NASA Reference Publication 1008

Lightning Protection of Aircraft

Franklin A. Fisher and J. Anderson Plumer

General Electric Company
Pittsfield, Massachusetts

Prepared for
Aerospace Safety Research and Data Institute
NASA Lewis Research Center

National Aeronautics
and Space Administration

Scientific and Technical
Information Office

1977

PREFACE

This book is an attempt to present under one cover the current state of knowledge concerning the potential lightning *effects* on aircraft and the means that are available to designers and operators to *protect* against these effects. The impetus for writing this book springs from two sources——the increased use of nonmetallic materials in the structure of aircraft and the constant trend toward using electronic equipment to handle flight-critical control and navigation functions. Nonmetallic structures are inherently more likely to be damaged by a lightning strike than are metallic structures. Nonmetallic structures also provide less shielding against the intense electromagnetic fields of lightning than do metallic structures. These fields have demonstrated an ability to damage or cause upset of electronic equipment.

Such concerns, when added to the continuing apprehension regarding the vulnerability of fuel systems to lightning, have led in the past decade to increased research into lightning effects on aircraft. The results of this research are contained in the technical reports and literature published by ourselves and by researchers in other laboratories who are also working on these problems. Conferences and symposiums have been held so that researchers could exchange ideas and information; there is a high degree of cooperation among all of those working towards the goal of complete safety-of-flight in the lightning environment.

The persons who can best *use* information on aircraft protection from lightning are the aircraft designers and operators, but generally they are not among those who produced this information. Moreover, they are often unaware of its existence, and they seldom have the background to distill from it the important facts that can and should be applied to achieve safer designs. The purpose of this book is to present the most important parts of this body of knowledge in a manner most useful to the designer and the operator.

This book is organized into seventeen chapters. In the first of these we review what lightning is and how it originates. The second chapter describes how the aircraft becomes involved with the lightning flash and why it is that aircraft do not produce their own lightning flashes, but may, we think, sometimes trigger natural ones. Chapter 3 considers how often and under what conditions aircraft have been struck, reviews avoidance procedures now in use by operators, and reviews their degree of success. We also take up the question of whether or not strikes could be totally avoided. The fourth chapter summarizes the various effects which have occurred when lightning has struck aircraft, giving the operator an idea of the direct and indirect effects which he may expect when his aircraft is "zapped."

Since our main purpose is to help the designer protect against those effects that may be hazardous, the remainder of the book is devoted to this purpose. Chapters 5, 6, and 7 deal with protection against the direct physical damage effects. Chapter 5 sets forth three philosophical steps which guide us in the design work that follows. Attention is also called to government standards or certification regulations which deal with aircraft lightning protection.

Chapter 6 reviews in some depth what is known about lightning effects on aircraft fuel systems and tells how to design protection against these effects. We give considerable attention to this subject and urge the designer to do likewise because of the serious consequences in the past of lightning effects on these systems. Chapter 7 deals with the protection of the aircraft from structural damage resulting from lightning, with emphasis on the nonmetallic materials—— materials that may be more vulnerable than the metal structures they are beginning to replace.

The remainder of the book deals with indirect effects. Chapter 8 introduces the reader to the basic mechanisms by which induced voltages occur in aircraft electrical circuits, and Chapters 9 through 14 treat these mechanisms in greater detail. Chapters 15 and 16 then consider the impact of induced voltages upon solid state electronic devices and tell how these devices may be protected. Here, concepts such as the *transient control level* philosophy are presented. Concepts of this kind may form the basis for future specifications that define the roles to be played by both the aircraft designer and the electronics equipment designer. Finally, in Chapter 17 we show how aircraft can be tested to determine their actual susceptibility to indirect effects and how equipment can be tested to determine its vulnerability or prove that protection design goals have been met.

To some extent, each chapter stands by itself and can be utilized without knowledge of the others. Dependencies often exist, however, among the lightning effects on structural, electrical, and fuel systems; for the most thorough understanding of lightning effects on any one of these systems, the reader is urged to read the entire book.

Part of the research upon which this book is based was conducted by ourselves, but a significant amount was conducted by researchers at other laboratories. We have referred to or incorporated that work frequently. Without it our present understanding of lightning effects——as well as this book——would not have been possible. Not all of the work conducted in this field could be referenced, of course, but we have carefully studied most of it. The work we reference is that which we consider to be definitive, and we have taken care to provide complete source details so that the reader can refer to them for additional information.

Even though much has now been learned about lightning effects on aircraft and how to design protection, there are still some lightning effects which are not fully understood. Examples of these are (1) the mechanisms by which lightning currents diffuse into interior structural members and conducting parts together with the extent to which this happens and (2) the effects of electromagnetic radiation from the lightning arc upon aircraft electrical and electronic systems. We have tried to identify these areas as they are encountered, and we caution the reader to remain alert for developments in these areas in the future.

While much of this book may appear oriented to the designer, there is

iv

much here of benefit to the operator as well. Familiarity with its contents will enable him to know what to expect, what not to expect, and why particular things happen when lightning strikes his aircraft. The book will show him where to look for damage after a strike, and, we hope, help him to better identify potential problem areas and to communicate them to researchers and designers in time to avert future problems.

The preparation of this material was supported by the Aerospace Safety Research and Data Institute, Lewis Research Center, National Aeronautics and Space Administration under Contract NAS3-19080. We wish to acknowledge the support of that organization. We also wish to acknowledge the support of H. V. Bankaitis, Solomon Weiss and, in particular, of Paul T. Hacker of that organization. It was Mr. Hacker who first suggested that a book of this nature be written.

Also, we deeply appreciate the help of Beryl I. Hourihan in digging through our files for materials and in preparing the rough draft; and the diligence, skill, and long hours applied by Catharine L. Fisher, who edited this book, bringing order and clarity to a sometimes confusing array of facts and figures.

F. A. Fisher
J. A. Plumer

April 1977

TABLE OF CONTENTS

TABLE OF CONTENTS

TABLE OF CONTENTS

TABLE OF CONTENTS

CHAPTER 1
THE LIGHTNING ENVIRONMENT

1.1 Introduction

The lightning flash originates with the formation of electrical charge in the air or, more commonly, clouds. The most common producer of lightning is the cumulonimbus thundercloud. Lightning, however, can also occur during sandstorms, snowstorms, and in the clouds over erupting volcanos. Lightning has even been reported to occur in clear air, though this phenomenon is rare and is possibly a result of lightning originating in conventional clouds beyond the observer's field of vision. Lightning originating in sandstorms and volcanic eruptions is not of serious concern to aircraft, but lightning associated with snowstorms occurs sufficiently often as to present a problem, not because its nature is different from lightning associated with thunderstorms but because it is apt to occur when it is unexpected.

The most common types of lightning are those involving the cloud and ground, called *cloud-to-ground lightning,* and lightning between charge centers within a cloud, called *intracloud lightning.* This latter is sometimes erroneously called *intercloud* or *cloud-to-cloud lightning.* True cloud-to-cloud lightning between isolated cloud centers is possible; however, what appears to be cloud-to-cloud lightning is often a spectacular manifestation of intracloud discharges.

Most research on lightning has centered on cloud-to-ground lightning; despite its importance to aircraft operation, much less information on the characteristics of intracloud lightning than on those of cloud-to-ground lightning exists, for a variety of reasons. The first is simply that intracloud lightning, unlike cloud-to-ground lightning, is largely hidden from direct observation. The second is that conducting research on the characteristics of lightning is often a labor of love requiring both extensive apparatus and extreme patience. Observing lightning from a fixed ground station is much easier and cheaper than observing lightning from a moving aircraft. The third is that most of the funding for research on lightning has come, directly or indirectly, from those who are concerned with the effects of lightning on electric power transmission and distribution lines, which are affected only by cloud-to-ground strokes.

The characteristics of lightning are discussed in the sections which follow. However, much of the material relative to the physics of lightning will be discussed primarily in terms of cloud-to-ground strokes for the reasons cited above and because frequently it is difficult to say for any given flash which type was involved. While aircraft may be involved with any of the three types of lightning, cloud-to-ground and intracloud lightning flashes are the most common types. But where aircraft design and operation are of concern, the type of flash makes little difference.

Figure 1.1 shows a generalized waveshape for the current flowing to ground from a typical negative cloud-to-ground flash and presents terms for the five main regions shown: the leader, the initial return stroke, an intermediate

1

current, a continuing current, and one or more restrikes. Each of these aspects will be discussed.

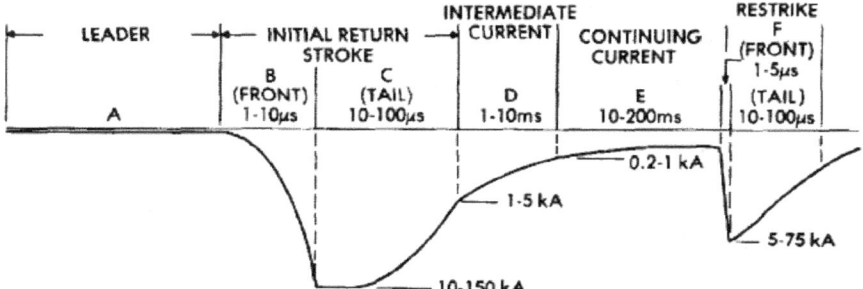

Figure 1.1 Generalized waveshape of current in negative cloud-to-ground lightning. (Note that the drawing is not to scale.)

1.2 Generation of the Lightning Flash

1.2.1 Generation of the Charge

The energy that produces lightning is assumed to be provided by warm air rising upwards into a developing cloud. As the air rises it becomes cooler, and at the dew point, the excess water vapor condenses into water droplets, forming a cloud. When the air has risen high enough for the temperature to drop to -40 °C, the water vapor will have frozen to ice. At lower elevations there will be many supercooled water drops that are not frozen, even though the temperature is lower than the freezing point. In this supercooled region, ice crystals and hailstones form.

According to one theory, the cloud becomes electrically charged by the following process (Reference 1.1). Some of the ice crystals which have formed coalesce into hailstones. These hailstones fall through the cloud gathering additional supercooled water droplets. As droplets freeze onto a hailstone, small splinters of ice chip off. Apparently, these splinters carry away a positive electrical charge, leaving the hailstone with a net negative charge. The vertical wind currents in the cloud carry the ice splinters into the upper part of the cloud, while the hailstone, being heavier, falls until it reaches warmer air, where some portion of it melts and the remainder continues to earth. Thus, the upper part of the cloud takes on a positive charge while the lower region takes on a negative charge. In some other manner, another smaller pocket of positive charge may be formed near the front of the base of the cloud and below the main body of negative charge.

Other theories have been proposed to account for the electrification of the cloud (References 1.2 to 1.7). All of them are based on experimentally observed evidence that the charge in the top of the cloud is positive. There may also be a body of positive charge near the front of the base of the cloud. The charge in the

2

rest of the cloud is negative. Figure 1.2 shows a typical cloud with the charge distributed as previously described. The cloud is moving to the left. The unbroken lines represent stream lines of air.

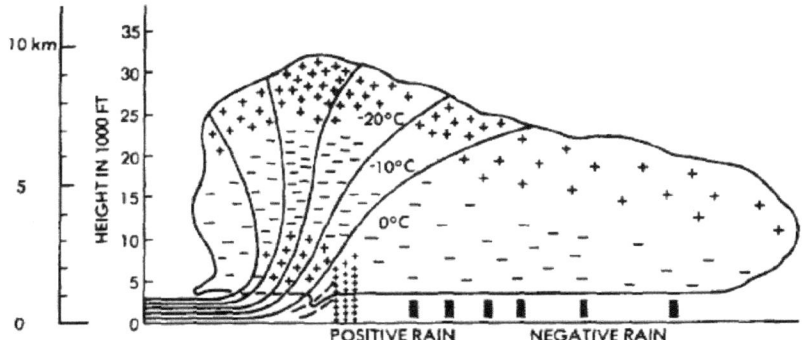

Figure 1.2 Generalized diagram showing distribution of air currents and electrical charge distribution in a typical cumulonimbus cloud.

The air currents and the electrical charges tend to be contained in localized cells and the cloud as a whole to be composed of a number of cells. A typical cloud might have the cell structure shown in Figure 1.3 (Reference 1.8). The electrical charge contained within a cell might appear as shown in Figure 1.4 (Reference 1.9). The temperature at the main negative-charge center will be about -5 °C and at the auxiliary pocket of positive charge below it, about 0 °C. The main positive-charge center in the upper cloud will be about 15 °C colder than its negative counterpart.

The lifetime of a typical cell is about 30 minutes. At its mature stage the cell as a whole will have a potential, with respect to the earth, of 10^8 to 10^9 volts (V). It will have a total stored charge of several hundred coulombs (C) with potential differences between positive- and negative-charge pockets again on the order of 10^8 to 10^9 V. The cell as a whole will have a negative charge.

1.2.2 Conditions on the Ground

As the cloud passes over a point on the ground, an electrical charge is attracted into the ground under the cloud. The average electric field at the surface of the ground will change from its fair-weather value of about 300 volts per meter (V/m) positive (air positive with respect to the earth) to as high as several thousand volts per meter. Generally, when a cloud is overhead, the field from cloud to earth will be negative, but when a localized positive region is overhead, the field may be positive. The potential gradient will be concentrated around sharp protruding points on the ground and can exceed the breakdown strength of the air, which has a nominal value of 30 000 V/cm. When the

3

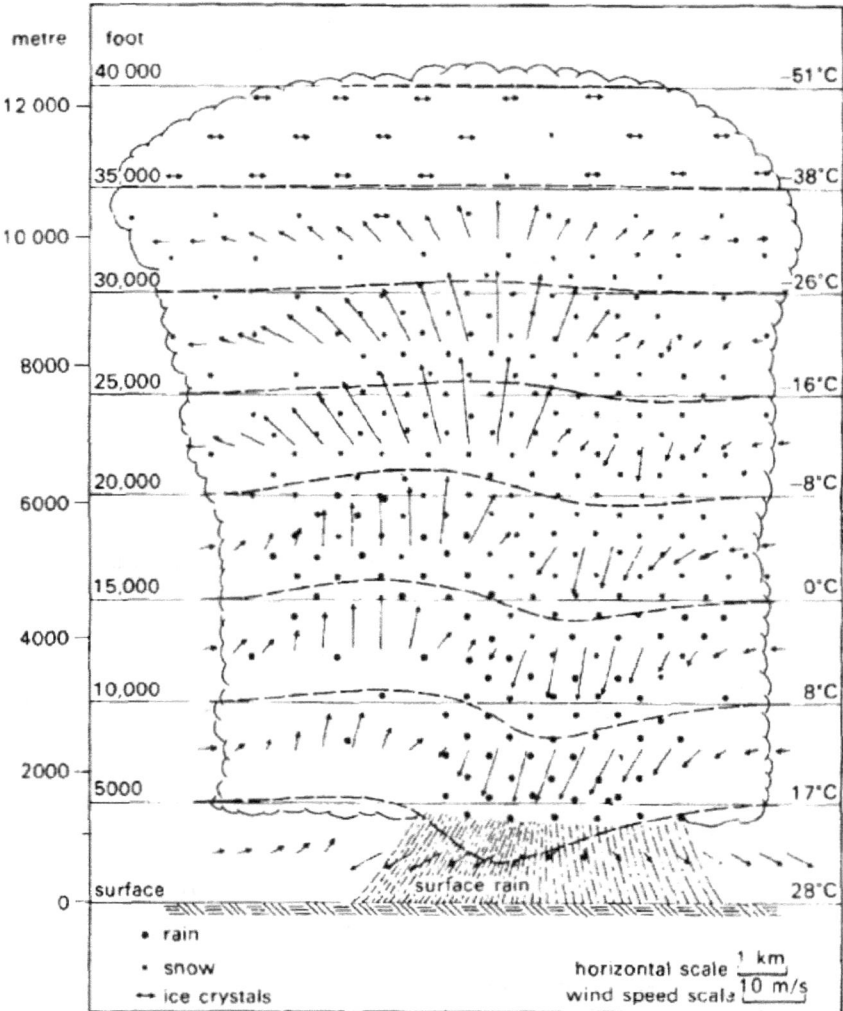

Figure 1.3 An idealized cross section through a thunderstorm cell in its mature stage. Key: ●, rain; *, snow; ↔, ice crystals.

breakdown strength of the air is exceeded, current into the air increases sharply and a bluish electrical discharge called *corona* forms around a point. This discharge is the St. Elmo's fire that appears on such places as masts of ships, aircraft wing and tail tips, and from trees and grass on high ground. The magnitude of the current from a single discharge point may range from 1 or 2 microamperes (μA) to as high as 400 μA.

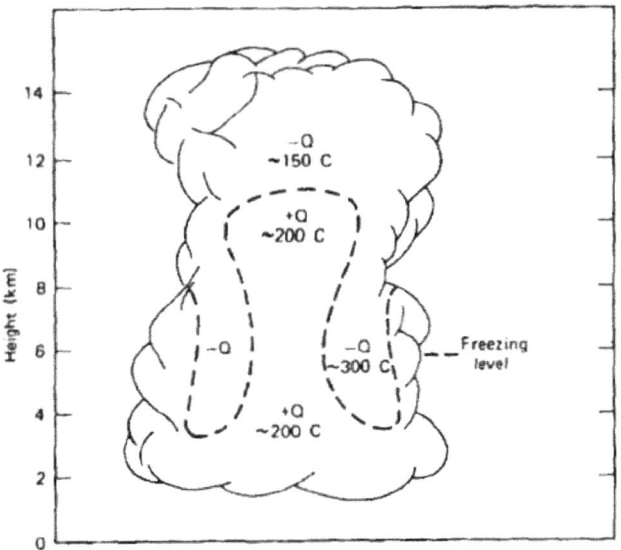

Figure 1.4 Estimated charge distribution in a mature thundercloud (after Phillips).

1.2.3 Development of the Leader

At some state in the electrification of the cloud, a discharge towards the earth takes place. It starts as a slow-moving column of ionized air called the *pilot streamer.* After the pilot streamer has moved perhaps 30 to 50 m, a more intense discharge called the *step-leader* takes place. This discharge lowers additional negative charge into the region around the pilot streamer, recharges it, so to speak, and allows it to continue for another 30 to 50 m, after which the cycle repeats.

A discharge propagating in this manner is called a *streamer discharge*; its development is illustrated in Figure 1.5. The streamer is initiated when a free electron is accelerated in a sufficiently high electrical field (a). An electron (b) so accelerated collides with neutral molecules of air (c), ionizes them, leaving them with a positive charge, and creates new electrons at a rate of a per unit length. The electrons, being much less massive than the positive ions, move under the influence of the electric field, leaving the positive ions behind. If the discharge continues to develop over the length a, there will be at the head of the discharge (d) a number of electrons given by

$$N = \exp \left(\int_0^\ell (a-\eta) \right) dx \qquad (1.1)$$

where a is the Townsend ionization coefficient and η is the attachment coefficient. In air at atmospheric pressure, electron multiplication can exist only where the field is higher than 25 kV/cm because only in this case is a greater than η.

5

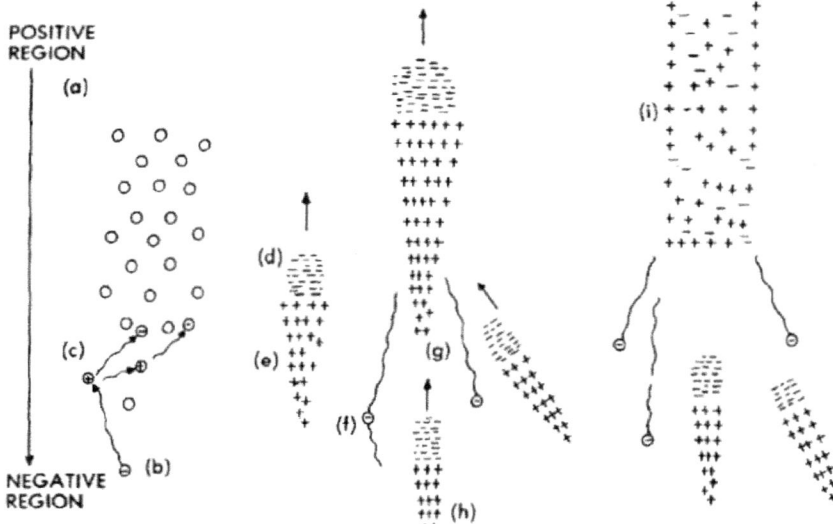

Figure 1.5 Stages in the development of a leader.

Left in the wake of these electrons will be positively charged ions (e). If the electric field is high enough, the initial avalanche will reach a critical size (approximately 10^8 electrons) for another avalanche of electrons (f) to be initiated by photoionization (g) from the initial discharge. The electric field that accelerates this secondary discharge is the sum of the initial electric field and that produced by the positive space charge left behind by the initial avalanche. Under the action of the total field, these successive avalanches reach the positive space charge (h), neutralize it, and leave a new positive charge a little farther on. With such a mechanism a positive charge moves step by step into the un-ionized air leaving behind it a partially ionized filament (i). This filament is a conductor, though at this stage of its development perhaps only a poor conductor.

The processes just described relate to a positive electric field. In a negative electric field a similar, though more complicated, phenomenon occurs. The initial avalanches seem to develop in the air farther ahead of the leader and to propagate both ways: into virgin air and back toward the more fully developed leader. The end result is much the same; behind the advancing head of the leader is an ionized column in this case with a predominance of negative charge and having at its center a more heavily conducting filament.

If the initial development of the leader takes place in the charged cloud, the developing streamer branches and begins to collect charge from its surroundings. Because it collects charge in this way, the streamer may be viewed as connected to the cloud and at the same potential as the cloud. As the head of the leader moves farther into the un-ionized air, charge flows down from the charged regions of the cloud, along the partially conducting filament and toward the head of the leader, thus tending to keep all parts of the leader at a very high potential. The amount of charge, q_0, lowered into the leader will be on the order of 2 to 20 x 10^{-4} C/m of length. A leader 5 km long would then have stored within it a charge of 1 to 10 C.

Since the potential of the leader is very high, there will be a high radial electric field along the leader. This field will be high enough to exceed the breakdown strength of the air, and secondary streamers will branch out radially away from the central filament. The filaments will branch out radially until the field strength at the edge of the ionized region falls to about 30 kV/cm.

It can be shown that the electric field strength at the edge of a cylinder containing a charge, q_0, per unit length is

$$E_r = \frac{1.8 \times 10^{10}\, q_0}{r} \qquad (1.2)$$

$$(V/m, C/m, m)$$

From this and the above breakdown strength of air, it can be deduced that the radius of the leader will be 1.2 to 12 m. At higher elevations the breakdown strength of air is less; hence the leader radius may be more.

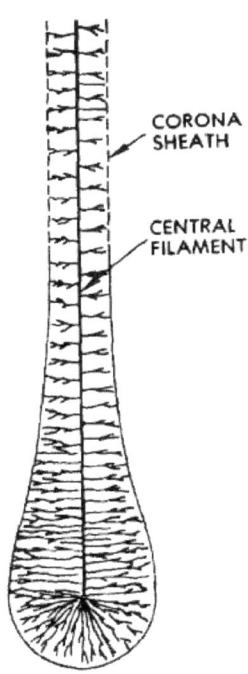

CORONA
SHEATH

CENTRAL
FILAMENT

Figure 1.6   The lightning leader as postulated by Wagner.

7

As postulated by Wagner (Figure 1.6 [Reference 1.10]), the head of the leader may have a larger diameter than that of the rest of the leader, though this is difficult to prove by photographs. The head of the leader, nevertheless, is generally visible because of the optical radiation associated with the extension of the electron avalanches. But once the growth ceases, the radiation stops; consequently, the corona sheath surrounding the central conducting filament is not visible.

The process can be studied in the laboratory, albeit on a smaller scale than that of natural lightning. Figure 1.7 shows typical phenomena observed during the breakdown of the air between electrodes about 10 m apart. The sketch is based on a series of short-duration (about 1 microsecond [μs]) photographs taken at intervals of about 25 μs with an image-converter camera. As the electrons are attracted out of the initially ionized region at the head of the leader, the conducting filament lengthens and the corona discharge at the head of the leader occurs farther on in previously un-ionized air.

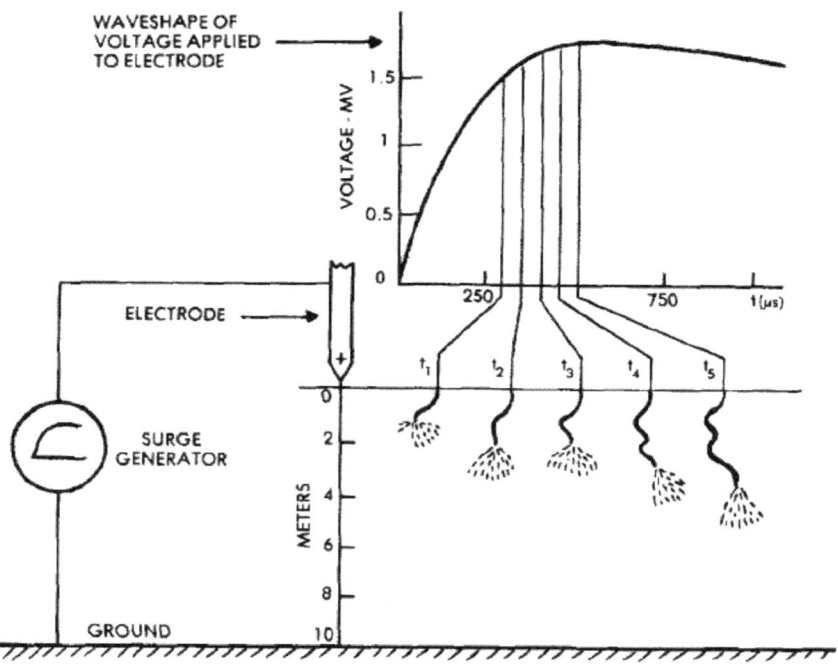

Figure 1.7 The development of a leader in the laboratory.

Photographs of actual lightning leaders may be taken with a Boys camera, a camera in which the film moves relative to the camera lens. An example of such a photograph is shown in Figure 1.8 (Reference 1.11). The leader is seen originating at the top left-hand corner of the picture and lengthening as time

8

increases. The bright line at the right of the picture is called the *return stroke* and will be discussed shortly.

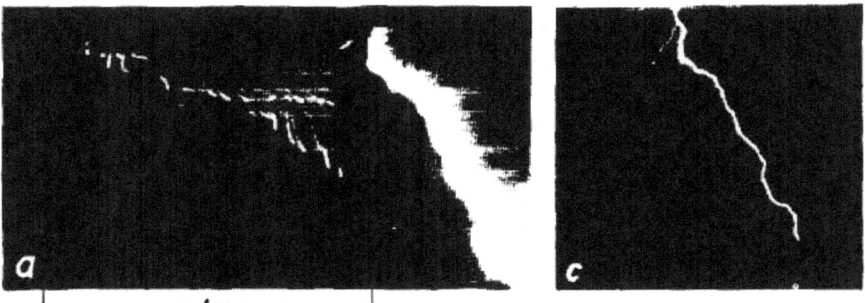

Figure 1.8 Boys camera photograph of a lightning leader.

From such photographs it has been learned that the leader advances at about 1 to 2 x 10^5 m/s, or 0.03 to 0.06% of the speed of light (Reference 1.12). In order that a charge of 2 to 20 x 10^{-4} C be deposited by a leader advancing at the rate of 1 x 10^5 m/s requires the average current in the leader, i_ϱ, to be 20 to 200 A. A current of this magnitude could be carried only in a highly conducting arc discharge, the assumed central conducting filament of the leader. Such an arc would have a diameter on the order of a few millimeters and an axial voltage gradient, g_ϱ, of about 5 x 10^3 V/m. A leader 4 km long would then have a voltage drop along its length of 2 x 10^7 V. The longitudinal resistance, R_ϱ, of the conducting filament g_ϱ/i_ϱ would then be in the range of 40 to 400 ohms per meter (Ω/m).

While of less importance as regards aircraft, it might be noted that leaders sometimes start at the ground and work their way toward the sky. This happens most frequently from tall buildings or towers, or from buildings or towers located atop hills. Generally, one can tell from the direction of the lightning flash branching whether the leader started at the cloud or at the ground: if the branching is downward (Figure 1.9), the leader originated at the cloud; if the branching is upward, the leader originated at the ground.

1.2.4 The Return Stroke

As the negatively charged step-leader approaches the ground, positive charge accumulates in the ground underneath it or, more accurately, negative charge is repelled away from the region under the leader. At some point the electric field strength around objects on the ground becomes sufficiently high that a streamer starts at the ground and works its way toward the downward-approaching leader. When the streamers meet, the conducting filament in the center of each streamer provides a low-impedance path so that the charge stored in the head of the leader can flow easily to ground. As the current in the central

9

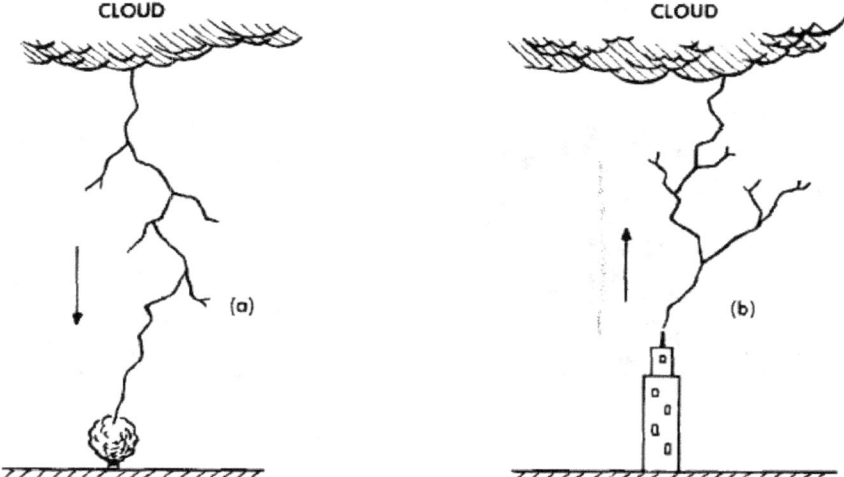

Figure 1.9   Leader direction as determined from direction of branching.
(a)   Downward-branching leader starts at cloud.
(b)   Upward-branching leader starts at ground.

filament rises from its initial current of a few tens of amperes to higher values, it gets hotter, its diameter expands, its longitudinal gradient decreases, and it becomes an even better conductor, which in turn allows even more current to flow in the arc. As the charge in the lower part of the leader is neutralized, the heavily conducting arc reaches higher into the charged leader channel. The head of the region in which this neutralization takes place moves upwards at a rate of roughly 100 000 km/s (or one-third the velocity of light) until it reaches the cloud. This heavily conducting region, called the *return stroke*, produces the intense flash normally associated with the lightning stroke. Some stages in the development of the return stroke are shown in Figure 1.10 (Reference 1.13).

The velocity with which this return stroke propagates, together with the amount of charge deposited in the leader channel, determines the amount of current developed in the return stroke. Let v be the velocity of the return stroke and q be the amount of charge deposited per unit length, dℓ, along the leader channel. Since

$$I = \frac{dq}{dt} \qquad (1.3)$$

and

$$v = \frac{d\ell}{dt} \qquad (1.4)$$

then

$$I = qv \qquad (1.5)$$

10

As a numerical example let

$$v = 10^8 \text{ m/s and } q = 10 \times 10^{-4} \text{ C/m}$$

$$I = 10 \times 10^{-4} \times 10^8 = 10 \times 10^4 = 100\ 000 \text{ A}$$

The velocity of the return stroke is not constant from one stroke to the next. It seems to vary with the magnitude of current that is ultimately developed. The relationship between current and velocity may be deduced either from theoretical concepts or experimentally. The relationship derived by Wagner (Reference 1.14) is shown in Figure 1.11. Considerations of the return stroke velocity are primarily of importance in studying the time history of the electric field produced by the lightning flash. The velocity, however, may affect the surge impedance of the lightning stroke, and thus the way that the stroke interacts with a metallic conductor like an aircraft.

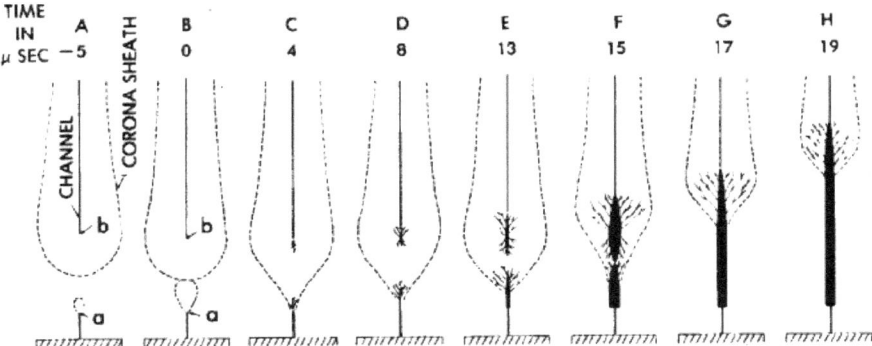

Figure 1.10   Stages in the development of the return stroke.

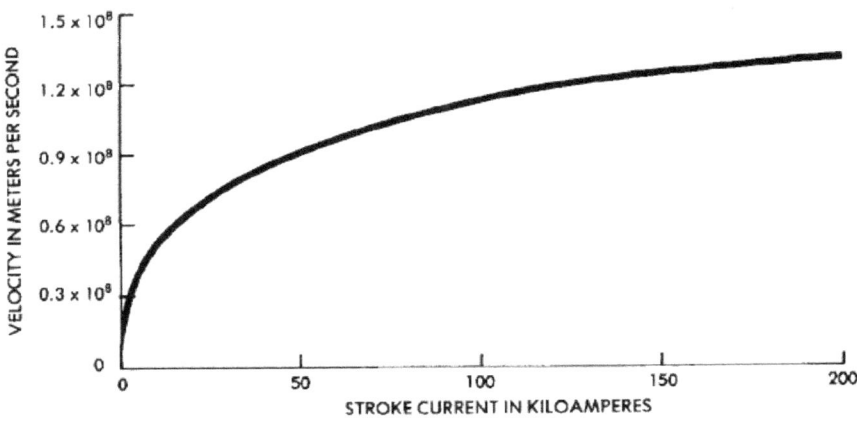

Figure 1.11   Relation between stroke current and velocity of return stroke.

11

The velocity of propagation of the return stroke is less than that of the speed of light for two basic reasons. The first reason involves the longitudinal resistance of the return stroke channel.

Some of the factors associated with this longitudinal resistance are shown in Figure 1.12. Central to the phenomenon is the fact that the current in the lightning channel must increase fairly rapidly from the 200 A (approximately)

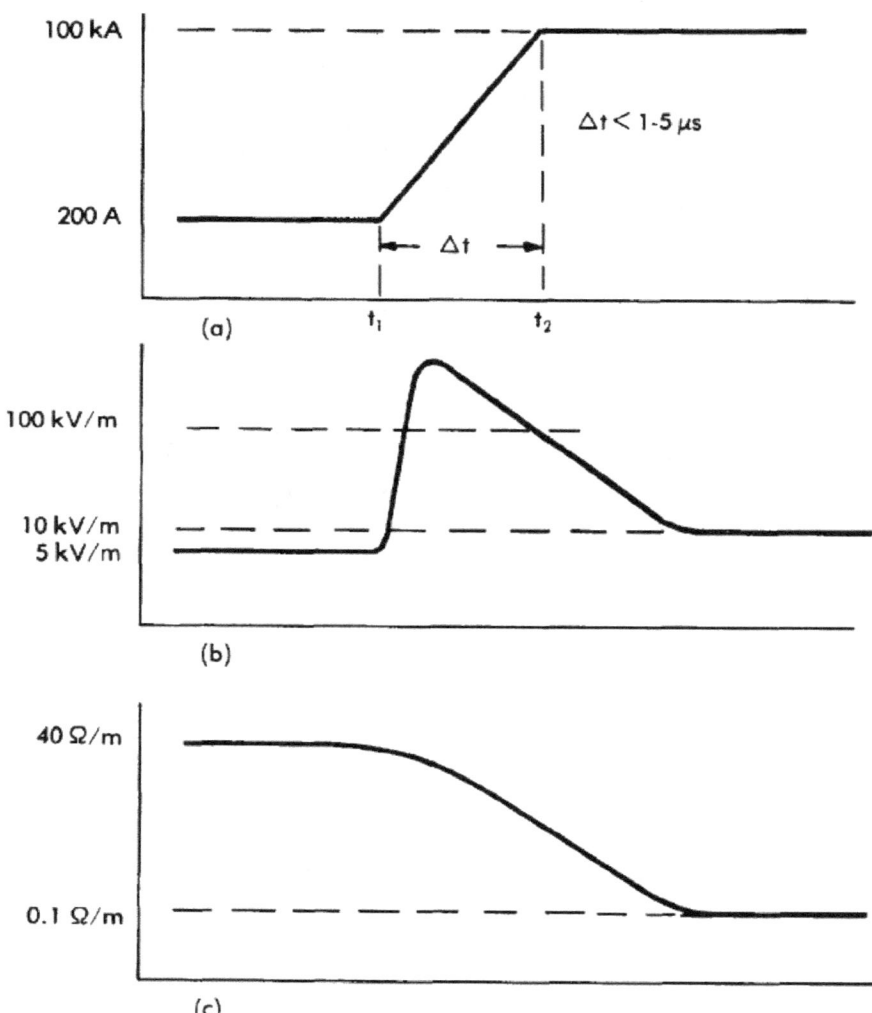

Figure 1.12 Phenomena associated with passage of the return stroke.
- (a) Current
- (b) Longitudinal voltage
- (c) Longitudinal resistance

12

current associated with the initial development of the leader to a current of perhaps 100 kA as the return stroke becomes fully developed. It is a characteristic of an arc channel discharge that if the current through the arc is increased, the arc channel expands in diameter, keeping a fairly constant current density across the channel. This channel cannot expand instantaneously, since energy must be put into the channel to cause the channel to heat up sufficiently to force it to expand. Accordingly, if the current through the arc channel is increased suddenly by a large magnitude, as in Figure 1.12(a), the longitudinal voltage gradient of the channel must suddenly increase. Since the rate at which energy is injected into the channel is the product of the current and the longitudinal voltage gradient, the increased longitudinal voltage gradient may be taken as the mechanism forcing the arc channel to get hot enough to expand to the diameter required to carry the high currents. It is not known what the maximum longitudinal voltage gradient would be in a lightning channel, but it is known from studies of arcs in laboratories that the gradient will fall to values on the order of 100 kV/m in a fraction of a microsecond. Presumably in a few microseconds, the channel diameter will have expanded to its final value, and the longitudinal voltage gradient will have decayed back toward values on the order of 5 to 10 kV/m. The longitudinal resistance, then, would fall from values on the order of 40 Ω/m to values on the order of a small fraction of an ohm per meter, in times on the order of a few microseconds.

This collapse of longitudinal resistance, however, is far from instantaneous. The initial resistance of the leader is sufficiently high to retard the development of the upward-going return stroke and hence reduce its velocity of propagation below that of the speed of light. Presumably, leaders which lead to the formation of high-amplitude lightning currents either have a sufficiently low longitudinal resistance to begin with or the longitudinal resistance is reduced to low values sufficiently fast by the high-amplitude return strokes that the longitudinal resistance presents less of an obstacle for the upward-going return stroke than it does for the flashes which involve lower peak currents.

An additional factor that affects the velocity of the return stroke and is the second reason that the velocity is less than the speed of light is shown in Figure 1.13. As explained earlier, the leader deposits in its wake a column of electrical charge with diameters on the order of several meters. At the center is a highly conducting core, which has a diameter of a few millimeters for the leader and which expands to a few centimeters during the passage of the return stroke. The inductance of this return stroke is determined by the diameter of the highly conductive central core, and the capacitance by the diameter of the column of electrical charge. The lightning stroke may then be modeled as shown in Figure 1.13(b), in which a highly conductive central conductor is fastened onto a series of projecting splines, much like the backbone of a fish. A better analogy might be to view the lightning flash as a piece of tinsel rope for decorating a Christmas tree: a central piece of string is surrounded by a tube of fine filaments projecting radially away from the central core. In either case, the radial filaments can carry a radial current, i_r, but cannot carry an axial current, i_a. Accordingly, the lightning return stroke has both a high capacitance and a high inductance per unit of length. In this respect it differs from a solid conductor of large diameter which, while possessing a high capacitance per unit length, simultaneously possesses a low inductance per unit length. It follows that the surge impedance, governed by the ratio of inductance to capacitance, is high while the velocity of propagation, governed by the product of inductance and capacitance, is less than that of the speed of light.

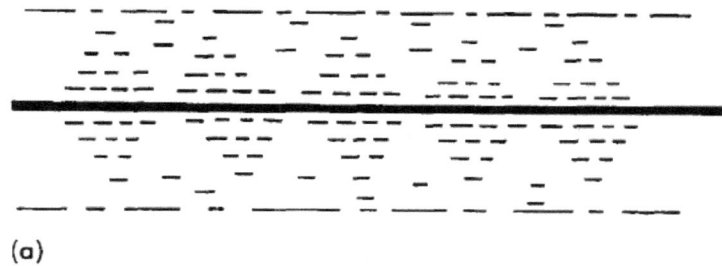

(a)

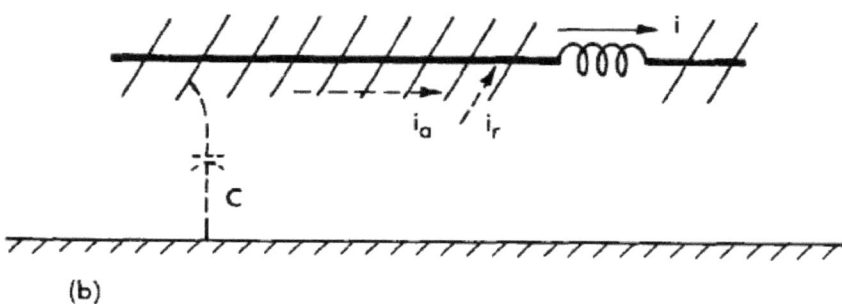

(b)

Figure 1.13 Effect of corona cloud on velocity of propagation.

    (a) Distributed charge surrounding a highly conductive central core

    (b) Highly conductive central conductor fastened onto a series of projecting splines

Wagner (Reference 1.15) concludes that the surge impedance of the lightning flash is of the order of 3 000 Ω for return strokes of large amplitude, 100 kA. This value is large compared to the surge impedance ($\approx$500 Ω) of a simple conductor in air and remote from a ground plane or other current return path.

The waveshapes of lightning-flash currents measured at ground level are reasonably well known, principally from the work of Berger (Reference 1.16). Typical waveshapes detailing the front of the initial return stroke are shown in Figures 1.14 and 1.15 (References 1.17 and 1.18). In all cases, the current is

14

seen to have a concave front, the current initially rising slowly but then increasing to a maximum current rate of change just before crest amplitude is reached. It may be speculated that the initial slowly changing portion of these current oscillograms (which, of course, were measured at ground level) represents the growth of an upward-going leader from the lightning tower reaching upwards to contact the downward-approaching lightning leader. It can also be speculated that the maximum rate of change of current, which occurs just before crest, is most representative of the rate at which the current can increase in the lightning channel as the return stroke passes one particular point in space. This is supported by the observation that subsequent strokes in a lightning flash, even measured at ground level, exhibit front times considerably faster than the rise time of the initial stroke in the flash.

The true front time of the leading edge of the return stroke as it passes a point remote from ground has probably never been measured. It, however, seems appropriate to assume that it will be faster than the leading edge of currents measured at ground level.

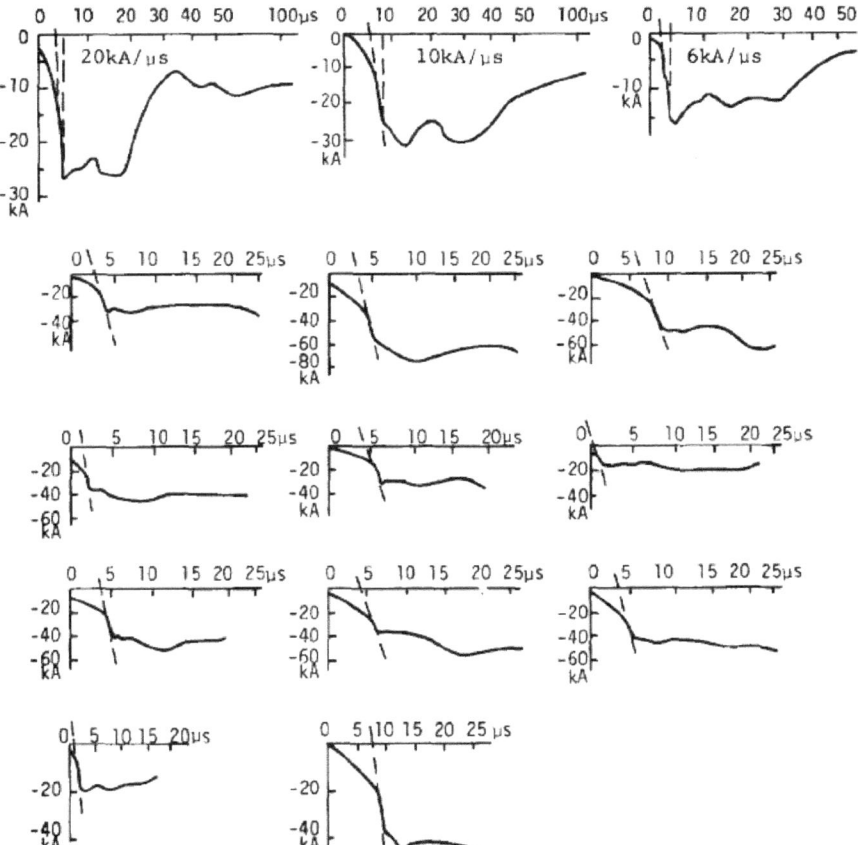

Figure 1.14 Front waveshapes of lightning currents as measured by Berger.

15

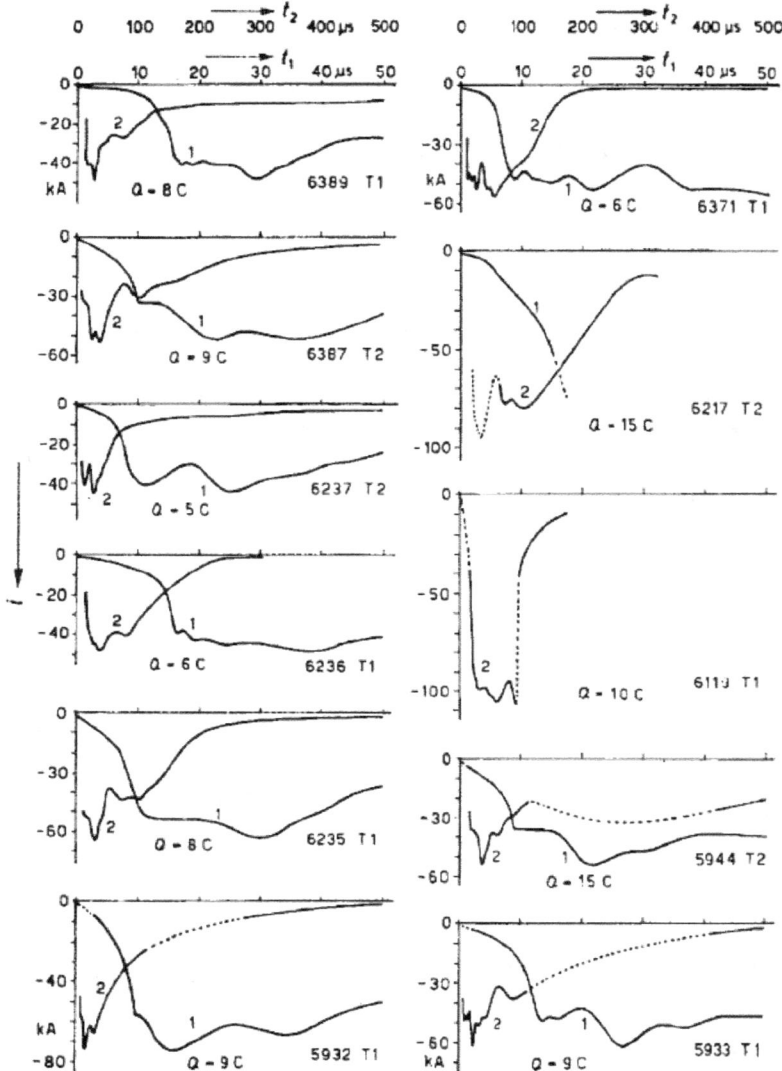

Figure 1.15 Current oscillograms from single strokes or first downward strokes.
1) Fast time-scale t_1; 2) Slow time-scale t_2. In osc. No. 6119 TI,
chopping may be caused by a flashover in the measuring equipment.

1.3 Further Development of the Lightning Flash

After the charge has been drained from the leader by the upwardly moving
return stroke, the current measured at the ground decays, though at a rate
slower than that at which the current rose to its peak. Oscillograms showing
typical decay times are shown in Figure 1.15 (Reference 1.18). The figure

16

displays the current on two different time scales, emphasizing the front and the tail. Some of the oscillograms showing the front are the same as those shown in Figure 1.14 (Reference 1.17).

As the return stroke approaches the cloud, it may encounter other branches of the leader, as shown in Figure 1.16. As it passes these branches, the charge stored in them will feed into the developing lightning stroke and momentarily increase the current. Eventually, the return stroke will reach the cloud. Our understanding of the phenomena occurring within the cloud is hindered by our not being able to see the phenomena, but we can infer some of the phenomena from measurements of the electrical radiation produced by the developing flash and from the usual behavior of the flash after the initial return stroke has passed. As the return stroke reaches into the cloud, it appears to encounter a much more heavily branched leader than it did in the air below the cloud. The return stroke can thus tap the charge diffused through a large volume of the cloud, rather than only the charge in the more localized leader. It would

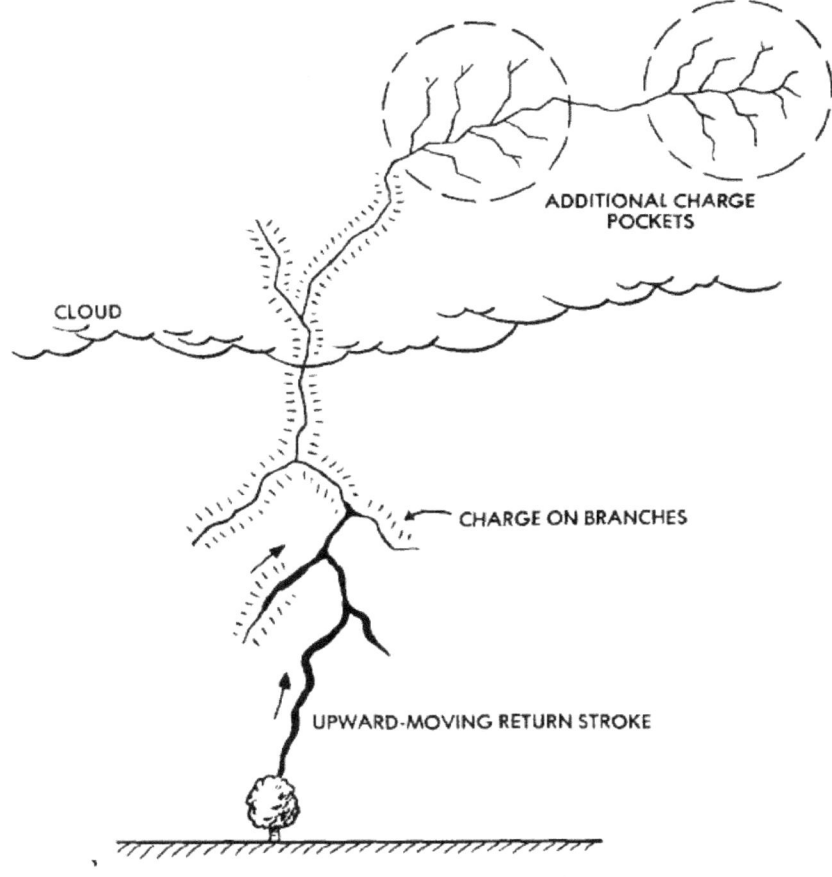

Figure 1.16 Further development of the flash.

17

appear to be during this period that the intermediate current (Component D of Figure 1.1) is developed. As the discharge continues to spread through the cloud, for times on the order of fractions of a second, currents on the order of a few hundred amperes continue to flow in the lightning flash. These are referred to as *continuing currents* (Component E of Figure 1.1). As one may expect, there is no clear-cut demarcation between the tail of the return stroke and the intermediate current, or between the end of the intermediate current and the start of the continuing current.

Eventually, and usually, the developing discharge within the cloud reaches into a different cell of the cloud or, at any rate, into a region where there is another localized body of electrical charge. At this stage there occurs what is called a *restrike* (Component F of Figure 1.1). The restrike starts with additional charge being lowered from the cloud to form a new leader, or, more properly, to recharge the central portion of the old leader. Presumably, because of the residual ionization in the channel, this charging process occurs smoothly, not in the step-by-step process by which the initial leader penetrates into the virgin air. Accordingly, this is called a *dart leader* instead of a step-leader. Unlike the initial step-leader, the dart leader is seldom branched. When the dart leader reaches ground level, a return stroke again occurs. The amplitude of this return stroke is again high, since the current comes from an intensely ionized region close to the ground. While the amplitude is usually not as high as that of the first return stroke, the current rises to crest more rapidly than does that of the initial return stroke, presumably because the upward leader from the ground does not have to propagate into virgin air.

1.4   Lightning Polarity and Direction

Most lightning flashes originate in the cloud and lower negative charge to earth. The question of direction of the lightning flash is sometimes confusing. With the intent of clarifying matters, the statement is sometimes made that lightning strikes upward and not downward. This is at least partially true; the return stroke that produces the high peak currents, thunder, and the highest intensity light, in fact does start near the ground and grow upward into the ionized channel previously established by the step-leader, thus tapping the charge in the step-leader. The step-leader, nevertheless, originated at the cloud. The source of energy is in the cloud, and the lower amplitude and longer duration currents have their origin in the charge stored in the cloud. Thus, in terms of the engineering definition of current, these flashes result in the direction of current flow from the earth to the cloud. This type is commonly called a *negative polarity flash*.

When tall buildings or mountain tops are involved, the lightning flash often does originate at the ground; the step-leader starts at ground level and propagates upwards into the cloud. Such flashes seem to be triggered by the high electric field concentrated around the top of the building or mountain. They may be recognized by the upward direction of branching, as mentioned earlier and shown in Figure 1.9(b). This type of flash therefore results in current flow from

18

the cloud to the ground and is called a *positive polarity flash*. Positive polarity flashes usually have lower peak currents than do flashes that originate at the cloud.

About 10% of all flashes are positive polarity flashes, and a fraction of these involve the highest peak currents and charge found associated with lightning. Examples of some of these strong positive flashes are shown in Figure 1.17 (Reference 1.19). The positive flashes typically have only one high current stroke; they lack the restrike phase generally noted on flashes of negative polarity.

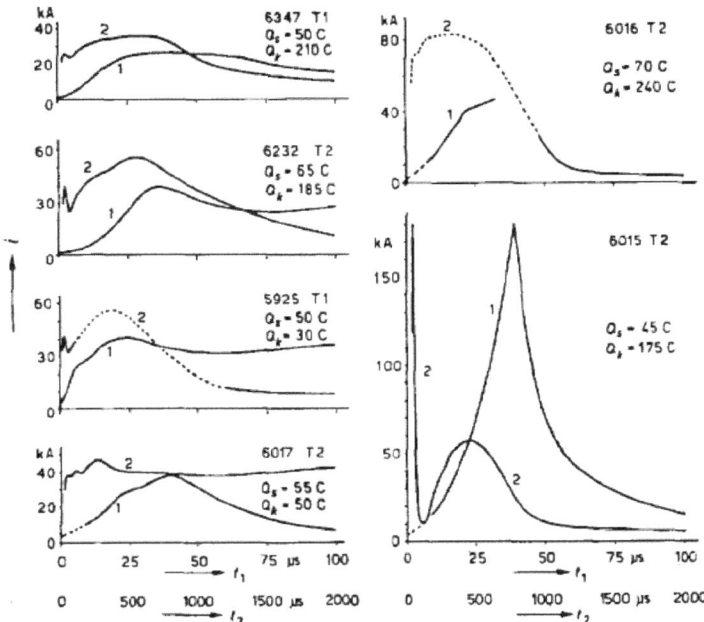

Figure 1.17 Examples of strong positive strokes. Currents are recorded on time-scales t_1 and t_2; Q_s electric charge (coulombs) within $2\ \mu s$ from the origin; Q_k electric charge (coulombs) in the continuing current after $2\ \mu s$.

1.5 Intracloud Flashes

Intracloud flashes occur between charge centers in the cloud. A distinguishing characteristic of intracloud flashes is that they seem to lack the intense return stroke phase typical of flashes to the ground, or at least that the electrical radiation associated with true intracloud discharges lacks the characteristics associated with the return stroke of cloud-to-ground flashes. Discharges between charge centers take place during cloud-to-ground flashes as well, and, to an observer within the cloud, it may be difficult to tell whether or not a flash to ground occurred.

19

With regard to aircraft the matter may be academic. Aircraft are struck underneath clouds by clear-cut cloud-to-ground flashes and by flashes within clouds. Based on the damage observed, the peak current sometimes is very high. Whether the high current was associated with the upper end of a cloud-to-ground flash or with a true intracloud flash makes little difference.

In temperate regions about two-thirds of all flashes are intracloud flashes. In tropical regions, where there is more lightning activity, the ratio is higher.

1.6 Measured Characteristics of the Lightning Flash

Lightning flashes are quite variable from one to another. Peak currents, total duration, waveshapes, number of strokes in the flash, charge transferred, etc., may all vary over wide limits, and only in general terms can one find a correlation between different parameters. Data on the characteristics of lightning are best presented in statistical terms, the mode that will be used in the following sections.

One item that needs to be emphasized is that virtually all the data on lightning comes from measurements made at ground level, and these measurements may be influenced by the growth of an upward leader. Very few measurements have been made of the amplitude and waveshape of lightning currents passing through aircraft. Most of the measurements that have been made were of strokes with lower peak currents and longer times to crest than those often observed at ground level. In part this may be explained by chance and in part by the fact that many of the flashes intercepted may have been intracloud flashes and not cloud-to-ground flashes. As noted earlier, intracloud flashes often lack the well-defined high-amplitude return stroke of cloud-to-ground flashes.

The best summary of the statistical characteristics of lightning is that compiled by Cianos and Pierce (Reference 1.20). They observed that many of these characteristics were nearly linear when plotted as a log-normal distribution. They then made a judgment as to the linear distribution that was the best fit to the experimentally observed data. The figures that follow are reproduced from their report.

Figure 1.18 (Reference 1.21) shows data on the peak current amplitude in lightning strokes. Regarding the damage that may be caused by lightning, this is one of the most important parameters. There are two curves shown, one for the first return stroke in a flash and one for subsequent return strokes. The first return stroke is generally of the highest amplitude. For engineering analysis, Cianos and Pierce have determined that subsequent return strokes may be represented as half the amplitude of the first return stroke. Marked on the curves are the amplitudes corresponding to the 2, 10, 50, 90, and 98% probabilities.

The peak value of the current is related to the explosive, or blasting, effect of lightning. It is also relative to the maximum voltage developed across ground resistance and hence to the risk of side flashes occurring in the vicinity of objects struck (or, related to the maximum voltage developed across loading resistors struck (or, related to the maximum voltage developed across bonding resistance and hence to the possibility of sparking at structural interfaces).

20

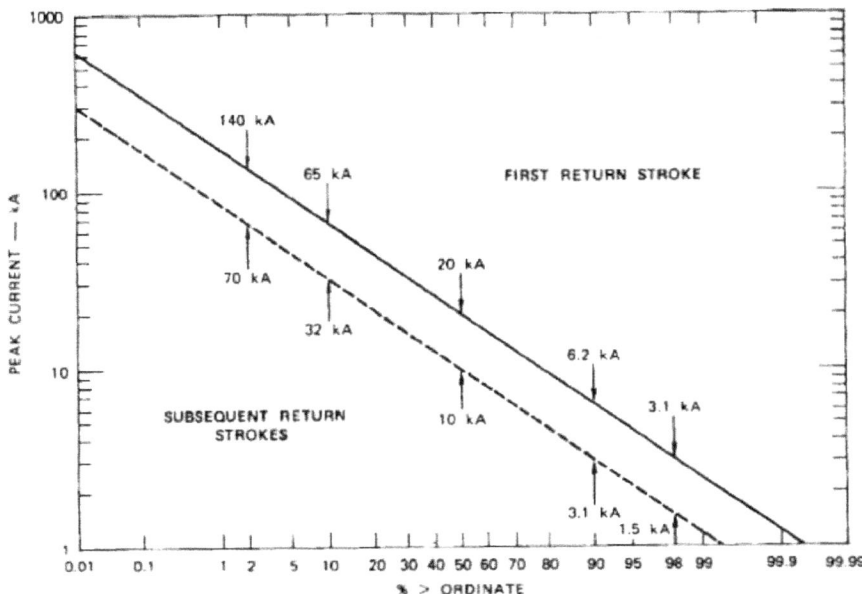

Figure 1.18 Distribution of peak currents for first return stroke and subsequent strokes.

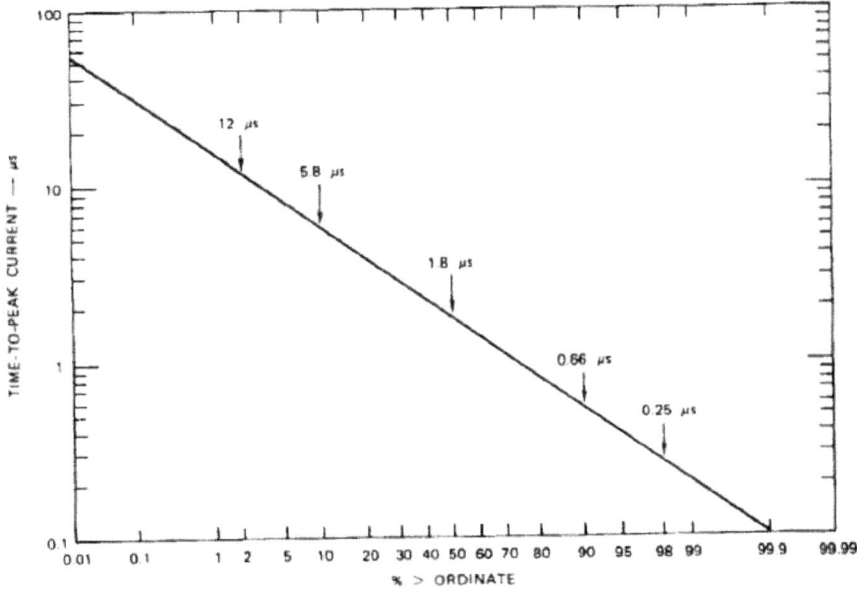

Figure 1.19 Distribution of time to peak current.

21

Figure 1.19 (Reference 1.22) gives a distribution of the time for the current to reach its peak amplitude. This time is subject to considerable interpretation for any particular lightning stroke, since there is seldom a clearly definable time at which the stroke starts. Lightning strokes typically have a concave front, starting out slowly and then rising faster as the current gets higher. Thus, the effective rate of rise of the lightning current is not directly obtained by dividing the peak current by the front time. The best summary of the effective rates of rise is given on Figure 1.20 (Reference 1.23). The rate of rise of a lightning current is an important factor in determining how much voltage is induced into electrical equipment, and in determining how many lightning conductors are needed and how they should be placed.

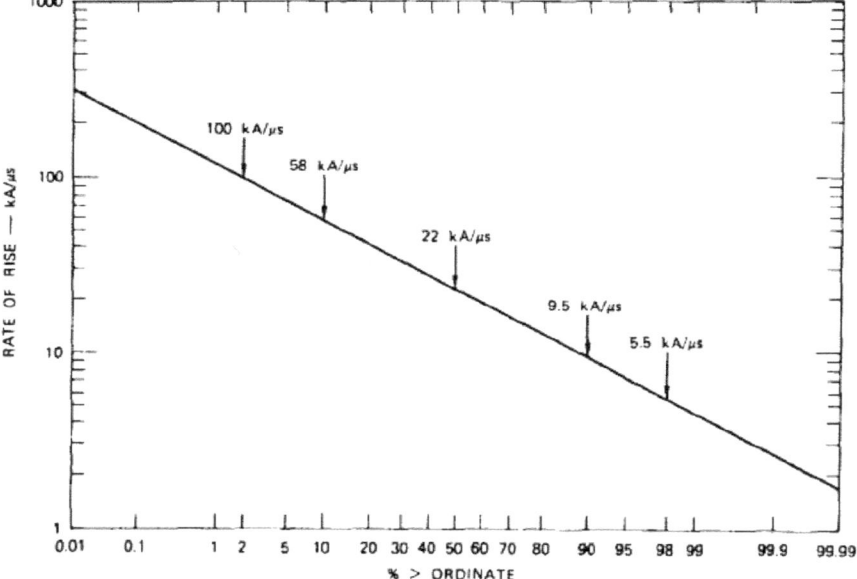

Figure 1.20 Distribution of rates of rise.

The duration of the stroke current affects the distance across which side flashes may develop, and affects how severely metal structures may be deformed by magnetic forces or the explosive liberation of energy. This distribution is shown in Figure 1.21 (Reference 1.24). The duration of the stroke, which is measured in tens of microseconds, should not be confused with the total duration of the lightning flash. The total duration, shown in Figure 1.22 (Reference 1.25), is frequently on the order of a second. The duration of the total flash is influenced by the number of return strokes in the flash (Figure 1.23 [Reference 1.26]) and the time interval between strokes (Figure 1.24 [Reference 1.27]).

Figure 1.25 (Reference 1.28) gives information on the total charge

22

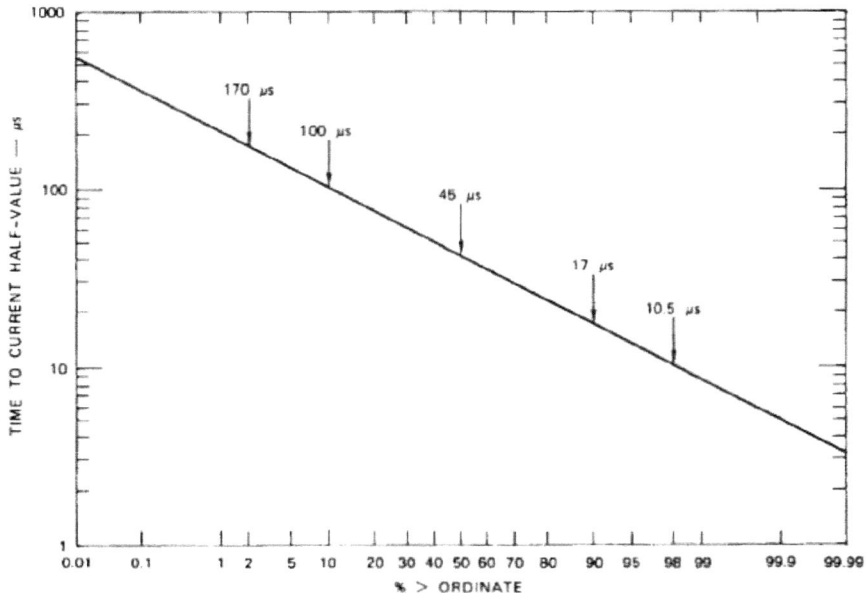

Figure 1.21   Distribution of time to current half value.

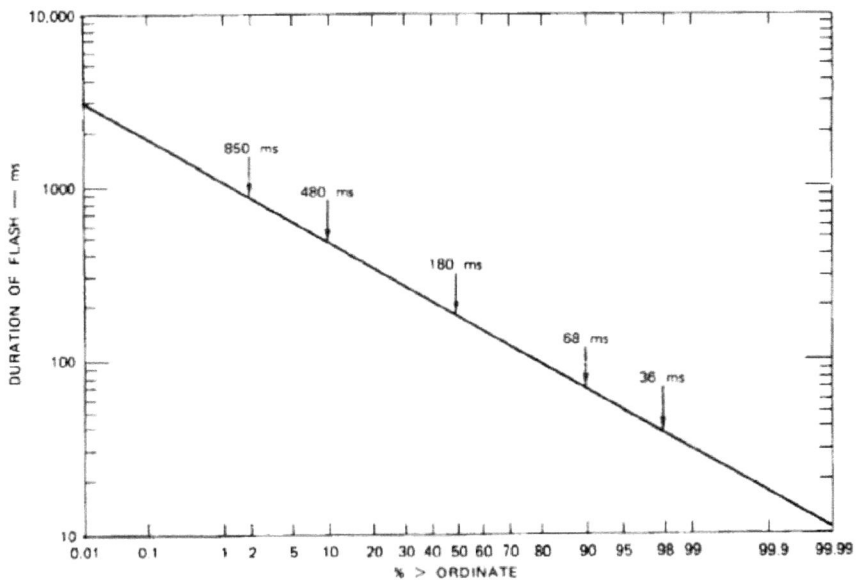

Figure 1.22   Distribution of duration of flashes to earth.

23

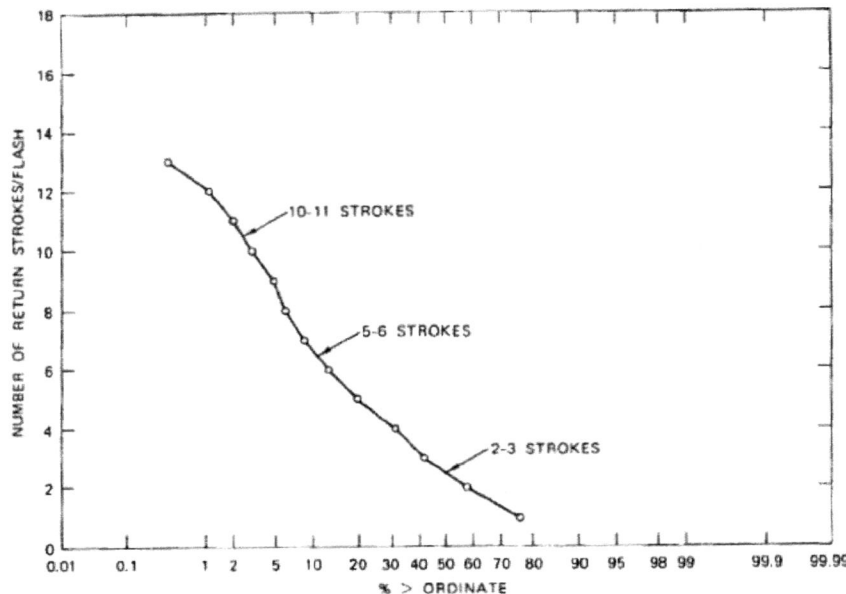

Figure 1.23 Distribution of the number of return strokes/flash.

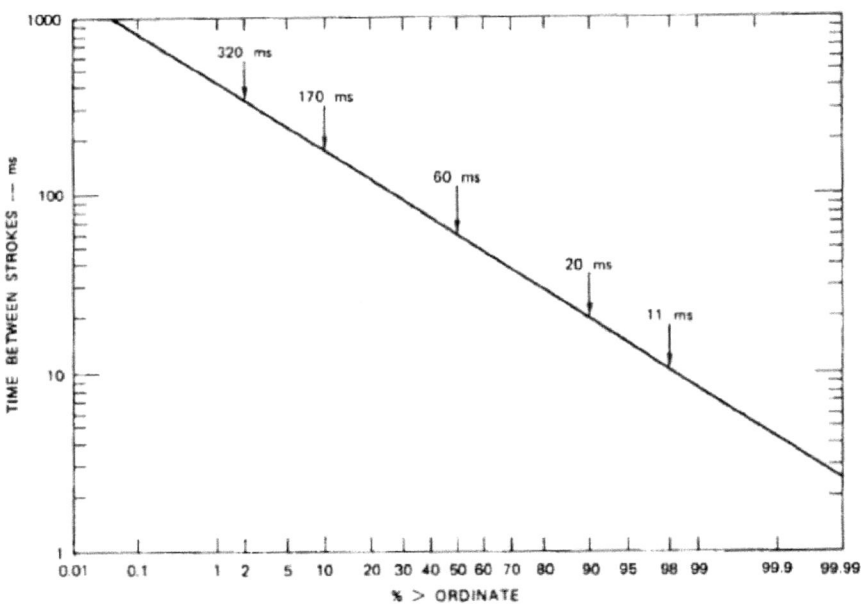

Figure 1.24 Distribution of time interval between strokes.

24

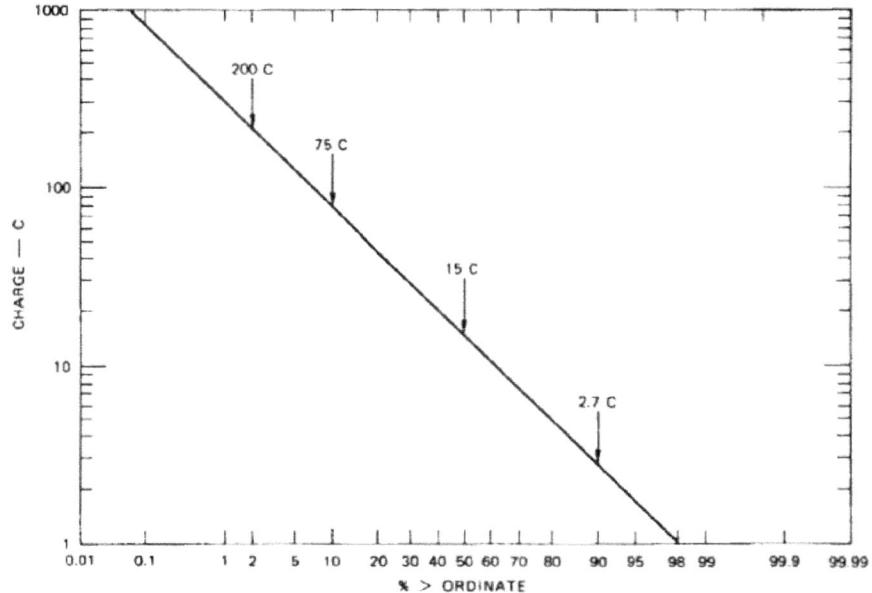

Figure 1.25 Distribution of charge/flash.

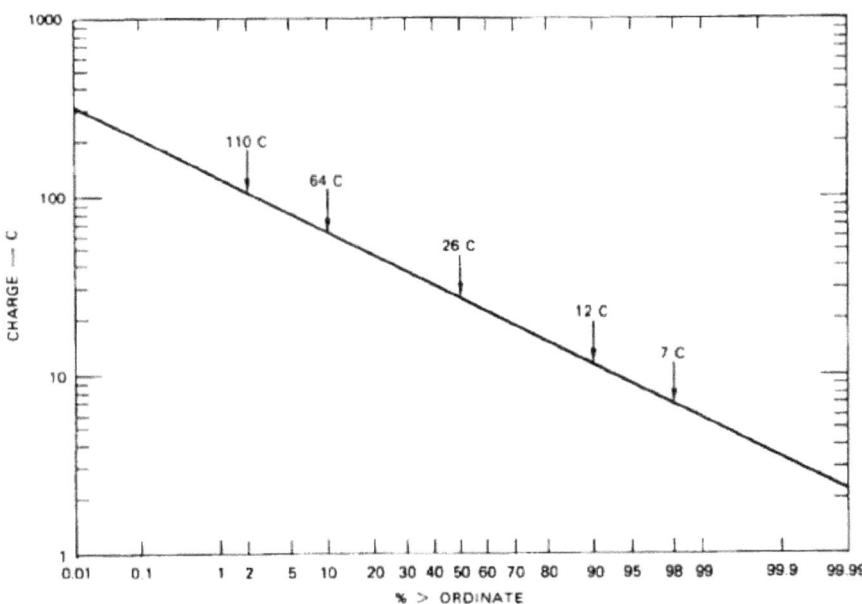

Figure 1.26 Distribution of charge in continuing current.

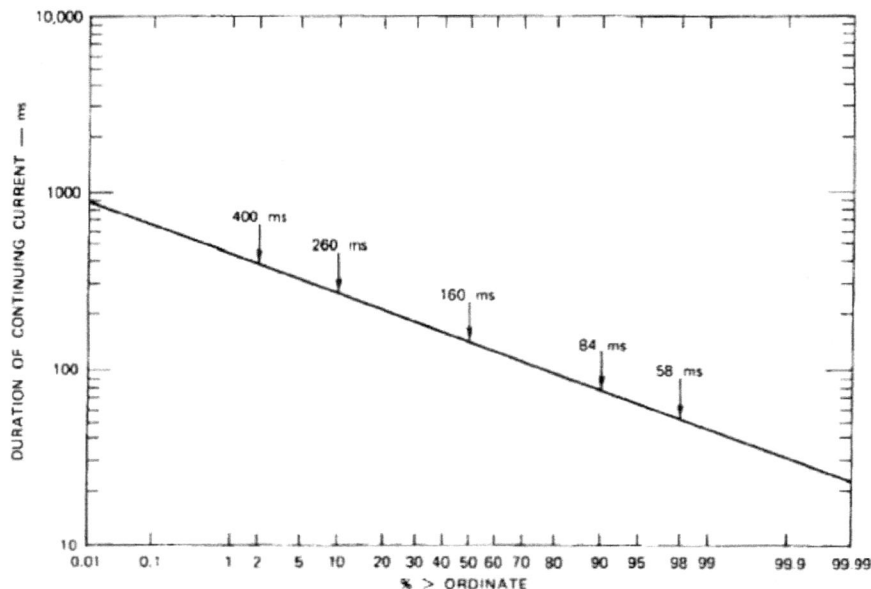

Figure 1.27 Distribution of duration of continuing currents.

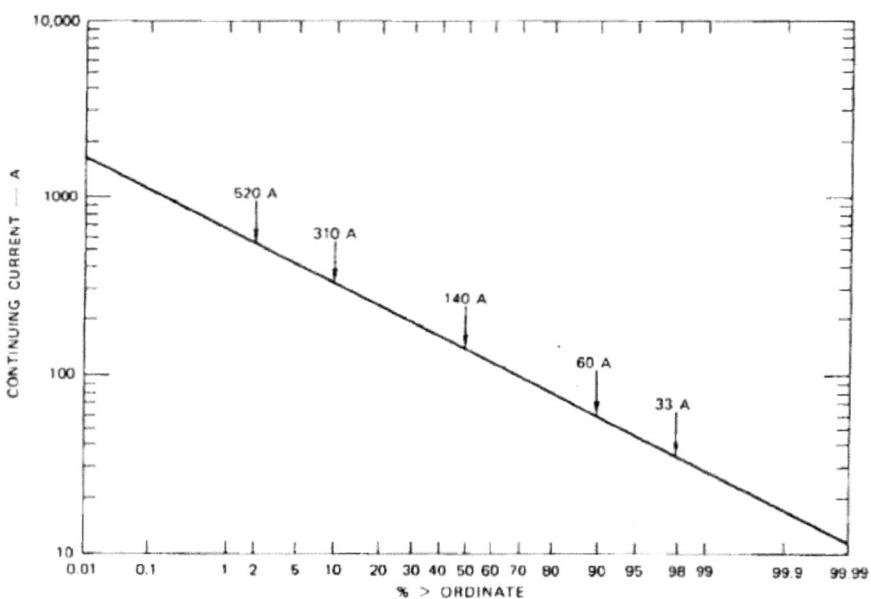

Figure 1.28 Distribution of amplitude of continuing current.

transferred in the flash. Little of the charge is transferred by any one stroke. Instead, most of it is transferred by the continuing currents. The total charge transfer and the amplitude and duration of the continuing currents largely govern the thermal effects of lightning. Data on the characteristics of these continuing currents are shown in Figures 1.26, 1.27, and 1.28 (References 1.29, 1.30, and 1.31).

1.7 Thunderstorm Frequency and Lightning-Flash Density

One of the major factors to consider in determining the probability of lightning damage is the number of lightning flashes to earth in a given area and for a given time. Since precise quantitative data do not exist (except at a few specifically instrumented structures), a secondary measure, the frequency of thunderstorms, is used.

For many years, weather bureau stations have recorded *thunderstorm days* (the number of days per year on which thunder is heard). This index, called the *isokeraunic level*, is shown for the continental United States in Figure 1.29 (Reference 1.32). It should be noted that the information so collected is of limited value for several reasons. First of all, no distinction is made between cloud-to-cloud discharges and cloud-to-ground flashes. Also, there is no allowance for the duration of a storm. A storm lasting an hour would be counted as heavily as one lasting several hours. A better indicator of lightning frequency would be thunderstorm hours per year. Some weather bureau records are now being made of thunderstorm hours, but not much data has yet been

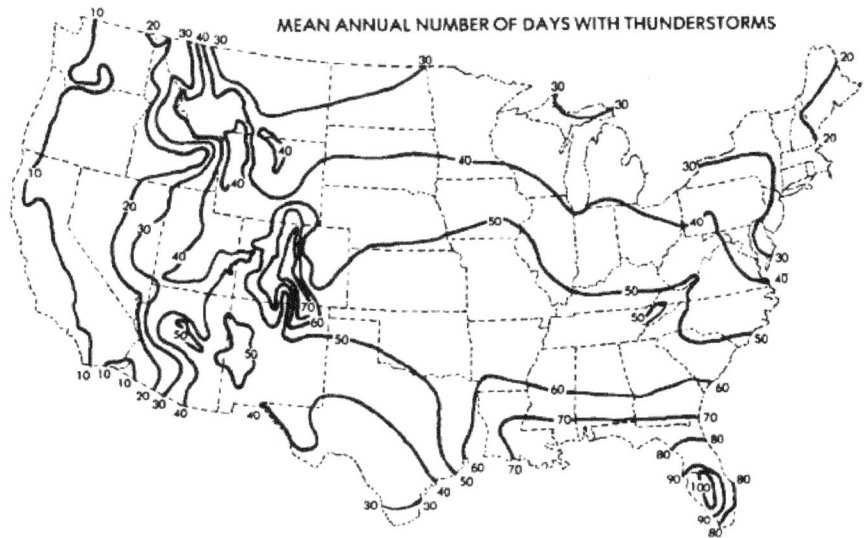

MEAN ANNUAL NUMBER OF DAYS WITH THUNDERSTORMS

Figure 1.29 Thunderstorm days (isokeraunic level) within the continental United States as reported by U.S. Weather Bureau.

27

accumulated. Despite its limitation, the isokeraunic level is broadly useful and can be correlated at least partially with lightning strokes to earth-based objects. Pierce has summarized some of the available data, shown in Figure 1.30 (Reference 1.33).

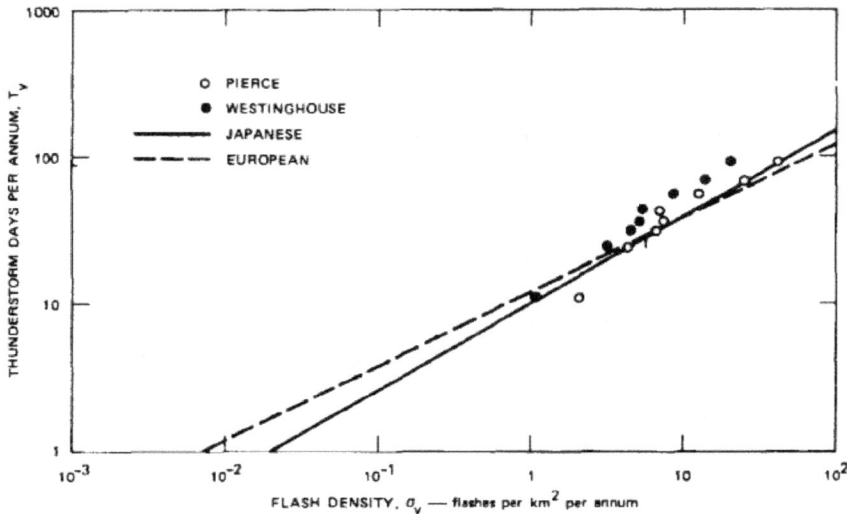

Figure 1.30 Relationship between annual thunderstorm day (T_y) and flash density (σ_y) values.

This flash density includes both flashes between clouds and flashes to ground. There is some evidence that the proportion of flashes that go to ground is related to the geographical latitude of the point under study. This relation would exist because the proportion of flashes to ground depends partly on the average height of clouds, and this, in turn, depends on the type of storm formation. Pierce (Reference 1.34) has proposed the relation shown in Figure 1.31).

The number of lightning flashes to any particular object depends strongly on the terrain upon which the object is situated. Objects on the crests of hills are more prone to be struck than are objects in valleys. All other things considered, however, the probability that a given object will be struck depends upon the area covered by that object. For objects flat upon the ground--for example, a long group of cables laid on the ground surface--the strike interception area can be taken as directly equal to the area covered. Figure 1.32 shows such a group of cables. Equating the stroke interception area to the actual area covered implies that the stroke interception probability is unity for stroke 1 and zero for stroke 2 falling somewhat to the side. This may not be strictly true, since the ultimate contact point of a lightning flash seems to depend upon the junction of the downcoming lightning leader and induced leaders which are drawn from the

28

ground when the electric field strength at the ground surface becomes high enough. Such induced leaders could well appear somewhat earlier from a group of cables lying on the ground than from the undisturbed ground surface.

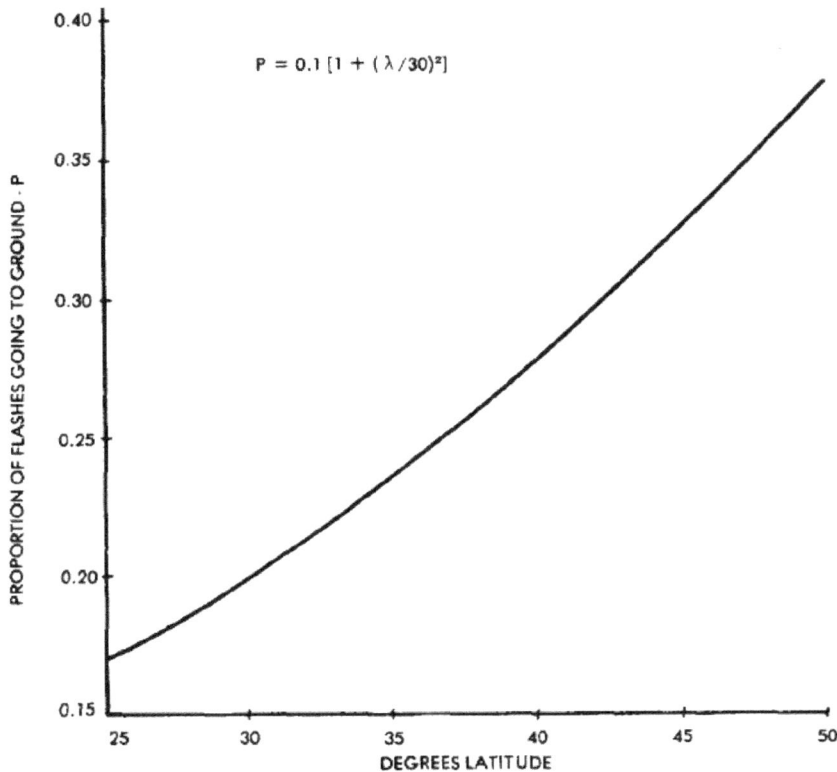

$$P = 0.1 \left[1 + (\lambda/30)^2\right]$$

Figure 1.31 Relationship between geographical latitude and proportion of flashes to ground.

If there is a protruding object on the ground, the stroke interception area depends upon the height of the object. While there is no uniformity of judgment on the effective stroke interception area of a protruding object, one can at least glean some information from the observed pattern of damage to objects near a tall building or on a building protected by lightning rods. It has long been observed that lightning very seldom strikes within a one-to-one cone of protection of a protruding object. The significance of this fact is shown on Figure 1.33. A lightning flash that would ordinarily continue directly to the ground at point r will instead be diverted to the protruding object. This implies that if the distance to point r is less than h, the probability of the protruding object being struck at a distance less than h is very little less than unity. It stands to reason also that the probability of striking the ground falls off from unity

29

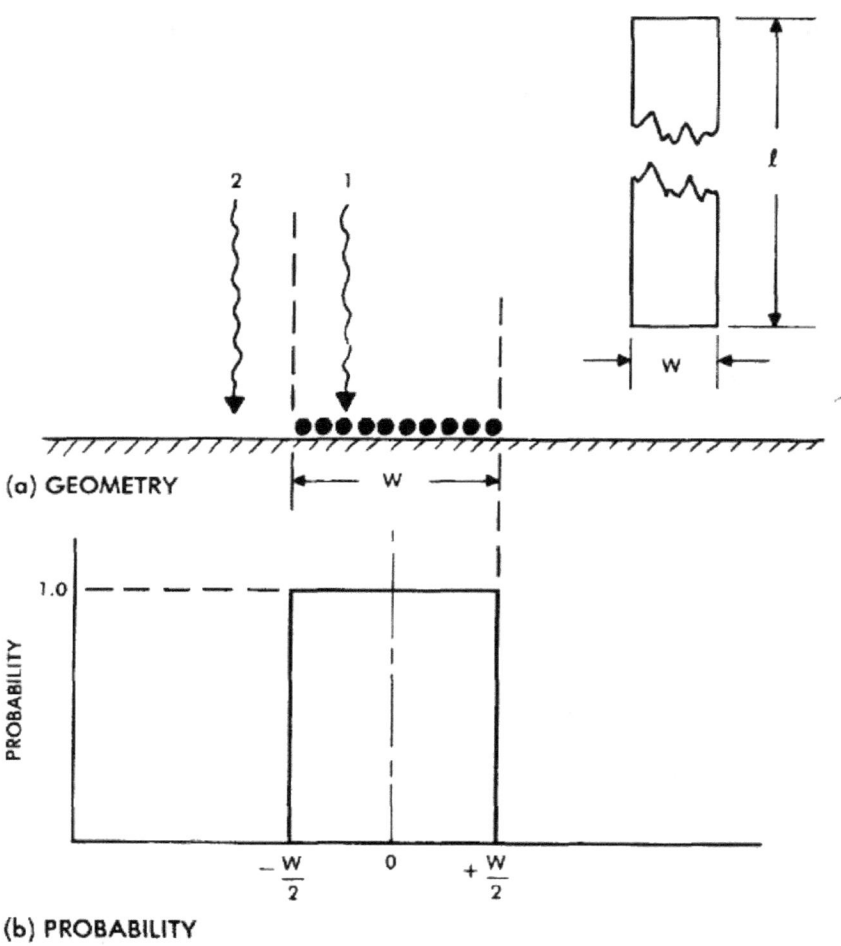

(a) GEOMETRY

PROBABILITY

1.0

$-\dfrac{W}{2}$ 0 $+\dfrac{W}{2}$

(b) PROBABILITY

Figure 1.32 Probability of striking a flat area.

only gradually as one moves farther and farther away from the protruding object. For purposes of calculations of lightning-stroke incidence rate, it can then be assumed that a structure of height h will intercept all flashes that would ordinarily strike the ground over a circle of radius 2h.

It is not as easy to determine analytically how often aircraft in flight will be struck. Field experience seems to be the only reliable guide. Commercial aircraft in regularly scheduled service in the United States are struck about once per year, frequently while in takeoff, landing, or holding patterns, and usually while flying at less than 15 000 ft altitude. Transport aircraft are seldom struck while at cruising altitudes and speeds, partly of course as a result of the fact that storms usually can be, and are, avoided. Aircraft constrained to operate at lower altitudes and along fixed corridors tend to be struck more often.

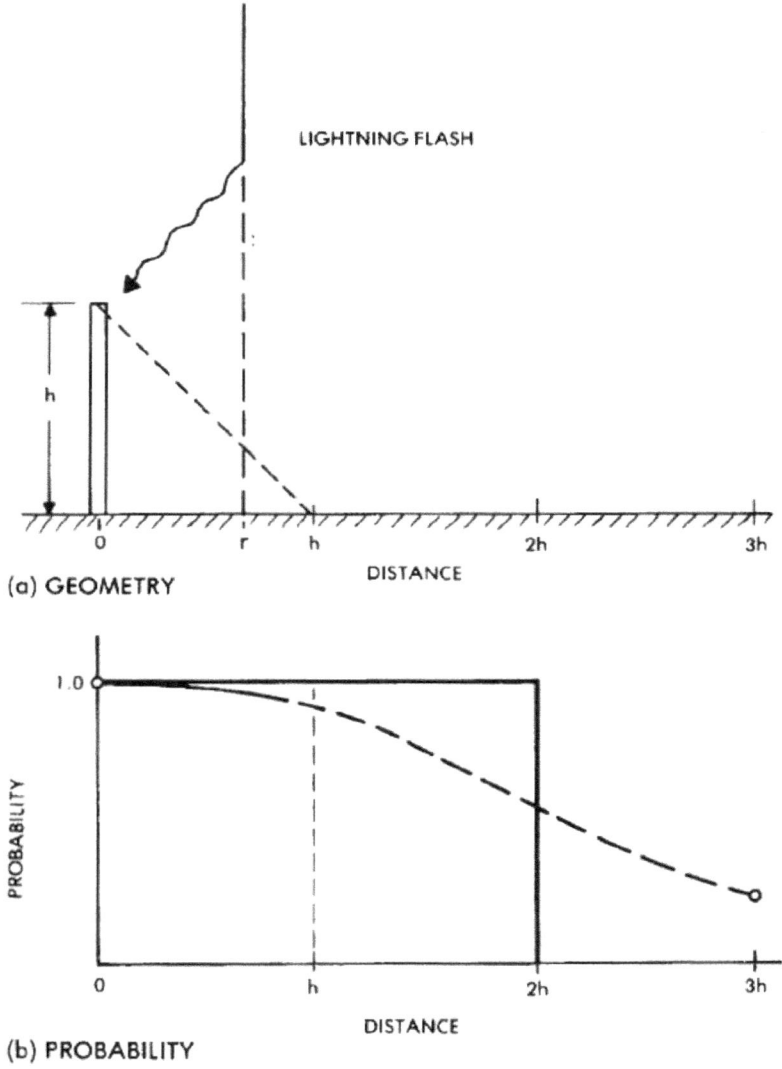

(a) GEOMETRY

(b) PROBABILITY

Figure 1.33 Probability of striking a protruding object.

Military aircraft tend to be struck less often than commercial aircraft, since they are flown in training flights, which are usually scheduled during good weather, much more often than in combat.

1.8 Engineering Models of Lightning Flashes

For purposes of analysis it is helpful to have models of the current flowing in both typical and severe lightning flashes. Pierce (Reference 1.35) gives several

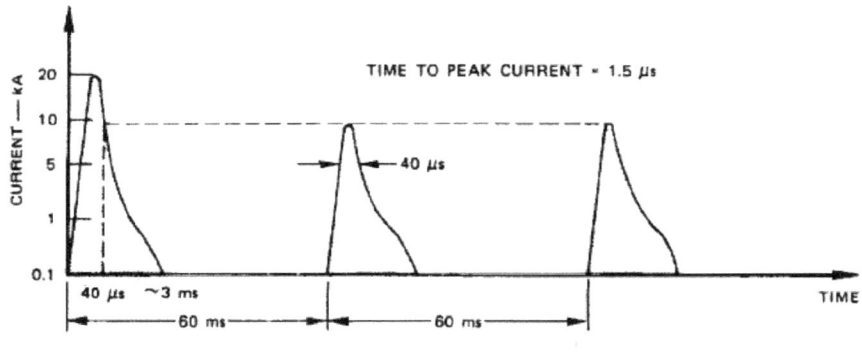

(a) FLASH WITHOUT ANY CONTINUING CURRENT STAGES

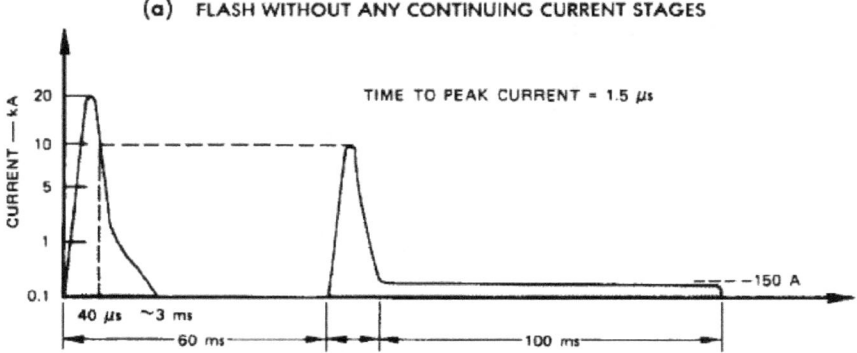

(b) FLASH WITH FINAL STAGE CONTINUING CURRENT

Figure 1.34 Time history of typical (basic) lightning models.

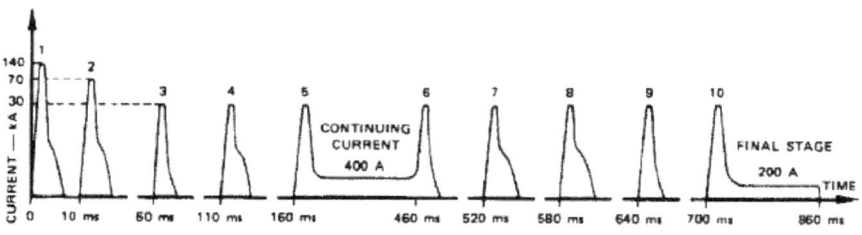

Figure 1.35 Time history of severe (basic) lightning model.

models, shown in Figures 1.34 and 1.35 (References 1.36 and 1.37). Other models may be devised. One that has received wide prominence in the aerospace field is the *Space Shuttle* Lightning Protection Criteria waveform (Figure 1.36 [Reference 1.38]). These, of course, are models for design purposes. As such they duplicate the effects (usually worst case effects) of lightning, but the

32

chances of any real lightning flash producing currents of this exact shape is virtually zero.

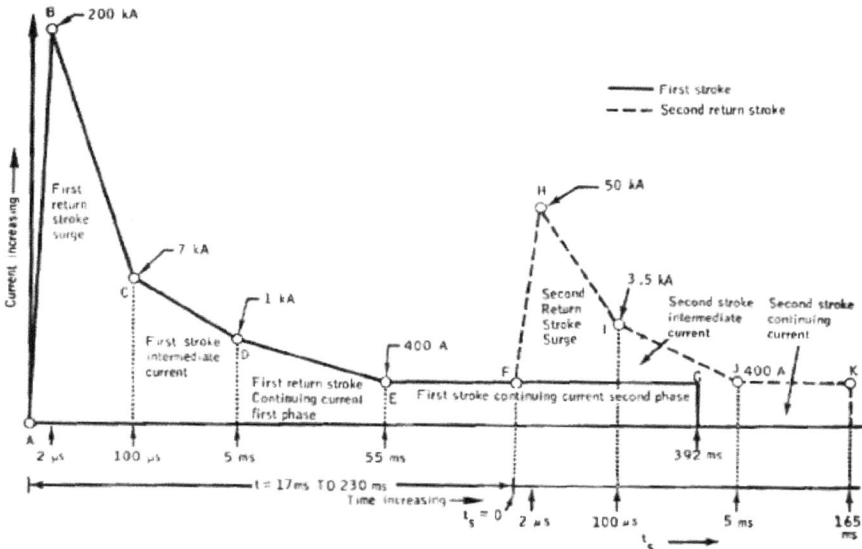

Figure 1.36 Diagrammatic representation of lightning model.
(Note that the diagram is not to scale.)

REFERENCES

1.1   Sir Basil Schonland, "Lightning and the Long Electric Spark," *Advancement of Science* 19 (November 1962): 306-313.

1.2   C. T. R. Wilson, "Some Thunderstorm Problems," *Journal of The Franklin Institute* 208, 1 (July 1929): 1-12.

1.3   C. T. R. Wilson, "Investigations on Lightning Discharges on the Electrical Field of Thunderstorms," *Proceedings of the Royal Society of London,* Series A, 221, the Royal Society of London (1920): 73-115.

1.4   Sir George Simpson, "The Mechanism of a Thunderstorm," *Proceedings of the Royal Society of London,* Series A, 114, the Royal Society of London (1927): 376-99.

1.5   E. A. Evans and K. B. McEachron, "The Thunderstorm," *General Electric Review,* September 1936, pp. 413-25.

1.6   G. C. Simpson, "Lightning," *Journal of the Institution of Electrical Engineers* 67, 395 (November 1929): 1269-82.

1.7   G. C. Simpson and F. J. Scrase, "The Distribution of Electricity in Thunderclouds," *Proceedings of the Royal Society of London,* Series A, 161, the Royal Society of London (1937): 309-52.

1.8   R. H. Golde, *Lightning Protection* (London: Edward Arnold, 1973), p. 6.

1.9   J. L. Marshall, *Lightning Protection* (New York: John Wiley and Sons, 1973), p. 12.

1.10   C. F. Wagner and A. R. Hileman, "The Lightning Stroke—II," *AIEE Transactions* 80, Part III, American Institute of Electrical Engineers, New York, New York (October 1961): 622-42: 623.

1.11   K. Berger and E. Vogelsanger, "Photographische Blitzuntersuchungen der Jahre 1955 ... 1965 auf dem Monte San Salvatore," *Bulletin des Schweizerischen Elektrotechnischen Vereins,* 14 (July 9, 1966): 599-620: 608.

1.12   K. Berger, "Novel Observations on Lightning Discharges: Results of Research on Mount San Salvatore," *Journal of The Franklin Institute* 283, 6 (June 1967): 478-525: 514.

1.13   Wagner and Hileman, "The Lightning Stroke—II," p. 629.

1.14   C. F. Wagner, "Lightning and Transmission Lines," *Journal of The Franklin Institute* 283, 6 (June 1967): 558-594: 560, and C. F. Wagner, "The Relation between Stroke Current and the Velocity of the Return Stroke," *IEEE Transactions on Power Apparatus and Systems* 82, American Institute of Electrical Engineers, New York, New York (October 1963): 609-17: 609.

1.15   C. F. Wagner and A. R. Hileman, "Surge Impedance and Its Application to the Lightning Stroke," *AIEE Transactions* 80, Part 3, American Institute of Electrical Engineers, New York, New York (February 1962): 1011-22.

1.16   See Berger and Vogelsanger, "Photographische Blitzuntersuchungen der Jahre 1955 ... 1965 auf dem Monte San Salvatore," *Bulletin des Schweizerischen Elektrotechnischen Vereins,* 14 (July 9, 1966): 599-620, and K. Berger, "Novel Observations on Lightning Discharges: Results of

Research on Mount San Salvatore," *Journal of The Franklin Institute,* 283, 6 (June 1967): 478-525.

1.17 Redrawn from C. F. Wagner, "Lightning and Transmission Lines," p. 562.

1.18 Berger, "Novel Observations," p. 492.

1.19 Berger, "Novel Observations," p. 494.

1.20 N. Cianos and E. T. Pierce, *A Ground-Lightning Environment for Engineering Use,* Technical Report 1, prepared by the Stanford Research Institute for the McDonnell Douglas Astronautics Company, Huntington Beach, California (August 1972).

1.21 Cianos and Pierce, p. 65.

1.22 Cianos and Pierce, p. 67.

1.23 Cianos and Pierce, p. 68.

1.24 Cianos and Pierce, p. 69.

1.25 Cianos and Pierce, p. 63.

1.26 Cianos and Pierce, p. 61.

1.27 Cianos and Pierce, p. 64.

1.28 Cianos and Pierce, p. 66.

1.29 Cianos and Pierce, p. 72.

1.30 Cianos and Pierce, p. 70.

1.31 Cianos and Pierce. p. 71.

1.32 U.S. Department of Commerce, National Oceanic and Atmospheric Administration, Environmental Data Service, National Climatic Center, Asheville, North Carolina.

1.33 Cianos and Pierce, p. 124.

1.34 E. T. Pierce, "Latitudinal Variation of Lightning Parameters," *Journal of Applied Meteorology,* 9 (1970): 194-95.

1.35 Cianos and Pierce, *A Ground-Lightning Environment,* pp. 75-83.

1.36 Cianos and Pierce, p. 76.

1.37 Cianos and Pierce, p. 82.

1.38 *Space Shuttle Program Lightning Protection Criteria Document,* JSC-07636, Revision A, National Aeronautics and Space Administration, Lyndon B. Johnson Space Center, Houston, Texas (November 4, 1975), p. A-5.

CHAPTER 2
AIRCRAFT LIGHTNING ATTACHMENT PHENOMENA

2.1 Introduction

Statistics on lightning strikes reported by aircraft pilots seem to indicate that no aircraft is likely to receive more than one or two lightning strikes in a year but that some types of aircraft receive more strikes than others. Compared with exposure to other hazards, such as hail, birds, and turbulence, which are also encountered during flight, the exposure of aircraft to lightning strikes seems relatively infrequent. Because of this relatively low incidence, inclusion of lightning protective measures in the designs of some aircraft has been considered unnecessary. The question also arises: If lightning strikes do in fact occur infrequently, can they be avoided altogether? Because some types of aircraft seem to experience more than their "fair share" of lightning strikes, a related question also arises: Why are some aircraft more vulnerable than others? To answer these questions a considerable amount of research on the effects of such factors as aircraft size, engine exhaust, and microwave radar emissions on lightning-strike formation has been undertaken during the past few years. Much of this research has been aimed at answering the question of whether or not an aircraft can in fact produce its own lightning strike, or *static discharge*, or if it can *trigger* an impending flash from a nearby cloud. While some of the findings are as yet inconclusive, others have provided definite answers to some of these questions. In this chapter we summarize what has been learned about the aircraft's influence on lightning-strike occurrence and dispel some other misconceptions about this phenomenon. In succeeding chapters we examine how other factors, such as flight and weather conditions, affect lightning strikes and conclude why it is important that lightning protection be incorporated into all new aircraft designs.

2.2 Aircraft Influence on Flash Formation

At the beginning of lightning-flash formation, when a stepped-leader propagates outward from a cloud charge center, the ultimate destination of the flash at an opposite charge center in another cloud or on the ground has not yet been determined. The difference of potential which exists between the stepped-leader and the opposite charge center(s) establishes an electrostatic force field between them, represented by imaginary equipotential surfaces. They are shown as lines in the two-dimensional drawing of Figure 2.1. The field intensity, commonly expressed in kilovolts per meter, is greatest where equipotential surfaces are closest together. It is this field that is available to ionize air and form the conductive spark which is the leader. Because the direction of electrostatic force is normal to the equipotentials and strongest where they are closest together, the leader is most likely to progress toward the most intense field regions.

If an aircraft happens to be in the neighborhood, it will assume the

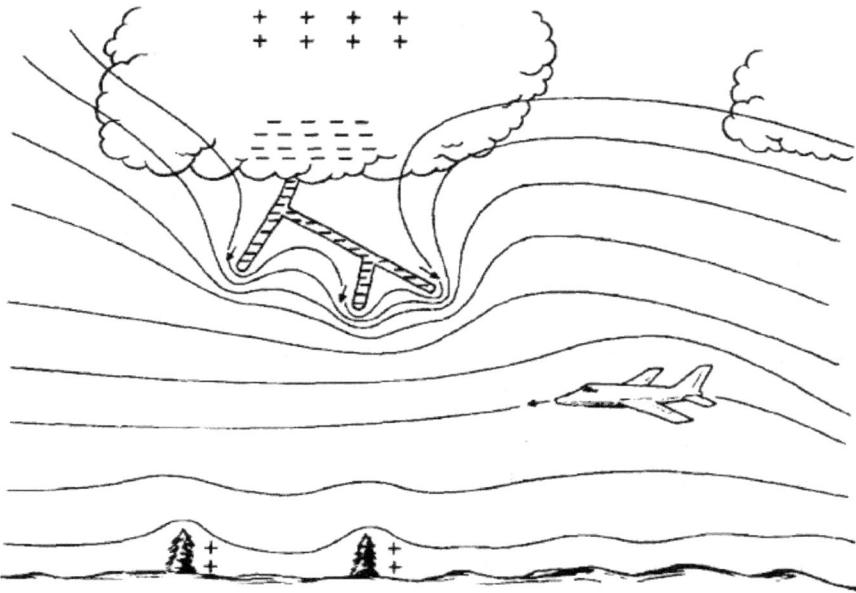

Figure 2.1 Aircraft influence on stepped-leader direction.

electrical potential of its location. Since the aircraft is a thick conductor and all of it is at this same potential, it will divert and compress adjacent equipotentials, thus increasing the electric field intensity in the vicinity of the aircraft, and especially between it and other charged objects, such as the leader. If the aircraft is far away from the leader, its effect on the field near the leader is negligible; however, if the aircraft is within several tens or hundreds of meters from the leader, the increased field intensity in between may be sufficient to attract subsequent leader propagation toward the aircraft. As this happens, the intervening field will become even more intense, and the leader will advance more directly toward the aircraft.

The highest electric fields about the aircraft will occur around extremities, where the equipotential lines are compressed closest together, as shown in Figure 2.2. Typically, these are the nose and wing and empennage tips, and also smaller protrusions, such as antennas or pitot probes. When the leader advances to the point where the field adjacent to an aircraft extremity is increased to about 30 kV/cm, the air will ionize and electrical sparks will form at the aircraft extremities, extending in the direction of the oncoming leader. Several of these sparks, called *streamers*, usually occur simultaneously from several extremities of the aircraft. These streamers will continue to propagate outward as long as the field remains above about 7 kV/cm (Reference 2.1). One of these streamers will meet the nearest branch of the advancing leader and form a continuous spark from the cloud charge center to the aircraft. Thus, when the aircraft is close enough to influence the direction of the leader propagation, it will very likely become attached to a branch of the leader system.

DIRECTION OF FIELD

↑

EQUIPOTENTIAL LINES

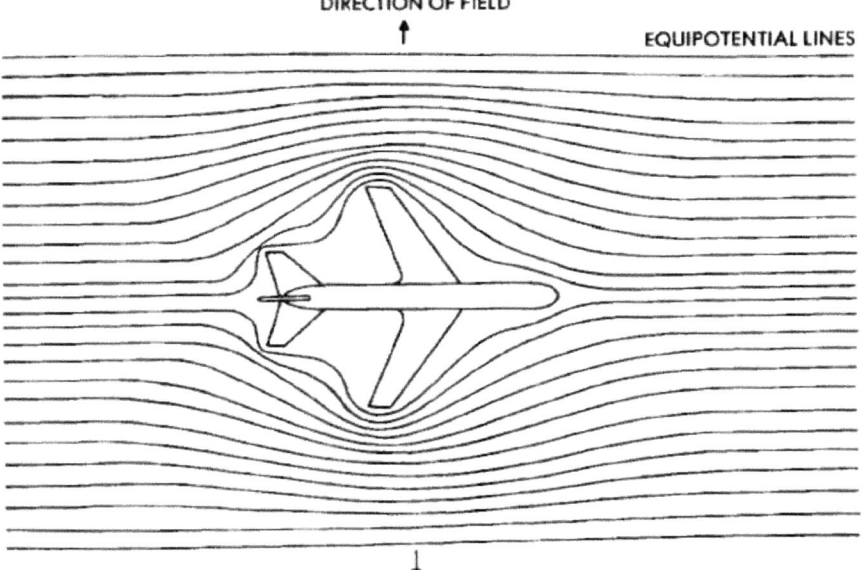

Figure 2.2 Compression of electric field around an aircraft.

When the aircraft is attached to the leader, some charge (free electrons) will flow onto the aircraft, but the amount of charge which can be taken on is limited by the aircraft size. The measure of the aircraft's ability to store charge is its *capacitance*. The capacitance of a complex object, such as an aircraft, is impractical to calculate but may be estimated by comparing it with the easily calculated capacitance of a sphere of equivalent surface area in an infinite field in which no other charge centers are present. The capacitance of such a sphere is given by

$$C = 4\pi\epsilon_r\epsilon_v a = 4\pi\epsilon_r\epsilon_v \sqrt{\frac{S}{4\pi}} \tag{2.1}$$

where

C = capacitance (farads [F])
a = radius (meters)
S = surface area (meters^2)
ϵ_v = absolute dielectric constant (=$10^{-9}/36\pi$ for vacuum)
ϵ_r = relative dielectric constant with respect to vacuum
   (= 1 for air)

39

For a typical transport aircraft with a total surface area of 3000 m^2, C is, from Equation 2.1

$$C = 4\pi \left(\frac{10^{-9}}{36\pi} \right) (1) \sqrt{\frac{S}{4\pi}} \qquad (2.2)$$

$$= \frac{10^{-9}}{9} \sqrt{\frac{3,000}{4\pi}} = \frac{15.45}{9} \times 10^{-9}$$

$$= 1.717 \times 10^{-9} = 1717 \text{ picofarads (pF)}$$

A similar calculation has been made by Schaeffer and Weinstock (Reference 2.2) for smaller fighter aircraft, of 420 pF, and by Moore (Reference 2.3) for helicopters, of 600 pF.

The amount of charge which can be stored by an object having a known capacitance is

$$Q = CV \qquad (2.3)$$

where

Q = charge (coulombs)
C = capacitance (farads)
V = object potential (volts)

Thus, before the amount of charge which can be stored on an aircraft can be found, it is necessary to determine the potential, V, which the aircraft can assume. For this purpose, it will be assumed that the potential of the entire aircraft is the same as that of a small sphere whose radius is similar to that of the sharp aircraft extremities from which streamers develop and from which charge begins to leave the aircraft. Assuming, for example, that this radius is 2 cm, the capacitance of a sphere of this radius is first determined directly from Equation 2.1.

$$C = 4\pi\epsilon_r\epsilon_v (0.02) \qquad (2.4)$$

$$\left(\frac{10^{-9}}{9} \right)(0.02) = 2.22 \text{ pF}$$

The electric field at the surface of this charged sphere (Reference 2.4) is given by

$$E = \frac{Q}{4\pi\epsilon_r\epsilon_v a^2} \qquad (2.5)$$

If E is assumed to be equal to 30 kV/cm, which is the ionization stress of air in a uniform field, and if Equation 2.3 is substituted in Equation 2.5 for Q, then

$$30 \text{ kV/cm} = \frac{CV}{4\pi\epsilon_r\epsilon_v a^2} \qquad (2.6)$$

40

The values of C and a of Equation 2.4 are then substituted into Equation 2.6 and the equation solved for V, as follows:

$$V = \frac{(30 \times 10^5)\,[(0.11 \times 10^{-9})\,(0.02)^2]}{2.22 \times 10^{-12}} \tag{2.7}$$

$$= \frac{0.00132 \times 10^{-4}}{2.22 \times 10^{-12}} = 0.000594 \times 10^8$$

$$= 59.4 \text{ kV}$$

Since the cloud and ground charge centers already establish an ambient field, 59.4 kV is the potential at which the aircraft must be relative to the ambient field for streamering to occur. If the capacitance of even a large aircraft is no more than about 2000 pF, as given by Equation 2.2, then the maximum amount of charge which the aircraft can retain is

$$Q = (2 \times 10^{-9})\,(59.4 \times 10^3) \tag{2.8}$$

$$Q = 118.8 \times 10^{-6} \text{ coulombs (C)}$$

$$Q \approx 100 \text{ microcoulombs } (\mu C)$$

This can be considered the amount of charge which a leader can deliver to the aircraft before streamers occur from opposite extremities. If additional charge flows onto the aircraft, more profuse streaming will occur, and from extremities of larger radii of curvature. In fact, the maximum charge which can be on the aircraft probably exceeds by up to 100 times the streamer initiation value. However, statistics on natural lightning characteristics show that a typical leader contains about 1 to 10 C (Reference 2.5), so there is still no room for any significant portion of this to accumulate on an aircraft. Thus, the aircraft merely becomes an extension of the path being taken by the leader on its way to an ultimate destination at a reservoir of opposite polarity charge. Streamers may propagate onward from two or more extremities of the aircraft at the same time. If so, the incoming leader will have split, and the two (or more) branches will continue from the aircraft independently of each other until one or both of them reach their destination. This process of attachment and propagation onward from an aircraft is shown in Figure 2.3.

When the leader has reached its destination and a continuous ionized channel between charge centers has been formed, recombination of electrons and positive ions occurs back up the channel, and this forms the high-amplitude return stroke current. This stroke current and any subsequent stroke or continuing current components must flow through the aircraft, which is now part of the conducting path between charge centers, as shown in Figure 2.4(a).

If another branch of the original leader reaches the ground before the branch which has involved the aircraft, the return stroke will follow the former, and all other branches will die out, as shown in Figure 2.4(b). No substantial

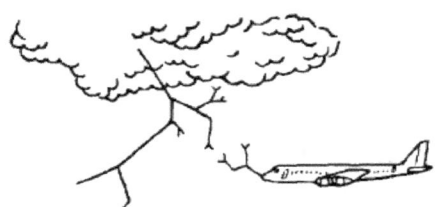

(a)

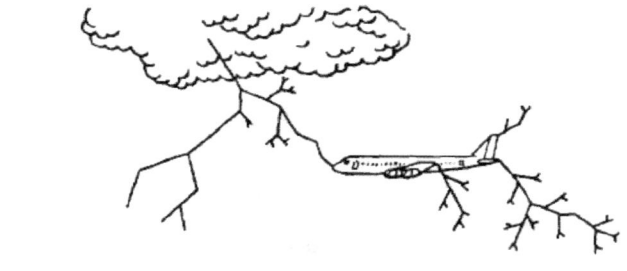

(b)

Figure 2.3 Stepped-leader attachment to an aircraft.
(a) Stepped-leader approaching aircraft
(b) Stepped-leader attachment and continued
propagation from an aircraft

currents will flow through the aircraft in such a case, and any damage to the aircraft will be slight. A still photograph of a downward-branching flash after completion of the main channel is shown in Figure 2.5. Several dying branches are evident in the photograph.

2.2.1 Precipitation Static

The foregoing analysis began with the assumption that the aircraft is at the potential of its position in an electric field established by the cloud and ground charge centers. In dry air this is correct, but if the aircraft is flying through dry

42

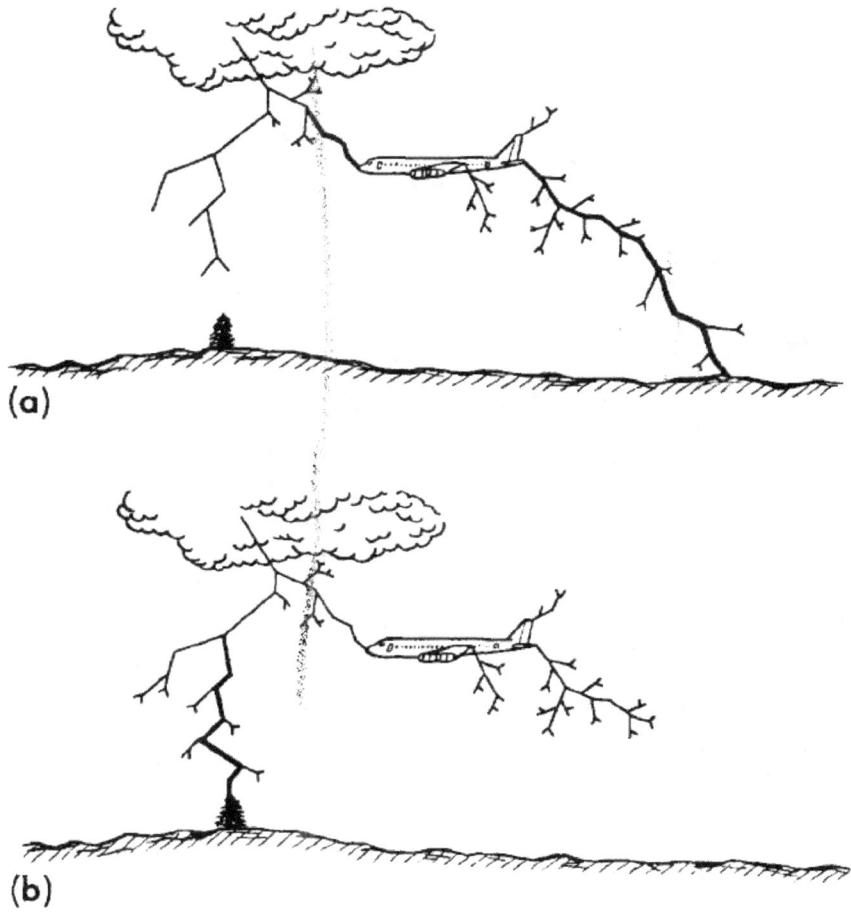

Figure 2.4 Return stroke paths.
(a) Return stroke through the aircraft
(b) No return stroke through the aircraft

precipitation in the form of sleet, hail, or snow, the impact of these particles on the aircraft will cause a charge to separate from the particle and join the aircraft, leaving the aircraft with a preponderance of positive or negative charge (depending on the form of precipitation), thereby changing its potential with respect to its surroundings. This phenomenon is known as *triboelectric charging* and has been extensively studied by Tanner and Nanevicz, and others (Reference 2.6). It is commonly referred to as *precipitation static*, or *P-static*.

The P-static charging process is easily capable of raising the aircraft to a potential of 50 kV, or more, with respect to its surroundings, a charge sufficient to cause ionization at sharp extremities. This ionization radiates broadband

43

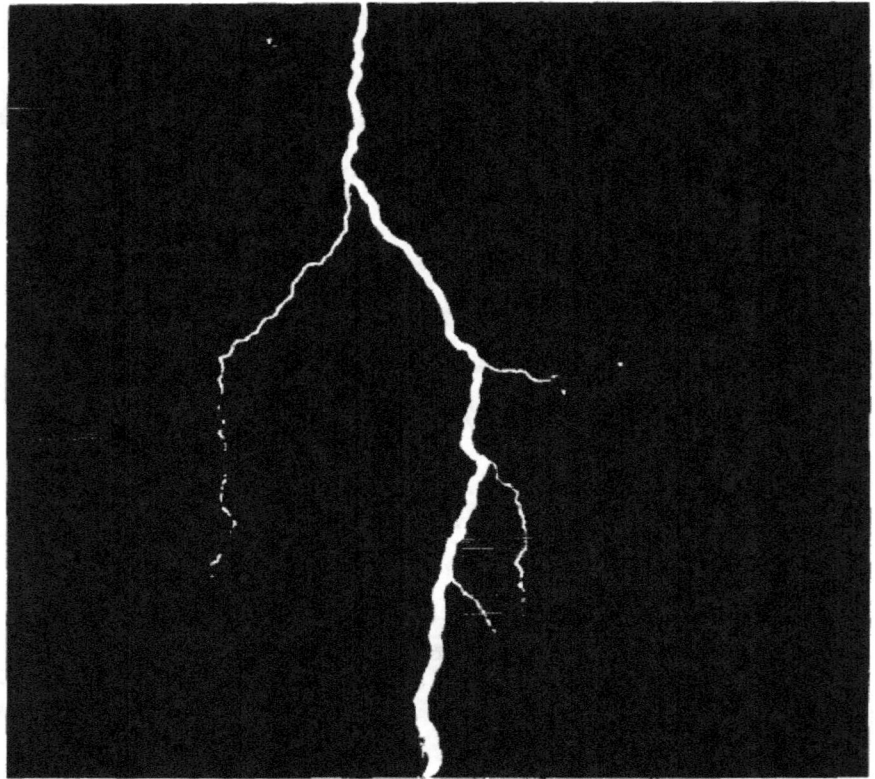

Figure 2.5 Downward-branching flash.

electromagnetic radiation (EMR) throughout the low- and high-frequency spectrum. This EMR is often received as interference, or *static*, by the aircraft communications or low-frequency automatic direction finding (LF-ADF) receivers. The radiated EMR spectrum from P-static discharging at a typical trailing edge is shown in Figure 2.6 (Reference 2.7).

The EMR results from a continuous series of minute streamer-like discharges of ionized air in the immediate (i.e., 10 cm) vicinity of sharp extremities. These discharges also produce a continuous ultraviolet glow visible at night and called *St. Elmo's Fire* or *corona*.

P-static discharging (corona) will occur initially from the sharpest extremities, where the surrounding field first reaches the ionization potential for air. If more P-static charge enters the aircraft than is bled off by the discharges at these extremities, the aircraft potential will increase until the field surrounding extremities of large radii also becomes intense enough to ionize air. Thus, as the aircraft potential increases, the radiated EMR from static discharging becomes more intense, and so does the associated static in communications receivers.

44

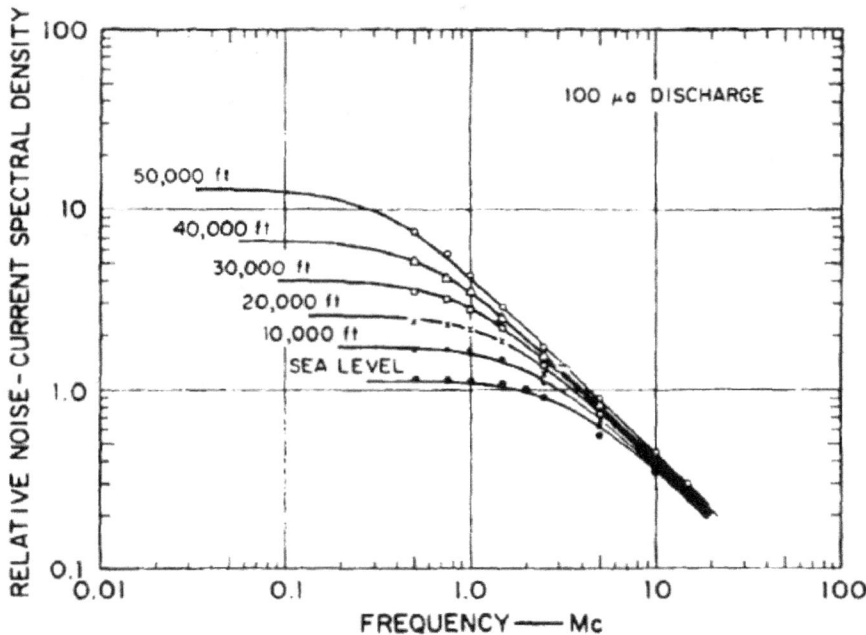

Figure 2.6 Normalized noise spectrum from trailing edge.

To reduce the level of P-static interference, it is therefore appropriate to reduce the amount of charge which can be stored on the aircraft, and thus its potential relative to its surroundings. This has been done (Reference 2.8) by attaching devices called *static dischargers* to the aircraft. These are brushes or sharp needles with very small radii of curvature and thus of low corona inception potential. They are most effective in draining charge from the aircraft when located at regions where the surrounding field stress is already high and where the airflow may readily carry away discharged electrons or ions. Thus, they are usually found on trailing edges of wing tips, empennage tips, and tail cones, as shown in Figure 2.7. While these dischargers are sometimes struck by lightning, they are not capable of either attracting or preventing lightning flashes, or of diverting them from other attachment points on the aircraft.

P-static persists for as long as the aircraft is being charged by impact with dry precipitation, such as sleet or snow. It is rarely reported in rain. It is thus a continuous phenomenon lasting from several seconds to many minutes. Once the aircraft has left such a region, the static in communications receivers quickly clears up, and the aircraft potential reverts to that of its surroundings—established again by its location in the ambient cloud-ground electric field. Because of its low capacitance, an aircraft cannot retain enough P-static charge to produce a startling flash of several meters (or more) in length or a loud report, such as is often heard when lightning strikes the aircraft. Nevertheless, pilots often report a "static discharge" from the aircraft, and proceed to describe the

45

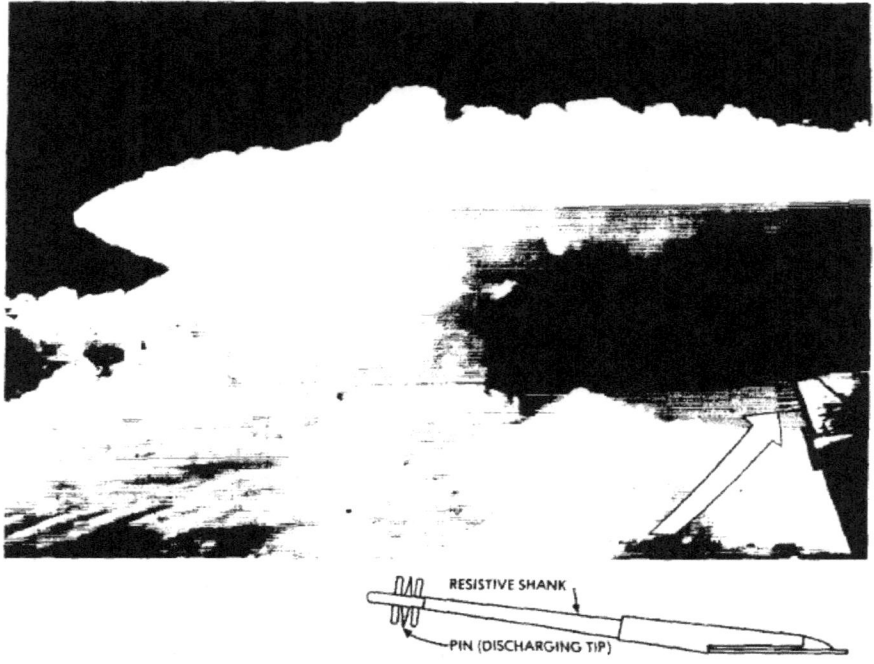

RESISTIVE SHANK

PIN (DISCHARGING TIP)

Figure 2.7 Precipitation static discharger on an aircraft.

symptoms of a lightning strike: i.e., a long, bright flash or spark extending outward from the aircraft, usually accompanied by a loud report, such as that which would be produced by a shotgun going off outside the cockpit.

Thus, the P-static process cannot contribute much to the formation of a lightning leader or to the process of leader attachment to an aircraft. It is true that an approaching streamer is most likely to attach to a point from which an opposing streamer has developed, but the intense field produced by an advancing leader would overcome that produced by the P-static process and draw streamers of its own from most locations where P-static discharges were occurring.

The P-static discharging process may be intensified when the aircraft is in a region where the ambient electric field is relatively intense, as is the case when a lightning flash is imminent. When the flash occurs (whether or not it intercepts the aircraft), the main charge centers are neutralized and the field collapses, thereby reducing the intensity of P-static discharging. This is the reason that pilots frequently report that P-static interference gradually intensifies until a lightning flash occurs and then diminishes instantaneously. This experience reinforces the pilot's impression that the flash is a sudden static discharge from the aircraft alone. For this reason P-static interference should be looked upon by pilots as an indicator that a lightning strike may be imminent (i.e., within a few seconds to a few minutes). P-static interference is, in fact, reported prior to the lightning strike in about half of the lightning-strike incidents described in recent airline lightning-strike reports (Reference 2.9).

46

2.2.2 Can an Aircraft Trigger a Lightning Strike?

A question often asked is If an aircraft cannot produce its own lightning flash, can it trigger a natural one? Stated another way the question might be Would the lightning flash have occurred if the aircraft were not present?

While there is insufficient scientific data upon which to base a conclusive answer to these questions, the following factors suggest that the aircraft does not often trigger a flash.

1. Aircraft often fly through electrified regions without being struck, while lightning flashes are occurring nearby.

2. The stepped-leader must begin from a charge source capable of furnishing it with several coulombs of charge. Thus, the potential (voltage) of this center, and the surrounding field intensity, would seem to be much greater than that about an aircraft, leaving the implication that, unless the aircraft is very close to the charge center, it can have little influence on the surrounding field or on the process of leader initiation.

3. Laboratory breakdown tests of long high-voltage air gaps, thought to be similar to lightning leader formation, show that initial ionization always begins at one of the electrodes and not from an object suspended in the gap (Reference 2.10). Such an object significantly influences the voltage level at which breakdown begins only if it is close enough to one electrode to influence the field about this electrode.

It is more probable that the aircraft does not become involved until after leader propagation has begun. If the leader happens to approach the aircraft, the field intensification produced by the presence of the aircraft becomes much more significant, and the leader may now be attracted to the aircraft.

There is some evidence (Reference 2.11) that jumbo jet (wide-body) aircraft do trigger their own flashes, but this is not yet conclusive, since accumulated flight hours are not yet nearly as great as those for conventional aircraft. If large-body aircraft are in fact triggering flashes, it is probably because their larger sizes make a more noticeable perturbation on the electric field near the cloud charge centers from which leaders begin.

The aircraft motion has little influence on the propagating leader because the aircraft is moving much slower, about 10^2 m/s, than the leader, which is advancing at 10^5 to 10^6 m/s. Thus, the aircraft appears stationary to the leader during the leader formation process.

Several other stimuli have been mentioned as possible causes of aircraft lightning strikes. These include engine exhaust and radiated electromagnetic energy (i.e., radar transmission).

2.2.2.1 Effect of Jet Engine Exhaust

There has been speculation that the hot jet-engine exhaust gases may contain a sufficient number of ionized particles to attract or trigger a lightning flash to the aircraft. This speculation has been heightened by the widely publicized launch of *Apollo 12*, which apparently triggered a lightning flash that struck the top of the vehicle when it had reached 1950 m and again at 4270 m. The flash exited from the vehicle exhaust plume. Subsequent studies by Nanevicz, Pierce, and Whitson (Reference 2.12) of this and other incidents in which a rocket was rapidly introduced into an intense electric field indicate that the exhaust plume does appear electrically conductive, making the rocket appear longer than its own physical length. An empirical study by Pierce (Reference 2.13) of documented strikes to tall grounded and airborne conductors concludes that there must also be a potential discontinuity between the conductor and the adjacent atmosphere of up to 10^6 V if the lightning leader is initiated from the conductor, and that the rapid discharge of hot ionized gas from the rocket engine may cause sufficient charge separation from the vehicle to increase its potential to 10^6 V or more with respect to its surroundings.

Shaeffer and Weinstock (Reference 2.14) have studied the conductivity of an aircraft jet-engine exhaust. In this case, ionized particles and free electrons in a jet exhaust originate in the combustion chamber as a result of chemical reactions taking place between the intake air and jet fuel. The ion concentration in a jet-engine exhaust has been measured by Fowler (Reference 2.15) to be between 5×10^6 and 3×10^7 particles per cubic centimeter (p/cm^3) and the free electron density deduced from this to be between 5×10^3 to 3×10^5 p/cm^3. The electron density in luminous rocket exhaust has been calculated by Pierce (Reference 2.16) to be 10^{12} p/cm^3, as has that in the tip of an advancing leader. Conversely, the free electron density in ambient air ranges from 10^0 to 10^3 p/cm^3. Evidently, then, the jet-engine exhaust is only slightly more ionized than the ambient air and much less so than the rocket exhaust, with the result that the jet exhaust would not be expected to have sufficient conductivity to initiate or attract a lightning leader. This conclusion is supported by aircraft lightning-strike incident reports, which indicate that engine tail pipes are not often lightning attachment points unless they are already located at an aircraft extremity, where the electric field would be intense from geometrical conditions alone.

There is also no evidence to suggest that jet aircraft are struck more often than piston-engined aircraft. Overall, the ability of the jet aircraft to operate at higher altitudes and spend less time climbing and descending to airports has probably rendered the jet less susceptible than its piston-engined predecessor to lightning strikes, which occur predominantly at low or intermediate altitudes. Recorded strikes which have hit jet engines almost always have terminated on the tail pipes of the aft- (tail-) mounted engines of conventional jet transport aircraft. However, a few recent reports of strikes to wide-body aircraft show evidence of strikes to the tail pipes of wing-mounted engines (Reference 2.17). The evidence to date is indirect, since no tell-tale burn marks have been found on the tail pipes themselves. Instead, electronic engine instruments with sensors mounted on the engine have experienced damage, and no attachment points have been found other than the flash entry point at a nose or wing tip. Lightning attachment points on jet tail pipes may be masked by exhaust deposits or simply be indiscernible because of the high melting temperature of tail pipe metals. If this is true, the increased volume and temperature of the jumbo jet exhaust may have become a region of sufficient ionization or weakened dielectric strength to divert a propagating leader into the exhaust channel or to enable the formation of outward-propagating streamers from the tail pipe, once the aircraft has been struck elsewhere by a leader. While this engine exhaust may divert an existing leader, it seems improbable that it could trigger a flash by itself.

Viewed another way, if the jet engine were leaving behind an ionized exhaust, it seems probable that an imbalance of charge on the aircraft would eventually result, even when flying in clear weather. As with P-static charging, this would raise (or lower) the aircraft potential with respect to its surroundings and cause static discharging and corona from sharp extremities, causing interference in radio equipment. It is well established that this does not happen.

2.2.2.2 Effects of Electromagnetic Radiation

Another suggestion has been made that an aircraft's radar may trigger or divert lightning strikes. This possibility was also investigated by Schaeffer and Weinstock (Reference 2.18), who show that the transmitted power level of microwave radiation necessary to produce an electric field capable of ionizing air is about 6.7×10^6 watts (W), which is far greater than that available from aircraft radars. They also point out that a radar would not be designed to ionize air because the energy required would reduce transmitted beam energy and because ionized air would cause an undesired radar return signal.

Aircraft lightning-strike incident reports also show no evidence of radar or other EMR having been involved in the lightning-strike formation. There are cases in which radomes are punctured, but clearly these are simple cases of dielectric breakdown of the plastic radome material, with flash termination on some airframe-grounded object inside the radome or on the radar antenna itself. The punctures have occurred whether or not the radar set was turned on. The addition of conductive diverter strips to the outside of the radome usually prevents these punctures by enabling lightning flashes to attach directly to a diverter.

2.3 Swept Strokes

After the aircraft has become part of a completed flash channel, the ensuing stroke and continuing currents which flow through the channel may persist for up to a second or more. Essentially, the channel remains in its original location, but the aircraft will move forward a significant distance during the life of the flash. Thus, whereas the initial entry and exit points are determined by the mechanisms previously described, there may be other, subsequent, attachment points that are determined by the motion of the aircraft through the relatively stationary flash channel. In the case of a fighter aircraft, for example, when a forward extremity such as the pitot boom becomes an initial attachment point, its surface moves through the lightning channel, and thus the channel appears to sweep back over the surface, as illustrated in Figure 2.8. This occurrence is known as the *swept-stroke* phenomenon. As the sweeping action occurs, the type of surface can cause the lightning channel to attach and dwell at various surface locations for different periods of time. If part of the surface, such as the radome, is nonmetallic, the flash may continue to dwell at the last metallic attachment point (aft end of the pitot boom) until another exposed metallic surface (the fuselage) has reached it; or the channel may puncture the nonmetallic surface and reattach to a metallic object beneath it (the radar dish). Whether puncture or surface flashover occurs depends on the amplitude and rate of rise of the voltage stress created along the channel, as well as the voltage-withstand strength of the nonmetallic surface and any air gap separating it from enclosed metallic objects. When the lightning arc has been swept back to one of the trailing edges, it may remain attached at that point for the remaining duration of the lighting flash. An initial attachment point at a trailing edge, of course, would not be subjected to any swept-stroke action.

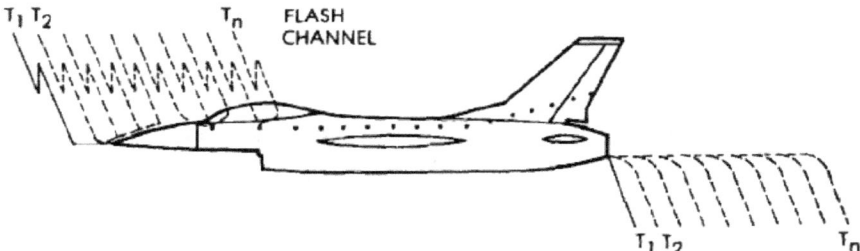

Figure 2.8 Typical path of swept-stroke attachment points.

The aircraft cannot fly out of, or away from, the channel. This is because the potential difference between charge centers (cloud and earth or another cloud) is sufficient to maintain a very long channel until the charges have neutralized each other and the flash dies. The aircraft is a very small (and highly conductive) part of the channel and cannot move sufficiently far away from the vicinity of the channel to become detached from it during the brief lifetime of a lightning flash.

50

2.4   Aircraft Lightning Attachment Zones

Since there are some regions on the aircraft where lightning will not attach, and others which will be exposed to attachment for only a small portion of the total flash duration, it is appropriate to define the zones on the aircraft surface which will be exposed to different components of the flash and therefore receive different types and degrees of effects. For purposes of aircraft fuel system protection, the Federal Aviation Administration (FAA) has defined in its advisory circular AC 20-53 (Reference 2.19) the following zones:

Zone 1.
(a)   All surfaces of the wing tips located within 18 inches of the tip measured parallel to the lateral axis of the aircraft, and surfaces within 18 inches of the leading edge on wings having leading edge sweep angles of more than 45 degrees
(b)   Projections such as engine nacelles, external fuel tanks, propeller disc, and fuselage nose
(c)   Tail group: within 18 inches of the tips of horizontal and vertical stabilizer, trailing edge of horizontal stabilizer, tail cone, and any other protuberances
(d)   Any other projecting part that might constitute a point of direct stroke attachment

Zone 2. Surfaces for which there is a probability of strokes being swept rearward from a Zone 1 point of direct stroke attachment. This zone includes surfaces which extend 18 inches laterally to each side of fore-and-aft lines passing through the Zone 1 forward projection points of stroke attachment. All fuselage and nacelle surfaces, including 18 inches of adjacent surfaces not defined as Zone 1, are included in Zone 2.

Zone 3. Surfaces other than those covered by Zones 1 and 2. Ignition sources in these areas would exist only in the event of streamering.

These definitions confine zone boundaries to the 18-inch distances described above. Even though they have proved adequate for protection of present-day transport category aircraft, it has been recognized that the lightning attachment zones of new aircraft designs in fact may not fit these rigid definitions. Therefore, an industry committee, designated special Task F of the Society of Automotive Engineers (SAE) Committee AE-4 on Electromagnetic Compatibility, has recently clarified the definition of these zones and broadened their application by removal of the arbitrary 18-inch description. The Task F recommended definitions are as follows (Reference 2.20):

Zone 1. Surfaces of the vehicle for which there is a high probability of direct lightning-flash attachment or exit

51

Zone 2. Surfaces of the vehicle across which there is a high probability of a lightning flash being swept by the airflow from a Zone 1 point of direct flash attachment

Zone 3. Zone 3 includes all of the vehicle areas other than those covered by Zone 1 and Zone 2 regions. In Zone 3 there is a low probability of any direct attachment of the lightning-flash arc, but Zone 3 areas may carry substantial amounts of electrical current by direct conduction between some pairs of direct or swept-stroke attachment points in other zones.

The Task F definitions further divide Zones 1 and 2 into A and B regions depending on the probability of the flash hanging on for any protracted period of time. An A-type region is one in which there is low probability that the arc will remain attached, and a B-type region is one in which there is a high probability that the arc will remain attached. Some examples of zones are as follows:

Zone 1A. Initial attachment point with low probability of flash hang-on, such as a nose
Zone 1B. Initial attachment point with high probability of flash hang-on, such as a tail cone

Zone 2A. A swept-stroke zone with low probability of flash hang-on, such as a wing mid-span
Zone 2B. A swept-stroke zone with high probability of flash hang-on, such as a wing trailing edge.

The Task F definitions are not in conflict with the FAA criteria but, rather, are broader in scope and consider the complete aircraft in addition to those areas of importance to the fuel system, which is the system primarily addressed by FAA AC 20-53.

In accord with the Task F criteria, typical lightning effects and test criteria (described in subsequent chapters of this book) have been categorized according to each of these zones. The actual boundaries of each of these zones on a particular aircraft can never be exactly determined, but they can be established with sufficient accuracy to enable adequate lightning protection to be designed for the aircraft.

Initial lightning attachment points (Zone 1A) can usually be established by inspection and comparison with other aircraft in-flight experience. Locations of the other zones can then be deduced from the locations of Zone 1A, in accordance with the definitions. Alternately, simulated lightning flashes can be fired in a laboratory to a model of the aircraft, and the Zone 1 (A and B) attachment points determined photographically; however, the validity of such scale model testing is somewhat questionable because of the nonlinear behavior of air breakdown phenomena. Model test results have shown good comparison with aircraft of conventional wing and empennage configuration but may be less capable of accurately predicting the attachment points of more modern designs of blended wing-fuselage construction or high wing sweep-back angles. An example of lightning-strike attachment zones determined from a model test of the Lockheed S-3A aircraft is shown on Figure 2.9.

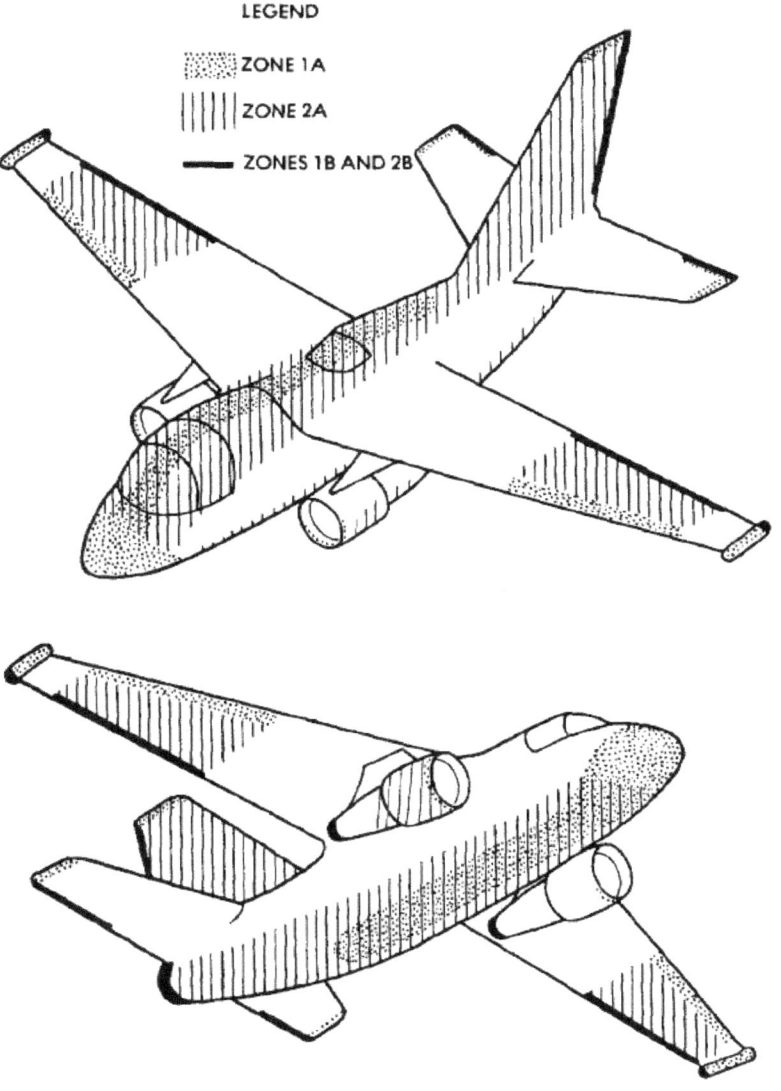

Figure 2.9 Lightning-strike attachment zones predicted from model test results.

REFERENCES

2.1 C. T. Phelps, "Field Enhanced Propagation of Corona Streamers," *Journal of Geophysical Research* 76, 24 (August 20, 1971): 5799-5806.

2.2 J. F. Shaeffer and G. L. Weinstock, *Aircraft Related Lightning Mechanisms,* Technical Report AFAL-TR-386, prepared by the McDonnell Aircraft Company, McDonnell Douglas Corporation, for the Air Force Avionics Laboratory, Air Force Systems Command, Wright-Patterson Air Force Base, Ohio (January 1973), p. 21.

2.3 K. A. Moore, "Precipitation Static Noise Problems on Operational Aircraft," *Lightning and Static Electricity Conference, 3-5 December 1968, Part II: Conference Papers,* AFAL-TR-68-290, Air Force Avionics Laboratory, Air Force Systems Command, Wright-Patterson Air Force Base, Ohio (May 1969), pp. 168-78.

2.4 M. A. Uman, *Lightning* (New York: McGraw-Hill, 1969), pp. 75-81.

2.5 R. Plonsey and R. E. Collin, *Principles and Applications of Electromagnetic Fields* (New York: McGraw-Hill, 1961), p. 50.

2.6 R. L. Tanner and J. E. Nanevicz, "An Analysis of Corona-Generated Interference in Aircraft," *Proceedings of the IEEE* 52, 1, Institute of Electronic and Electrical Engineers, New York, New York (January 1964): 44-52.

2.7 Tanner and Nanevicz, "Corona-Generated Interference," p. 47.

2.8 Cumulonimbus with cyclonic circulation at low levels to its southwest, photographed from U.S. Weather Bureau DC-6B at 20,000 ft. near Topeka, Kansas, 1750 CST, 21 April 1961.

2.9 B. I. Hourihan, *Data from the Airlines Lightning Strike Reporting Project, June 1971 to November 1974,* Summary Report GPR-75-004, High Voltage Laboratory, Environmental Electromagnetics Unit, Corporate Research and Development, General Electric Company, Pittsfield, Massachusetts (March 1975), p. 5.

2.10 R. H. Golde, *Lightning Protection* (London: Edward Arnold, 1973), p. 25.

2.11 E. T. Pierce, "Triggered Lightning and Some Unsuspected Lightning Hazards," *American Association for the Advancement of Science, 138th Annual Meeting, 1971,* Stanford Research Institute, Menlo Park, California (January 1972), pp. 14-28.

2.12 J. E. Nanevicz, E. T. Pierce, and A. L. Whitson, *Atmospheric Electricity and the Apollo Series,* Note 18, Stanford Research Institute, Menlo Park, California (June 1972).

2.13 Pierce, "Triggered Lightning."

2.14 Shaeffer and Weinstock, *Aircraft Related Lightning Mechanisms,* pp. 33-45.

2.15 R. T. Fowler, *Ion Collection by Electrostatic Probe in a Jet Exhaust,* graduate thesis, Air Force Institute of Technology, Wright-Patterson Air Force Base, Ohio (June 1970).

2.16 E. T. Pierce, *Atmospheric Electric and Meteorological Environment of*

Aircraft Incidents Involving Lightning Strikes, Special Interim Report 1, Stanford Research Institute, Menlo Park, California (October 1970).

2.17 Hourihan, *Data from the Airlines*, pp. 40-41.

2.18 Shaeffer and Weinstock, *Aircraft Related Lightning Mechanisms*, pp. 65-66.

2.19 *Protection of Aircraft Fuel Systems Against Lightning*, Federal Aviation Agency Advisory Circular AC 20-53, Federal Aviation Administration, Department of Transportation, Washington, D.C. (October 6, 1967), pp. 2-3.

2.20 *Lightning Test Waveforms and Techniques for Aerospace Vehicles and Hardware*, Society of Automotive Engineers, Warrendale, Pennsylvania (5 May 1976).

CHAPTER 3
LIGHTNING STRIKE EXPERIENCE

3.1 Introduction

The atmospheric and flight conditions under which aircraft have been struck by lightning have been of interest since the beginning of powered flight because lightning and other thunderstorm effects, such as turbulence and icing, are to be avoided if possible. To learn about these conditions, various lightning-strike incident reporting projects have been implemented. Beginning in 1938, the Subcommittee on Meteorological Problems of the National Advisory Committee for Aeronautics (NACA) prepared and distributed a sixteen-page questionnaire to airlines and the Armed Forces (Reference 3.1). Pilots filled out one of these questionnaires after each lightning-strike incident and forwarded it to the NACA subcommittee for analysis. This questionnaire was evidently too long for widespread use, however, and, although it could have been shortened, was discontinued by 1950. Nevertheless, the program provided important data for the first time on the meteorological conditions prevailing when strikes occurred and the resulting effects on the aircraft. For one thing, the data obtained from the NACA program showed that some lightning-strike conditions are common to many incidents.

Subsequent programs were conducted by the Lightning and Transients Research Institute (LTRI) (Reference 3.2) and the FAA (Reference 3.3). Projects currently in operation are being conducted by Plumer and Hourihan of General Electric (Reference 3.4), Anderson and Kröninger of South Africa (Reference 3.5), Perry of the British Civil Aviation Authority (Reference 3.6), and Trunov of the USSR National Research Institute for Civil Aviation (Reference 3.7). Strike incidence data, based largely on turbojet or turboprop aircraft, is usually summarized according to the following categories:

- Altitude
- Flight path (i.e., climb, level flight, descend, etc.)
- Meteorological conditions
- Outside air temperature
- Lightning-strike effects on the aircraft

Altitude, flight path, meteorological conditions, and air temperature are topics discussed in this chapter. Lightning-strike effects on aircraft will be discussed in the succeeding chapter.

3.2 Altitude and Flight Path

Figure 3.1 shows the altitudes at which these reporting projects show aircraft are being struck, as compared with a typical cumulonimbus (thunder) cloud. The turbojet and turboprop data from the four summaries are in close agreement. For comparison, the data from the earlier piston aircraft survey of Newman (Reference 3.8) are also presented. Cruise altitude for jet aircraft is considerably higher (10 km) than that of earlier piston aircraft, which flew at about 4 to 5 km; yet Figure 3.1 shows the altitude distribution of lightning-strike incidents to be nearly the same. This fact indicates (1) that there are more lightning flashes to be intercepted below about 6 km than above this altitude, and (2) that very likely jet aircraft are being struck at lower than cruise altitudes: that is, during climb, descent, or hold operations. Flight regime data obtained from the jet projects shown in Figure 3.2 (Reference 3.9) confirm this.

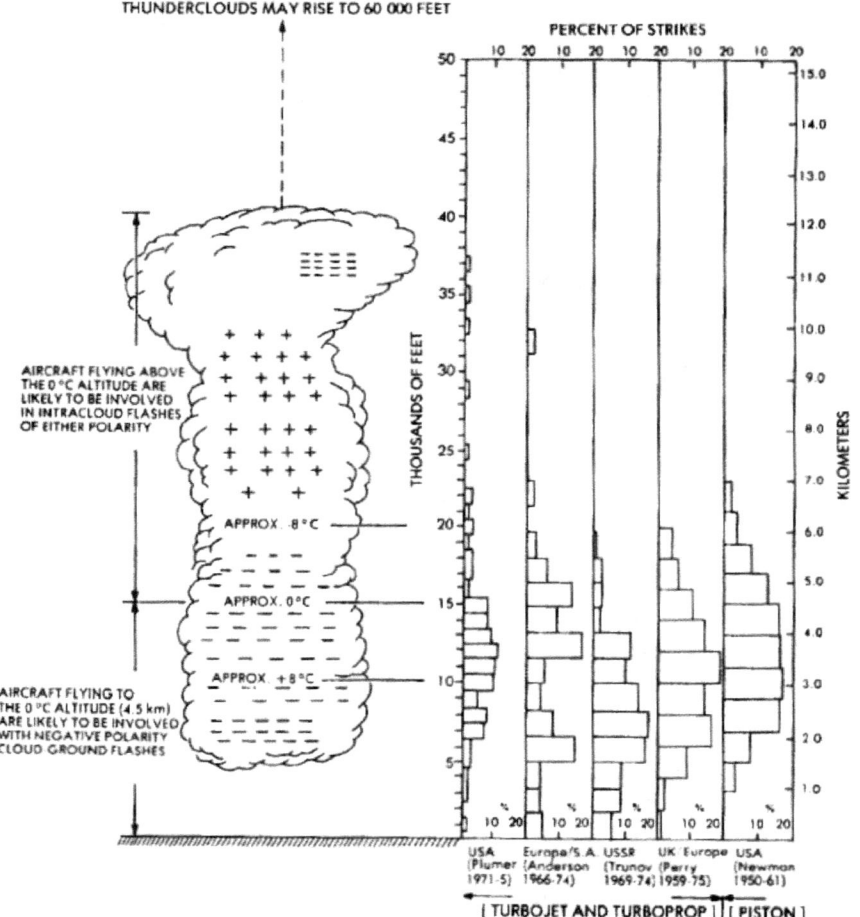

Figure 3.1 Aircraft lightning-strike incidents vs altitude.

If the strike altitudes shown in Figure 3.1 are compared with the electrical charge distribution in the typical thundercloud shown in the figure, it is evident that strikes which occur above about 3 km result from intracloud flashes between positive and negative charge centers in the cloud (or between adjacent clouds), whereas strikes below about 3 km probably result from cloud-to-ground flashes. Strike incidents occurring above 6 km are rare because of the absence of

58

concentrated charge centers at the higher altitudes and because aircraft at these altitudes can more easily divert around thunderclouds than can aircraft at lower altitudes.

FLIGHT CONDITION

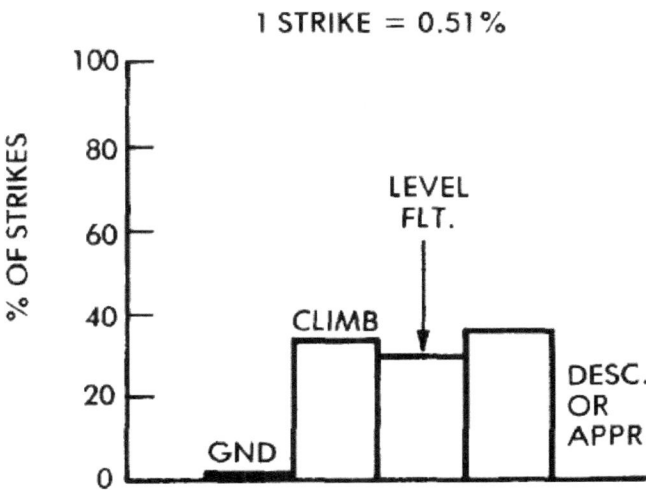

Figure 3.2 Flight conditions when struck.

3.3 Synoptic Meteorological Conditions

Data discussed thus far might imply that an aircraft must be within or beneath a cloud to receive a strike and, since electrical charge separation is brought about by precipitation, that most strikes would occur when the aircraft is within a cloud in regions of precipitation. Strike incident reports show that these conditions often do exist, but other lightning strikes occur to aircraft in a cloud when there is no evidence of precipitation nearby, or even to aircraft flying in clear air a supposedly safe distance from a thundercloud. FAA and airline advisory procedures instruct pilots to circumvent thunderclouds or regions of precipitation evident either visibly or on radar, but strikes to aircraft flying 25 miles from the nearest radar returns of precipitation have been reported. Occasionally a report of a "bolt from the blue," with no clouds anywhere around, is received. It is highly improbable, however, that these reports are correct because it does not seem possible for electrical charge separation of the magnitude necessary to form a lightning flash to occur in clear air. In most well-documented incidents, a cloud is present somewhere (i.e., within 25 miles) when the incident occurs.

Perhaps of most interest to aircraft operators are the area weather conditions which prevailed at the time of reported strikes. There is no universal

59

data bank for this type of data, but a summary has been made by H. T. Harrison (Reference 3.10) of the synoptic meteorological conditions prevailing for 99 United Air Lines lightning-strike incidents occurring between July 1963 and June 1964. Table 3.1 lists the synoptic type and the percentage of incidents ($\approx$ number of cases) occurring in each type. Examples of the most predominant synoptic conditions are presented in Figure 3.3 (a), (b), (c), (d), and (e) (Reference 3.11).

Table 3.1 SYNOPTIC TYPES INVOLVED WITH 99 ELECTRICAL DISCHARGES JULY 1963 TO JUNE 1964

Synoptic type	Percentage
Airmass instability	27
Stationary front	18
Cold front	17
Warm front	9
Squall line or instability line	9
Orographic	6
Cold LOW or filling LOW	5
Warm sector apex	3
Complex or intense LOW	3
Occluded front	1
Pacific surge	1

Harrison has summarized this data by saying that any conditions which will cause precipitation may also be expected to cause electrical discharges (lightning), although he adds that no strikes were reported in the middle of warm front winter snowstorms. Data from the projects of Plumer and Perry (Reference 3.12) presented in Figure 3.4 (Reference 3.13) show that lightning strikes to aircraft in the United States and Europe occur most often during the spring and summer months, when thunderstorms are most prevalent.

It is also important to note that many strike incidents have been reported where no bona fide thunderstorms have been visually observed or reported. L.P. Harrison (Reference 3.14), for example, reports that in 1946 active thunderstorms were manifested in less than half of the 150 strike incidents which he studied. Later, from July 1963 to June 1964, United Air Lines flight crews reported 99 cases of static discharges or lightning strikes in flight (Reference 3.15). Correlation of these incidents with weather conditions prevailing in the vicinity and in the general area at the time of strike gave the results shown in Table 3.2.

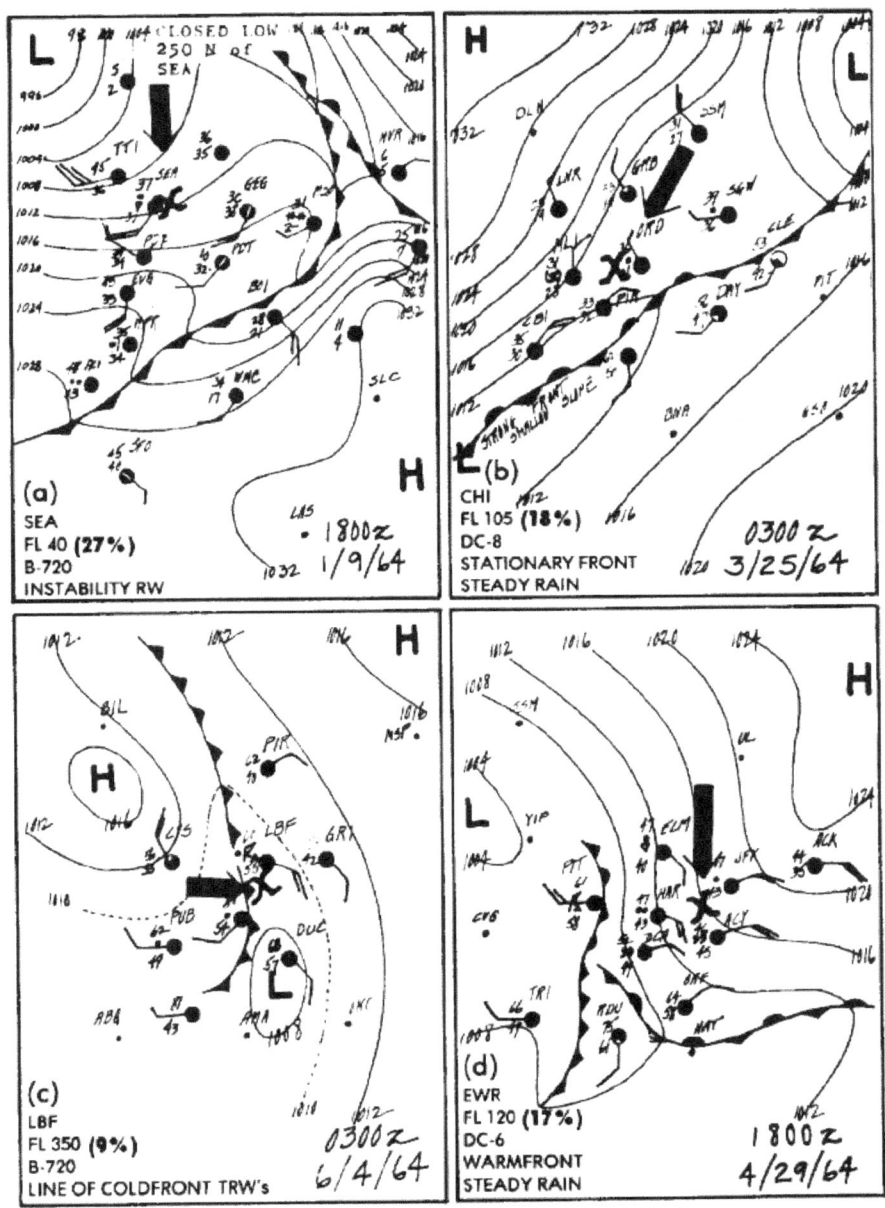

Figure 3.3 Examples of most frequent synoptic meteorological conditions when aircraft have been struck. Tip of arrow indicates position of aircraft when struck (continued).

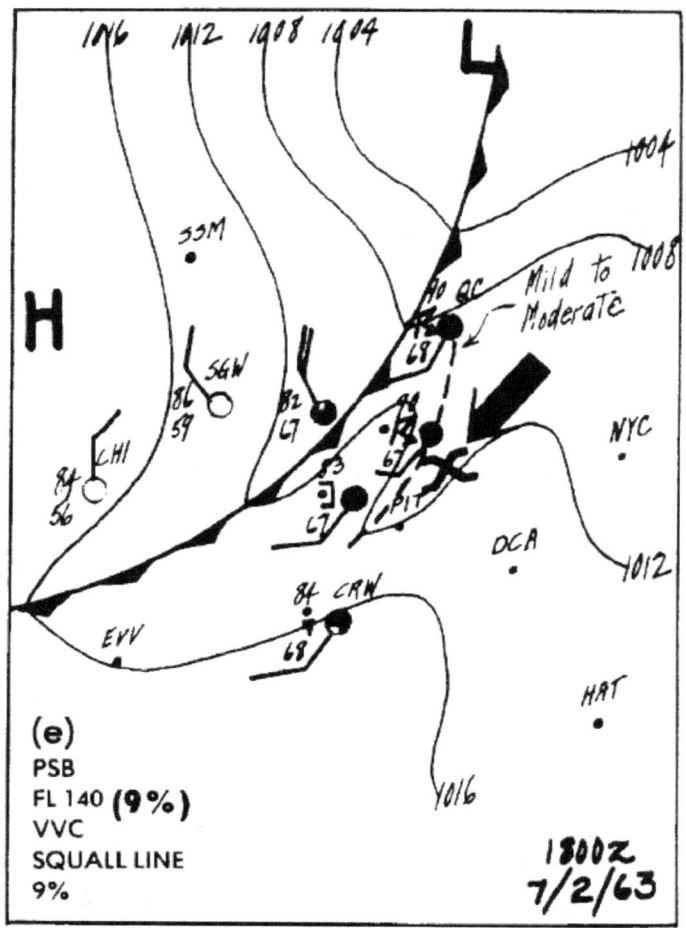

Figure 3.3 (Continued).

Table 3.2 PERCENTAGES OF STRIKE INCIDENTS VS
REPORTED THUNDERSTORMS

Thunderstorms Reported in Vicinity	33%
Thunderstorms Reported in General Area	24%
No Thunderstorms Reported	42%

3.4   Immediate Environment at Time of Strike

Figures 3.5, 3.6, and 3.7 show the immediate environment of the aircraft
at the times of the 214 strikes reported in the project of Plumer and Hourihan

(Reference 3.16). In over 80% of the strikes reported, each aircraft was within a cloud and was experiencing precipitation and some turbulence. With precipitation present, turbulence is to be expected, since vertical air currents acting against precipitation are thought to be the cause of electrical charge separation.

Incident reports also show that most aircraft strikes occur when an aircraft is in the freezing level of 0 °C. Figure 3.8 (Reference 3.17) from Newman's project (Reference 3.18), for example, shows the distribution of lightning strikes to aircraft as a function of outside air temperature. Freezing temperatures (and below) are thought to be required for the electrical charge separation process to function. Of course, strikes to aircraft at temperatures higher than +10 °C have occurred when the aircraft was close to (or on) the ground, where the ambient air temperature may be as high as about +25 °C.

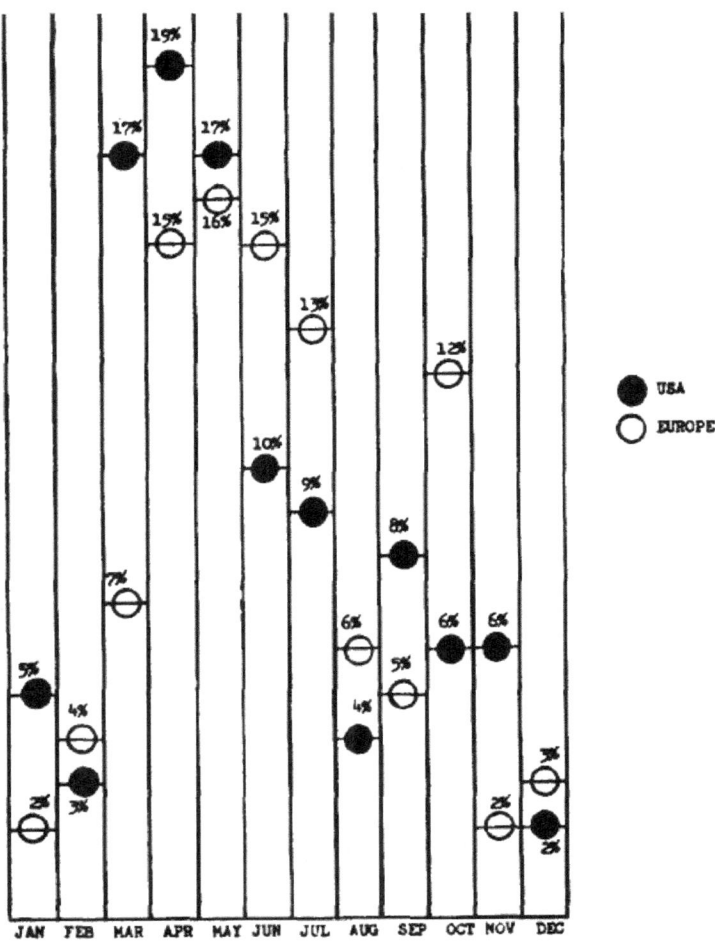

Figure 3.4 Occurrence of lightning strikes relative to months, in Europe and the United States.

63

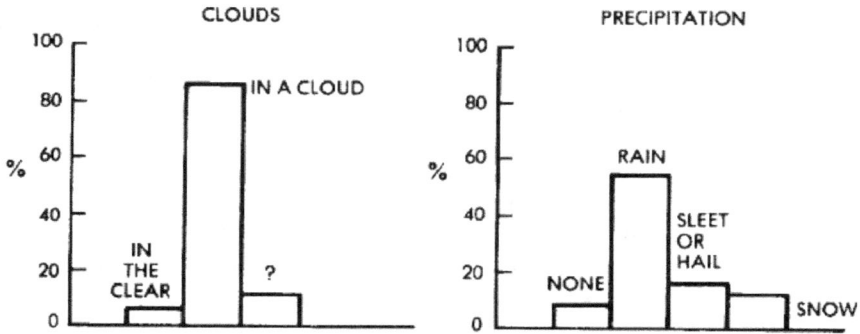

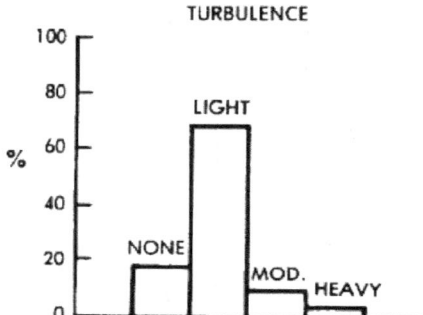

Figures 3.5, 3.6, 3.7   Environmental conditions at time of strike.

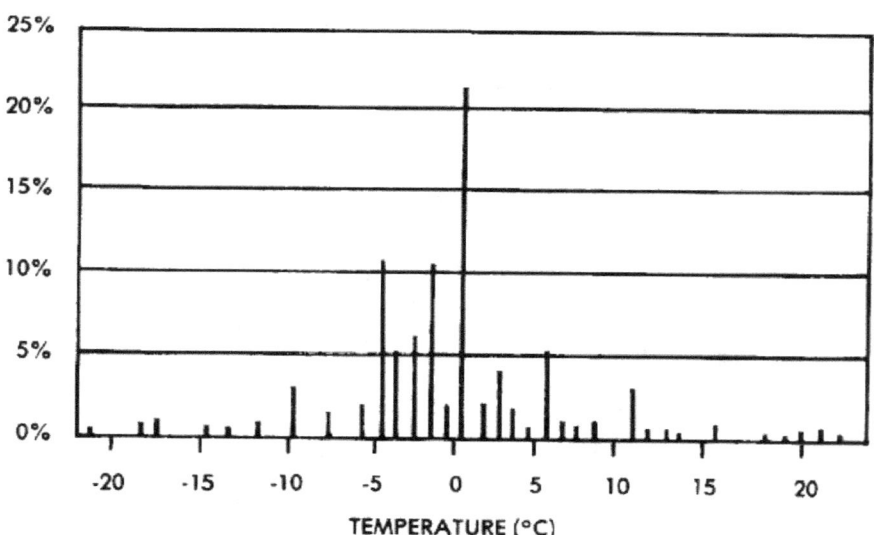

Figure 3.8   Lightning strikes to aircraft as a function of temperature.

3.5 Thunderstorm Avoidance

Clearly, whenever it is possible to avoid the severe environments which thunderstorms present, it is desirable to do so. For, even if the aircraft is adequately protected against lightning effects, the turmoil caused by wind and precipitation in or near thunderstorms presents a serious hazard to safe flight. Consequently, the operating procedures of commercial airlines and those of other air carriers strongly advise against penetration of thunderstorms.

In attempts to avoid thunderstorm regions, pilots use three indicators:

- Visual sighting of thunderclouds (cumulonimbus) in daytime and of lightning at night
- Airborne radar patterns of precipitation areas
- Ground-base radar patterns, if available, relayed by Air Traffic Control (ATC) to aircrews as instructions for thunderstorm avoidance.

Methods of thunderstorm avoidance in common use are, in order of preference

- Circumvention of thunderclouds, ideally by 25 miles or more
- Flying over the tops of thunderclouds
- Flying beneath the bases (bottoms) of thunderclouds.

Obviously, when other conditions are equal, the degree to which any of these measures is successful depends on the accuracy of the information received by the pilots.

Aside from visual observation (which has obvious limitations), the most common method of detecting thunderstorms is using airborne weather radar. Radar, however, cannot detect clouds themselves; it can detect only the associated precipitation (if any), which is capable of producing an echo; only the rain that may be present in the cloud will produce a radar echo.

A typical C-band airborne weather radar presentation of a thunderstorm (cumulonimbus cloud) with active precipitation and frequent lightning is shown in Figure 3.9 (Reference 3.19). The pictures were taken during a research project carried out by Beckwith of United Air Lines to determine the weather detection capability of airborne radar. The photographs shown were taken during a United Air Lines flight from Chicago to Denver on August 3, 1960.

Figure 3.9 (Reference 3.19) shows the northern end of a line of severe thunderstorms, developed from a cold front, in Illinois. A detour to the north was planned and successfully executed with the aid of this radar presentation. The flight remained in clear and generally smooth air while making the detour. The strong echos were easily detected with a slight upward tilt of the radar antenna to eliminate ground clutter.

It is possible to obtain more information regarding the intensity of a storm by use of the contour circuit provided on most radars. This circuit provides a means of eliminating any reflected signal the intensity of which is above a certain level. Figure 3.10 (Reference 3.20) shows the same radar return as that in Figure 3.9 (Reference 3.19) one minute later with the contour circuit employed.

The thin, distinct outlines which now appear in place of the original echo indicate a narrow boundary across which the intensity of rain varies from no rain (outside the white outline) to intense rain (inside the white outline). This change in return intensity is called the *rain gradient*; and the narrower the white outline of the return, the more abrupt, or steeper, is the gradient.

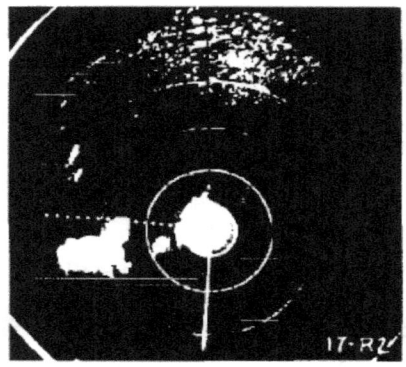
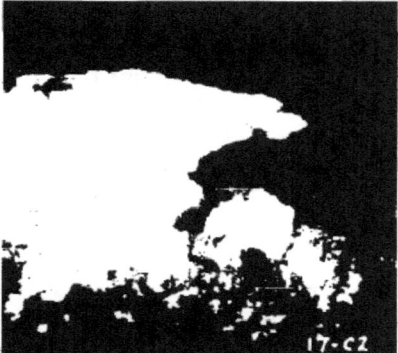

TIME: 0026Z
ALTITUDE: 28 000 FT.
RANGE MARKS: 10-MILE
ANT.: 1½° UP
HEADING: 255° TRUE
　　STRONG ECHOES HERE ARE THE N END OF A LINE OF THUNDERSTORMS WHICH WERE PRO-DUCING SEVERE WIND AND LIGHTNING DAM-AGE BELOW. LESS THAN ONE HOUR EARLIER, HAILSTONES OF GOLF BALL SIZE WERE RE-PORTED IN THE AREA OF THE NEAREST LARGE ECHO. STORM PICTURED ON PHOTO 17-C2.

TIME: 0028Z
ALTITUDE: 28 000 FT.
　　SOUTH OF DUBUQUE, IOWA, LOOKING SE TO S AND W EDGE OF THUNDERSTORM LINE. THIS CUMULONIMBUS WAS BUILT TO AN ESTI-MATED ALTITUDE OF 45 000 FEET. ECHO ON PHOTO 17-R2 CORRESPONDS TO THIS VISUAL.

Figure 3.9 Radar presentation and subsequent photograph of a thunderstorm.

The amount of electric charge separation and lightning activity is known to be directly related to the degree of interaction between precipitation and vertical air currents (turbulence). Further, the severity of turbulence is also related to the amount of temperature difference that exists between different masses of air. Thus, turbulence and electrical activity are likely to exist at well-defined boundaries, such as those indicated by steep rain gradients on contoured radar. These boundaries derived from contour radar are often used in planning a detour.

Considering the variable nature of thunderstorms and the limited information as to their whereabouts and severity available to pilots, it is not surprising that there are varying opinions as to what detour distance is adequate to avoid turbulence and lightning. Primarily, a pilot is advised to use distances commensurate with his radar's specific capability.

The specifications and policies of one of the major airlines (Reference 3.21) follow.

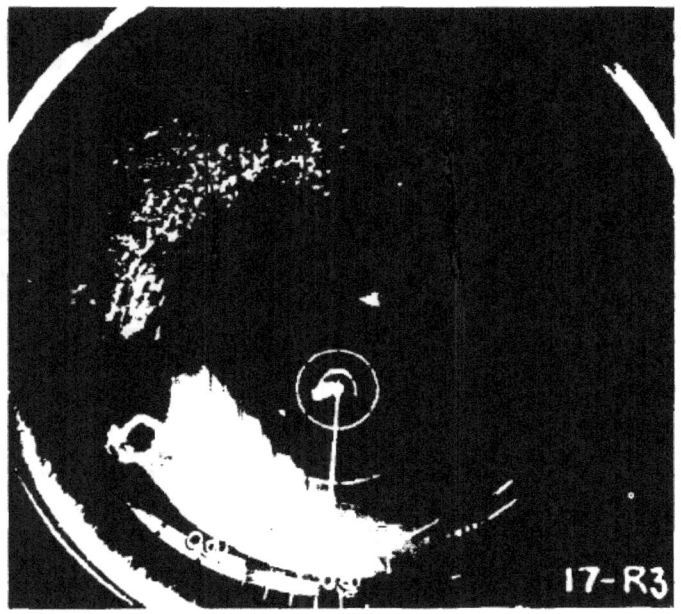

TIME: 0027Z
ALTITUDE: 28 000 FT.
RANGE MARKS: 10-MILE
ANT.: 1° UP
HEADING: 255° TRUE
CONTOUR: ON

Figure 3.10 Same as Figure 3.9 but with contour to show very steep rain gradient of each cell.

Specifications:
> *Wavelength:* 3.2 cm (X-Band) through 5.6 cm (C-Band)
> *Antenna Size:* 12-inch or larger (X-Band); 25-inch or larger (C-Band)
> *Power (Peak):* 10 kW (X-Band) or higher; 75 kW (C-Band) or higher

Policies:
1. When the temperature at flight level is 0 °C or higher, avoid all echoes exhibiting sharp gradients by 5 nautical miles.
2. When the temperature at flight level is less than 0 °C, avoid all echoes exhibiting sharp gradients by 10 nautical miles.
3. When flying above 23 000 feet avoid all echoes, even though no sharp gradients are indicated, by 20 nautical miles.

Weather radar, however, is not a foolproof means of detecting and avoiding thunderstorms because situations exist in which radar is not capable of distinguishing a thunderstorm return from ground or other precipitation returns in the same vicinity. Such a case is illustrated in Figure 3.11 (Reference 3.22).

67

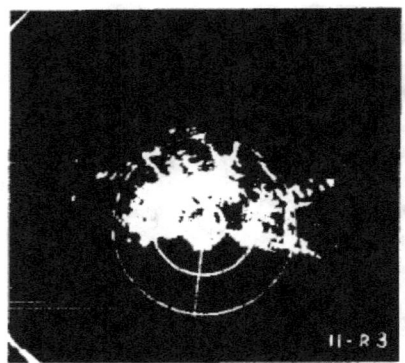 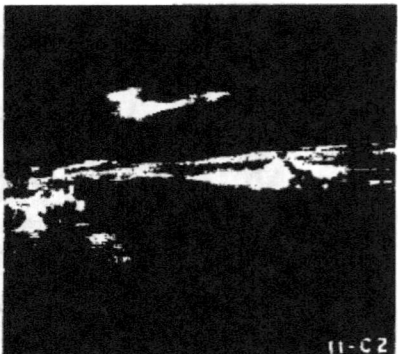

TIME: 0153Z
ALTITUDE: 29 000 FT.
RANGE MARKS: 25-MILE
ANT.: 0°
HEADING: 060° TRUE. OVER ST. GEORGE, UTAH.
    A THUNDERSTORM ECHO IS NOT DISTIN-GUISHABLE FROM TERRAIN AT THIS OR OTHER ANTENNA SETTINGS TO MATCH THE VISUAL SIGHTING AT 9:30 BEARING IN PHOTO 11-C2.

TIME: 0155Z
ALTITUDE: 29 000 FT.
    DECAYING THUNDERSTORM BUILT TO ABOVE FLIGHT ALTITUDE FOR WHICH NO MATCHING ECHO WAS VISIBLE IN SCOPE PIC-TURE 11-R-3.

Figure 3.11 Thunderstorms not distinguishable on radar scope.

In this case, returns from the ground *(ground clutter)* obscured the return from the storm. However, if ground clutter does not obscure a storm return and an aircraft is successful in avoiding all thunderstorms by the recommended distances of up to 25 miles, the severe turbulence associated with thunderstorms is usually also avoided. Nevertheless, lightning flashes may extend farther outward from the storm center than does turbulence and for this reason are not as easily avoided. Indeed, there are several reports each year of aircraft receiving strikes "in the clear" 25 or more miles from the nearest storm. That lightning flashes can propagate this distance is evident from ground photographs of very long, horizontal flashes.

A decaying thunderstorm, moreover, may not present a distinctive radar echo. Sometimes this type of storm becomes embedded in expanding anvils or cirrus clouds in such a way that it is not visible. In-flight measurements conducted by the Air Force and the Federal Aviation Administration, and reported by Fitzgerald (Reference 3.23), indicate that thunderstorms in their early stages of dissipation have sufficient charge to cause a few lightning discharges if a means of streamer initiation becomes available; an aircraft entering such a region may initiate, or trigger, such a flash. Thus, in normal IFR operations in regions where an active thundercloud is merged with decaying thunderclouds and other cloudy areas, those diversions from the normal course that are taken to avoid the active cloud may redirect flight through a decaying area, where a lightning strike is possible.

Attempts have been made from time to time to develop an airborne instrument capable of warning pilots of an impending lightning strike and providing information to the pilot for use in avoidance. Most such instruments are based on the principle of detecting the ambient electric field which would exist when a lightning flash is imminent (Reference 3.24). None of these instruments have been successfully tested in an aircraft, and because of the apparent wide variation of electric field magnitudes and directions which may exist at the aircraft surface just before a strike occurs, the prospects for success seem remote. The situation is further complicated by the problems of field interpretation and translation into advisory information to aid the pilot in deciding on an avoidance maneuver.

Perhaps the most effective warning of imminent strikes available to flight crews is that which is readily available—the buildup of static discharging and (at night) St. Elmo's Fire (corona). Static discharging causes interference (instability) in Low Frequency Automatic Direction Finding (LF-ADF) indicators, or audible "hash" in most communications receivers. St. Elmo's Fire is visible at night as a bluish glow at aircraft extremities where the discharging is occurring. Pilot responses to lightning strikes (which may also be called *static discharges* or *electrical discharges*) vary. Typical answers by pilots of one airline (Reference 3.25) to the question "Do you have any recommendations for avoiding electrical discharges?" were as follows:

"From cruise speed, a reduction of 25% to 30% in airspeed will often allow the static buildup to stabilize at a lower maximum and dissipate rather than discharge. These buildups are generally accompanied by a buzz-type static on VHF [very high frequency] and ADF [automatic direction finding] and a random swinging of the ADF needles though I have observed the ADF needles to hold a steady error of up to 90° as the static level stabilized at or near its peak, generally just prior to the discharge or beginning of dissipation."

"Climb and descend through the freezing level as quickly as possible."

"Avoid all precipitation. I know of no way to predict accurately where a discharge will occur."

"Slow down to minimum safe speed, change altitude to avoid temperature of 20 °F to 35 °F."

"Not without excessive detour, both route and altitude."

69

"The static discharges I have encountered have built up at a rate which would preclude any avoidance tactics (3 to 15 seconds)."

"No, I have never known when to expect this until just prior to the discharge."

"No, not in the modern jets. Once the static begins the discharge follows very quickly."

"All information received at the ... training center applicable to static discharges and their avoidance has been completely accurate and helpful."

"No, hang on!"

"Lead a clean life."

Thus, there is a wide divergence of pilot opinion regarding the best way to avoid lightning strikes. However, from this and many other sources, it is possible to list the symptoms most often present just prior to experiencing a lightning strike, and the actions (if any) which most pilots take to reduce the possibility of receiving a strike.

A lightning strike is imminent when a combination of some of the symptoms which follow is present.

Symptoms:
1. Flight through or in the vicinity of the following:
   Unstable air
   Stationary front
   Cold front
   Warm front
   Squall line
2. Within a cloud
3. Icy types of precipitation
4. Air temperature near 0 °C
5. Progressive buildup of radio static
6. St. Elmo's fire (when dark)
7. Experiencing turbulence
8. Flying at altitudes between 1.5 and 4.5 km (5000 and 15 000 ft); most prevalent: 3.35 km (11 000 ft)
9. Climbing or descending in the vicinity of airports

Actions:

1. Circumvent areas of heavy precipitation.
2. Reduce speed (or rpm with piston-engine aircraft).
3. Change altitude to avoid temperature near 0 °C.
4. Turn up cockpit lights.
5. Have one pilot keep eyes downward.

Since air traffic congestion often precludes circumvention of precipitation and since diversion often poses hazards, avoidance, while desirable, is neither a dependable nor an adequate means of protecting the aircraft against lightning strikes. The aircraft, therefore, must be designed to safely withstand lightning strike effects.

3.6 Frequency of Occurrence

The number of lightning strikes which actually occur as compared with flight hours for piston, turboprop, and pure jet aircraft is tabulated in Table 3.3 based on the data of Newman (Reference 3.26) and Perry (Reference 3.27). From this data it follows that an average of one strike can be expected for each 3000 hours of flight for any type of commercial transport aircraft, with slightly greater exposure expected for piston aircraft only. The probable reason for this difference is that piston aircraft are limited to lower cruise altitudes, where lightning flashes are more prevalent.

Unlike commercial airlines, military and general aviation aircraft need not adhere to strict flight schedules or congested traffic patterns around metropolitan airports. The result is that these aircraft do not experience as many strikes as do commercial aircraft, as is evident from Table 3.4, which shows U.S. Air Force experience for the years 1965 to 1969.

Comparison of the experience reported in Tables 3.3 and 3.4 indicates that the probability of commercial aircraft being struck is anywhere from 10 to 200 times greater than for U.S. Air Force aircraft for the same number of flight hours. The discrepancy is most certainly a result of variations in flight operations, rather than a result of physical differences in the types of aircraft flown. Whereas the Air Force may curtail flight operations in adverse weather, commercial air lines usually continue operations, and their aircrafts' exposure to lightning is further increased in bad weather by traffic congestion and holding requirements near airports.

Statistics such as these, which apply to a broad category of aircraft and include data from a variety of different operators in varying geographic locations, may be misleading, however. For example, whereas Table 3.4 shows that there is an average of 99 000 flying hours between lightning strikes to U.S. Air Force fighter-type aircraft, the strike experience in Europe is known to be about 10 times more frequent than strike experience in the U.S. and in most other parts of the world. Weinstock and Shaeffer (Reference 3.28) report 10.5 strikes per 10 000 hours for certain F-4 models flying in Europe, which rate is about 5 times greater than the world-wide exposure rate for these aircraft. A similar situation pertains to commercial aircraft operating in Europe, as indicated by Perry's summary of United Kingdom and European strike data (Reference 3.29), for example. This unusually high lightning-strike exposure seems to result both from the high level of lightning activity in Europe as compared with that in many other regions and from the political constraints placed on flight paths in this multinational region.

There are several trends in commercial and general aviation which are likely to cause even greater exposure of aircraft everywhere to lightning strikes in the future:

- More time in holding patterns as a result of increased traffic at major airports
- Increases in the number of intermediate stops along former nonstop routes, resulting in more time in descent-hold-climb patterns at lower altitudes
- Increasing use of radar and other navigation aids in general aviation aircraft, permitting IFR flight under adverse weather conditions.

These factors warrant continued diligence in the design and operation of aircraft with respect to the possible hazards which lightning may present.

Table 3.3   INCIDENCE OF REPORTED LIGHTNING STRIKES
TO COMMERCIAL AIRCRAFT

	Newman (1950 – 1961)		Perry (1959 – 1974)		TOTALS		
	Strikes	Hours	Strikes	Hours	Strikes	Hours	No. hours per strike
Piston	808	2 000 000	–	–	808	2 000 000	2475
Turboprop	109	415 000	280	876 000	389	1 291 000	3320
Pure Jet	41	427 000	480	1 314 000	521	1 741 000	3340
ALL	958	2 842 000	760	2 190 000	1718	5 032 000	2930

Table 3.4   INCIDENCE OF LIGHTNING STRIKES TO
U.S. AIR FORCE AIRCRAFT

Aircraft type vs mean hours between lightning strikes					
	1965	1966	1967	1968	1969	Average per year
Bomber	55 500	48 000	47 900	73 000	28 000	50 480
Cargo	68 000	140 000	112 000	124 000	76 000	104 000
Fighter	141 000	105 000	112 000	65 000	73 000	99 200
Trainer	246 000	378 000	500 000	224 000	130 000	295 600

REFERENCES

3.1   L. P. Harrison, *Lightning Discharge to Aircraft and Related Meteorological Conditions,* Technical Note 1001, National Advisory Committee for Aeronautics, Washington, D.C. (May 1946).

3.2   M. M. Newman and J. D. Robb, *Aircraft Protection from Atmospheric Electrical Hazards,* ASD Technical Report 61-493, L and T Report 374, Lightning and Transients Research Institute, Minneapolis, Minnesota (December 1961).

3.3   "Lightning Strike Survey Report for the Period of January 1965 through December of 1966," *Federal Aviation Agency Report of the Conference on Fire Safety Measures for Aircraft Fuel Systems,* Appendix II, Department of Transportation, Washington, D.C. (December 1967).

3.4   B. I. Hourihan, *Data from the Airlines Lightning Strike Reporting Project, June 1971 to November 1974,* Summary Report GPR-75-004, High Voltage Laboratory, Electromagnetics Unit, Corporate Research and Development, General Electric Company, Pittsfield, Massachusetts (March 1975).

3.5   R. B. Anderson and H. Kröninger, "Lightning Phenomena in the Aerospace Environment: Part II, Lightning Strikes to Aircraft," *Proceedings of the 1975 Conference on Lightning and Static Electricity at Culham Laboratory, England, 14-17 April 1975,* Session I: Fundamental Aspects and Test Criteria, the Royal Aeronautical Society of London (December 1975).

3.6   J. A. Plumer and B. L. Perry, "An Analysis of Lightning Strikes in Airline Operation in the USA and Europe," *Proceedings of the 1975 Conference on Lightning and Static Electricity at Culham Laboratory, England, 14-17 April 1975,* Session IV: Aircraft Applications, the Royal Aeronautical Society of London (December 1975).

3.7   O. K. Trunov, "Conditions of Lightning Strikes in Air Transports and Certain General Lightning Protection Requirements," *Proceedings of the 1975 Conference on Lightning and Static Electricity at Culham Laboratory, England, 14-17 April 1975,* Session IV: Aircraft Applications, the Royal Aeronautical Society of London (December 1975).

3.8   Newman and Robb, *Aircraft Protection,* pp. 97, 99.

3.9   Hourihan, *Data from the Airlines,* p. 11.

3.10   H. T. Harrison, *UAL Turbojet Experience with Electrical Discharges,* UAL Meteorological Circular No. 57, United Air Lines, Chicago, Illinois (January 1, 1965), pp. 27-48.

3.11   H. T. Harrison, *UAL Turbojet Experience,* pp. 37, 39, 47, 43, 30.

3.12   Plumer and Perry, "An Analysis of Lightning Strikes," pp. 2, 10.

3.13   Plumer and Perry, "An Analysis of Lightning Strikes," p. 10.

3.14   L. P. Harrison (1946) summarized by H. T. Harrison in *UAL Turbojet Experience,* p. 4.

3.15   H. T. Harrison, *UAL Turbojet Experience,* p. 10.

3.16   Hourihan, *Data from the Airlines,* p. 11.

73

3.17 Newman and Robb, *Aircraft Protection*, p. 99.

3.18 See Newman and Robb, *Aircraft Protection*, Appendix I: Lightning Strike Statistics Extended to Jet Aircraft.

3.19 W. B. Beckwith, *The Use of Weather Radar in Turbojet Operations*, UAL Meteorological Circular No. 53, United Air Lines, Denver, Colorado (April 1, 1961), pp. 34, 35.

3.20 Beckwith, *The Use of Weather Radar*, p. 35.

3.21 H. R. Hoffman and G. W. Peckham, *The Use of Airborne Weather Radar*, United Air Lines, Denver, Colorado (1968), p. 12.

3.22 Beckwith, *The Use of Weather Radar*, p. 25.

3.23 D. R. Fitzgerald, "Probable Aircraft Triggering of Lightning in Certain Thunderstorms," *Monthly Weather Review*, December 1967, pp. 835-42.

3.24 G. A. M. Odam, *An Experimental Automatic Wide Range Instrument to Monitor the Electrostatic Field at the Surface of an Aircraft in Flight*, Technical Report 69218, Royal Aircraft Establishment, Farnborough Hants, England (October 1961).

3.25 Harrison, *UAL Turbojet Experience*, pp. 74-77.

3.26 Newman and Robb, *Aircraft Protection*, p. 100.

3.27 Plumer and Perry, "An Analysis of Lightning Strikes," pp. 1-2.

3.28 J. F. Shaeffer and G. L. Weinstock, *Aircraft Related Lightning Mechanisms*, Technical Report AFAL-TR-72-386, prepared by the McDonnell Aircraft Company, McDonnell Douglas Corporation, for the Air Force Avionics Laboratory, Air Force Systems Command, Wright-Patterson Air Force Base, Ohio (January 1973).

3.29 Plumer and Perry, *"An Analysis of Lightning Strikes,"* p. 2.

CHAPTER 4
LIGHTNING EFFECTS ON AIRCRAFT

4.1 Introduction

"*We had just taken off from Presque Isle, Maine, and had been in cruise power for 50 minutes, when a large thunderhead cumulus was observed directly on course. Lightning could be seen around the edges and inside the thunderhead. All cockpit lights were on and the instrument spotlight was full on, with the door open. I had just finished setting the power and fuel flows for each engine. As the ship approached the thunderhead, there was a noticeable drop in horsepower and the airplane lost from 180 MPH airspeed to 168 MPH, and continued to lose airspeed due to power loss as we approached the thunderhead. . . . A few seconds before the lightning bolt hit the airplane all four engines were silent and the propellers were windmilling. Simultaneous with the flash of lightning, the engines surged with the original power. The lightning flash blinded the Captain and me so severely that we were unable to see for approximately eight minutes. I tried several times during this interval to read cockpit instruments and it was impossible. The First Officer was called from the rear to watch the cockpit. Of course, turbulent air currents inside the cumulus tossed the ship around to such an extent that, had the airplane not been on auto-pilot when the flash occurred and during the interval of blindness by the cockpit occupants, the ship could have easily gone completely out of control. The Captain and I discussed the reason for all four engines cutting simultaneously prior to the lightning flash and could not explain it, except for the possibility of a magnetic potential around the cumulus affecting the primary or secondary circuit of all eight magnetos at the same time.*" First Officer N.A. Pierson's experience on a flight from Presque Isle, Maine, to the Santa Maria Islands on July 9, 1945 (Reference 4.1).

It wasn't long after the beginning of powered flight that aircraft began being struck by lightning—sometimes with catastrophic results. The early wooden aircraft with metal control cables and guy wires were not capable of conducting lightning-stroke currents of several thousand amperes or more. Wooden members and even the control cables exploded or caught fire. Even if severe structural damage did not occur, pilots were frequently shocked or burned by lightning currents entering their hands or feet via control pedals or the stick. Sometimes fuel tanks caught fire or exploded. These effects, coupled with the air turbulence and precipitation also associated with thunderstorms, quickly taught pilots to stay clear of stormy weather.

With the advent of all-metal aircraft, most of the catastrophic effects were eliminated, but thunderstorms continued to be treated with respect. Nonethe-

75

less, because a few bad accidents attributed to lightning strikes continued to happen, in 1938 the Subcommittee of Aircraft Safety, Weather and Lightning Experts was formed by the National Advisory Committee for Aeronautics (NACA) to study lightning effects on aircraft and determine what additional protective measures were needed. Dr. Karl B. McEachron, Director of Research at the General Electric High Voltage Laboratory, was a key member of this committee, and during its twleve-year existence he performed the first man-made-lightning tests on aircraft parts. During and subsequent to this period, other organizations, such as the U.S. National Bureau of Standards, the University of Minnesota, and the Lightning and Transients Research Institute, also began to conduct research into lightning effects on aircraft.

For a long time the physical damage effects at the point of flash attachment to the aircraft were of primary concern. These included holes burned in metallic skins, puncturing or splintering of nonmetallic structures, and welding or roughening of movable hinges and bearings. If the attachment point was a wing tip light or an antenna, the possibility of conducting some of the lightning current directly into the aircraft's electrical circuits was also of concern. Today, these and other physical damage effects are called the *direct effects*. Since present-day military and commercial aircraft fly IFR (Instrument Flight Rules) in many kinds of weather, protective measures against direct effects have been designed and incorporated into these aircraft so that hazardous consequences of lightning strikes are rare.

In recent years it has become apparent that lightning strikes to aircraft may cause other effects, or *indirect effects*, to equipment located elsewhere in the aircraft. For example, the operation of instruments and navigation equipment has been interfered with, and circuit breakers have popped in electric power distribution systems when the aircraft has been struck by lightning. The cause of these indirect effects are the electromagnetic fields associated with lightning currents flowing through the aircraft. Even though metallic skins provide a high degree of electromagnetic shielding, some of these fields may penetrate through windows or seams and induce transient voltage surges in the aircraft's electrical wiring; these surges in turn may damage electrical or electronic equipment.

To date, few aircraft accidents can be attributed positively to the indirect effects of lightning, but there are two trends in aircraft design which threaten to aggravate the problem unless new protective measures are developed and utilized. The first of these trends is the increasing use of miniaturized, solid state components in aircraft electronics and electric power systems. These devices are more efficient, lighter in weight, and far more functionally powerful than their vacuum tube or electromechanical predecessors, and they operate at much lower voltage and power levels. Thus, they are inherently more sensitive to overvoltage transients, such as those produced by the indirect effects of lightning.

The second trend is the increasing use of reinforced plastics and other nonconducting materials in place of aluminum skins, a practice that reduces the electromagnetic shielding previously furnished by the conductive skin. This reduced shielding may greatly increase the level of surges induced in wiring not protected by other means. Because electronic systems were being increasingly depended upon for safety of flight, the National Aeronautics and Space Administration, the Federal Aviation Agency, and the Department of Defense initiated research programs, beginning in 1967, to learn how to measure or predict the levels of lightning-induced voltages and how to protect against them. A considerable amount of research has followed these initial programs.

Since the indirect effects originate in the aircraft's electrical wiring, their consequences may show up anywhere within the aircraft, such as at equipment locations remote from the lightning-flash attachments. The direct effects, on the other hand, occur primarily at the points of arc attachment. This comparison is illustrated in Figure 4.1.

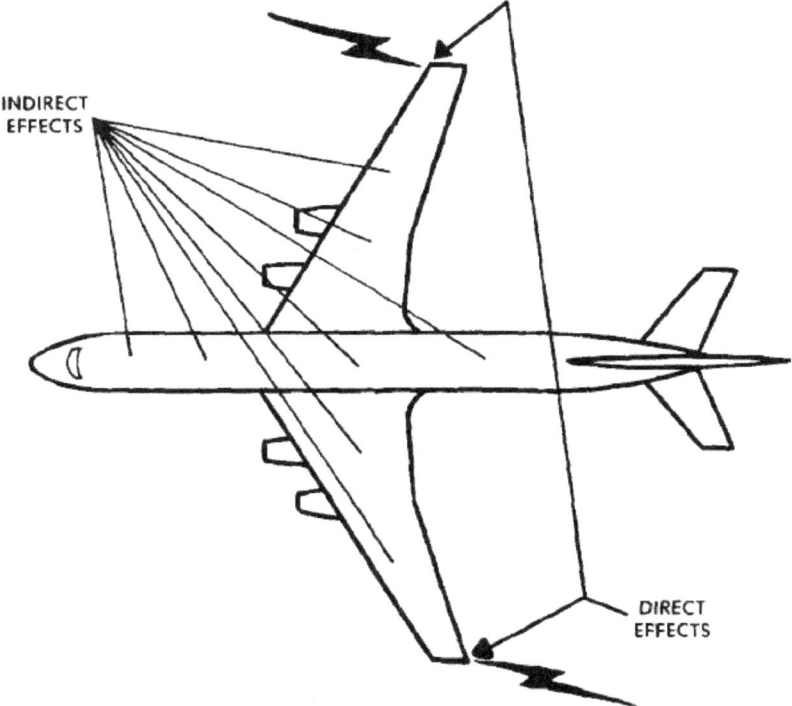

Figure 4.1 Areas of direct and indirect effects.

Before discussing the techniques for protecting aircraft against either type of lightning effect, it is worthwhile to review some of the common examples of each which occur on today's aircraft. The purpose of this review is to remind the aircraft designer of areas needing particular attention and to alert pilots to what to expect when lightning strikes occur in flight. Detailed discussion of the causes of each effect, including the lightning-flash characteristic most responsible and its quantitative relationship to damage severity, is deferred until protective measures are discussed, beginning with Chapter 5.

4.2 Direct Effects on Metal Structures

Metal structures include the outer skins of the aircraft together with internal metallic framework, such as spars, ribs, and bulkheads. Because lightning currents must flow between lightning entry and exit points on an aircraft and because these currents tend to spread out as they flow between attachment points, using the entire airframe as a conductor, the aluminum with which most of these structures are fabricated provides excellent electrical conductivity. As a result, the current density at any single point in the airframe is rarely sufficient to cause physical damage between entry and exit points. Only if there is a poor electrical bond (contact) between structural elements in the current flow path is there likely to be physical damage. On the other hand, where the currents converge to the immediate vicinity of an entry or exit point, there may be a sufficient concentration of magnetic force and resistive heating to cause damage. Damage at these points is further compounded by the lightning arc, from which intense heat and blast forces emanate. Discussion of individual effects follows.

4.2.1 Melting and Burnthrough

If a lightning arc touches a metal surface for a sufficient time, melting of the metal will occur at the point of attachment. Common evidences of this are the successive pit marks often seen along a fuselage or empennage, as shown in Figure 4.2 (Reference 4.2) or the holes burned in the trailing edges of wings or empennage tips, as shown in Figure 4.3 (Reference 4.3). Most holes are melted in skins of 1 mm (0.040″) thick, or less, except at trailing edges, where the lighting arc may hang on for a longer time and enable holes to be burned through much thicker pieces. Since a relatively large amount of time is needed for melting to occur, the continuing currents are the lightning-flash components most conducive to melting and burnthrough. Melting or burnthrough of skins is usually not a safety-of-flight problem unless this occurs in an integral fuel tank skin.

4.2.2 Magnetic Force

Metal skins or structures may also be deformed as a result of the intense magnetic fields which accompany concentrated lightning currents near attachment points. It is well known that parallel wires with current traveling in the same direction are mutually attracted to each other. If the structure near an attachment point is viewed electrically as being made up of a large number of parallel conductors converging to this attachment point, then as lightning current flows from the point, forces occur which tend to draw these conductors closer together. If a structure is not sufficiently rigid, pinching or crimping may occur, as shown in Figure 4.4 (Reference 4.4). The amount of damage created is

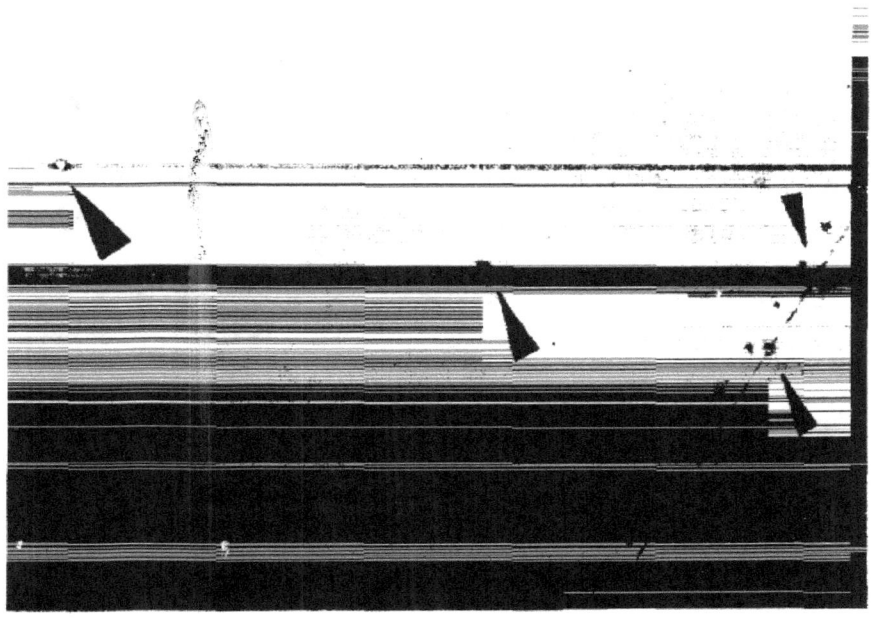

Figure 4.2 Successive pit marks extending backward from leading edge of vertical stabilizer.

proportional to the square of the lightning-stroke current amplitudes and is directly proportional to the length of time during which this stroke current flows. Thus the high amplitudes of return stroke and intermediate stroke are the lightning-flash components most responsible for magnetic force damage.

Besides the main airframe, other parts which may be damaged by magnetic forces include bonding or diverter straps, pitot probes, or any other object which may conduct lightning-stroke currents. Magnetic force damage is usually not, by itself, significant enough to require abortion of a flight, and may not even be detected until the aircraft is on the ground. However, since overstress or severe bending of metals is involved, aircraft parts damaged by this phenomenon are not often repairable.

4.2.3   Pitting at Structural Interfaces

Wherever poor electrical contact exists between two mating surfaces, such as a control surface hinge or bearing across which lightning currents may flow, melting and pitting of these surfaces may occur. In a recent incident, for example, the jackscrew of an inboard trailing edge flap of a jet transport was so

79

Figure 4.3 Hole burned in trailing edge corner of ventral fin.

damaged by a lightning flash that the flap could not be extended past 15°. Since this jackscrew is located on the inboard side of the flap, the flash must have reached it after sweeping along the fuselage from an earlier attachment point near the nose, as shown on Figure 4.5. Instead of continuing to sweep aft along the fuselage, the flash apparently hung on to the jackscrew long enough to melt a spot on it. The event did not cause difficulties in landing the aircraft, and the damage, in fact, was not discovered until after the aircraft was on the ground. The damage, however, was extensive enough that the jackscrew had to be replaced.

It should be noted that the jackscrew in this instance was not an initial, or Zone 1, attachment point (See Figure 2.9). It subsequently became an attachment point only by being in the path of a swept stroke and is therefore in Zone 2B, a swept-stroke zone with high probability of hang-on.

A second illustration of pitting is the damage caused to the seals of the hydraulic jack operating the tail control surfaces of another jet transport aircraft. In this case the jack was shunted by a jumper of adequate cross section to carry lightning-stroke currents but of excessive length, which caused most of the current to flow through the lower inductance path directly through the jack body and across the seals, resulting in leakage of hydraulic fluid.

Earlier aircraft, especially those with nonmetallic skins, experienced more troublesome consequences from lightning currents, nearly all of which had to flow through the control cables. In these cases, lightning currents entering a

80

Figure 4.4 Example of magnetic pinch effect at lightning attachment points.

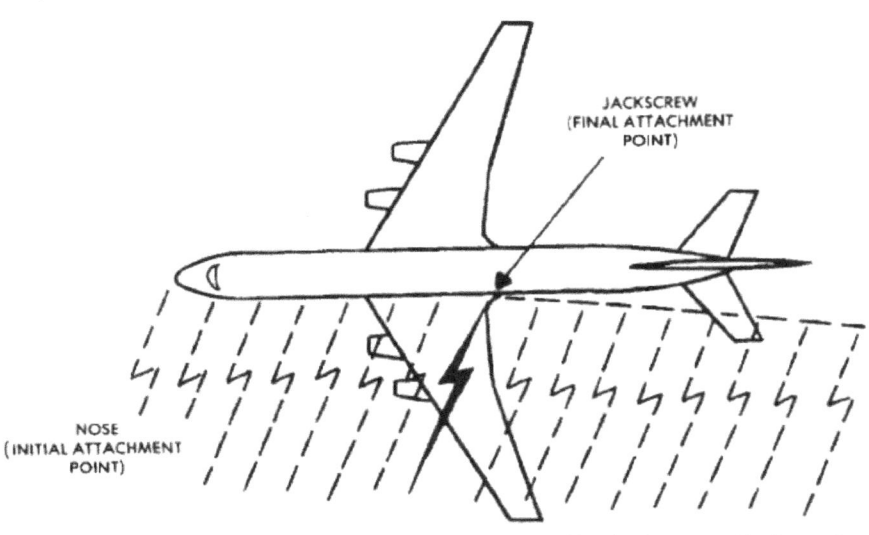

Figure 4.5 Swept-stroke attachment and inboard flap jackscrew attachment.

control surface have been conducted all the way through the aircraft via the control cables, sometimes with very damaging results. There is at least one case on record (Reference 4.5) of a wooden glider lost as a result of lightning current which entered the aircraft at the left aileron pulley support and then flowed via the aileron cables to the right pulley, from which point it exited the aircraft. The cables were disintegrated, and the wreckage indicated that extensive damage to the wooden airframe had occurred before impact, probably as a result of the cables exploding inside. There are no records of loss of a powered aircraft from this effect, since these aircraft have nearly always been of metallic construction.

The high-amplitude stroke currents are primarily responsible for pitting, but continuing currents may also contribute to this damage, as illustrated by the jackscrew incident.

4.2.4 Resistive Heating

The glider accident mentioned above is also an example of another direct effect: resistive heating of conductors. When the resistivity of a conductor is too high or its cross-sectional area too low for adequate current conductance, lightning currents flowing in it may deposit appreciable energy in the conductor and cause an appreciable temperature rise. Since the resistivity of most metals increases with temperature rise, a given current in a heated conductor will deposit more energy than it would in an unheated, less resistant conductor; this process in turn increases the conductor temperature still further.

Resistive energy deposition is proportional to the lightning current action integral ($\int i^2 dt$), and for any conductor there is an action integral value at which the metal will melt and vaporize, as shown in Figure 4.6. The result is the

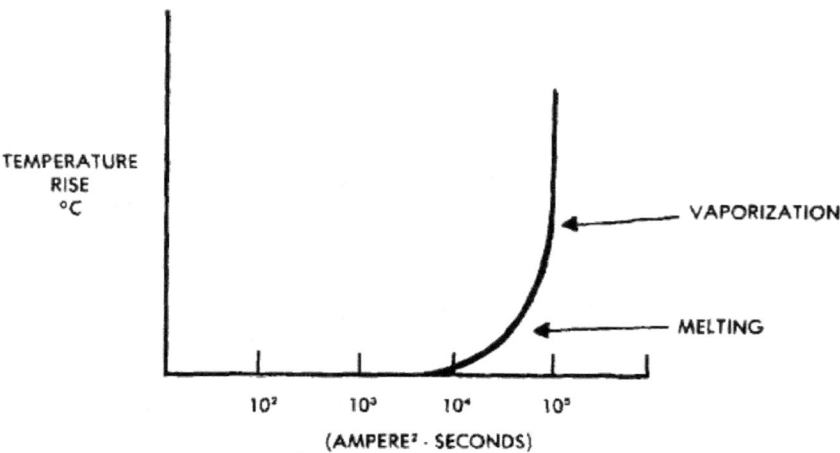

Figure 4.6 $\int i^2 dt$ vs temperature rise in a conductor.

82

exploding conductor apparent in the glider incident described in Section 4.2.3.

Other consequences of resistive heating and explosive vaporization of conductors are shown in Figures 4.7 and 4.8 (References 4.6 and 4.7). The damage is usually most severe when the exploding conductor is within an enclosure, which contains the explosion until the pressure has built up to a level sufficient to rupture the container.

Figure 4.7 Lightning damage to radome—probably as a result of exploding pitot tube ground wire.

4.2.5 Shock Wave and Overpressure

When a lightning-stroke current flows in an ionized leader channel (as when the first return stroke occurs), a large amount of energy is delivered to the channel in 5 to 10 μs, causing the channel to expand with supersonic speed. Its temperature has been measured by spectroscope techniques to be 30 000 °K and the channel pressure (before expansion) about 10 atmospheres (Reference 4.8). When the supersonic expansion is complete, the channel diameter is several centimeters and the channel pressure is in equilibrium with the surrounding air. Later, the channel continues to expand more slowly to the equilibrium situation of a stable arc. The cylindrical shock wave propagates radially outward from the center of the arc, and, if a hard surface is intercepted, the kinetic energy in the shock wave is transformed into a pressure rise over and above that in the shock wave itself. This results in a total overpressure of several times that in the free

83

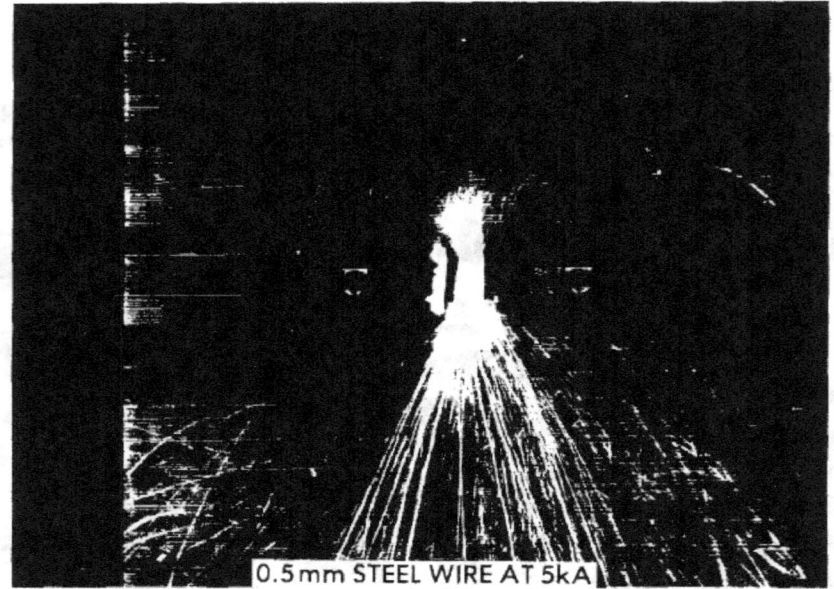

0.5mm STEEL WIRE AT 5kA

Figure 4.8   Resistive heating and explosive vaporization of conductors.

shock wave at the surface. Depending on the distance of the arc from the surface, overpressures can range up to several hundred atmospheres at the surface, resulting in implosion-type damage, such as that shown in Figure 4.9 (Reference 4.9). The arc does not have to contact the damaged surface but may simply be swept alongside it, as was evident in the case shown in Figure 4.9. Apparently a return stroke or restrike occurred as the tip of the propeller passed just below the leading edge of the wing, positioning a cylindrical shock wave horizontally beneath the wing, as in Figure 4.10. This hypothesis by Hacker (Reference 4.10) is supported by scorching of the paint on the imploded panels, an indication of a nearby heat source.

If an arc is contained inside a structure, such as would occur when a nonmetallic assembly is punctured, its overpressure may cause additional damage to the structure. This may have been responsible for some of the damage to the radome shown in Figure 4.7.

Other examples of shock wave implosion damage include cracked or shattered windshields and navigation light globes. Modern windshields, especially those aboard transport aircraft, are of laminated construction and evidently of sufficient strength to have avoided being completely broken by arc blast and overpressures. Broken windshields resulting from a lightning strike, however, are considered a possible cause of the crash of at least one propeller-driven aircraft. (Reference 4.11).

84

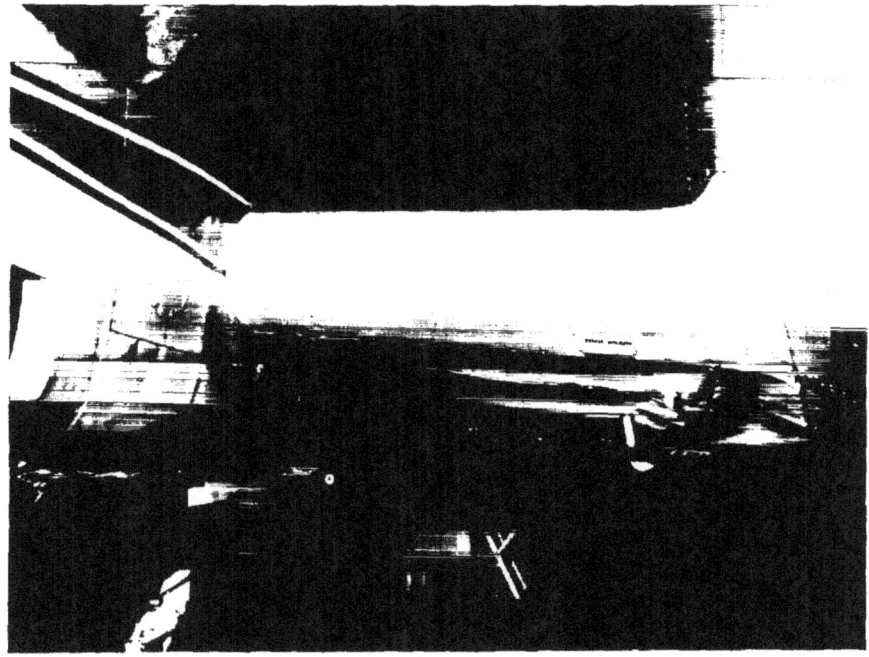

Figure 4.9 Implosion damage from lightning-flash overpressure. Flash swept aft beneath wing from propeller.

4.3 Nonmetallic Structures

Early aircraft of wood and fabric construction would probably have suffered more catastrophic damage from lightning strikes had it not been for the fact that these aircraft were rarely flown in weather conducive to lightning. The all-aluminum aircraft which followed were able to fly in or near adverse weather and receive strikes, but because aluminum is an excellent electrical conductor, severe or catastrophic damage from lightning was rare. There is a trend again, however, toward use of nonmetallic materials in aircraft construction. These include fiber-reinforced plastics and polycarbonate resins, which offer improvements in cost and performance. However, some of these materials have begun to appear at aircraft extremities, such as nose, wing and empennage tips, and access door covers, where structural loads are moderate but where lightning strikes frequently attach. Often the nonmetallic material is used to cover a metallic structure. If this material is nonconducting, such as is the case with fiberglass, electric fields may penetrate it and initiate streamers from metallic objects inside. These streamers may puncture the nonmetallic material as they propagate outward to meet an oncoming lightning leader. This puncture begins as a pinhole, but, as soon as stroke currents and accompanying blast and shock waves follow, much more damage occurs. An example of a puncture of a fiberglass-honeycomb radome is shown on Figure 4.11 (Reference 4.12). In this case a

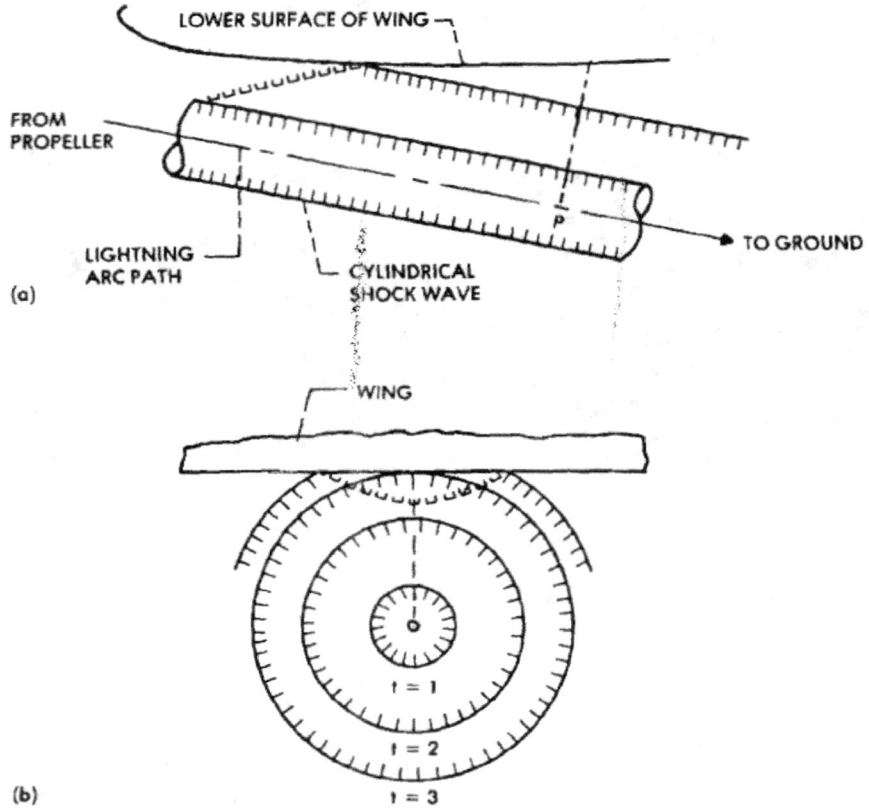

SHOCK WAVES

ⲧⲧⲧⲧⲧⲧⲧⲧ INCIDENT

ⲛⲛⲛⲛⲛⲛⲛ REFLECTED

LOWER SURFACE OF WING

FROM
PROPELLER

TO GROUND

LIGHTNING
ARC PATH

CYLINDRICAL
SHOCK WAVE

(a)

WING

t = 1

t = 2

(b)   t = 3

Figure 4.10 Orientation of lightning path and shock wave with respect to lower
side of aircraft wing shown in Figure 4.9.

(a) Chordwise plane

(b) Fore-aft plane: perpendicular to lightning path at point P of part (a).

streamer evidently propagated from the radar dish or some other conductive
object inside the radome, puncturing the fiberglass-honeycomb wall and rubber
erosion protection boot on its way to meet an oncoming lightning leader. Most
of the visible damage was done by the ensuing stroke current.

Other materials, such as boron- and graphite-reinforced composites, do
have some electrical conductivity, and, because of this, their behavior with
respect to lightning is considerably different from that of nonconductive

86

Figure 4.11　Puncture of a fiberglass-honeycomb radome.

materials. At the present time, boron and graphite composites are only beginning to see in-flight service in Zone 1 and Zone 2 regions (Figure 4.5), where lightning attachments may occur; because of widespread concern regarding possible lightning vulnerability, these zones have been conservatively protected with conductive strips or coatings. No reports exist as yet of composite parts damaged by natural lightning. Simulated lightning tests which have been performed on composites in the laboratory, however, have shown (Reference 4.13) that unprotected composites are likely to be vulnerable. The reason is that there is sufficient conductivity in the reinforcing fibers or filaments to prevent electric field penetration and puncture of the composite but not enough conductivity to safely carry away the lightning flash currents, which in this case tend to flow into the fibers or filaments themselves. When carrying even minute portions of the total flash current, these poor conductors overheat, and damage to themselves or to the surrounding resin matrix results. An example of what happens to the filaments in a boron-reinforced composite is shown in Figure 4.12. The lightning current entered the composite at the damaged areas shown and flowed to a conducting plate at one end of the panel. There is little or no damage to the resin matrix, but the damage to the reinforcing filaments results in loss of overall material strength (Reference 4.14).

Because of their greater electrical conductivity and other differences,

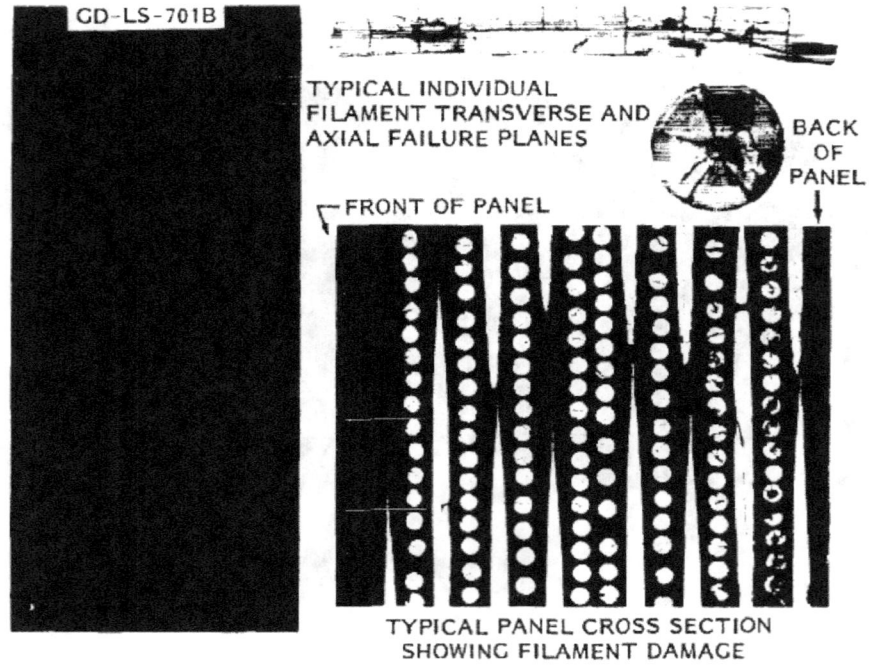

GD-LS-701B

TYPICAL INDIVIDUAL
FILAMENT TRANSVERSE AND
AXIAL FAILURE PLANES

BACK
OF
PANEL

FRONT OF PANEL

TYPICAL PANEL CROSS SECTION
SHOWING FILAMENT DAMAGE

Figure 4.12 Fifty-six kiloampere simulated lightning current damage to boron
composite laminate panel. Seventeen-ply, 0°-90°-0° orientation.

graphite fibers can withstand more lightning current than can boron fibers. At
higher current levels the graphite fibers remain intact but become hot enough to
boil and ignite the plastic resin. Figure 4.13 (Reference 4.15) shows a small
"popsicle stick" graphite composite laminate through which 10 kA of lightning
current have been passed. Nearly all of the resin has burned away, and the
graphite fibers are left in disarray. As a whole, however, graphite-reinforced
composites are able to withstand higher amounts of lightning current without
damage than can boron-reinforced composites. Often some damage to either
material can be accepted or repaired, but sometimes a lightning-damaged
composite part will need replacement.

Transparent acrylics or polycarbonate resins are often utilized for canopies
and windshields. These materials are usually found in Zone 1 or Zone 2
locations, where either direct or swept-lightning flashes may occur. Most of the
polycarbonates are very good insulators, however, and so will successfully resist
punctures by lightning or streamers. The electric field will penetrate them and
induce streamers from conducting objects inside, but these streamers are not
usually able to puncture a polycarbonate. Thus, fighter pilots beneath polycarbonate canopies have often reported electric shocks indicative of streamering
off their helmets, but the current levels involved have not been harmful because
the streamers have not come in direct contact with the lightning flash. Leaders
approaching the outside of a canopy travel along its surface to reach a metallic

88

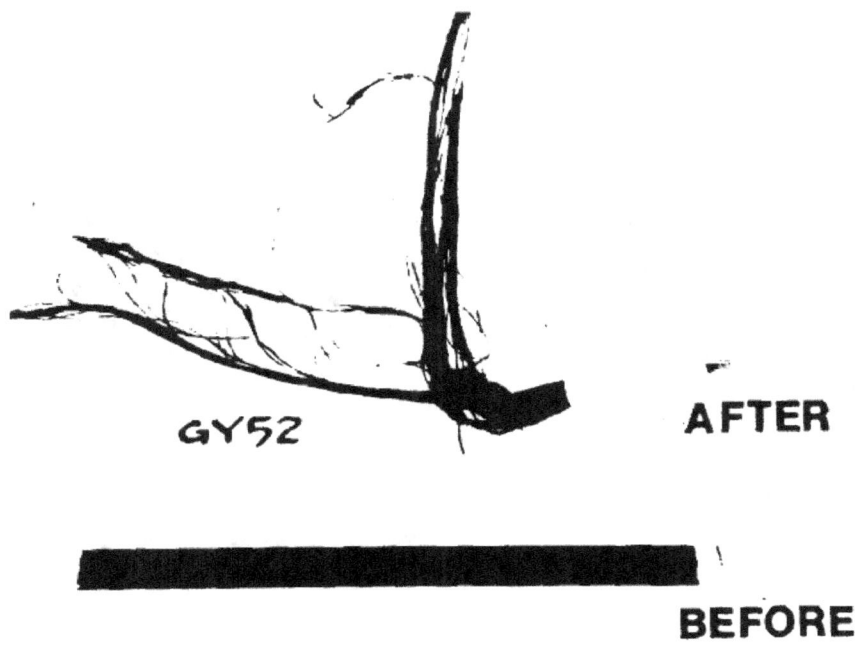

GY52

AFTER

BEFORE

Figure 4.13 Unidirectional graphite composite sample before and after conducting 10 kA of lightning current end-to-end.

skin, or those initially attached to a forward metal frame may be swept aft over a canopy until they reattach to an aft metallic point. Sometimes this occurrence will leave a scorched path across the canopy, as shown in Figure 4.14 (Reference 4.16). Scorches like this can usually be polished away.

While harmless to a canopy itself, flashes passing just outside frequently cause electrical shock or flash blindness to the pilot. In at least one case shock or blindness to the pilot caused him to lose control of the aircraft at low altitude and resulted in a fatal accident.

In addition to the direct effects described in the preceding paragraphs, replacement of metallic skins with nonmetallic materials removes the inherent protection against electromagnetic field penetration that is an important by-product of aluminum skins. Electrical wiring and electronics components enclosed inside nonmetallic skins are therefore likely to be much more susceptible to the indirect effects of lightning than those inside metallic skins unless specific measures are taken to reduce this susceptibility.

4.4 Fuel Systems

Potentially, aircraft fuel systems represent the most critical lightning

Figure 4.14 Evidence of lightning attachment to canopy fastener and scorching of canopy.

hazard to flight safety. An electric spark produced by only 0.2 millijoule (mJ) of energy is sufficient to ignite a propagating flame in a near stoichiometric mixture of hydrocarbon fuel and air (Reference 4.17); yet lightning-flash currents may deposit several thousand joules of energy in an aircraft.

There are several jet and turbojet transport accidents on record which have been attributed to lightning ignition of fuel. Although the exact location of ignition in each case remains obscure, the most prevalent opinion is that lightning ignited fuel vapor at the wing tip vent outlets of these aircraft (References 4.18 and 4.19). It is also possible that sparking occurred somewhere inside a fuel tank as lightning currents flowed through the aircraft. The inflight loss of at least two military aircraft also has been attributed to lightning ignition of fuel, and there is a report of a lightning strike igniting fuel in another military aircraft parked on the ground (Reference 4.20).

In addition to the direct effects described above, there are several instances in which indirect effects have evidently accounted for ignition of fuel. Lightning-induced voltages in aircraft electrical wiring are believed to have resulted in sparks, for example, across a capacitance-type fuel probe or some other electrical object inside fuel tanks of several military aircraft, resulting in loss of external tanks in some cases and the entire aircraft in others. Capacitance-type fuel probes are designed to preclude such occurrences, and laboratory tests (Reference 4.21) have shown that the voltage required to spark a typical capacitance-type probe is many times greater than that induced in fuel gauge circuits by lightning. However, other situations involving unenclosed circuits, such as externally mounted fuel tanks, exist wherein induced voltages may be much higher than those found in circuits completely enclosed by an airframe.

The accidents mentioned above prompted extensive research into the lightning effects on and protection of aircraft fuel systems. Improved bonding, lightning-protected filler caps and access doors, active and passive vent flame suppression devices, flame-retardant foams, and safer (i.e., less flammable) fuels are examples of developments which have resulted from this research. In addition, government airworthiness requirements now include lightning protection for aircraft fuel systems and specify requirements and tests that must be passed to demonstrate compliance prior to aircraft certification. As a result of these safety measures, lightning strikes present fewer hazards to the fuel systems aboard modern transport aircraft than to those of older aircraft, and properly certificated aircraft may expect to experience lightning strikes with no adverse effects on fuel systems. Continued changes in airframe designs and materials, however, make it mandatory that care and diligence in fuel system lightning protection not be relaxed in the future.

4.5 Electrical Systems

If an externally mounted electrical apparatus, such as a navigation lamp or antenna, happens to be at a lightning attachment point, protective globes or fairings may shatter and permit some of the lightning current to enter associated electrical wiring directly.

In the case of a wing tip navigation light, for example, lightning may shatter the protective globe and light bulb. This may in turn allow the lightning arc to contact the bulb filament so that lightning currents may flow into the electrical wires running from the bulb to the power supply bus. Even if only a fraction of the total lightning current enters the wires, they may be too small to conduct the thousands of amperes involved and thus be melted or vaporized. The accompanying voltage surge may cause breakdown of insulation or damage to other electrical equipment powered from the same bus. At best, the initial component affected is disabled, and, at worst, enough other electrical apparatus is disabled along with it to require evacuation of the crew and loss of the aircraft. There are many examples of this effect, involving both military and civil aircraft. Externally mounted hardware most frequently involved includes navigation lights, antennas, windshield heaters, pitot probe heaters, and, in earlier days, the trailing long-wire antennas that were deployed in flight for high-frequency radio communications. The latter were quite susceptible to lightning strikes, and, since these wires were too thin to conduct the ensuing currents, they were frequently burned away. The high-frequency radio sets feeding these antennas were also frequently damaged, and cockpit fires were not uncommon.

Damage may be increased when an electrical assembly is mounted on nonmetallic portions of the airframe because some lightning current may have to use the assembly ground wire as a path to the main airframe. That the resulting damage can be extensive was exemplified by a recent strike to a small single-engine aircraft with fiberglass wing tips which included fuel tanks, the type pictured in Figure 4.15. The details of this incident will illustrate several of the strike effects described in this section.

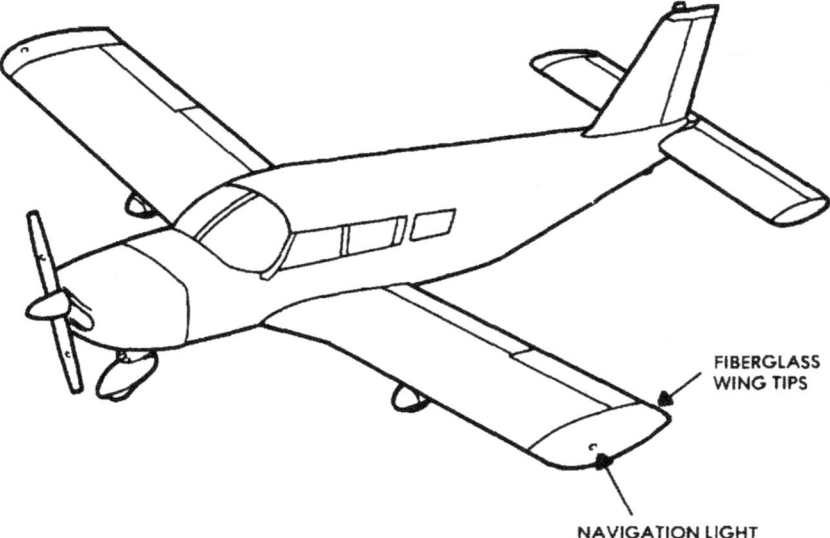

Figure 4.15 General avaiation aircraft with plastic wing tips.

This aircraft, flying at about 900 m (3 000 ft), was experiencing light rain and moderate turbulence when it was struck by lightning. The pilots had seen other lightning flashes in the vicinity before their aircraft was struck, and embedded thunderstorms had been forecast enroute, but there had been no cells visible on the air traffic control (ATC) radar being used to vector the aircraft, which, of course, had no weather radar of its own.

The strike entered one wing tip and exited from the other. It sounded to the pilot reporting like a rifle going off in the cabin, and the cabin immediately filled with smoke. Other effects follow.

- The No. 1 VHF communication set burned out.
- Seventy-five percent of the circuit breakers were popped, of which only 50% could be reset later.
- The left wing tip fuel tank quantity indicator was disabled.
- The right main fuel tank quantity indicator was badly damaged.
- Several instrument lights were burned out.
- The navigation light switch and all lights were burned out.

The aircraft, nevertheless, was able to land at a nearby airport. Subsequent inspection showed extensive damage to the right and left wing tips and to their electrical wiring. The attachment points and direct effects are pictured in Figure 4.16 (A-F) (Reference 4.22) and are represented by a diagram in Figure 4.17. The evidence suggests that the flash included two or more strokes separated by a few milliseconds of continuing current. Assuming, for purposes of explanation, that the original lightning flash approached the right wing tip, the probable sequence of events was as follows: the initial point of attachment was the right wing tip navigation light housing (Figure 4.16 [A]). Current from this stroke entered the housing ground wire and exploded both sections of it on the way to the right outboard metallic rib, as evidenced by the absence of these wires and the blackened interior shown in Figure 4.16(B). Current continued through the airframe to the left outboard rib and out the sender unit ground wire to the sender unit, the base of which is shown in Figure 4.16(C). From there, the current followed the filler cap ground braid and exited the aircraft at the filler cap (Figure 4.16[D]). The current exploded the sender unit ground wire but not the heavier filler cap ground braid, which was only frayed. Sparks undoubtedly occurred inside the fuel tank along the ground braid and between the filler cap and its receptacle, but the fuel-air mixture in the ullage of these half-full tanks was probably too rich to support ignition.

Blast forces from stroke No. 1 at the right navigation light housing also shattered the lamp globe and bulb, as shown in Figure 4.16(A). This shattering allowed a portion of the first stroke current to enter the right navigation light power wire, exploding it between the lamp and the outer rib, where the current jumped to the rib and continued through the rest of the airframe to the left sender unit ground wire.

Lightning current flowing in the navigation lamp power wire elevated its voltage to several thousand volts with respect to the airframe, a voltage high enough to break down the insulation at the outer rib feed-through point, as shown in Figure 4.17. Until breakdown occurred here (a few microseconds after the first stroke began), the wire was at sufficiently high voltage to break down the insulation to the neighboring sender wire. This breakdown occurred all along the wire inside the right wing. The portion of current arcing into the sender wire caused a large voltage to build up across the right wing tip tank fuel gauge magnet inductance, to which this wire connects. This voltage in turn sparked over the gap between the gauge terminal and the nearest grounded housing wall, the arcing badly damaging the gauge unit. While the left navigation light power wire was also exploded, it is probable that this did not occur until the second stroke.

Since the aircraft was moving forward, the entry and exit points of the second stroke were farther aft on both wing tips than the points of the first stroke. Since no other metallic components were present aft of the first stroke entry point on the right wing tip, the second stroke punctured a hole in the fiberglass trailing edge and contacted the metallic outboard rib, as shown in Figure 4.16(E). As shown in Figure 4.17, current from this stroke proceeded

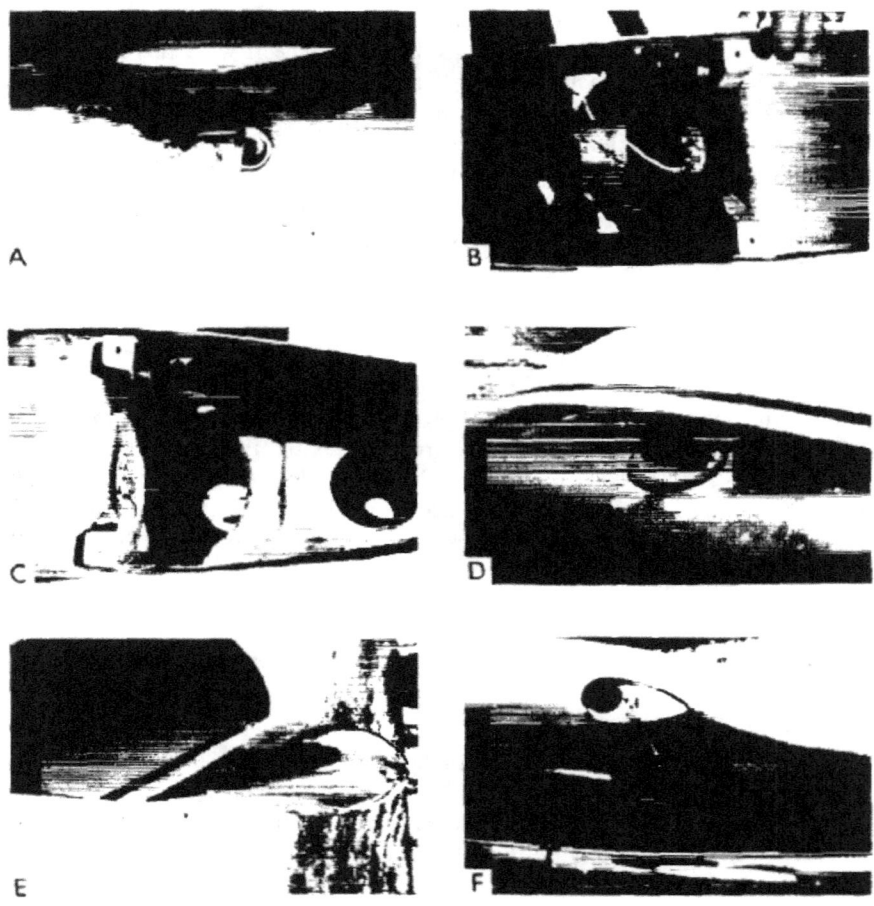

Figure 4.16 Attachment points and direct effects on plastic wing tips.

through the airframe to the left wing tip, where by this time the stroke had swept aft adjacent to the navigation lamp (shown in Figure 4.16[F]), from which point the stroke current exited. Current from stroke No. 2 thus probably arced between the left outer rib to the navigation lamp power wire (the ground wire having been vaporized by the first stroke), which it followed to the lamp housing. The power wire was vaporized by the second stroke current flowing in it.

Both left and right navigation lamp power wires were connected together in the cabin and to both the 12 V dc bus and the tail light. The voltage and current surges which entered the lamp power wires inboard of the outer rib feedthroughs were also conducted to the tail light, burning it out, and to the 12 V dc bus. The surge on the bus, of course, was immediately imposed on all of the other electrical equipment powered from this bus, or *all* of the electrical equipment in this aircraft. Arcing undoubtedly occurred in a number of

94

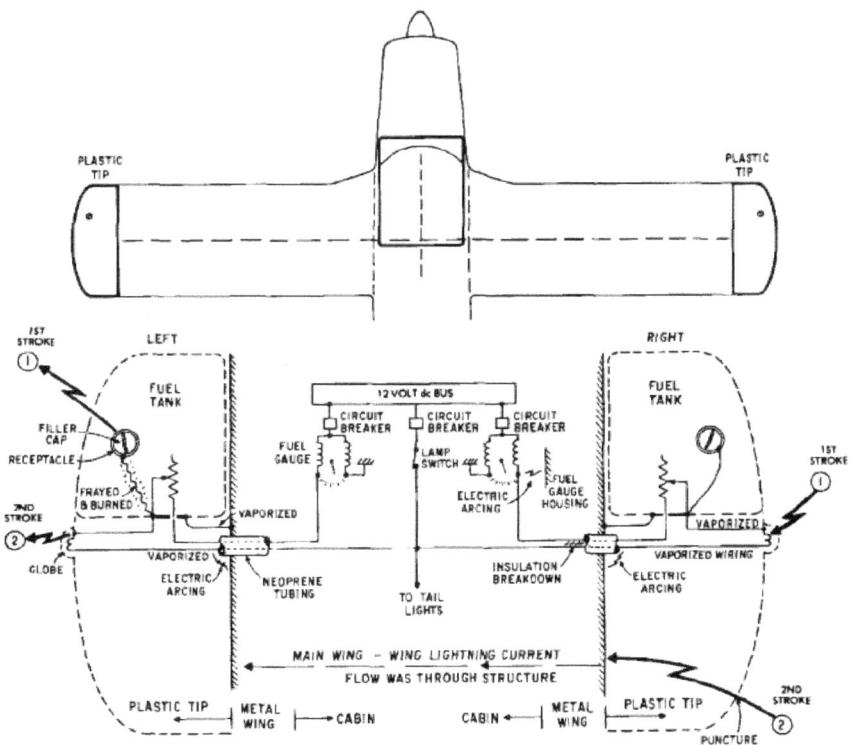

Figure 4.17 Plastic wing tips and associated electric circuits and locations of lightning effects.

components, causing circuit breakers to pop. Because circuit breakers, however, react much too slowly to prevent passage of a lightning surge, at least one piece of equipment (the No. 1 VHF communication set) and several instrument lamps were burned out.

There have been several similar incidents (References 4.23 and 4.24), and together these have stimulated the design and verification of protective measures (Reference 4.25) for general aviation aircraft with fiberglass components such as these wing tip fuel tanks. Many of these aircraft, however, are still flying without adequate protection.

The foregoing incident is an example of how a change in materials can increase the vulnerability not only of the airframe but also of other systems which previously had the inherent protection of conventional aluminum skins. The lightweight lamp and sender unit electrical wires were quite adequate for an installation in which a metal skin was available to carry away lightning currents, but they are woefully inadequate when used unmodified inside a plastic wing tip, where they become the only conducting path available to lightning currents trying to enter the main airframe.

95

4.6 Engines

With the exception of a few incidents of temporary malfunction similar to the incident reported in the introduction to this chapter, there have been no reports of adverse lightning effects on reciprocating engines. Metal propellers and spinners have been struck frequently, of course, but effects have been limited to pitting of blades or burning of small holes in spinners, as shown in Figures 4.18 and 4.19 (References 4.26 and 4.27). Lightning currents must flow through

Figure 4.18 Lightning-strike damage to a propeller.

propeller blade and engine shaft bearings, but these are massive enough to carry these currents with no harmful effects. Wooden propellers, especially ones without metal leading edges, could probably undergo more damage, but these are seldom used on aircraft which fly in weather conditions where lightning strikes occur.

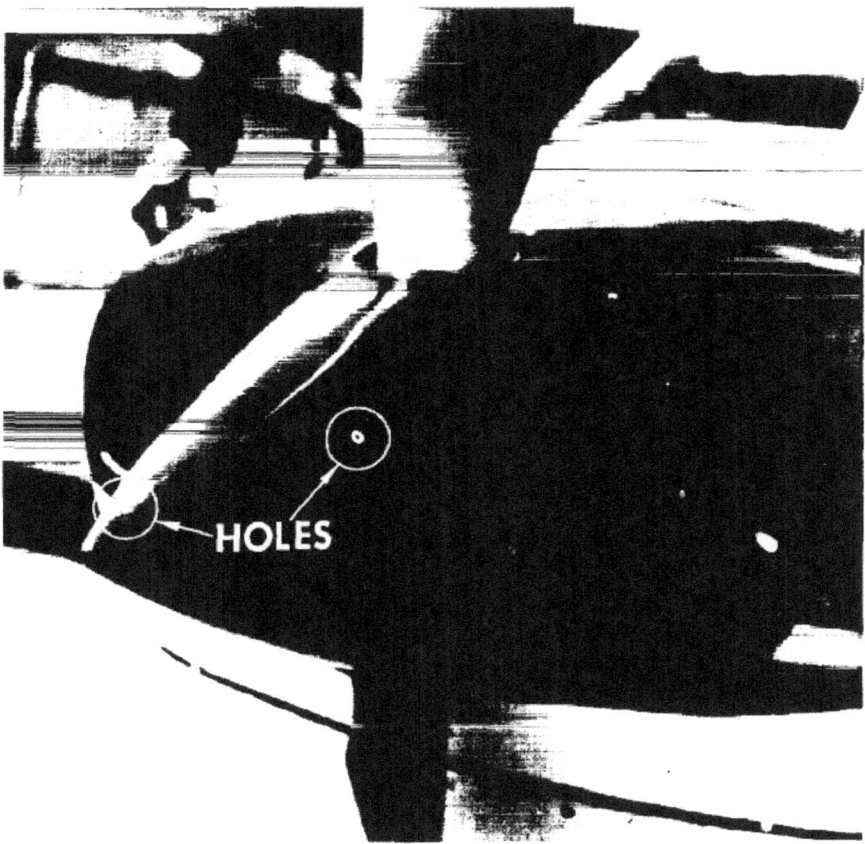

Figure 4.19 Lightning-strike damage to a spinner.

Reported lightning effects on turbojet engines show that these effects also are limited to temporary interference with engine operation. Flameouts, compressor stalls, and roll-backs (reduction in turbine rpm) have been reported after lightning strikes to aircraft with fuselage-mounted engines. This type includes military aircraft with internally mounted engines and fuselage air intakes, or other military and civil aircraft with engines externally mounted on the fuselage. There have been no attempts to duplicate these events with simulated lightning in a laboratory, and there has been no other qualitative analysis of the interference mechanism; however, it is generally believed that

97

these events result from disruption of the inlet air by the shock wave associated with the lightning arc channel sweeping aft along a fuselage. This channel may indeed pass close in front of an engine intake, and if a restrike occurs, the accompanying shock wave is considered sufficient to disrupt engine operation. The steep temperature gradient may also be important. These effects have been reported as occurring more often on smaller military or business jet aircraft than on larger transport aircraft. Thus, smaller engines are probably more susceptible to disrupted inlet air than are their larger counterparts.

In some cases a complete flameout of the engine results, while in others there is only a stall or roll-back. There is no case on record, however, in which a successful restart or recovery of the engine to full power was not made while still in flight. Perhaps because of this, together with the impracticality of a laboratory simulation, there has been little research into the problem. Nevertheless, operators of aircraft with engines or inlets close to the fuselage should anticipate possible loss of power in the event of a lightning strike and be prepared to take quick corrective action.

There are no reports of lightning effects on wing-mounted turbojet engines, since lightning strikes do not often occur near the inlets of these engines, and there are no reports of power loss of turboprop engines as a result of lightning strikes.

4.7 Indirect Effects

Even if the lightning flash does not directly contact the aircraft's electrical wiring, strikes to the airframe are capable of causing voltage and current surges in this wiring which may be damaging to aircraft electronics.

The mechanism whereby lightning currents induce voltages in aircraft electrical circuits is illustrated in Figure 4.20. As lightning current flows through an aircraft, strong magnetic fields which surround the conducting aircraft and change rapidly in accordance with the fast-changing lightning-stroke currents are produced. Some of this magnetic flux may leak inside the aircraft through apertures such as windows, radomes, canopies, seams, and joints. Other fields may arise inside the aircraft when lightning current diffuses to the inside surfaces of skins. In either case these internal fields pass through aircraft electrical circuits and induce voltages in them proportional to the rate of change of the magnetic field. These magnetically induced voltages may appear between both wires of a two-wire circuit, or between either wire and the airframe. The former are often referred to as *line-to-line voltages* and the latter as *common-mode voltages.*

In addition to these induced voltages, there may be resistive voltage drops along the airframe as lightning current flows through it. If any part of an aircraft circuit is connected anywhere to the airframe, these voltage drops may appear between circuit wires and the airframe, as shown in Figure 4.20. For metallic aircraft made of highly conductive aluminum, these voltages are seldom significant except when the lightning current must flow through resistive joints

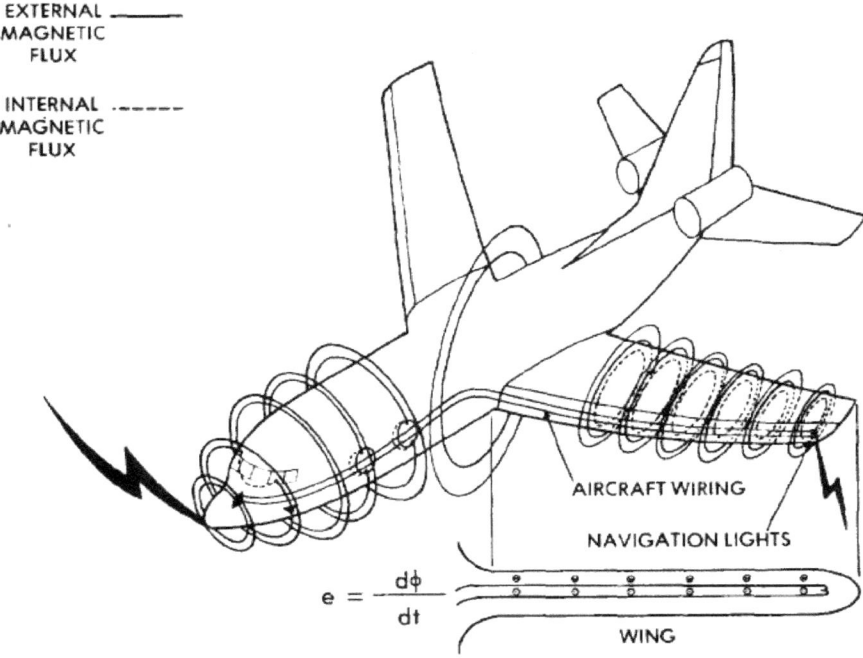

EXTERNAL _____
MAGNETIC
FLUX

INTERNAL . _ _ _ _
MAGNETIC
FLUX

AIRCRAFT WIRING

NAVIGATION LIGHTS

$$e = -\frac{d\phi}{dt}$$

WING

Figure 4.20 Magnetic flux penetration and induced voltages in electrical wiring.

or hinges. However, the resistance of titanium is 10 times that of aluminum, and that of composite materials many hundred times that of aluminum, so the resistive voltages in future aircraft employing these materials may be much higher.

Upset or damage of electrical equipment by these induced voltages is defined as an *indirect effect*. It is apparent that indirect effects must be considered along with direct effects in assessing the vulnerability of aircraft electrical and electronics systems. In situations like that of the light aircraft described in Section 4.5, the direct effects are clearly the most severe. Other aircraft exist, however, whose electrical systems are well protected against direct effects but not so well against indirect effects.

Until the advent of solid state electronics in aircraft, indirect effects from external environments, such as lightning and precipitation static, were not much of a problem and received relatively little attention. No airworthiness criteria are available for this environment. There is increasing evidence, however, of troublesome indirect effects. Incidents of upset or damage to avionic or electrical systems, for example, without evidence of any direct attachment of the lightning flash to an electrical component are showing up in airline lightning-strike reports. Table 4.1 summarizes the reports of interference or outage of avionic or electrical equipment reported by a group of U.S. airlines for the period June 1971 to November 1974 (Reference 4.28).

99

Table 4.1 EVIDENCE OF INDIRECT EFFECTS IN COMMERCIAL AIRCRAFT
(214 strikes)

	Interference	Outage
HF communication set	—	5
VHF communication set	27	3
VOR receiver	5	2
Compass (all types)	22	9
Marker beacon	—	2
Weather radar	3	2
Instrument landing system	6	—
Automatic direction finder	6	7
Radar altimeter	6	—
Fuel flow gauge	2	—
Fuel quantity gauge	—	1
Engine rpm gauges	—	4
Engine exhaust gas temperature	—	2
Static air temperature gauge	1	—
Windshield heater	—	2
Flight director computer	1	—
Navigation light	—	1
ac generator tripoff	(6 instances of tripoff)	
Autopilot	1	—

The incidents reported in Table 4.1 occurred in 20% of the total of 214 lightning-strike incidents reported during the period. U.S. military aircraft have had similar experience. This experience is probably a result of the increasing sensitivity of miniaturized solid state electronics to transient voltages, a trend which necessarily would not have posed a problem in older, less sophisticated equipment. In any one incident, only a few electronic components are affected; others are not. Yet laboratory tests (Reference 4.29) have shown that lightning-induced voltages appear in all aircraft electric wiring at once. Thus it is evident that surges reach higher values in some circuits than in others or that some electronics are less tolerant of such surges than others.

Indirect effects were evident in about 20% of all the incidents reported, while outages were reported in only about 10% of all incidents. Since severe strokes also occur in only about 10 to 20% of all flashes, it is probable that a severe stroke may be required to cause noticeable effects.

While the indirect effects are not presently a major safety hazard, there are four trends in aircraft design and operations which could increase the potential problem. These include the following:

- Increasing use of plastic or composite skin
- Further miniaturization of solid state electronics

- Greater dependence on electronics to perform flight-critical functions
- Greater congestion in terminal airways, requiring more frequent flight through adverse weather conditions at altitudes where lightning strikes frequently occur.

Design of protective measures against direct effects is the subject of the next three chapters. Design of protective measures against indirect effects is treated in the remainder of this book.

REFERENCES

4.1 Answer to questionnaire *Effects of Electrical Phenomena upon Airplanes in Flight,* National Advisory Committee for Aeronautics, Washington, D.C. (July 7, 1945).

4.2 U.S. Air Force photograph.

4.3 National Aeronautics and Space Administration photograph.

4.4 National Aeronautics and Space Administration photograph.

4.5 B. L. Perry, "British Researches and Protective Recommendations of the British Air Registration Board," *Lightning and Static Electricity Conference, 3-5 December 1968, Part II, Conference Papers,* Technical Report AFAL-TR-68-290, Air Force Avionics Laboratory, Air Force Systems Command, Wright-Patterson Air Force Base, Ohio (May 1969), pp. 81-103: 96.

4.6 U.S. Air Force photograph.

4.7 U.S. Air Force photograph.

4.8 M. A. Uman, *Lightning* (New York: McGraw-Hill, 1969), p. 230.

4.9 Paul T. Hacker, *Lightning Damage to a General Aviation Aircraft— Description and Analysis,* NASA TN-7775, National Aeronautics and Space Administration, Lewis Research Center, Cleveland, Ohio (September 1974).

4.10 Hacker, *Lightning Damage.*

4.11 *Report of the Investigation of an Accident Involving Aircraft of U.S. Registry, NC 21789, Which Occurred near Lovettsville, Virginia, on August 31, 1940,* Civil Aeronautics Board, Washington, D.C. (1941).

4.12 United Air Lines photograph.

4.13 L. C. Walko, *Current Status of Composites Vulnerability to Lightning,* SRD-74-090, Corporate Research and Development, General Electric Company, Schenectady, New York (August 1, 1974).

4.14 F. A. Fisher and W. M. Fassell, *Lightning Effects Relating to Aircraft, Part I: Lightning Effects on and Electromagnetic Shielding Properties of Boron and Graphite Reinforced Composite Materials,* Technical Report AFAL-TR-72-5, Air Force Avionics Laboratory, Air Force Systems Command, Wright-Patterson Air Force Base, Ohio (January 1972), pp. 48-80.

4.15 From studies of A. P. Penton, J. L. Perry, and K. J. Lloyd. See "Fundamental Investigations of High Intensity Electric Current Flow, Processes and Resultant Damage in Advanced Composites," *1970 Lightning and Static Electricity Conference, 9-11 December,* sponsored jointly by the Air Force Avionics Laboratory and the Society of Automotive Engineers, Air Force Systems Command, Wright-Patterson Air Force Base, Ohio (December 1970), pp. 253-297.

4.16 General Electric Company photograph.

4.17 *Fire Protection Research Program for Supersonic Transport,* APL TDR-64-105, Air Force Aero Propulsion Laboratory, Wright-Patterson Air Force Base, Ohio (October 1964), pp. 111-113.

4.18 *Aircraft Accident Report: Boeing 707-121, N709PA, Pan American World Airways, Inc., near Elkton, Maryland, December 8, 1963,* File No. 1-0015, Civil Aeronautics Board, Washington, D.C. (February 25, 1965).

4.19 *Aircraft Accident Report: TWA Lockheed 1649A near Milan, Italy,* File No. 1-0045, Civil Aeronautics Board, Washington, D.C. (November 1960). English translation of report by Italian Board of Inquiry.

4.20 C-141 on-ground incident.

4.21 J. A. Plumer, "Lightning-Induced Voltages in Electrical Circuits Associated with Aircraft Fuel Systems," *Report of Second Conference on Fuel System Fire Safety, 6 and 7 May 1970,* Federal Aviation Administration, Washington, D.C. (1970), pp. 171-92.

4.22 General Electric Company photographs.

4.23 Don Flagg, "Night Flight," *Aero Magazine,* January/February 1972, pp. 18-21.

4.24 Page Shamberger, "Learning About Flying the Hard Way," *Air Progress,* February 1971, p. 64.

4.25 J. A. Plumer, *Guidelines for Lightning Protection of General Aviation Aircraft,* FAA-RD-73-98, Federal Aviation Administration, Washington, D.C. (October 1973).

4.26 National Aeronautics and Space Administration photograph.

4.27 National Aeronautics and Space Administration photograph.

4.28 J. A. Plumer and B. L. Perry, "An Analysis of Lightning Strikes in Airline Operation in the USA and Europe," *Proceedings of the 1975 Conference on Lightning and Static Electricity at Culham Laboratory, England, 14-17 April 1975,* Session III: Aircraft Applications, the Royal Aeronautical Society of London (December 1975).

4.29 J. A. Plumer, F. A. Fisher, and L. C. Walko, *Lightning Effects on the NASA F-8 Digital Fly-By-Wire Airplane,* NASA CR 2524, prepared by the High Voltage Laboratory, Environmental Electromagnetics Unit, Corporate Research and Development, General Electric Company, Pittsfield, Massachusetts, for the National Aeronautics and Space Administration, Lewis Research Center, Cleveland, Ohio (March 1975), p. 138.

CHAPTER 5
DIRECT EFFECTS PROTECTION

5.1 Introduction

Successful protection of an aircraft against the direct effects of lightning depends upon protection of each of its various systems or components that may be susceptible either to direct lightning attachment or to current flow between lightning attachment (entry and exit) points. Components located in different sections of the aircraft are likely to experience different degrees of susceptibility to lightning, and they may be vulnerable to different components of the lightning flash. In this chapter we present the basic steps which should be followed in establishing lightning protection by discussing the varieties of lightning susceptibility and the designing of suitable protective measures against arc entry and current flow-through damage. We explain how the lightning-strike zones and lightning current environments are established, since environmental conditions in the zones are those under which specific protective measures must perform. We then call attention to those airworthiness regulations which apply to lightning protection. In subsequent chapters we present specific protection techniques for aircraft fuel and structural systems exposed to the direct effects of lightning strikes.

5.2 Basic Steps in Protection Design

5.2.1 Establishment of the Lightning Susceptibility

Design of successful protection usually involves taking the following basic steps. As described in Chapter 1, lightning strikes commonly attach to an aircraft nose, wing tips, vertical fin tip, horizontal stabilizer tips, and to other appendages, such as propellers, pitot booms, antennas, or pylon-mounted external stores. These are the places where the hot lightning arc attaches directly to the aircraft and thus the places of initial concern as far as direct effects are concerned. However, lightning currents must flow through the aircraft skin and structural members between these arc entry and exit points, and it is possible for these currents to damage components along the way if safe conduction paths are not provided. Thus, the lightning protection designer's job is not completed when he has adequately treated only the attachment points. Instead, he must also look throughout the aircraft at other components that might be vulnerable in one way or another to direct effects from current flow through the aircraft.

The susceptibility (degree of exposure to lightning effects) of components depends greatly on their location on or within the aircraft. Thus it is helpful to utilize the lightning-strike zone definitions of Chapter 2 to establish the lightning environment which each particular component experiences. Once this has been established, the vulnerability (if any) of the component to its own environment can be determined, and protective measures designed if necessary.

The lightning attachment zone definitions proposed by SAE Special Task F (Reference 5.1) are repeated below:

Zone 1A: An initial attachment point with a low probability of flash hang-on, such as a leading edge

Zone 1B: An initial attachment point with a high probability of flash hang-on, such as a trailing edge

Zone 2A: A swept-stroke zone with a low probability of flash hang-on, such as a wing mid-span

Zone 2B: A swept-stroke zone with high probability of flash hang-on, such as a wing inboard trailing edge

Locations of these zones can be determined either from comparison with actual lightning attachment points on existing aircraft or by laboratory tests in which a scale model of the aircraft is subjected to simulated lightning strikes. Perry (Reference 5.2) has shown the locations of actual lightning attachment points on several types of propeller and turbojet aircraft in use in the United Kingdom and Europe. His strike patterns for three aircraft are shown on Figures 5.1, 5.2, and 5.3 (Reference 5.3). From these figures it is evident that initial attachment points (Zone 1A and 1B) are at or very near the extremities, such as nose, wing tips, or empennage tips. If an initial attachment point is at a trailing edge, the arc must hang on there until the flash dies naturally; thus more damage may occur. On the other hand, if the initial attachment point is a forward extremity, such as the nose, the aircraft can fly through the flash channel, allowing the arc to reattach at subsequent points along the line of flight, as evident from the lines of successive burn marks along the fuselage in the three figures. Thus, forward initial attachment points are commonly in Zone 1A and the subsequent burn marks in Zone 2A. Once in a while, the flash may remain attached to a forward-most extremity or to a small protrusion, such as an antenna or even a rivet head along the fuselage. In such cases the attachment point becomes a B-Zone point with high probability of hang-on. Since most lightning flashes persist for one-fourth of a second or more (up to about 1 second), the aircraft may move forward its entire length, or more, in this period of time, and its entire length may be exposed to swept-lightning flashes. Its entire length must therefore be considered to be in Zone 2.

If a new aircraft design is of the same general configuration (that is, relative position and size of fuselage, wings, empennage, etc.) the general location of its lightning attachment zones can be determined by comparison with known attachment points on similar aircraft. Knowledge of the exact boundaries between one zone and another is important, however, and these boundaries must often be determined by a laboratory test in which a full-size replica of the aircraft assembly in question (such as a wing tip or nose radome) is subjected to simulated lightning flashes from various directions. Knowledge of these zone boundaries is particularly important in designing protection for wing fuel tanks, for example. Recognizing that laboratory tests may not be practical, the Federal Aviation Administration in its advisory circular on protection of aircraft fuel systems against lightning defined Zone 1 as including "[a]ll surfaces

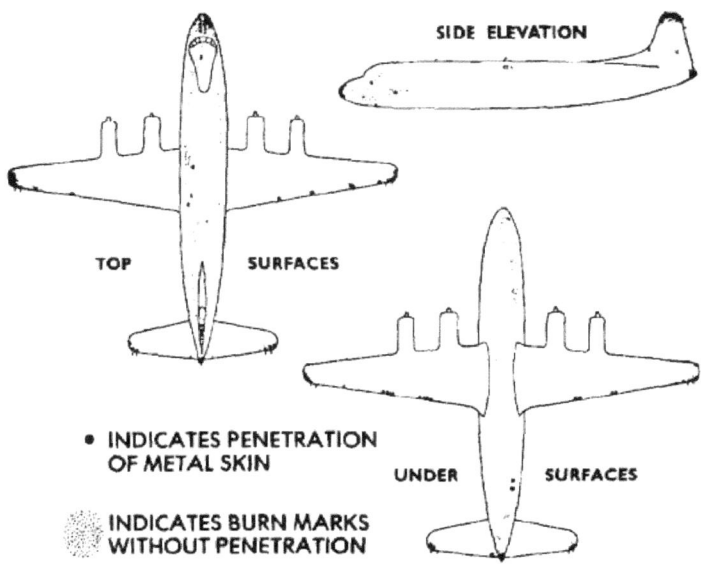

SIDE ELEVATION

TOP SURFACES

• INDICATES PENETRATION
OF METAL SKIN

UNDER SURFACES

INDICATES BURN MARKS
WITHOUT PENETRATION

Figure 5.1   Position of lightning strikes on Viscount aircraft — March 1959 to June 1964.

of the wing tips located within 18 inches of the tip measured parallel to the lateral axis of the aircraft, and surfaces within 18 inches of the leading edge on wings having leading edge sweep angles of more than 45 degrees." Also included are "[p]rojections such as engine nacelles, external fuel tanks, propeller disc, and fuselage nose"; in the tail group all surfaces "within 18 inches of the tips of horizontal and vertical stabilizer, trailing edge of horizontal stabilizer, tail cone, and any other protuberances," as well as "[a]ny other projecting part which might constitute a point of direct stroke attachment."

FAA defines Zone 2 as including "[s]urfaces for which there is a probability of strokes being swept rearward from a Zone 1 point of direct stroke attachment. This zone includes surfaces which extend 18 inches laterally to each side of fore-and-aft lines passing through the Zone 1 forward projection points of stroke attachment. All fuselage and nacelle surfaces, including 18 inches of adjacent surfaces, not defined as Zone 1 are included in Zone 2" (Reference 5.4). (FAA does not subdivide Zone 1 or Zone 2 into A and B zones as recommended by SAE Task F.)

Thus, the establishment of the lightning attachment zones for aircraft of conventional configuration is relatively straightforward and can be done with reasonable accuracy. Prediction of the lightning attachment zones for new aircraft of unconventional design is not as easily accomplished by reference to inflight data or to the FAA 18-inch criteria. Such an aircraft might look like the one in Figure 5.4 (Reference 5.5). The gradual blending of fuselage and wing lines makes identification of the regions of maximum electric field stress (and lightning attachment) less easy to distinguish. With the exception of the pointed

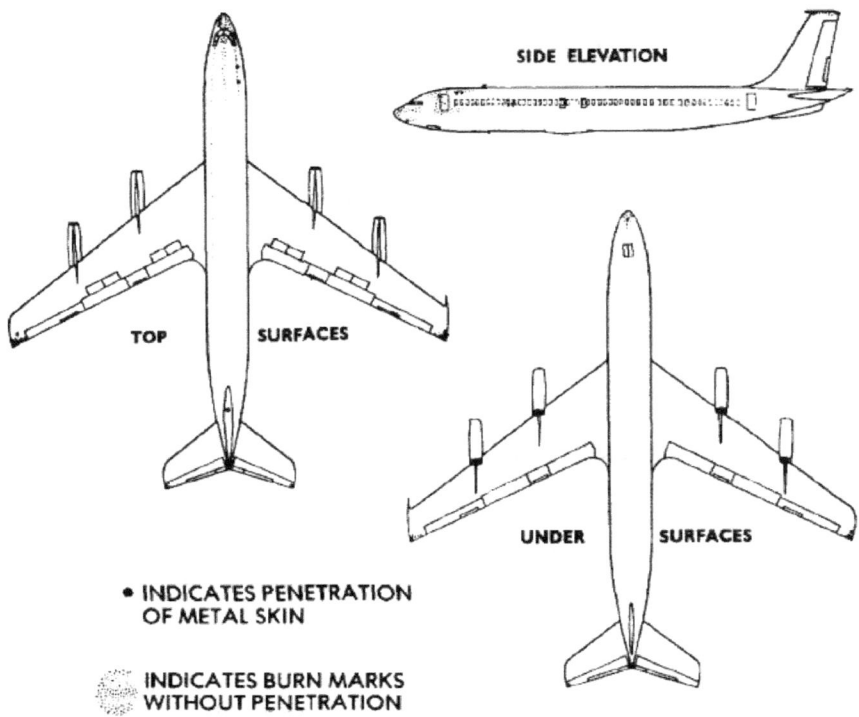

SIDE ELEVATION

TOP   SURFACES

UNDER   SURFACES

• INDICATES PENETRATION
  OF METAL SKIN

INDICATES BURN MARKS
WITHOUT PENETRATION

Figure 5.2 Position of lightning strikes on Boeing 707 aircraft – January 1962
to December 1967.

nose and vertical wing tips, the regions of highest electric field stress are less evident on this aircraft than on the conventional aircraft of Figures 5.1 through 5.3. Thus, in this case strike attachment tests on a scale model in a laboratory are desirable. The SAE Task F report (Reference 5.1) describes how this test should be performed.

5.2.2 Protection of Exposed Systems or Components in Zones 1 and 2 Against Arc Entry Damage

The most obvious protection task is to protect systems or components located in Zone 1 or Zone 2 from damage that results from direct arc attachment. In some cases, complete protection against burning and erosion is impractical, since the consequences of permitting some of this damage to occur are of a maintenance nature only. In other cases, such as those involving integral fuel tank skins, it is very important that complete protection be designed and applied.

108

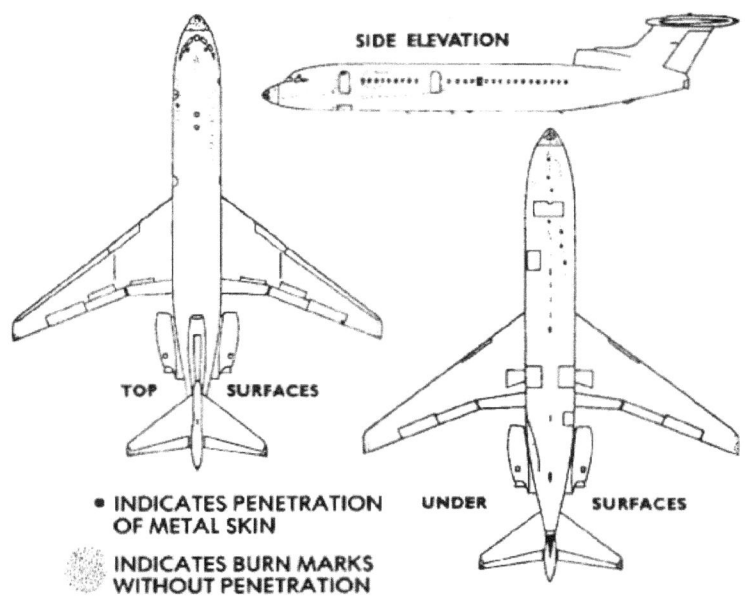

SIDE ELEVATION

TOP SURFACES

UNDER SURFACES

● INDICATES PENETRATION OF METAL SKIN

INDICATES BURN MARKS WITHOUT PENETRATION

Figure 5.3 Position of lightning strikes on Trident aircraft — May 1964 to June 1968.

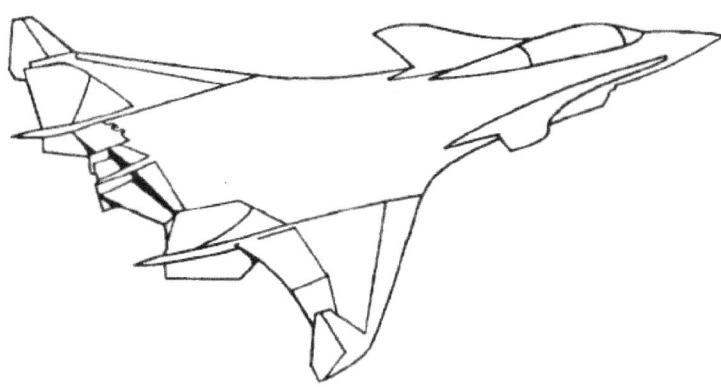

Figure 5.4 Possible aircraft configuration of the future.

5.2.3 Provision of a Safe, Controlled Path for Lightning Currents

Since the aircraft is part of the lightning current path between two external charge centers, lightning currents must flow through the airframe between attachment points. In most cases aluminum skins and structural elements provide an excellent conductive path without the need of additional provisions for lightning currents. Nonmetallic materials, however, such as fiber-reinforced plastics are finding increased use in newer aircraft. At present, use of these materials has been limited to outer wing tips or radomes, for example; therefore, in these cases additional conductive paths need only be added across or through the nonmetallic section. However, an aircraft made entirely of fiberglass has been fabricated. In this case positive steps must be taken to provide the necessary conducting path(s) for lightning currents to flow through the entire aircraft.

5.3   The Lightning Environment as Related to Attachment Zones

For guidance in the lightning protection design and qualification testing of aerospace vehicles and hardware, the Society of Automotive Engineers Committee AE4, Special Task F, has formulated an idealized representation of the current components of a severe lightning flash. This representation incorporates the important aspects of both positive and negative flashes. Four current components, A, B, C, and D, represent the four current characteristics found in most lightning flashes. These are shown together on Figure 5.5 (Reference 5.6). This model represents a very severe flash wherein intensities are exceeded less than 1% of the time. It may be used to define the environment that aircraft systems and components must withstand or against which they must be protected. At the present time, this model is not included in any government specification or standard dealing with lightning protection; however, the model is expected to be incorporated into forthcoming revisions of most of these requirements. The lightning environment described in Figure 5.5 is equal in severity to the lightning currents which are described in present specifications, or greater than such currents. The model is also in general agreement with new

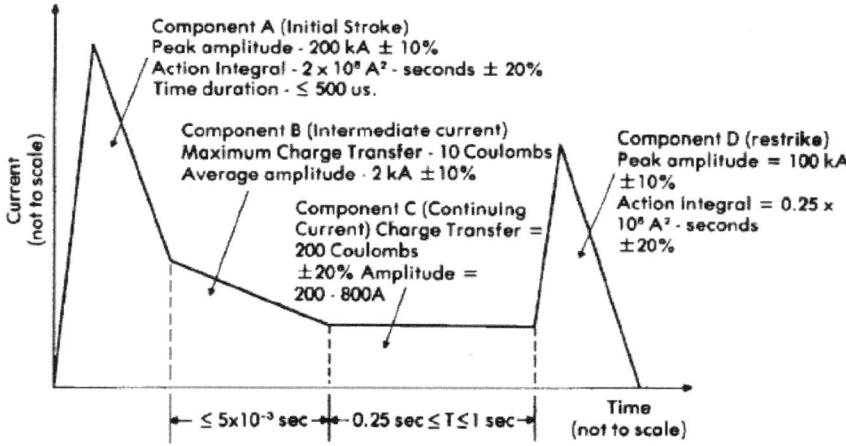

Figure 5.5 Current test waveform components for evaluation of direct effects.

lightning protection requirements being formulated in the United Kingdom and Europe (Reference 5.7).

In order to design protection for a particular system or piece of hardware, its susceptibility to the various lightning current characteristics must be known.

110

For this purpose SAE Task F has also defined the lightning current component(s) likely to strike the aircraft in each of the four lightning attachment zones. The results are shown on Table 5.1. Further references will be made to these Task F criteria during discussion of protection techniques in succeeding chapters of this book.

Table 5.1 LIGHTNING CURRENT COMPONENTS
EXPERIENCED IN EACH ZONE

Current Component Zone	A First Return Stroke	B Intermediate Stroke	C Continuing Current	D Restrike
1A: initial attachment point with low probability of flash hang-on	X	X		
1B: initial attachment point with high probability of flash hang-on	X	X	X	X
2A: swept-stroke zone with low probability of flash hang-on		X	X*	X
2B: swept-stroke zone with high probability of flash hang-on		X	X	X
3: low probability of direct attachment	X		X	

*Only if flash hang-on time is greater than 5 ms

5.4 Applicable Airworthiness Regulations

There are a number of government airworthiness regulations that apply specifically to aircraft lightning protection. These are applicable in the United States and other countries and generally must be complied with as part of the aircraft certification requirements. Table 5.2 lists the regulations presently in effect. These are concerned primarily with fuel system protection, since this system has presented the most serious hazard in the past. Other areas such as nonmetallic structures and aircraft electronics systems are of increasing concern, however, and are beginning to appear in updated requirements. Applicable portions of the regulations in Table 5.2 will be referred to and discussed in succeeding chapters.

Table 5.2   GOVERNMENT AIRWORTHINESS REQUIREMENTS AND RELATED DOCUMENTS PERTAINING TO LIGHTNING PROTECTION OF AIRCRAFT

Document Identification	Title	Applicable Paragraphs	Country – Agency
Federal Aviation Regulations, Part 25	*Airworthiness Standards: Transport Category Airplanes*	25.581, 25.954	US Federal Aviation Administration (FAA) Dept. of Transportation
Advisory Circular AC 20-53 6 October 1967	*Protection of Aircraft Fuel Systems Against Lightning*	All	US Federal Aviation Administration (FAA) Dept. of Transportation
MSC-07636 June 1973	*Space Shuttle Lightning Protection Criteria Document*	All	US National Aeronautics and Space Administration (NASA) Lyndon B. Johnson Space Center
MIL-B-5087B (ASG) 15 October 1964	*Military Specification: Bonding, Electrical, and Lightning Protection, For Aerospace Systems*	3.3.4, 6.3.1, 6.3.9	US Air Force, Navy
MIL-C-38373A (ASG) 26 February 1969	*Military Specification: Cap, Fluid Tank Filler*	3.5.1, 4.6.14	US Air Force
MIL-A-9094D (ASG) 17 March 1969	*Military Specification: Arrester, Lightning, General Specification For*	All	US Air Force

Table 5.2 GOVERNMENT AIRWORTHINESS REQUIREMENTS AND RELATED DOCUMENTS PERTAINING TO LIGHTNING PROTECTION OF AIRCRAFT (CONT.)

Document Identification	Title	Applicable Paragraphs	Country – Agency
MIL-I-83456 (USAF) 2 June 1976	*Military Specification: Installation of Segmented Lightning Diverter Strips on Aircraft Radomes, General Specification For*	All	US Air Force
BCAR D4-6 1 November 1963	*British Civil Airworthiness Requirements Sub-Section D4 – Design and Construction Chapter D4-6 Electrical Bonding and Lightning Discharge Protection (Aircraft)*	All	Civil Aeronautics Board, London
BCAR G4-6 7 November 1975	*British Civil Airworthiness Requirements Sub-Section G4 – Design and Construction Chapter G4-6 Electrical Bonding and Lightning Discharge Protection (Rotocraft)*	All	Civil Aeronautics Board, London
TSS Standard No. 8-6 29 May 1973	*Electrical Bonding and Lightning Discharge Protection (Supersonic Transport)*	All	Anglo-French SST Organization

113

REFERENCES

5.1 *Lightning Test Waveforms and Techniques for Aerospace Vehicles and Hardware*, Society of Automotive Engineers, Warrendale, Pennsylvania (5 May 1976).

5.2 B. L. Perry, "British Researches and Protective Recommendations of the British Air Registration Board," *Lightning and Static Electricity Conference, 2-3 December 1968, Part II, Conference Papers*, AFAL-TR-68-290, Air Force Avionics Laboratory, Air Force Systems Command, Wright-Patterson Air Force Base, Ohio (May 1969), pp. 81-103.

5.3 Perry, "British Researches and Protective Recommendations," pp. 90, 92, 91.

5.4 *Protection of Aircraft Fuel Systems Against Lightning*, Federal Aviation Agency Advisory Circular AC 20-53, Federal Aviation Administration, Department of Transportation, Washington, D.C. (6 October 1967), p. 2.

5.5 Rockwell International designed research vehicle, *Aviation Week and Space Technology*, November 17, 1975, p. 38.

5.6 *Lightning Test Waveforms and Techniques for Aerospace Vehicles and Hardware*, p. 9.

5.7 J. Phillpott, "Simulation of Lightning Currents in Relation to Measured Parameters of Natural Lightning." *Proceedings of the 1975 Conference on Lightning and Static Electricity at Culham Laboratory, England, 14-17 April 1975*, Session I: Fundamental Aspects and Test Criteria, the Royal Aeronautical Society of London (December 1975), pp. 6-7.

CHAPTER 6
FUEL SYSTEM PROTECTION

6.1 Introduction

Because of the hazardous nature of most fuels and the potential for catastrophe if ignition should occur, the design of adequate lightning protection for the aircraft fuel system is probably the most important lightning protection job to be accomplished. This is the system that government airworthiness certification requirements stress most heavily, for the cause of the most tragic of lightning-related aircraft accidents has been attributed to the vulnerability of fuel systems.

Elements of the fuel system are typically spread throughout much of an aircraft and occupy much of its volume. They include the fuel tanks themselves, as well as associated vent and transfer plumbing, and electrical controls and instrumentation. Careful attention must be paid to all of these elements if adequate protection is to be obtained.

The main objective of fuel system protection is of course to prevent any direct or indirect source of ignition of the fuel by lightning. Accomplishment of this goal is quite challenging when it is remembered that thousands of joules of energy must be conducted through the airframe when the aircraft is struck by lightning, that every metallic structural element in and on the aircraft is involved to some degree in this conduction process, and that a tiny spark of $\sim 2 \times 10^{-4}$ joule may be all of the energy that need be released inside a fuel tank to ignite a fire or initiate an explosion.

Prevention of fuel ignition from lightning must be accomplished by at least one of the following approaches:

(a) removing one or more of the requirements for combustion
(b) eliminating all sources of ignition.

In this chapter we review typical aircraft fuel flammability characteristics and possible ways to control fuel flammability and ignition. Research in the field of fuel system protection distinguishes the areas of adequate knowledge and practice from areas of inadequate knowledge and practice. Because current guidelines for controlling lightning-related ignition derive in part from this research, we refer in some detail to several important studies, by way of background, before presenting guidelines for effective protection of fuel systems.

6.2 Fuel Flammability

Ignition of fuel cannot occur until some of the fuel has become vaporized and mixed in a compatible proportion with oxygen or air. Thus, the ability of fuels to vaporize and mix with air in typical aircraft fuel tanks is of primary concern.

The flammability of the vapor space in a fuel tank varies according to the concentration of evaporated fuel in the available air. Reducing the fuel-to-air ratio below a definite minimum value produces a vapor and air mixture that is too lean in fuel to burn. Likewise, there is a maximum fuel/air ratio which, when exceeded, results in a vapor space mixture too rich in fuel to be flammable. In between these extremes there is a range of mixtures that will burn. When only equilibrium conditions are considered, the particular fuel/air ratio that can exist is determined by the temperature and altitude of the fuel system. The temperature determines the quantity of the fuel by controlling its vapor pressure, and the altitude determines the quantity of air. Therefore, by a suitable combination of temperature and altitude, under equilibrium conditions, the ullage of a fuel tank can be made either flammable or nonflammable by varying combinations of temperature and altitude.

Much research into the fuel vapor and flammability characteristics that exist within aircraft fuel systems has been accomplished by many organizations concerned with aircraft propulsion and the resulting need to safely transport large quantities of fuel aloft in a vehicle with severe space and weight limitations. This research has included laboratory investigations of the flammability characteristics of the fuel vapors that exist within laboratory-type containers as well as those that exist during flight. Perhaps the most comprehensive laboratory work is that of Nestor (Reference 6.1), who has paid particular attention to the behavior of fuels in aircraft tanks and flight environments. Nestor shows that, as a result of the wide variety of temperatures, pressures, and motions that can exist in flight, there are correspondingly wide variations in the amount of fuel that can exist in the vaporized state. One reason for this variation is that, in addition to fuel already present as a vapor, the aircraft motion and vibration also cause the liquid to be dispersed in the vapor space in the form of mists and spray. An ignition source may preheat these droplets of fuel to a point where sufficient vapors are evolved to support a flame, even if insufficient vapor has been released from the bulk fuel surface itself. The heat from the developing flame then causes more fuel to vaporize and continue to feed and enlarge the flame. Temperature and altitude, however, are the primary characteristics that relate to flammability.

The relationship of temperature and altitude to tank vapor flammability is illustrated by the *flammability envelope*. A typical flammability envelope is shown in Figure 6.1 (Reference 6.2).

Turbine engine fuels are placed in two broad categories based on their distillation temperature ranges: wide-cut turbine engine fuels and aviation kerosenes. Aircraft with reciprocating engines are commonly fueled with 100 octane gasoline. Turbine- (jet-) powered aircraft fueled in the United States and in some foreign countries utilize aviation kerosenes. These include Jet A and Jet A-1 fuels. U.S. military aircraft and commercial transports in some other foreign countries are fueled with wide-cut fuels, such as Jet B or JP-4, which are kerosene-type fuels with a higher percentage of volatile components. The flammability limits for the turbine fuels as determined in the laboratory by Nestor are presented in Figure 6.2 (Reference 6.3).

The flammability envelopes are defined as lines beyond which no flammable points were observed in the laboratory tests. In Figure 6.2 these lines

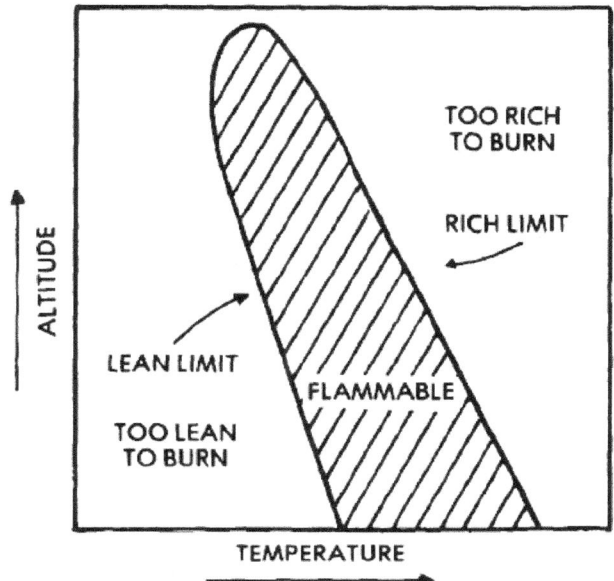

Figure 6.1 Typical flammability envelope of an aircraft fuel.

have been drawn to enclose those data points at which ignition and flame propagation occurred but exclude some of the data points where only ignition occurred with no flame propagation. This results in a random scatter of ignition points about the lines, as shown by Nestor in Figure 6.2. The flammability envelopes of the aviation kerosene-type fuels (Jet A and Jet A-1) differ from the wide-cut fuels (JP-4 and Jet B) primarily in their temperature ranges. The aviation kerosenes have an average lean limit at sea level of about 38 °C and a rich limit of approximately 80 °C. For the wide-cut fuels, the lean and rich limits at sea level are about -20 °C and 10 °C, respectively.

Figure 6.3, taken from a Lockheed study on flammability of advanced flight vehicles (Reference 6.4) shows the flammability envelopes of aviation gasolines (AVGAS) along with a number of turbine fuels. The Reid vapor pressure of each of these fuels is also presented, and here the relationship between this parameter and volatility becomes apparent. Fuels with higher vapor pressures will release sufficient vapor at lower temperatures to reach a flammable mixture, thus lowering the flammability envelope as compared with the envelope of a fuel with a lower vapor pressure, which must be at a higher temperature to release sufficient vapor to form an ignitable mixture.

The temperature vs altitude profiles for standard and tropical atmospheres are also presented in Figure 6.3 to show the mean air temperature at various flight altitudes.

There are important factors which may alter the flammability limits of the vapor inside an aircraft fuel tank from those shown in Figures 6.2 and 6.3, however. One of these is the mixing of one type of fuel with a small amount of

117

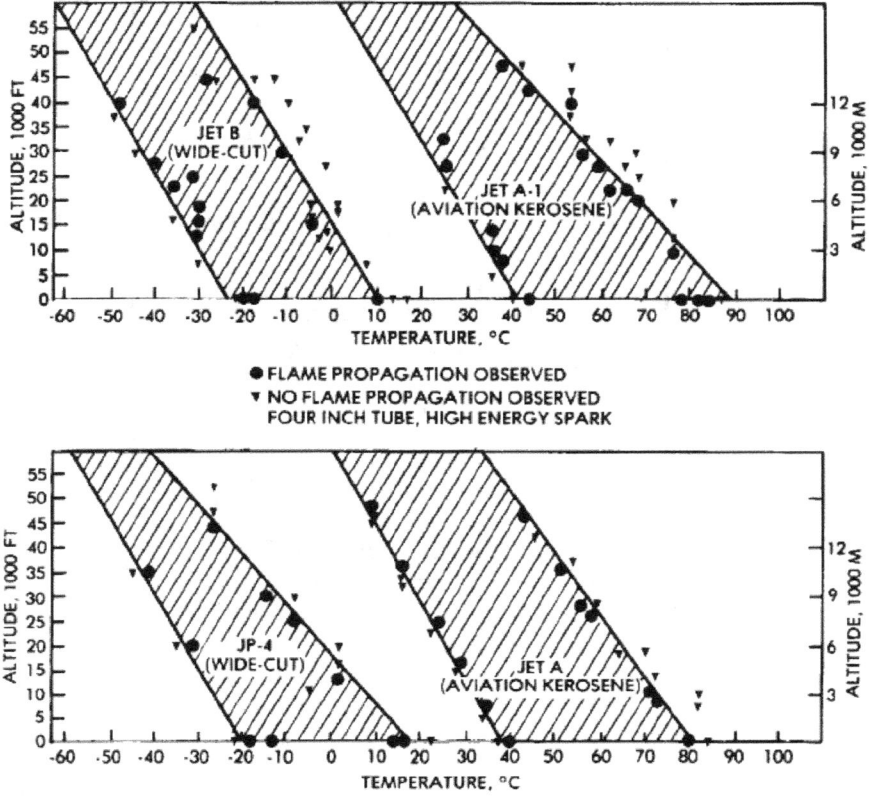

Figure 6.2 Flammability limits for typical aircraft turbine fuels as a function of altitude and temperature.

another. Figure 6.4 (Reference 6.5) shows, for example, the flammability limits determined by Nestor of a mixture proportionately comprised of 85% Jet A and 15% Jet B fuels. Such a mixture might have occurred if the aircraft had originally been fueled with Jet B and later refueled with Jet A. The flammability envelope of the resulting mixture has been lowered such that it encompasses the altitudes and temperatures where most lightning strikes to aircraft have occurred.

Agitation is another way that the flammability envelopes might be altered from those shown in Figure 6.2, for the latter are valid only when the mixture is stabilized. Agitation of a fuel or spray from a pump or pressurized fuel line can result in extending the lower temperature flammability limit below the curves shown in Figures 6.2 and 6.4. Thus, even though Figures 6.2 and 6.4 indicate that the vapor-air mixture in the tanks of an aircraft fueled with Jet A would be too lean to support combustion, there can be no assurance that this is always so, since agitation of the fuel may occur in flight.

A number of methods have been developed to provide more positive

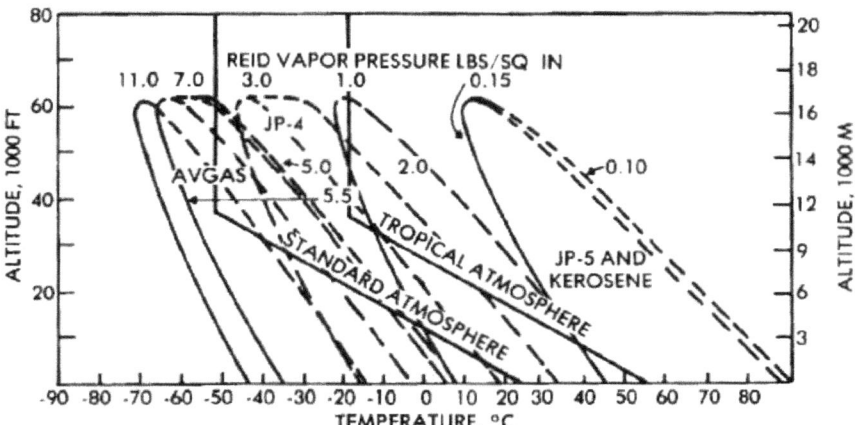

Figure 6.3 Flammability envelopes of fuels of different volatility.

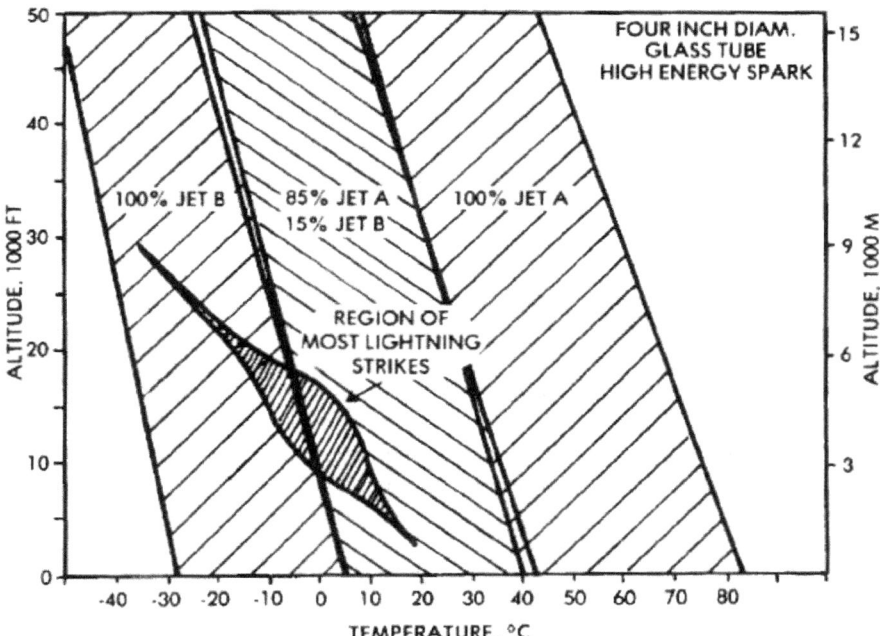

Figure 6.4 Flammability envelope of the fuel blend 85% Jet A/15% Jet B, and altitude/temperature envelope enclosing most lightning strikes.

inerting of the ullage in aircraft fuel tanks——inerting sufficient to inhibit the enlargement and propagation of flames away from a point of ignition, so that an ignition source could occur yet not result in a catastrophic fire or explosion.

119

These methods include inerting of the fuel tank ullage with gases, such as nitrogen or carbon dioxide, and mechanical subdivision of the tank volume with reticulated foams. Nitrogen-inerting systems have been developed and flight tested on the USAF C-141 and C-135 (Reference 6.6) among others, but they have not been used extensively on any fleet of commercial aircraft because of excessive cost and weight penalties. Inerting systems have been used on several military aircraft for protection against fuel ignition resulting from enemy gunfire.

Even if inerting systems were to come into common use, it would be advisable to take positive steps to assure that no ignition source could occur in the event of a lightning strike because there may never be complete assurance that a flammable mixture did not exist somewhere in the fuel system. Therefore, while the use of low-volatility fuels such as Jet A or while the inclusion of inerting systems to reduce the flammability of the atmosphere in a fuel tank will undoubtedly increase flight safety in the event of a lightning strike, these measures should not be relied upon to prevent all in-flight fires or explosions resulting from lightning. Positive steps must still be taken to prohibit ignition sources within the fuel system.

6.3 Sources of Ignition

While there have been a number of in-flight fuel tank explosions undoubtedly caused by lightning strikes, the exact location or source of ignition has not been positively identified in any of these accidents. This failure is perhaps a result of the fact that so small an amount of energy in an electrical spark is required to cause ignition that such a spark would leave little or no other evidence of itself (like pitting of a metal surface). It may be possible that ignition has begun outside of the fuel system itself—for example, at a fuel vent outlet—and then propagated inside, although there has been no proof of this.

Laboratory studies involving simulated lightning strikes to fuel tanks or portions of an airframe containing fuel tanks have demonstrated several possible ignition mechanisms, and investigations of the accidents involving fuel tank fires and explosion have raised other possibilities. Possible ignition sources can be divided into the following two broad categories.

Thermal Ignition Source
1. Direct contact of the lightning arc with the fuel-air mixture, as at a vent outlet
2. Hot spot formation or complete meltthrough of a metallic tank skin by lightning arc attachment

Electrical Ignition Sources
1. Electrical sparking between two pieces of metal conducting lightning current, such as poorly bonded sections of a fuel line or vent tube
2. Sparking from an access door or filler cap (which has been struck) to its adapter assembly
3. Sparking among elements of a capacitive-type fuel quantity probe,

caused by lightning-induced voltages in the electrical wires leading to such a probe

The balance of this chapter considers each of the above ignition sources and methods of their elimination.

6.4 Vent Outlets

The temperature of a lightning arc far exceeds that required to ignite a flammable fuel-air mixture; therefore, any direct contact of the lightning flash with such a mixture must be considered an ignition source. Since fuel tank vents are the primary means by which a flammable fuel vapor can be exposed to the outside of an aircraft, a considerable amount of research has been undertaken to evaluate the possibilities of lightning ignition of fuel vent vapors. One of the first studies of this possibility is that of the Lockheed-California Company and the Lightning and Transients Research Institute in 1963 for the National Aeronautics and Space Administration (Reference 6.7). In this program, fuel-air concentrations in the vicinity of aircraft vent outlets of several configurations were measured and mapped under various conditions of tank vapor fuel/air ratio and effluent velocity. The tests were run in a wind tunnel producing an airflow of up to 100 knots. Mast vents discharging into wakes and free air streams were tested, as well as flush vents discharging into boundary layers, as shown in Figure 6.5 (Reference 6.8). The study showed that a vent discharging into a free stream exhibits the greatest dilution and thus the smallest flammable region, but that flammable mixtures can exist in the immediate vicinity of each type of vent outlet.

A typical mixture concentration profile aft of a flush vent outlet is shown in Figure 6.6 (Reference 6.9). Depending upon the richness of the original effluent, dilution of it by the addition of air to 30%, or more, might well lean it out of the flammability envelope. Thus, dilution to a nonflammable mixture probably occurs farther than about one outlet diameter away from the outlet. This finding suggests that a lightning strike or associated streamer must occur very near to the edge of a vent outlet for an ignition to occur.

The above prediction was confirmed in the second part of the 1963 Lockheed-LTRI study, which included application of simulated lightning strikes to a flush vent outlet with flammable fuel-air mixtures exiting from it into an airstream moving at 100 knots, both with and without flame arresters installed. These flame arresters, which were being evaluated as possible protective devices, consist of a parallel bundle of small-diameter metal tubes inserted into the vent line near the vent outlet, as pictured in Figure 6.7(a) (Reference 6.10). An alternate construction utilized a series of baffles extending into the fuel vapor flow, as shown in Figure 6.7(b). The object in either case is to quench (extinguish) the flame by cooling it, and the small tubes, or baffles, are intended to accomplish this by providing additional surface area to conduct heat away from the flame. The test results showed that it was possible to get ignition of the vent effluent in the 100-knot airflow only when the lightning arc was delivered

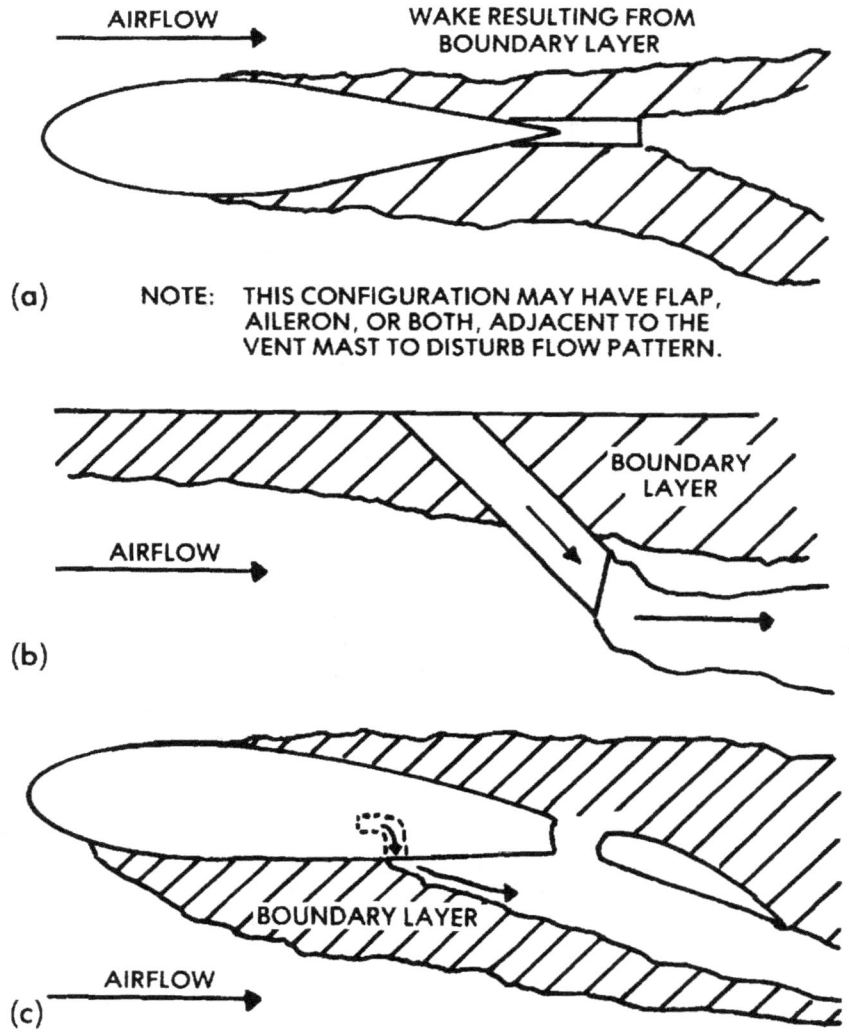

AIRFLOW
WAKE RESULTING FROM
BOUNDARY LAYER

(a) NOTE: THIS CONFIGURATION MAY HAVE FLAP,
AILERON, OR BOTH, ADJACENT TO THE
VENT MAST TO DISTURB FLOW PATTERN.

BOUNDARY
LAYER
AIRFLOW
(b)

BOUNDARY LAYER
AIRFLOW
(c)

Figure 6.5   Three general classes of fuel vent exits.

directly to the lip of the vent outlet. Arcs delivered to spots as little as 2.5 cm away did not ignite the effluent. The results also showed that flames ignited by the strikes to the outlet could propagate inward through the flame arresters when these were installed near the vent outlet. While the flame arresters did extinguish some flames ignited at the outlet, they did not stop all such flames. Evidently the intense heat and blast pressures of the lightning arc act to heat the arrester and force the flame through.

As a result of the crash of a Boeing 707 aircraft near Elkton, Maryland, on December 8, 1963, after being struck by lightning (Reference 6.11), another

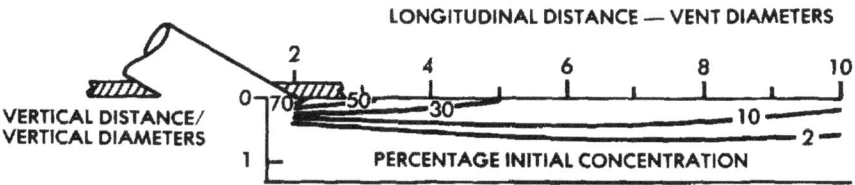

BOUNDARY LAYER THICKNESS 0.50 VENT DIAMETERS
VENT EXIT VELOCITY 0.10 X FREE STREAM VELOCITY

Figure 6.6 Typical profile of vent effluent in air stream aft of a flush vent outlet. (Profiles vary according to boundary layer thickness, effluent and air stream velocities, etc.)

investigation of the possibility of lightning-initiated fuel tank vent fires was undertaken by Bolta and others of the Atlantic Research Corporation with the support of Newman and others of the Lightning and Transients Research Institute (Reference 6.12). This work was sponsored by the FAA and focused on the Boeing 707 wing tank and vent system, with the objective of determining the conditions under which ignition of fuel vent effluent allows flames to propagate back through the vent duct and surge tank and from there into the reserve tank. Another objective was the evaluation of various protective measures, including flame arresters and flame-extinguishing systems.

Unlike the earlier Lockheed program, the vent was tested in still air, the rationale being that, if an ignitable effluent is assumed to be at the outlet, attachment of the lightning arc to the lip of the vent outlet is the governing factor in obtaining ignition. The ignitable mixture was a 1.15 stoichiometric mixture of propane and air.

Thermocouples were installed along the vent line to determine the time at which a flame passed by, thus enabling calculation of flame front velocities. The latter information was important because an automatic extinguishing system under consideration depended for its operation on sufficient time elapsing between the initial sensing of a flame at a vent outlet and the activation of an extinguisher in the surge tank located about 1 m down the vent line. The simulated lightning tests confirmed the earlier result--that a strike must occur very close to the vent outlet for ignition to occur. Average flame velocities of up to 45 m/s were recorded between the vent outlet and the surge tank when simulated lightning strokes of 175 kA and 1.5×10^6 ampere²-seconds (A²-s) were applied. These strokes simulated a severe return stroke such as that which might be experienced if the vent outlet were located in Zone 1A or 1B. Other tests were run with the high current delivered through a 10 cm aluminum foil tape which would explode as the current passed through it. The resulting flames reached higher velocities (up to 126 m/s), but the exploding tape may have caused the higher flame speeds, creating conditions more severe than those in a natural case, where an arc alone is the ignition source.

When lower amplitude currents of 44 kA and only 0.001×10^6 A²-s were

123

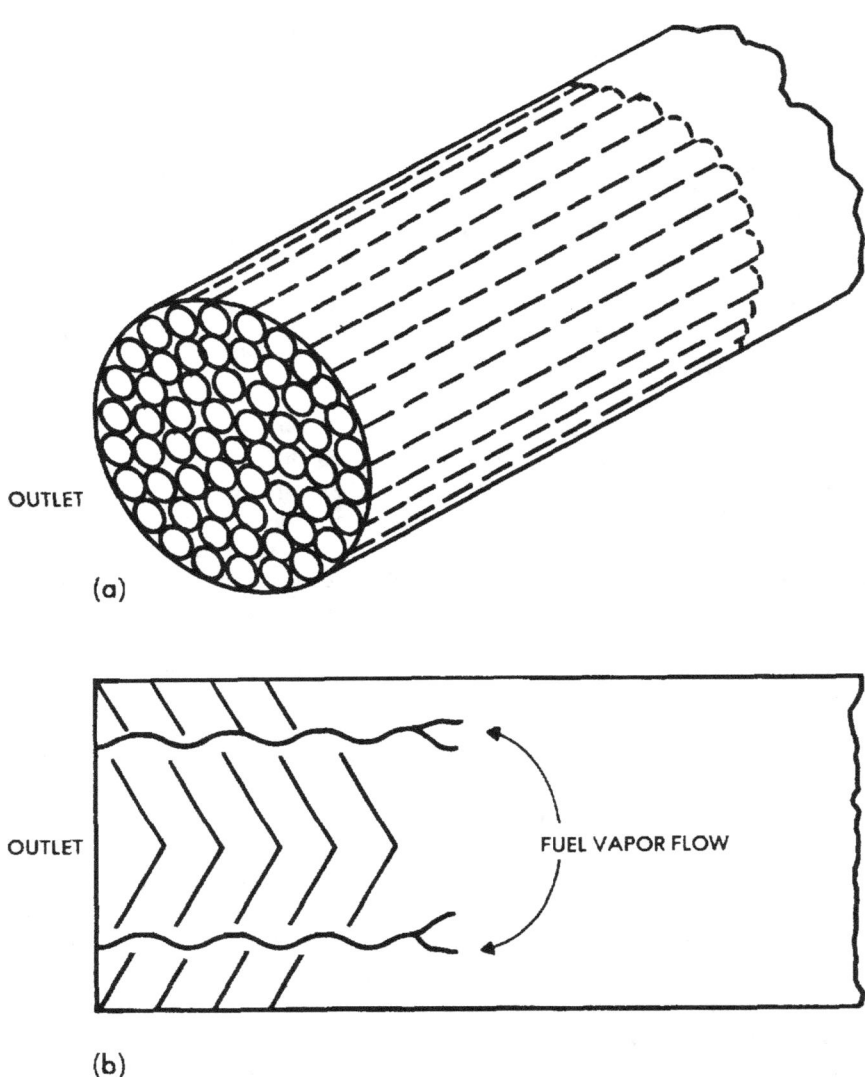

OUTLET

(a)

OUTLET

FUEL VAPOR FLOW

(b)

Figure 6.7 Flame arrester configurations.
    (a) Tubular construction
    (b) Baffle construction

applied, ignitions still occurred, but the highest average flame velocity was 17.4 m/s, which indicates that the intensity of the lightning discharge is one of the factors that affect the flame velocity. The 44 kA current is described by Bolta as a "high-voltage" discharge (Reference 6.13) because it was applied with a Marx-type, high-voltage generator, but the ignitions must have occurred from the ensuing 44 kA current discharge, since the 13 cm or 30 cm air gap between discharge electrode and vent outlet was too short to simulate the in-flight environment of the aircraft in a high electric field and allow formation of significant prebreakdown streamers at the vent outlet.

Various flame arresters fabricated from corrugated aluminum and stainless steel, a ceramic material, and various copper screens were tested. None of these arresters was capable of stopping a flame ignited by a simulated stroke current when the arrester was installed near the outlet of the vent line, but arresters wound from corrugated stainless steel 1.27 cm or 2.54 cm deep did stop the flames when these arresters were installed about 1 m upstream from the vent outlet near the surge tank (as shown in Figure 6.8 [Reference 6.14]), even when these flames traveled in the vent tube at an average speed as high as 122 m/s (the highest measured). Other arresters made of screens and ceramics did not stop flames ignited by the simulated strokes.

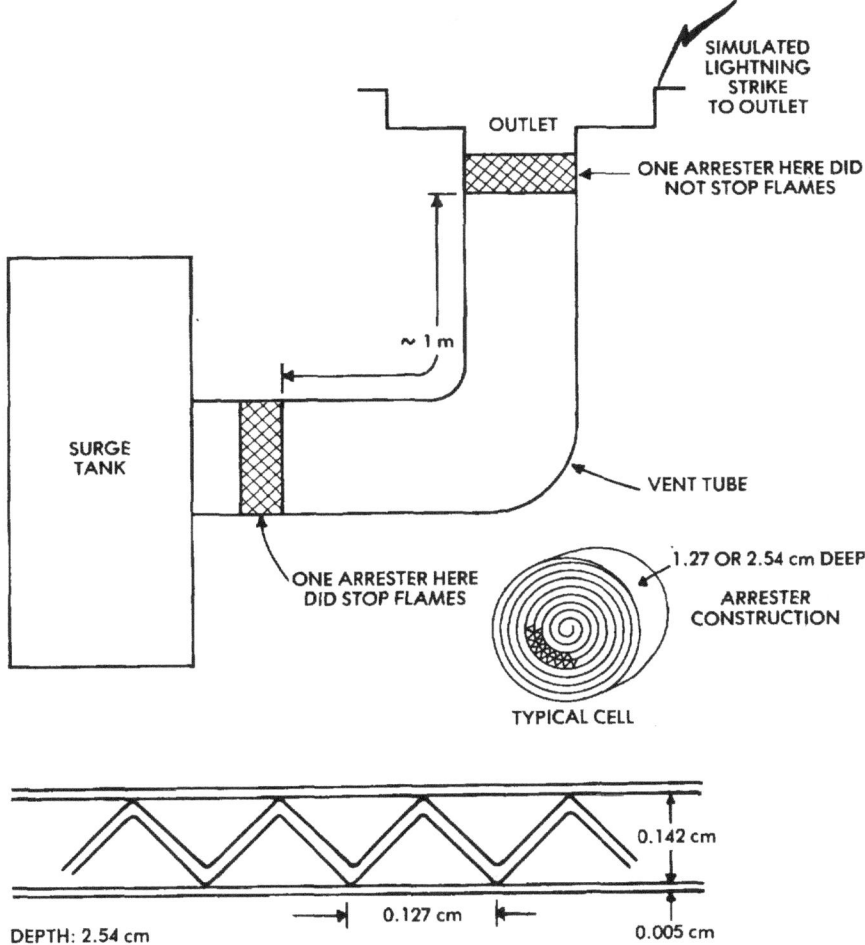

Figure 6.8 Successful flame arrester installation in simulated B-707 vent tube.

A flame-suppression system developed by Fenwal, Inc. (Reference 6.15) for industrial applications was also tested in this program. This system consisted of a fast-acting sensor for detecting the presence of a flame and a set of canisters containing a quantity of liquid suppressing agent for release into the surge tank by an electric detonator. When the detector sees the light of the flame, it sends a pulse to the detonator, which disperses the extinguishant into the tank within a few milliseconds of flame sensing, before the flame itself has reached the tank. This system effectively suppressed those flames which traveled slowly enough (30 m/s or less) to give the system time (18 ms) to react, but did not stop those, of course, which had gone past the surge tank by the time the extinguishant was released. In the latter case the extinguishant might still extinguish the fire in the surge tank, but the flame front on its way through interconnecting vent lines to the fuel tanks would have passed out of the reach of the extinguishant.

The Atlantic Research-LTRI program also included an investigation of arc plasma propagation into the vent line, which was inconclusive because of instrumentation difficulties. The question of icing of the successful flame arrester was also considered (Reference 6.16), resulting in the conclusion that unacceptable icing would occur only when the worst combination of atmospheric and flight conditions existed. This conclusion, however, is based on analysis only and should be verified by flight tests if this arrester or any other is to be incorporated into an aircraft fuel system.

6.4.1 Airflow Velocity Effects

Lightning attachment to the lip of a vent outlet was assumed to be possible in the Atlantic Research-LTRI and Lockheed programs reviewed in the preceding paragraphs. No attempt, however, was made in these programs to establish whether or not this phenomenon could in fact occur to an aircraft in flight or, if it could, how often. Yet neither the Elkton airplane nor any other airplane known to have been struck by lightning has shown evidence of lightning attachment to a vent outlet, and the outlet in the B-707, while located near the wing tip, is not located *at* the very tip of the wing, where lightning is known to attach. It is therefore improbable that flush-mounted vents, such as those on the B-707, will receive direct strikes. These questions then arise: Could an arc sweep across the outlet from another point and ignite the effluent?; if so, How close to the outlet must such an arc pass to cause ignition?; and Is it possible for streamering to be induced at the vent outlet and ignite an effluent? Answers to these questions were sought by Newman and others, who undertook an experimental program (Reference 6.17) during 1966 and 1967. Simulated lightning strikes were delivered to a B-707 wing tip and vent assembly to learn more about these possibilities, as well as the degree to which air flowing past at realistic flight speeds would make ignition unlikely even if an arc did in fact become attached to the outlet.

Whereas in earlier programs ignitions were obtained nearly 100% of the time when the vent outlets were in still air, Newman found that ignition of a 1.5

stoichiometric propane-air mixture by a 48 kA, 0.009×10^6 A^2-s simulated direct stroke occurred only once in 34 shots with a 90-knot (46 m/s) airflow over the outlet and not at all after 200 shots in a 200-knot (100 m/s) airflow. When strokes of longer duration were swept across the vent outlet by the 90-knot windstream, the effluent was ignited 11 times out of 15 runs, but when the airflow was increased to 200 knots, only 2 ignitions occurred in 46 runs.

Nearly all of Newman's tests were performed under the most vulnerable effluent condition, which was found to be a 1.5 m/s flow out of the vent outlet, such as might exist when an aircraft is climbing. Since more than half of all reported lightning strikes occur when the aircraft is either in level flight or descending and since most aircraft climb at well over 90 knots, the probability of an in-flight ignition from a *direct* stroke to a vent outlet must be remote. Newman's investigation, however, showed that a flash *sweeping* across the vent outlet might have a greater (2 in 46) chance of igniting an effluent, even under climb conditions at the more realistic speed of 250 knots. This result demonstrates the importance of locating vent outlets away from both direct and swept-stroke zones on the aircraft.

6.4.2 Explosive Ignitions

In the program just described, Newman and his colleagues conducted a test (Reference 6.18) in which a stroke to the vent outlet produced indications of unusually high flame velocities and severe deformation of the vent outlet, indicating much higher pressures than normal. They cite a similar case in another program in which flames traveling in excess of 300 m/s (1000 ft/s) were actually measured. The implication of these findings is serious because an arrester or surge tank protection system capable of extinguishing the lower velocity flames may not be able to stop flames traveling as fast as 300 m/s.

Kester and others (Reference 6.19) attempted to reproduce such speeds in a 14 cm simulated vent line, shown in Figure 6.9, but did not measure flame velocities higher than 20 m/s (65 ft/s) in this system, even when severe, 180 kA, 1.0×10^6 A^2-s strokes were applied. These velocities were, in fact, comparable to those measured in the Atlantic Research-LTRI program of 1964. Kester and his colleagues also reported one explosive ignition, when a stroke of 195 kA was delivered to the vent outlet. It was found that the 195 kA stroke current had induced a voltage in instrument wiring sufficiently high to spark over the insulation around several pressure probes inside the vent line. The vent outlet and parts of the surge tank were badly deformed, even though these parts were made of 0.64 cm (0.25 in) steel. Again, much higher than usual pressures were indicated. The explosion in the Kester program serves as a warning that a similar consequence might conceivably result from inductive voltages in an actual fuel system unless, by means of design, care is taken to eliminate such situations.

The question of whether or not these explosive ignitions could occur in actual fuel tank vent systems was of such importance that the Federal Aviation Administration undertook yet another study of flame propagation in vent

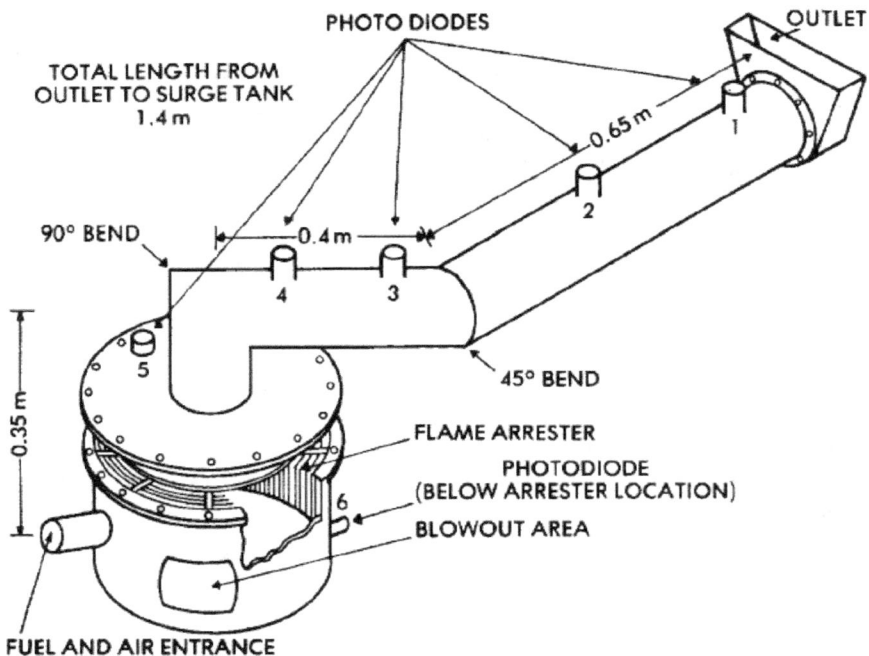

Figure 6.9 Simulated vent system — flames up to 20 m/s.

systems. The work, conducted by Gillis (Reference 6.20), expanded upon earlier research by including the study of flame behavior in the long vent lines leading inboard from the surge tanks to the fuel tanks, which comprise a typical, complete, vent system (Figure 6.10) (Reference 6.21). Gillis did not use simulated lightning arcs for an ignition source but, instead, discharged 100 J of electrical energy into a spark plug at the vent outlet. This is much less energy than would be released by a lightning arc of the same length. Nevertheless, Gillis recorded flames (Reference 6.22) that had accelerated to 300 m/s (1000 ft/s) far inboard (shown in Figure 6.10) when the aircraft was in a climb condition.

When fresh air was made to flow into the vent outlet, as in a descent condition, supersonic velocities like this also occurred as near to the surge tank as probes Pl and P2 in Figure 6.10. These supersonic flames expanded some of the vent lines from a rectangular to an elliptical cross section, a result similar to those occurring in the earlier programs of Newman and Kester.

The total number of authentic tests run by Gillis was 13, of which 11 resulted in flame velocities of 150 m/s or higher. The occurrence of such high speeds is perhaps best explained in Gillis' own words (Reference 6.23):

> When an explosive gas is confined in a channel and ignited, the flow induced by the thermal expansion of the gas in the combustion wave is restricted by the channel wall. Consequently, the flow attains much higher velocities than under conditions of free expansion in an

open flame and flame and flow commonly augment each other by a feedback mechanism as follows: stream turbulence, however slight it may be initially, produces a wrinkling of the combustion wave surface; the resulting increase of surface increases the amount of gas burning per unit time, namely, the flow of gas in the channel; this in turn produces more turbulence and hence, increased wrinkling of the wave, and so on, so that the progress of the combustion wave becomes nonsteady and self-accelerating. In addition, the burning velocity increases as the unburned gas ahead of the flame is preheated and precompressed by the compression waves that are generated by the mass acceleration in the combustion wave. The compression wave is initially a comparatively weak pressure wave, which is overtaken and reinforced during its travel by numerous other pressure waves originating in the combustion zone. The coalescence of these pressure waves into a strong shock front in a configuration which is dead-ended can result in a reflection of the shock wave back toward the combustion zone. The effect of the passage of this reflected shock wave through the combustion wave is similar to the effect of a sudden release of pressure by a rupture of a diaphragm. A rarefaction wave propagates backward into the unburned gas and a jet of unburned gas develops which penetrates deeply into the burned gas. The shear between burned and unburned gas in this flow configuration produces extreme turbulence so that a sudden large increase in the burning rate occurs.

In three of the 13 tests mentioned above, localized pressures of sufficient intensity to distort 3- to 5-foot sections of the duct were developed in the rectangular vent duct. A subsequent hydrostatic pressure test of a 1 m section of similar duct showed that a pressure of approximately 475 psig was required to produce similar distortion. This pressure exceeds the structural limitations of typical aircraft fuel tank and vent structures.

Gillis concluded that the surge tank located just inboard of the vent outlet is a factor contributing to the high flame speeds because, when a flame reaches it from the vent outlet, the pressure permitted to build up in it serves as a force to drive flames rapidly on down the vent lines towards the fuel tanks. This creates turbulence in these ducts, which further serves to accelerate flames down the ducts. Gillis' work is perhaps the most convincing demonstration to date that flames traveling at sonic velocities and with damaging overpressures can occur in typical transport aircraft vent systems. Since flame arresters or STP systems reliably capable of stopping such flames are not yet available. the importance of preventing any source of ignition within or near the vent system is very clear.

6.4.3 Summary of Vent Flame Research and Protection Considerations

Table 6.1 summarizes the ignition and flame velocity results for each of

129

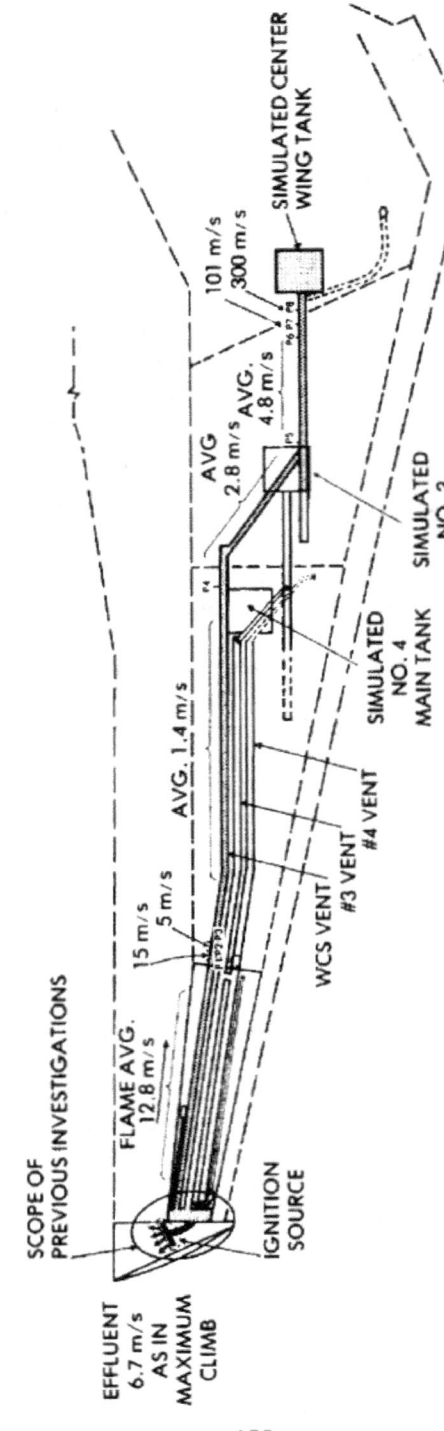

SCOPE OF PREVIOUS INVESTIGATIONS

EFFLUENT 6.7 m/s AS IN MAXIMUM CLIMB

IGNITION SOURCE

FLAME AVG. 12.8 m/s

15 m/s

5 m/s

AVG. 1.4 m/s

WCS VENT

#3 VENT

#4 VENT

SIMULATED NO. 4 MAIN TANK

AVG 2.8 m/s

AVG. 4.8 m/s

SIMULATED NO. 3 MAIN TANK

101 m/s

300 m/s

SIMULATED CENTER WING TANK

Figure 6.10 Flame speeds in mock-up vent system.

130

Table 6.1 SUMMARY OF RESULTS OF SIMULATED LIGHTNING STRIKES TO FUEL VENT SYSTEMS

(STROKES TO LIP OF VENT OUTLETS ONLY. NO IGNITIONS WERE OBTAINED FROM STROKES AWAY FROM VENT OUTLET.)

Program	Simulated Lightning		Airstream Velocity (knots)	Attachment Point	Results
	Amplitude (kA)	Action Integral (10^6 A²s)			
1963 Lockheed-LTRI	100	0.069	100 (50 m/s)	Lip of vent outlet	100% ignitions and flames propagating with and without flame arrester installed at vent outlet.
1964 Atlantic Research-LTRI (Bolta, et al)	175	1.5	0	Lip of vent outlet	100% ignitions and flame propagating up to 45 m/s, and at 126 m/s when ignited by an exploding foil. Flame arrester stopped these flames when installed 1 m upstream from outlet.
1964 Atlantic Research-LTRI (Bolta, et al)	44	0.001	0	Lip of vent outlet	100% ignitions and flames propagating up to 17.4 m/s (no arrester).
1966 LTRI (Newman, et al)	48	0.009	90 (46 m/s)	Lip of vent outlet	1 ignition and flame propagation out of 34 direct strokes to vent outlet (no arrester).
1966 LTRI (Newman, et al)	48	0.009	200 (100 m/s)	Lip of vent outlet	0 ignitions out of 200 direct strokes to vent outlet.
1966 LTRI (Newman, et al)	58	0.172	90 (45 m/s)	Swept access outlet	11 ignitions out of 15 swept strokes across the vent outlet.
1966 LTRI (Newman, et al)	58	0.172	200 (100 m/s)	Swept access outlet	2 ignitions out of 46 swept strokes across vent outlet.
1966 LTRI (Newman, et al)	58	0.172	250 (130 m/s)	Swept access outlet	No ignitions out of 2 swept strokes across vent outlet.
1966 Dynamic Science-GE (Kester, et al)	195	1.0	0	Lip of vent outlet	100% ignitions and flames propagating up to 20 m/s.
1969 Fenwal (Gillis)	100 J spark		0	Lip of vent outlet	11 ignitions and flame velocities over 150 m/s in. 13 authentic tests. Several at sonic velocity (300 m/s).

131

the research programs just discussed. While not all of the answers to lightning-related vent flame questions are in hand, a number of important conclusions and protection considerations can be drawn from the research.

1. As of this writing there has been no positive evidence that a natural lightning strike ever ignited a vent effluent on a transport-type aircraft, although there have been several in-flight explosions after lightning has struck wing tips at places within a meter or so of the vent outlet.

2. For ignition to occur, a lightning-type arc must attach directly to the edge of a vent outlet or within a few centimeters of it, or must be swept over a vent outlet. Neither of these situations has been reported to date. Most aircraft vent outlets are located in conformance with FAA Advisory Circular AC 20-53 (Reference 6.24). This guideline says that vent outlets should not be located in lightning-strike Zone 1 or 2 and defines these zones by the "18 inch" rule, which, if interpreted literally, are as illustrated in Figure 6.11(a). If the points defined by the "fore and aft line passing through the Zone 1 forward projection points of stroke attachment" are taken to be at the outboard tips only, then Zone 1 is restricted to a small area and Zone 2 may extend inboard only a little more than 18 inches (0.46 m) from the tip. If the tip is sharp, strikes are indeed most likely to attach only to the edge, and Zone 1 would be limited to a small distance (several centimeters or so) inboard. However, if the tip has a large radius of curvature, such as that illustrated in Figure 6.11, it is prudent to utilize a more conservative interpretation and consider Zone 1 to extend inboard the full 18 inches (0.46 m).

If the radius of curvature of the outboard tip is greater than 18 inches, the lightning attachment points will be even more scattered over the surface, and Zone 1 should be considered to extend inward a distance equal to the actual radius of curvature.

When nonmetallic skin materials are used on the wing tips, it is probable that direct strikes will occur to the edge of the outermost metallic element, such as a rib or spar, instead of to the nonmetallic tip itself. Therefore, Zone 1 must include not only the nonmetallic tip but also an area extending 18 inches inboard from the outermost metallic structure, as shown in Figure 6.11(b). If there is a metallic diverter assembly mounted on the outboard edge of a nonmetallic tip, this diverter instead of the outboard rib (for example) may receive strikes and thus permit Zone 1 to begin at the outboard edge, as in Figure 6.11(a). Metal diverter bars intended for this purpose have been installed on nonmetallic wing tips, and subsequently failed, however, in their intended purpose, with strikes circumventing the diverter and either puncturing the nonmetallic skin or attaching to an exposed metallic item inboard of the diverter. The most common reason for this failure is that the diverter has been coated with a tough, erosion-resistant paint or coating. Moreover, most of these paints also have good electrical insulating capability, which inhibits formation of the electrical streamers necessary for the diverter to intercept oncoming lightning flashes. Greater reliability for a diverter intended to protect a nonmetallic tip assembly or to enable location of a vent outlet farther outboard

132

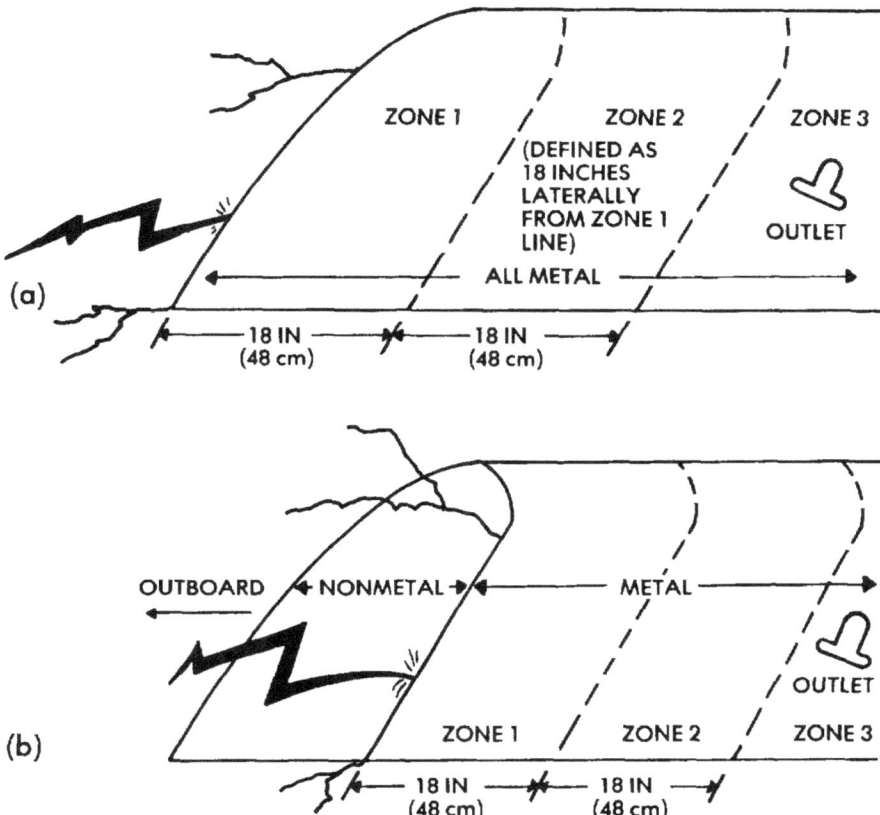

Figure 6.11 Vent outlets must be located in Zone 3. With nonmetallic tips, the 18-inch Zone 1 and Zone 2 areas must be measured from the outboard metallic structure and not from the nonmetallic tip.
(a) Lightning-strike zones on all-metal wing tip
(b) Lightning-strike zones on metal wing with nonmetal tip

can be assured if diverters are subjected to such high-voltage attachment tests as those described in *Lightning Test Waveforms and Techniques for Aerospace Vehicles and Hardware* (Reference 6.25). Since the authenticity of this test has not been absolutely established, it is nevertheless prudent to locate a vent outlet in Zone 3 as defined more conservatively in Figure 6.11(b).

3. Flame arresters of the corrugated steel type shown in Figure 6.8 have been the most effective of those experimented with in stopping flames. The research to date has not turned up a completely successful arrester, however.

Flame arresters are most satisfactory when located some distance away from the vent outlet, so that blast forces in the immediate vicinity of the lightning arc will not propel flames through the arrester. The most successful location seems to be at the surge tank end of the vent outlet tube, as shown in

Figure 6.8. A flame arrester at this location will certainly reduce the possibility of flames entering the surge tank, although there is no assurance that the arrester will stop all flames here. And if an arrester is located anywhere in the vent system, the possibility of its free air passage becoming blocked with ice must be evaluated, preferably by in-flight tests under the expected environmental conditions.

4. Surge tank protection (STP) systems, the purpose of which is to sense flames initiated at vent outlets and extinguish them before they reach fuel tanks, are available and should be considered if the vent outlet must be located in or close to Zone 1 or 2. Since the time required from sensing of the flame to dispersion of the extinguishant is several milliseconds, there is a possibility that flames traveling at sonic velocities will outrace the STP system. Thus, while an STP system will unquestionably improve overall safety, it must not be relied upon to provide absolute protection of the vent system.

5. Ninety-degree bends in the vent lines should be avoided because they expand the turbulence and surface area associated with propagating flames and thereby increase the velocity of propagation. Instead, straight or smoothly curved ducts should be utilized to minimize the possibility of explosive flame propagation.

6. Location of the fuel vent outlet in an ascertained Zone 3 area will provide the highest degree of protection of any of the methods described above. A recessed or flush outlet is greatly to be preferred over a protruding tube outlet, which could become a source of corona or streamering when the aircraft is flying in a strong electric field.

6.5 Fuel Jettison and Drain Pipes

On some aircraft provision exists for dumping (jettisoning) fuel overboard. Often this provision consists of a pipe extending into the airstream from a fuel tank, as shown in Figure 6.12. A normally closed, electrically operated valve is inserted into the pipe, so that it is unlikely a flame could travel past this valve into the fuel tank. Even if the pipe were struck by lightning while jettisoning fuel, it is improbable that flames could propagate through the full pipe into the tank.

A more likely hazard——if not guarded against in design——is that electrical sparks will occur from lightning currents flowing across a poor electrical bond between the pipe and the fuel tank wall, since the wall is frequently coated with an insulating, corrosion-resistant paint. Bonding jumpers installed across such joints may be adequate to equalize static charge differentials which sometimes occur in fuel systems, but these usually have too much inductance to prevent some lightning current from breaking down insulation and sparking at the interface between the pipe and tank wall. Clearly the best way is to make the shortest path also the conductive one, as shown in Figure 6.13. This may be done by making the faying surfaces of clean, uncoated metal and/or by providing bare metal-to-metal contact via the rivets or bolts. If there is doubt

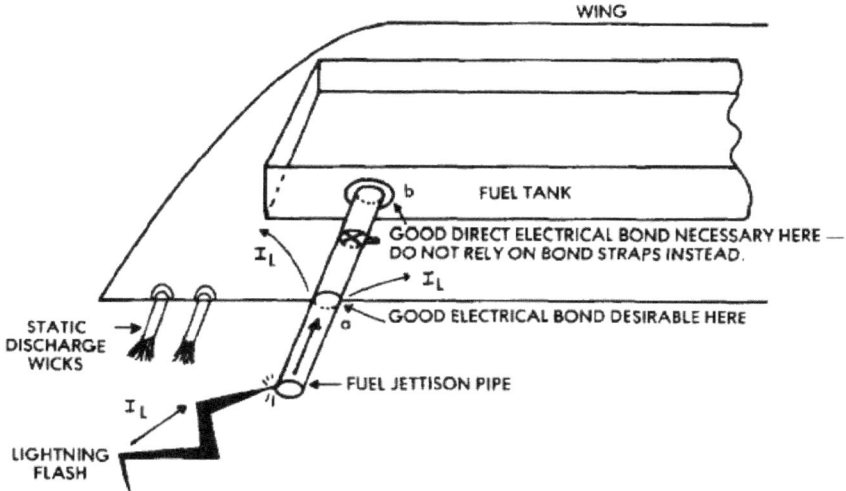

Figure 6.12 Fuel jettison pipe—a possible lightning attachment point.

- Good electrical bond between pipe and airframe desirable at point a, rather than forcing all lightning current to enter fuel tank wall at b.
- In every case a good electrical bond must be provided at b.

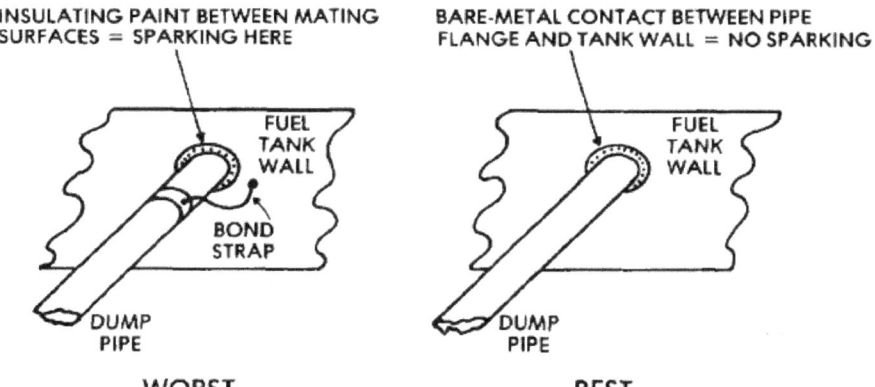

Figure 6.13 Bonding of fuel dump pipes to fuel tanks.

- Bond straps have too much inductance to prevent sparking across the shorter path.

about the adequacy of a particular bond, the bond should be tested with simulated lightning currents to assure that sparking does not occur at the interface. The test currents to be applied should be those recommended in *Lightning Test Waveforms and Techniques for Aerospace Vehicles and Hardware* (see Reference 6.25) for the particular lightning-strike zone in which the jettison or drain pipe is located.

6.6   Hot Spot and Hole Formation in Integral Tank Skins

Integral tank skins are those in which fuel is in direct contact with the outside skin of the aircraft. Tanks of this type are commonly found in the wings of transport and some general aviation aircraft, and in the wings and fuselages of modern fighter aircraft. External fuel tanks of the type carried on pylons or wing tips by military aircraft are also of the integral type.

If integral tank skins are located in direct or swept-lightning attachment zones, the possibility of hot spot formation or meltthrough of the skin must be protected against. For the aluminum skins in common use, making the skin thick enough to dissipate the heat from the lightning arc before the skin can be melted completely through will assure protection because ignition will occur only with complete penetration and the resulting arc contact with the fuel vapor. This phenomenon depends on the fact that the ignition temperature of most hydrocarbon fuel vapors ($\sim$1300 $^{\circ}$C) is higher than the melting temperature of aluminum ($\sim$500 $^{\circ}$C); thus, ignition will not occur unless the skin is melted completely through to the fuel vapor by the hot lightning arc, which can reach upwards of 30 000 $^{\circ}$C. On the other hand, the melting temperature of titanium ($\sim$1700 $^{\circ}$C) and stainless steel skins is higher than the fuel ignition temperature, so that, where this type is concerned, the skin need not be melted completely through for ignition to occur.

The amount of lightning current required to erode or melt holes in aircraft skins has long been of interest, first, for the purpose of estimating how much lightning current actually was involved in the damage sustained by aircraft in flight and, second, for determining the minimum skin thickness required to prevent meltthrough of an integral fuel tank skin. Quantitative relationships between electric arcs and typical aircraft skin metals were reported in 1949 by Hagenguth (Reference 6.26), who made laboratory measurements of the amount of continuing current necessary to melt holes of various sizes in various metals. He found a nearly linear relationship to exist between the amount of charge (Q) delivered to an arc attachment spot and the amount of skin metal melted away. And, although he reported that "the type of metal appears to have very little influence" (Reference 6.27) (a fact later found to be untrue), he determined that the size of a hole melted in a metal sheet of a given thickness can be expressed by the following two equations:

$$A = 25.3 \, Q \left(\frac{t}{2.54}\right)^{-0.9} \text{ for } 0 < t < 0.089 \text{ cm} \qquad (6.1)$$

$$A = 245 \, Q \left(\frac{t}{2.54}\right)^{-1.54} \text{ for } 0.089 < t < 0.381 \text{ cm} \qquad (6.2)$$

where   A = area of hole melted (square millimeters)

136

Q = charge (coulombs) delivered to the point by the arc
t = thickness of metal sheets (centimeters)

Results of Hagenguth's laboratory tests are presented in Figure 6.14 (Reference 6.28). By comparing the size of holes produced by actual lightning strikes with those produced in the laboratory, Hagenguth was able to estimate the amount of charge that had entered the aircraft through damaged parts, such as those shown in Figure 6.15.

Taken at face value, the Hagenguth equations would predict a hole of some size (A) through skin metals of any thickness (t) up to 0.381 cm (0.15 in)

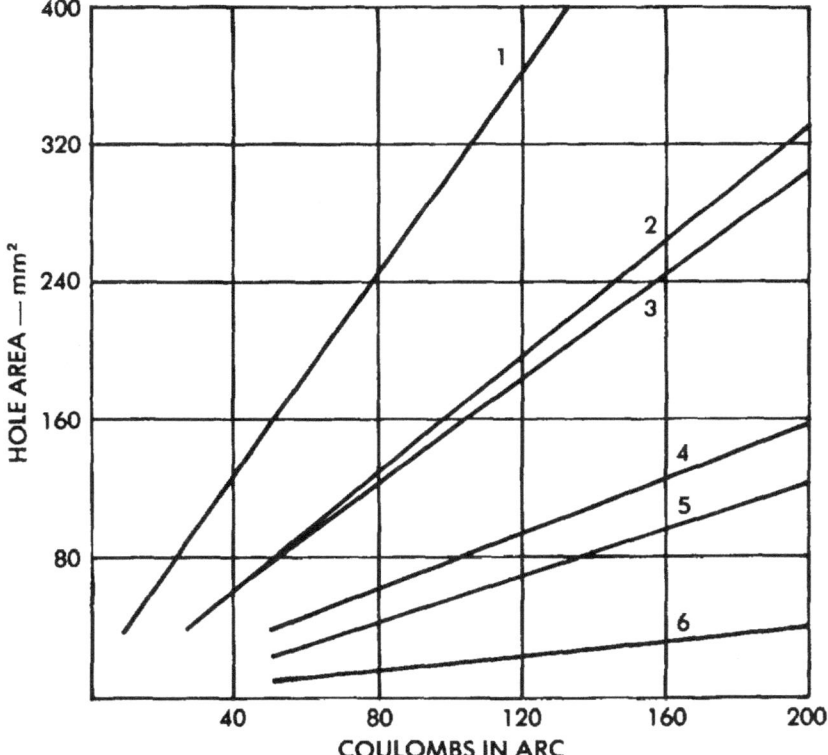

Figure 6.14 Relation between coulombs in the arc and the size of hole burned in metal sheets. Arc current amplitude ranges between 50 and 1000 A.

1. Stainless steel — 10 mil (0.254 mm)
2. Galvanized iron — 15 mil (0.381 mm)
3. Copper — 20 mil (0.508 mm)
4. Stainless steel — 40 mil (0.16 mm)
5. Aluminum — 51 mil (1.2954 mm)
6. Aluminum — 100 mil (2.54 mm)

137

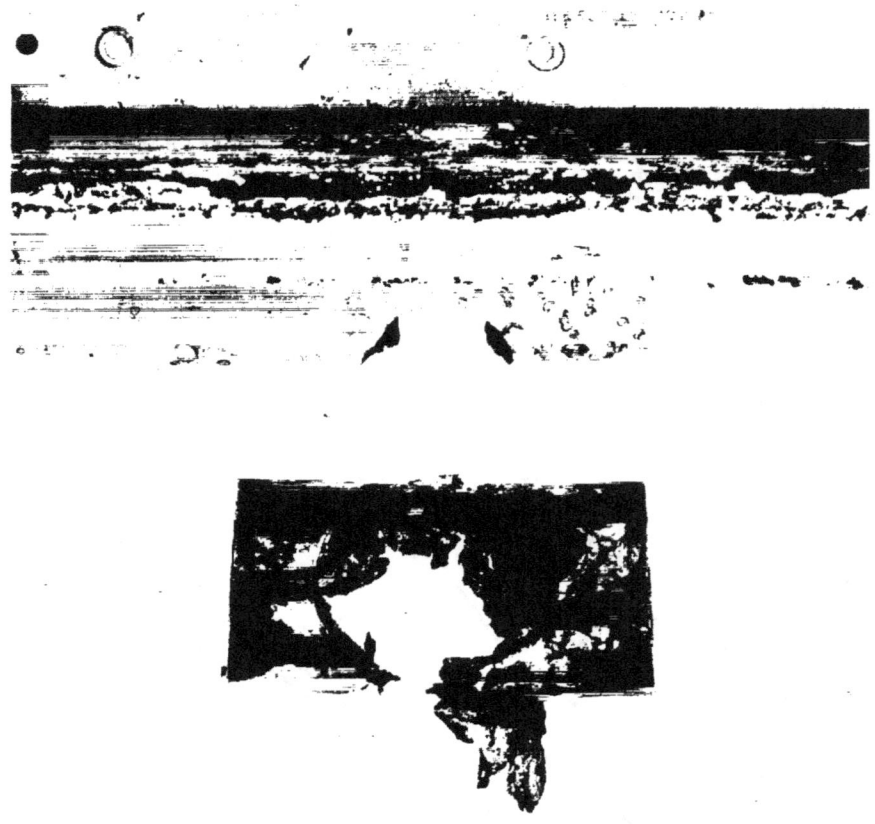

Figure 6.15 Natural lightning damage sustained by Airplane NC25629 about August 14, 1941. Top views.

Upper: trailing edge of left wing; material — 24st Alclad, 0.89 cm (0.35 in) thick. Note numerous small pitted spots.

Lower: rudder tip bow; material — 17st aluminum alloy, 2.54 cm (1 in) O.D. tubing, 0.07 cm (0.028 in) wall thickness.
Coulombs estimated from laboratory tests on metal sheets and tubing.

Wing edge:	large hole	137 C	Rudder bow:	300 C
	small hole	6 C		
	150 pits	150 C		
	total	293 C		

no matter how small an amount of charge (Q) is delivered, but it can be concluded logically that, if the charge and resultant heating were too low,

138

damage would be limited to partial meltthrough and no hole would be formed. Robb, Hill, Newman, and Stahmann reported this result in 1958 (Reference 6.29), based on tests in which a three-component simulated lightning current consisted of the following:

1	100 kA	0.069×10^6 A^2-s	Return stroke
2	5 kA	0.172×10^6 A^2-s	30 C intermediate stroke
3	200 kA	0.040×10^6 A^2-s	200 C continuing current

This current was applied to commonly used samples of 0.051 cm (0.020 in), 0.102 cm (0.040 in), 0.163 cm (0.064 in), 0.206 cm (0.081 in), and 0.259 cm (0.102 in) aluminum skin materials. Robb reported that this discharge, which is about as severe as a similar combination of SAE (Reference 6.30) current components D, B, and C (in that order, which would be specified for tests in Zone 2B), did not puncture the skins which were 0.206 cm or greater. Kester, Gerstein, and Plumer confirmed this result by showing that over 200 C (delivered at 288 A) are required to melt through 0.163 cm aluminum (Reference 6.31).

The three-component discharge including a 200 C continuing current would almost certainly appear only at a trailing edge (Zone 2B) after being swept there from an initial attachment point somewhere on a leading edge or forward extremity. Fuel is rarely in direct contact with trailing edges of aircraft or within direct stroke zones (Zone 1B), where extended flash hang-on is likely; yet airworthiness certification criteria, such as those listed in Table 5.1, generally require that integral fuel tank wall thicknesses be at least 0.203 cm (0.080 in) wherever lightning strikes are expected.

Robb also observed (Reference 6.32) that swept-flash pit marks from several centimeters to about 60 cm (2 ft) apart are commonly found, and that a slow, worst case aircraft velocity of 62 m/s (140 mph, 205 ft/s) would result in the lightning arc dwelling for 9.7 ms at each point.

$$t_{dwell} = \frac{0.6 \text{ m}}{62 \text{ m/s}} = 9.7 \text{ ms} \qquad (6.3)$$

If this is so, the amount of charge delivered to each dwell point across a typical 3 m wing surface by a lightning flash similar to Robb's three-component discharge would be as shown in Figure 6.16 (Reference 6.33).

Since the amount of charge entering any of the Zone 2A points in Figure 6.16 is considerably less than the 230 C applied by Robb to melt through 0.163 cm aluminum skins, it might be argued that a skin thinner than 0.203 cm would be adequate to prevent meltthrough. On the other hand, it has been found that the rate of charge delivery (current amplitude) is also important in determining whether sufficient energy is available to melt completely through, or not, and that smaller amounts of charge will also cause meltthrough of 0.163 cm aluminum if the amounts are delivered at a sufficiently high rate. Since any meltthrough of aluminum skins must be considered a possible source of fuel

139

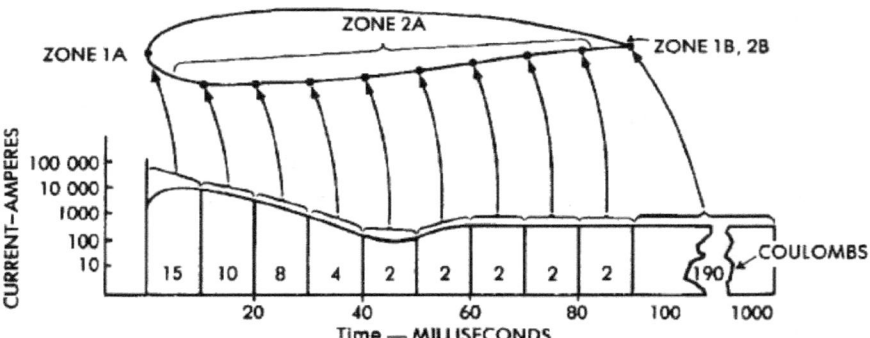

Figure 6.16 Charge delivered at successive dwell points across a wing surface.
● Only a small amount of charge enters Zone 2A dwell points.

ignition, it is imperative that the minimum amount of charge and current within the range of possible lightning characteristics be established for proposed integral fuel tank skins. This has been termed by Plumer the *coulomb ignition threshold* (Reference 6.34).

6.6.1 Coulomb Ignition Thresholds for Clean Skins

The amounts of charge and current required to melt through aluminum and titanium skins of various thicknesses and to cause fuel ignition were reported by Brick in 1968 (Reference 6.35) and by Oh and Schneider in 1972 (Reference 6.36) to depend heavily on current amplitude as well as charge. Whereas earlier work had shown that over 200 C, when delivered by a current of 200 A, were necessary to burn through 0.206 cm aluminum skins, the work of Brick, Oh, and Schneider shows that only about 10 C, when delivered by a current of 500 A, are required to melt completely through a 0.206 cm aluminum skin; and in laboratory tests as little as 2 C, when delivered by about 130 A, will melt a hole completely through 0.102 cm of aluminum. Oh and Schneider's meltthrough thresholds for these and other skin thicknesses are shown in Figure 6.17 (Reference 6.37). The close (0.24 to 0.48 cm) proximity of their test electrode to the skins may have restricted natural movement of the arc on the surface of the skin and been responsible for these unusually low coulomb ignition thresholds.

Work by Kester, Gerstein, and Plumer (Reference 6.38) with an L-shaped electrode spaced 0.64 cm above the skin, permitting greater arc movement, showed that 20 C or more, when delivered at 130 A, are required to burn through a 0.102 cm aluminum skin. Since a natural lightning arc is not restricted by an electrode, it is probable that the aluminum skin puncture threshold curves of Figure 6.17 are conservative, at least for unpainted surfaces. On the other hand, the electrical insulating properties of most paints tend to restrict arc movement and so would concentrate the heating effects at a smaller point, decreasing the amount of thermal energy required to melt completely through.

140

Oh and Schneider have similarly determined the coulomb ignition thresholds for titanium skin materials of various thicknesses, as shown in Figure 6.18 (Reference 6.39). They and other researchers have confirmed that, since the melting point of titanium is higher than the fuel ignition temperature, it is not necessary for a hole to be melted completely through for ignition to occur. For ignition to occur only a hot spot need be formed on the inside surface of a titanium skin. The lower thermal conductivity of titanium prevents rapid heat transfer away from the arc attachment point and accounts for the generally lower coulomb ignition thresholds than those for aluminum.

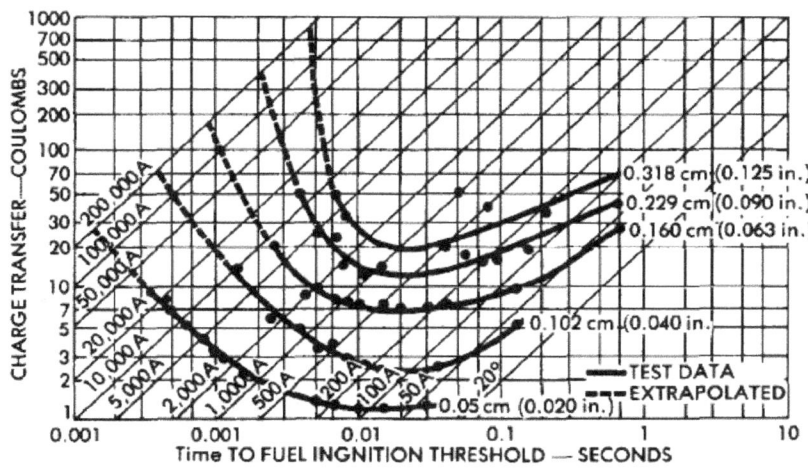

Figure 6.17 Coulomb meltthrough and ignition threshold for aluminum skins.

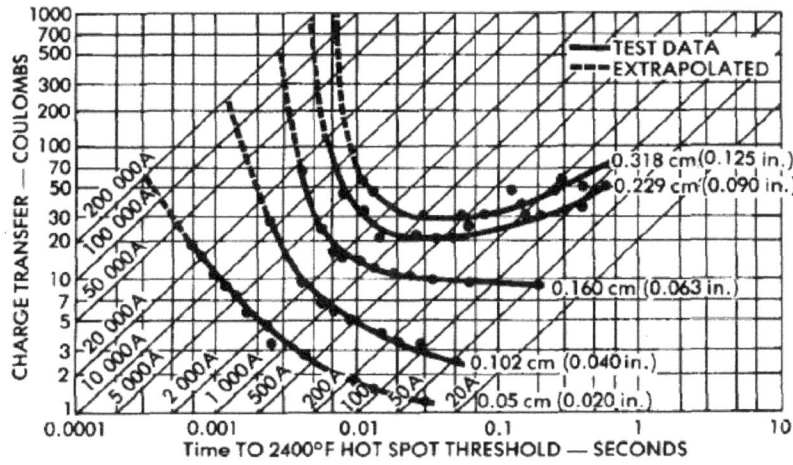

Figure 6.18 Coulomb hot spot and ignition thresholds for titanium skins.

Coulomb ignition thresholds for stainless steel have not been studied as thoroughly as have those for aluminum and titanium, but they have been determined for currents of 80 to 130 A by Kester, Gerstein, and Plumer (Reference 6.40) to be 4.4 C for 0.051 cm (0.02 in) skins, 12 C for 0.102 cm (0.04 in), and 41 C for 0.152 cm (0.06 in) stainless steel skins. A comparison of these thresholds with those found by Kester and his colleagues for aluminum and titanium skins of the same thicknesses is shown in Table 6.2. The times required for ignition to occur in several of the tests reported in Table 6.2 are listed in Table 6.3 (Reference 6.41).

Table 6.2   COULOMB IGNITION THRESHOLDS FOR VARIOUS SKIN
METALS. 1.5 STOICHIOMETRIC PROPANE-AIR MIXTURE

Material Thickness	Titanium (Ti-8Al-1MO-IV)	Aluminum (2024T-3)	Stainless steel (Series 304)
	Coulombs	Coulombs	Coulombs
0.051 cm (0.020 in)	5.5	not determined	4.4
0.102 cm (0.040 in)	5.75	20.0	12.0
0.152 cm (0.060 in)	100/125	46.0	41.0

Table 6.3   DISCHARGE AND IGNITION TIMES AT
COULOMB IGNITION THRESHOLD

Thickness	Coulomb	Dis. time (ms)	Ign. time (ms)	Coulomb	Dis. time (ms)	Ign. time (ms)	Coulomb	Dis. time (ms)	Ign. time (ms)
0.051 cm (0.020 in)	6.0	56	44				4.4	48	125
							*8.0	55	105
0.102 cm (0.040 in)	8.0	62	72	22.0	130	300			
				*37.0	300	100			
0.152 cm (0.060 in)	107.0	950	900						

*Above threshold level

The charges and times given in Tables 6.2 and 6.3 for ignition to commence are somewhat greater than those associated with the lowest coulomb ignition thresholds (bottoms of the curves) of Figures 6.17 and 6.18 because the

142

currents (rates of charge flow into the skins) of 80 to 130 A were generally lower in the tests of Tables 6.2 and 6.3 than the 100 to 1000 A currents which produced the lowest coulomb ignition thresholds of Figures 6.17 and 6.18. But since intermediate currents (as in component B) and continuing currents (as in component C) of from a few amperes to 1000 A are common in natural lightning, both data sources must be considered authentic. Clearly, the lower currents must dwell longer at a single attach point to produce the same damage and possibility of ignition that higher currents produce in shorter times. An illustration by Plumer (Reference 6.42) of the resulting wide range of charge and dwell time combinations necessary to melt through aluminum skins and ignite fuels beneath is shown in Figure 6.19.

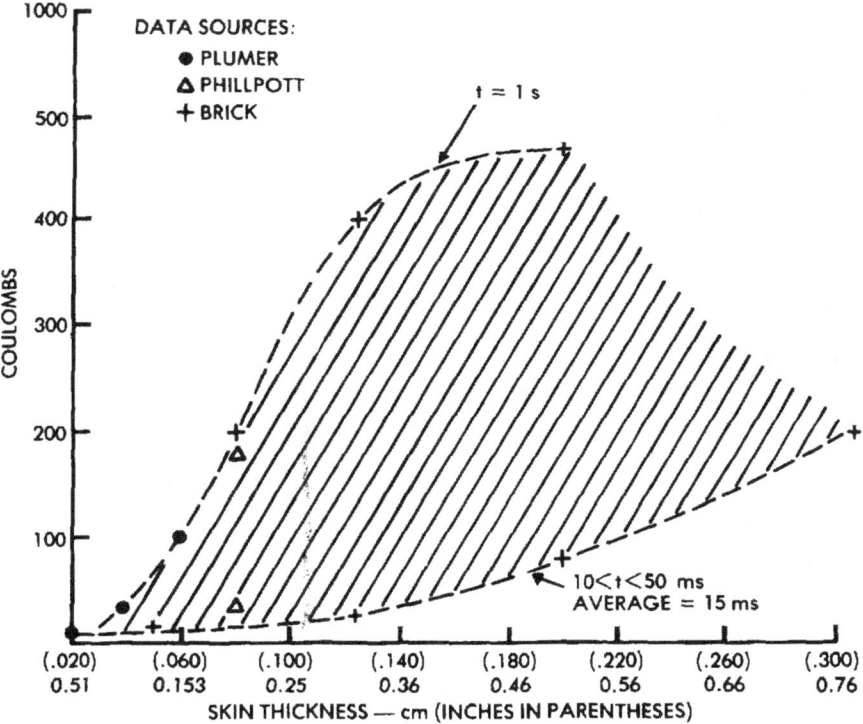

Figure 6.19 Range of minimum combinations of lightning dwell time and charge necessary to ignite fuels beneath bare aluminum skins.

6.6.2 Dwell Times

The data of Figures 6.17 and 6.18 and of Tables 6.2 and 6.3 can be used to determine the possibility of meltthrough or hot spot formation sufficient to cause ignition if the amount of time a lightning arc may dwell at a particular point and the amplitude of the current that flows into this point are known. Figure 6.20 (Reference 6.43) shows a time exposure photograph of an actual lightning arc attaching to successive dwell points along the radome-mounted pitot boom of a fighter aircraft, and Figure 4.2 shows what similar points look like along a typical aircraft surface.

Knowledge of dwell times at swept-stroke (Zone 2A) attachment points like these is of most importance because integral fuel tank skins are often found in these zones. Existing regulatory documents (Table 5.5) have specified that aluminum skins in lightning-strike zones should be at least 0.203 cm (0.080 in) thick to withstand meltthrough because the simulated lightning current of Figure 6.16 would not melt through this thickness of aluminum, even if all of this current is delivered to one point. The regulatory documents also imply that other materials should be of thicknesses capable of withstanding an equivalent lightning current.

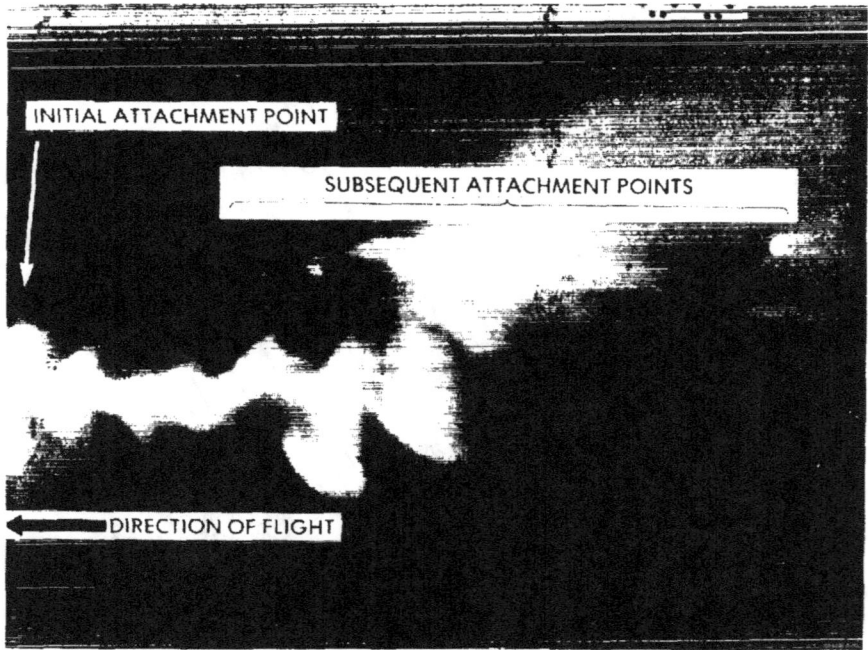

Figure 6.20 Time exposure of a lightning arc at successive dwell points.

Until recently, structural design demands also made it necessary to use skins of at least 0.203 cm, so the need for this thickness for lightning protection was not challenged. But in some recent designs, including the designs of supersonic transports, thinner skins would have met structural requirements and have permitted a savings in weight and cost were it not for the lightning

144

protection requirement for a 0.203 cm thickness. Therefore, Brick, Oh, and Schneider (Reference 6.44) studied dwell times of 400 A decaying arcs blown by the exhaust from a wind tunnel over aluminum and titanium skin panels with several surface treatments to see how long such an arc might actually dwell at one point before reattaching to the next. Wind speeds of 67 m/s (150 mph) and 112 m/s (250 mph), representing approach speeds for typical aircraft, were utilized. The test current they used was representative of lightning *continuing* current, and would have delivered an average charge of about 0.2 C/ms of dwell. The researchers reported that the arcs dwelled for 2 ms or less on uncoated surfaces of both metals, and for 4.8 ms on an anodized aluminum surface. At 0.2 C/ms of dwell these dwell points would have received 0.4 and 0.96 C, respectively. By Figures 6.17 and 6.18 this would not have been enough charge to melt through even the 0.05 cm thicknesses of either metal. If the lightning current amplitude experienced in Zone 2A never exceeded 400 A, this result would indicate that skins as thin as 0.05 cm would be safe. The problem is, of course, that the occurrence of higher currents has now been recognized as possible in Zone 2A.

At the end of the dwell at one point, the arc will reattach at a point farther aft on the aircraft, as shown in Figure 6.21 and explained in Section 2.3. During the dwell times the aircraft would be moving, and the distance covered would be

$$D = v \, t_d \qquad\qquad (6.4)$$

Where D = distance arc is drawn along the aircraft surface (m)
v = aircraft velocity (m/s)
t_d = dwell time (s)

Thus, at velocities of 67 and 112 m/s and with the dwell times of 2.0 and 4.8 ms reported by Brick, spacings between successive dwell points would be as given in Table 6.4.

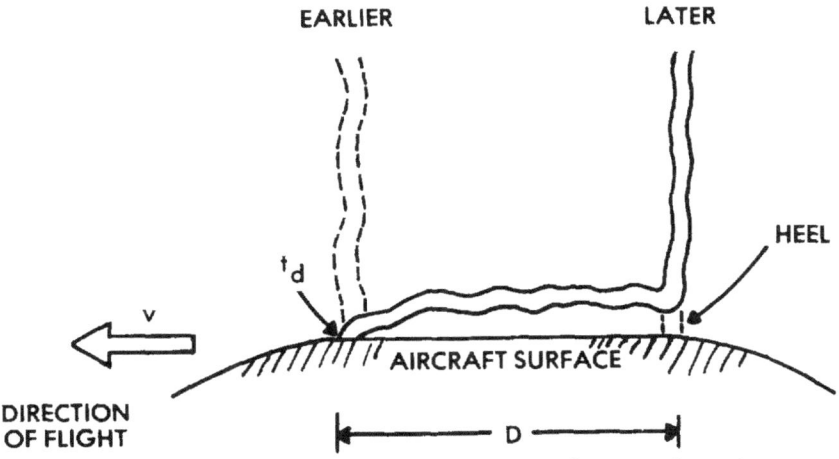

Figure 6.21 Basic mechanism of swept-stroke reattachment.

145

Table 6.4   DWELL POINT SEPARATIONS FOR TYPICAL LIGHTNING
DWELL TIMES AND AIRCRAFT VELOCITIES

Dwell Times (t_d)	Aircraft Velocities (v)	
	67 m/s	112 m/s
2.0 ms Unpainted surface	D = 13.4 cm (5.3 in)	D = 22.4 cm (8.8 in)
4.8 ms Painted surface	D = 32.2 cm (12.7 in)	D = 53.8 cm (21.1 in)

Because spacings of this order are often observed along aircraft surfaces, the dwell times predicted by the Brick tests are realistic. The later work of Oh and Schneider (Reference 6.45) basically confirms these results for uniform airflow conditions but shows that conditions which cause the airstream to leave the surface may force the arc to dwell longer at the last attachment point before the airflow is diverted. Oh and Schneider also demonstrate that higher aircraft velocities result in shorter arc dwell times, since the arc is stretched greater distances, allowing sufficient voltage to build up along it to break down the insulation at its heel (see Figure 6.21) at an earlier time.

Validity of the wind tunnel technique for simulating swept strokes along metallic surfaces has sometimes been questioned because the airstream is used to blow the arc along the surface of the test object, whereas in flight both the air and the arc are stationary. It might be expected that the moving air would carry the arc along at an equal velocity, but in practice this doesn't quite happen. In addition, there must be cooling of the arc by the faster moving air, which allows it to build voltage that in turn might create a new attachment point sooner than had cooling not occurred. Also, the behavior of the upper terminus of the arc as it moves along the electrode probably has *some* effect on the behavior of its lower terminus at the test object. The scope of these effects and the extent to which they cause departure from an authentic simulation of natural lightning has not been investigated.

A more realistic simulation would undoubtedly result if the test object, like an aircraft in flight, could be moved through a stationary arc. Plumer (Reference 6.46) attempted this by driving a wing tip fuel tank beneath a high-voltage electrode at a velocity of 15.5 m/s (35 mph) A flash was triggered when the approaching tank, carried atop a truck, sufficiently closed the airgap between the electrode and ground. Limitations of this technique were the low velocity and the low (4 A) test current amplitude. Nonetheless, subsequent in-flight lightning strikes to two fuel tanks of this type have confirmed the occurrence of the predicted attachment points and breakdown paths. Despite the lower velocity and current amplitude, the arcs in Plumer's tests also dwelled for times of between 1 and 4 ms on unpainted aluminum surfaces, results closely parallel to those of the wind tunnel work. These parallel conclusions seem to indicate that the arcs in the wind tunnel tests may not have moved as fast as the wind itself, and, also, that current *amplitude* has relatively little to do with dwell time.

Robb, Stahmann, and Newman (Reference 6.47) have utilized the wind tunnel technique to determine arc dwell times on various painted or coated surfaces. Most of these coatings are electrically insulating and thus require that the arc be further lengthened to allow the greater voltage buildup necessary to puncture the insulation and form the next attachment point. The reasonable agreement of test methods with the in-flight strike evidence makes it possible to show, with some confidence, in Table 6.5 the dwell times to be expected on several surfaces commonly found in swept-stroke zones (Zone 2A).

Table 6.5 LIGHTNING DWELL TIMES ON TYPICAL AIRCRAFT SURFACES IN ZONE 2A

Surface Type	Aircraft Velocities		
	15.5 m/s (35 mph)	58 m/s (130 mph)	103 m/s (230 mph)
Aluminum and titanium unpainted	[1] 1 to 4 ms	[2] 2.0 ms	[2] 1.0 ms
Aluminum anodized		[4] 4.8 ms	[2] 2.6 ms
Aluminum painted		[3] up to 20 ms	

Note 1 — Plumer (See Reference 6.46)
Note 2 — Oh and Schneider (See Reference 6.45)
Note 3 — Robb, Stahmann, and Newman (See Reference 6.47)
Note 4 — Brick, Oh, and Schneider (See Reference 6.44)

There are many different coatings and paints of various thicknesses found on aircraft surfaces, and the effect of many of these on arc dwell time has not yet been determined. It may therefore be advisable to perform swept-stroke tests on metallic skins by considering the particular coatings for a particular design. Guidance for performance of such tests is given in Section 4.3 of the SAE report *Lightning Test Waveforms and Techniques for Aerospace Vehicles and Hardware* (May 1976).

Once the expected dwell times are established, next it will be necessary to determine the amplitude of current that might flow during these periods.

147

6.6.3 Current Amplitudes

A wide range of current amplitudes appears in typical lightning flashes. These are discussed in Chapter 1; Figure 1.1, for example, shows that most of the charge in a flash is delivered by continuing currents at a rate of about 200 A and by intermediate currents at a rate of about 5 kA. The higher amplitude strokes are usually of too short a duration to deliver appreciable charge. Therefore, emphasis should be placed on the intermediate currents, which can occur to surfaces in Zone 2A and which can deliver large amounts of charge at appreciable current amplitudes. Section 4.4.2.3 of the SAE report specifies that current components B, C, and D should be applied in Zone 2A, with D applied first, since it is a restrike, which would be expected to create a new dwell point. Components B and C follow for whatever dwell time is expected — or for 50 ms if the dwell time is unknown and has not been determined from swept-stroke tests. This current is shown in Figure 6.22. If a dwell time of less than 5 ms is expected, an average current of 2 kA should be applied for the actual dwell time only.

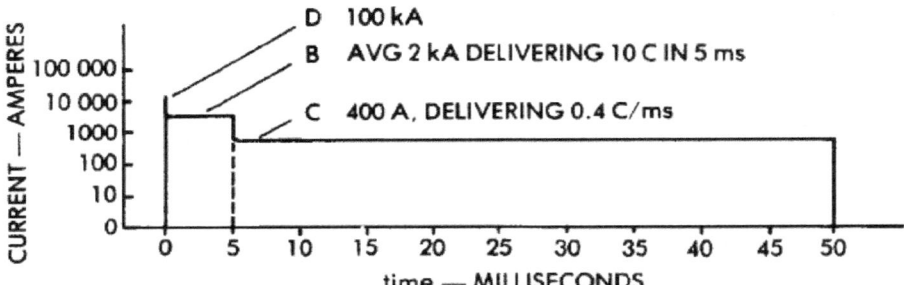

Figure 6.22 Current and charge expected at a Zone 2A dwell point.
- Delivers 10 C in first 5 ms at any dwell point
- Delivers .04 C/ms thereafter (From Figure 3.3 SAE report)
- Delivers 400 A for all 50 ms if dwell time not known (Drawn with straight lines for explanation purposes only)

The current amplitudes of Figure 6.22 are based on the SAE study of natural lightning data and in-flight damage reported by aircraft operators over the years. As such, these current amplitudes are a good representation of what to expect in a severe natural lightning flash, the type exceeded only 1 or 2% of the time. It is recognized, however, that more severe currents, like those which sometimes occur in a positive polarity, cloud-to-ground flash, could appear at Zone 2A dwell points. An example of such flashes, recorded by Berger (Reference 6.48), is shown in Figure 6.23.

The current measurement was made at an instrument tower on the ground, and the long, upward-moving leader which extended from this tower to the cloud base accounted for the (approximate) 11.5 ms which elapsed before the

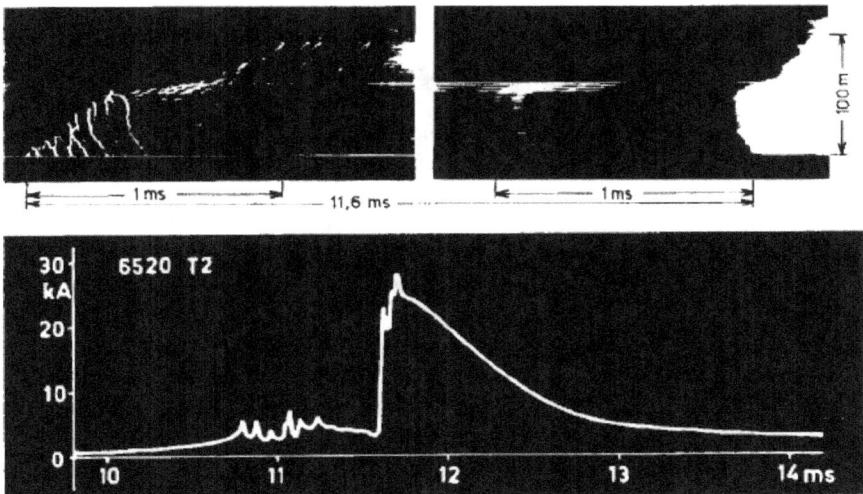

Figure 6.23 Positive flash of type 4.
Above: photograph on fast-moving film
Below: oscillogram of stroke current

return stroke appeared at the ground. Not all of this delay, therefore, would have occurred had the measurement been made on an aircraft in flight intercepted by this flash. But the several milliseconds which still would have elapsed between initial leader attachment to a forward extremity of the aircraft and the occurrence of the return stroke may have allowed the arc to move over Zone 2A. The positive polarity stroke of Figure 6.23 would have delivered about 20 C to this dwell point during the *minimum* of 2 ms of time it would have dwelled there. According to the data of Figure 6.17 meltthrough of a 0.16 cm (0.063 in) aluminum skin is possible under these conditions. There are reports of holes melted in 0.102 cm (0.040 in) skins from in-flight strikes in Zone 2A areas, but holes in 0.16 cm aluminum in Zone 2A have not been reported, a fact that demonstrates the apparent rarity of the high-energy positive polarity flash in Zone 2A areas. Because of the lack of positive evidence that these flashes have struck aircraft in critical zones, they have not been fully accounted for in present test current waveforms.

6.6.4 Determination of Skin-Thickness Requirements

With the establishment of methods to predict the lightning current amplitude and charge, as well as dwell time, sufficient information is available to utilize the charts of Figures 6.17, 6.18 or Tables 6.2 and 6.3 to determine the skin thicknesses which will or will not be melted through.

For example, let us assume that a bare aluminum skin is planned for an integral tank in Zone 2A, and that it is desired to determine how thick this skin should be to prevent meltthrough. Further, let us say this aircraft will fly at

149

velocities as low as 58 m/s (130 mph). From Table 6.5 the expected dwell time for this unpainted skin would be 2 ms. From Figure 6.22, an average of 2 kA would flow into the dwell point during this period, delivering 4 C of charge. According to Table 6.17 these parameters intersect at a point about half-way between the coulomb ignition threshold curves for 0.051 cm (0.020 in) and 0.102 cm (0.040 in) aluminum skins, indicating that 0.102 cm is the thinnest skin that should be considered. Since there would be little margin of safety if a skin of this thickness were actually used, it would be prudent to select instead the next larger thickness—0.16 cm (0.064 in).

If one notes that the airworthiness certification criteria of Table 5.6 generally call for an even thicker skin—0.203 cm (0.080 in)—in lightning-strike zones, one must also remember that these criteria make no distinction between painted and unpainted surfaces, and that dwell times on painted surfaces may be considerably longer than the 2 ms experienced on bare aluminum skins.

Different dwell points, different coatings, and different thicknesses will cause correspondingly different dwell times. Consider, for example, the longest dwell time, 20 ms, recorded by Robb, Stahmann, and Newman (Reference 6.47) for a painted surface. During this period the current of Figure 6.21 would deliver 16 C. These parameters intersect at a point just *above* the 0.229 cm (0.090 in) curve in Figure 6.17, indicating that even the 0.203 cm (0.080 in) thickness required by the certificating agencies is insufficient to prevent ignition where certain paints are used.

Therefore, if paints must be used, it is advisable to perform swept-stroke tests to establish the actual dwell time. The procurement or certificating agencies have generally been willing to accept test data in support of an alternate skin thickness to the aforementioned 0.203 cm (0.080 in) thickness, and such tests may lead to the use of either thinner or thicker skins for ensuring adequate protection.

6.6.4.1 Titanium and Stainless Steel Skins

By using the chart of Figure 6.18, it is possible to determine titanium skin thicknesses in a manner similar to that for determining aluminum skin thicknesses. The coulomb ignition threshold for titanium occurs when the back side of the skin reaches 1320 °C (2400 °F), a temperature sufficient to ignite a fuel-air vapor. Because titanium will not melt at this temperature, no hole will be formed when the coulomb ignition threshold is reached.

No charts of the type in Figure 6.17 and 6.18 have been generated for stainless steel or other metals, since, with the exception of certain space vehicles, these materials seldom appear as integral tank skins. Tables 6.2 and 6.3, while not accommodating as wide a variation in parameters, provide some guidance.

Certain other materials, such as fiber-reinforced plastics (composites) behave differently from metallic materials under lightning stress. These are treated separately in Section 6.2 and Chapter 7.

150

A comparison of the physical and electrical properties of titanium with those of aluminum has been made by Kosvic, Helgeson, and Gerstein (Reference 6.49) and is presented in Table 6.6.

Table 6.6 PHYSICAL AND ELECTRICAL PROPERTIES OF TITANIUM AND ALUMINUM

Property	Titanium	Aluminum
Density ρ (g/cm^3)	4.5	2.7
Thermal conductivity k (cal/s cm $^\circ$K)	0.040	0.37
Specific heat C_p (cal/g $^\circ$K)	0.13	0.23
Thermal diffusivity α (cm^2/s) = k/ρ C_p	0.07	0.60
Electrical Resistivity (μ Ω-cm)	199	6.3
Fusion point ($^\circ$K)	1950	930
Heat of fusion (cal/g)	77	93
Vaporization point ($^\circ$K)	3550	2750
Heat of vaporization ΔH_{vap} (cal/g)	2140	2580

As noted in this table, titanium has a markedly lower thermal conductivity and higher resistivity than has aluminum. These properties have given rise to the belief that the backside of titanium skins would reach the coulomb ignition threshold with a lower lightning current amplitude and charge transfer than those necessary for aluminum, raising the possibility that titanium skins might have to be thicker than aluminum skins to provide the same degree of lightning protection. But, at the same time, it was evident that structural requirements with titanium could be met with skins thinner than the 0.203 cm (0.080 in) which was necessary to meet strength requirements with aluminum. The question of lightning effects on titanium skins has also been of concern because this material is used in swept- or delta-shaped wings in some supersonic aircraft, and the swept-stroke zones may extend over internal fuel tanks in such shapes.

Therefore, several in-depth studies of titanium behavior under simulated lightning arcs have been conducted. One of the most important of these is the work of Kofoid (Reference 6.50), who conducted both experimental and analytical studies of the lightning current amplitude and dwell times necessary to reach the ignition temperature (coulomb ignition threshold) on the inside surfaces of titanium skins. Kofoid's experimental results closely follow an empirical heat diffusion equation:

$$I = \frac{d^2}{t_c} (480) \qquad (6.5)$$

where $\quad$ I $\ =\ $ lightning current amplitude (A) (assumed constant)

$\qquad\quad$ d $\ =\ $ skin thickness (cm)

$\qquad\quad$ t_c $\ =\ $ time to reach ignition temperature on inside surface (s)

151

The constant (480) is empirical. Curves according to Equation 6.5 are plotted in Figure 6.24, together with test data for simulated lightning continuing and intermediate currents of up to about 1900 A.

The experimental data adhere closely to the curves, and both are also in good agreement with the data of Figure 6.18 and Tables 6.2 and 6.3. Figure 6.24

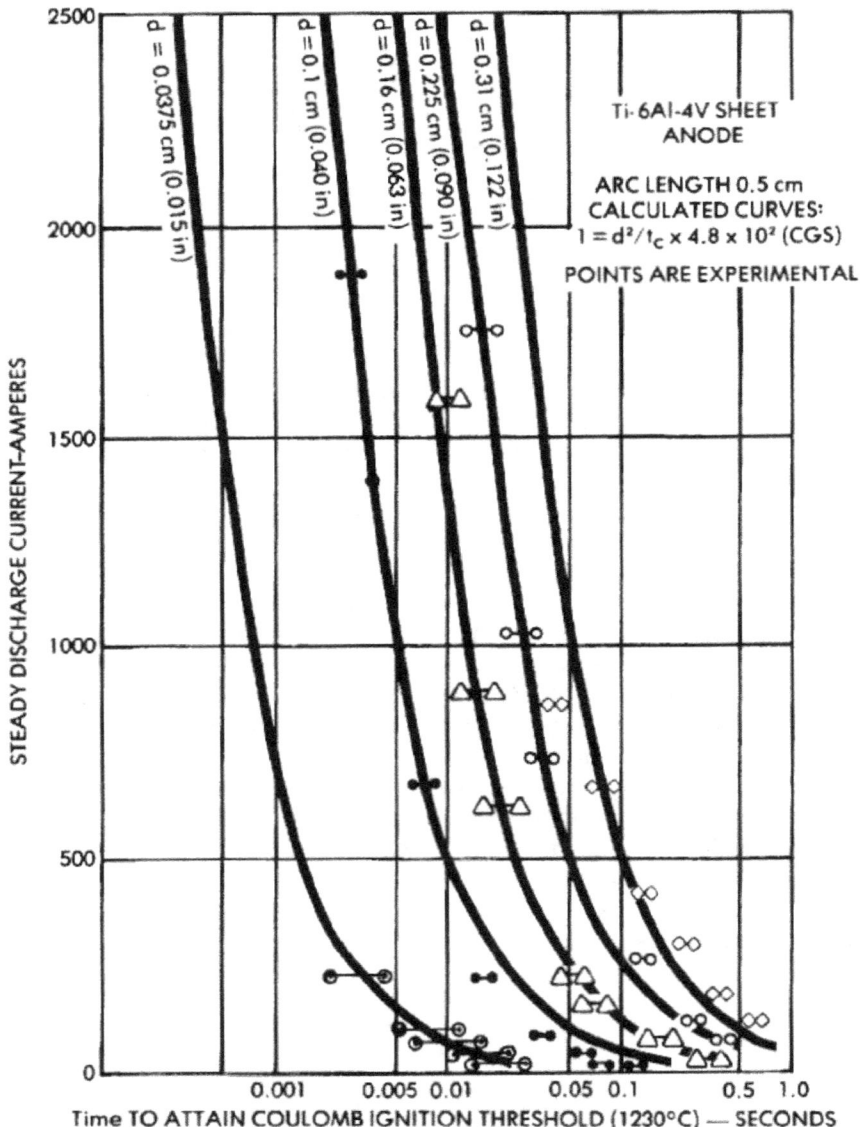

Figure 6.24 Time to reach coulomb ignition threshold vs constant discharge current amplitude.

(Reference 6.51) shows that the higher the lightning current amplitude, the shorter the time, t_c, required to reach the coulomb ignition threshold, and that for thicknesses as great as 0.102 cm (0.040 in), only 5 to 6 C of charge are necessary to reach this threshold if the charge is delivered in 10 ms or more by a current of 500 A or less. This means that if a titanium skin is proposed for use in Zone 2A, a lightning arc should not be permitted to dwell for more than about 5 ms at any one point on the surface. According to the dwell time data of Table 6.5, this means, in turn, that a 0.102 cm (0.040 in) skin may not have its outside surface painted. Anticorrosion paints or sealants on the inside surface of a titanium skin may ease the problem by providing thermal insulation between the fuel and the hot spot, but the degree of actual protection thus obtained should be determined by laboratory tests before such coatings are judged reliable in a design.

6.6.4.2 Ways to Reduce Dwell Time

The foregoing discussion makes clear the importance of lightning arc dwell time in establishing whether or not ignition from a strike to a certain skin is likely to occur. And dwell time is perhaps the only lightning characteristic over which the integral fuel tank designer may have any control. The objectives, of course, should be to have as short as possible a dwell time at any one spot and to spread the arc attachment among many different spots. A bare metallic external finish will best achieve these aims, since most paints and other coatings act to concentrate the attachment at more widely separated points for correspondingly longer times than does a bare surface.

If a paint must be used, lightning dwell times may be reduced by making the paint partially conductive. Robb and others (Reference 6.52) have demonstrated that aluminum powder is effective in increasing the conductivity of polyurethane paints, thereby increasing the ability of the arc to reattach to new points as the aircraft surface moves beneath the arc. Since no parametric data relating dwell time to amount of additive are yet available, it is advisable to determine in the laboratory the degree of improvement afforded by particular combinations.

6.6.5 Integral Fuel Tanks in Trailing Edges

If an integral tank skin is the aft-most trailing edge of an aircraft wing tip or other structure (Zones 1B, 2B), metal skins of almost any thickness will be melted completely through because the flash is likely to hang on there for most of its life. This conclusion derives from the data presented; the application follows. Figure 1.25 shows that 99% of all flashes deliver 200 C or less, and SAE current component C reflects this in its requirement for transfer of 200 C in 1 s or less. Figure 6.17 shows that this much charge would melt through aluminum skins up to 0.8 cm (0.313 in) thick, and Figure 6.18 shows that it would produce the fuel ignition temperature of 1320 °C (2400 °F) on the inside surface of titanium skins over 0.318 cm (0.125 in) thick. Therefore, because of the excessive flash hang-on times, it is not safe to have fuel in direct contact with trailing edge skins of practical thicknesses. If it is desired to have fuel in trailing edges, an extended, solid section capable of sustaining up to about 8 cm^3 of melting at an attachment point about 2.5 cm wide should be provided, as shown, for example, in Figure 6.25.

Externally mounted or tip tanks of the cylindrical variety may be protected by closing out the aft-most volume of the tank (as shown in Figure 6.26), so that if meltthrough occurs no fuel vapor will be contacted.

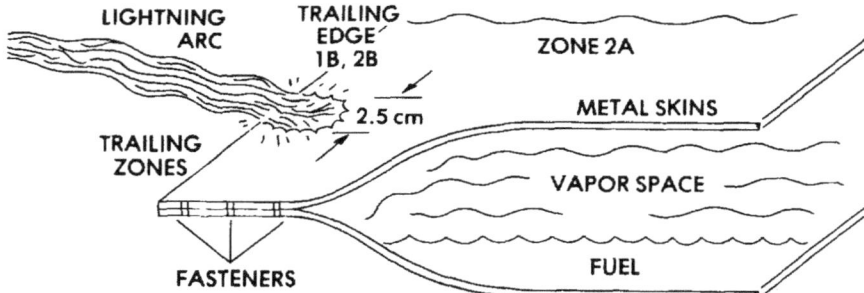

Figure 6.25 Trailing edge construction to avoid fuel ignition as a result of extended flash hangon.
- Volume of at least 8 cm^2 available for melting away
- Fuel or vapor does not contact hot arc or 1320 °C hot spot

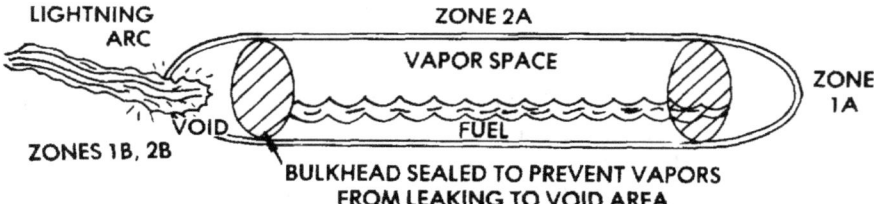

Figure 6.26 Trailing edge construction to avoid fuel ignition from extended flash hang-on.
- Hole melts into aft void area, but no vapor is present to ignite.
- Void sections removable for replacement
- Forward void area not as necessary because leading edge dwell times are short

6.7 Integral Fuel Tanks with Nonmetallic Skins

In some cases nonmetallic materials such as fiberglass-reinforced plastics (FRP) are used instead of metals for integral fuel tank skins. For example, the light aircraft of Figure 4.15 has FRP wing tips which include fuel tanks containing aviation gasoline; these tanks have suffered several in-flight explosions when lightning strikes occurred. The fact that the fuel in these tanks is in direct contact with the inside surface of the tank skin and the fact that portions of this skin are located in Zones 1A and 2A undoubtedly accounts for these explosions.

The basic differences between the lightning effects which may occur to metallic and nonmetallic skins result from the fact that nonmetallic skins are not electrically conductive. This means that the meltthrough or hot spot ignition mechanisms associated with attachment of the lightning arc to the skin will not occur. Instead, the lightning flash will either

- Puncture the skin and attach to an electrically conductive object inside the skin (tank), path (a) on Figure 6.27, or
- Divert around the nonmetallic skin and attach to an adjacent metallic skin or other object, path (b).

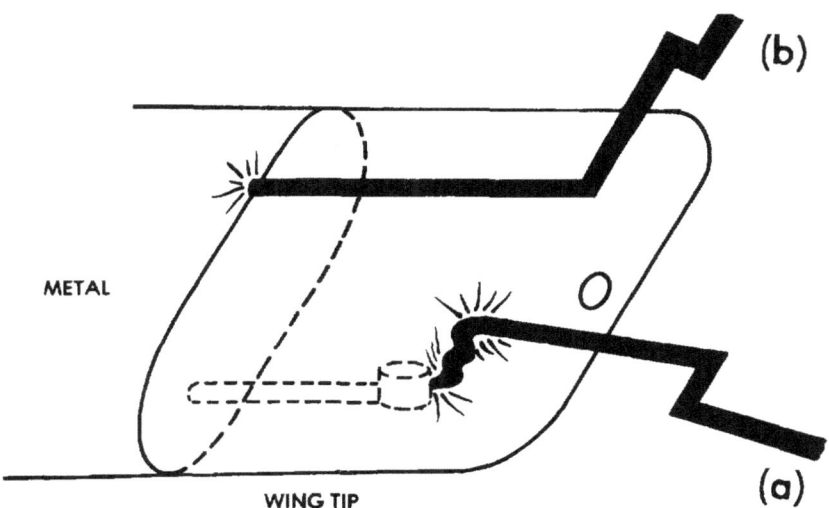

Figure 6.27 Possible lightning-flash attachment to nonconducting skins.

If the flash reaches the skin but remains on its outside surface, path (b), it is not likely that the skin will be burned through or that its inside surface will become hot enough to ignite fuel. The reason for this is that none of the current is being electrically conducted into the skin, and the hot arc does not lie close enough to or long enough against the skin to burn through it. In fact, its effects are usually limited to burning of external paints, if present, or to slight singeing of the nonmetallic skin.

The occurrence of punctures may be explained by the basic lightning attachment mechanism described in Chapter 2. An approaching leader induces streamers from conducting extremities on the aircraft—including objects located beneath nonconducting skins. These streamers are often restricted by the insulating skin, but if the electric field producing them is strong enough, the skin

155

may suffer dielectric breakdown (puncture) and allow the streamer to reach the approaching leader. This is most likely to occur when there is no external conducting object nearby from which another streamer could propagate and reach the leader first. Obviously, if a puncture occurs through an integral fuel tank wall, fuel is placed in direct contact with the lightning arc, and ignition is probable.

It is believed possible that, even if a puncture does not occur, an internal streamer can ignite fuel by itself. Protection against ignition therefore requires that two basic criteria be met:

1.   Punctures be prevented
2.   Internal streamers be prevented.

These criteria are, of course, in addition to other requirements such as elimination of ignition from strikes to vent outlets, or sparking across joints discussed elsewhere in this chapter.

Puncture of nonmetallic skins can be prevented, or at least greatly minimized, by placement of conductive diverters on the outside surface of the tank. These diverters must be placed close enough to each other or to other conductors to prevent an internal streamer from puncturing the skin. In theory this means that the diverters must be close enough together to greatly reduce the internal electric field and prevent internal streamer formation in the first place, because such streamers, once formed, will intensify the field about themselves as they propagate outward. Prevention of internal streamers by diverters is called electrostatic shielding, and successful accomplishment of this function is necessary to prevent skin punctures and, equally important, to prevent internal streamers from forming and becoming a source of ignition themselves. Types of diverters which have proved successful and criteria for placement and spacing of them are given in Chapter 7 (Structures Protection). Tests which may be used to verify diverter adequacy are also described in Chapter 7.

Because streamers form most readily from sharp conducting edges and corners, the number of metallic parts inside a nonmetallic fuel tank should be minimized wherever possible, and those that remain should be located as far from nonmetallic skins as possible and designed with smooth, rounded edges instead of sharp points. These concepts are illustrated in Figures 6.28 and 6.29. Also metallic parts should be located as far from nonmetallic skins as possible.

If a diverter protection system such as that shown in Figure 6.28 is used, the diverter should not be painted or covered with any other nonconducting material. The electrical insulation provided by most aircraft paints is nearly as good as that provided by the fiberglass skin itself, so the paint will mask the diverter and greatly reduce its effectiveness. Painting of the fiberglass skin itself, of course, is to be encouraged, as the added insulation provided by the paint will further reduce the probability of skin puncture.

Plumer (Reference 6.53) has successfully tested a diverter arrangement of the configuration shown in Figure 6.27, and similar arrangements have been flown on some light aircraft with nonmetallic wing tip tanks.

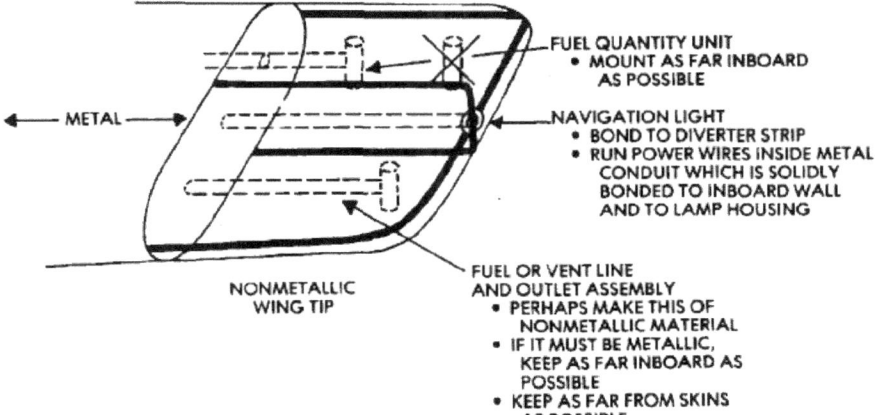

Figure 6.28 Location of metallic parts within nonmetallic fuel tanks.

6.7.1 Other Methods of Protecting Nonmetallic Tanks

If a diverter arrangement is not practical for some reason, a conductive coating may be applied to the tank instead. Such a coating may be a metal foil or a paint heavily doped with metal or carbon particles. The major disadvantage of these coatings is that a relatively large portion of the coating may be melted or burned away when the flash attaches to it. A coating would have to be prohibitively thick (one approaching the thickness of conventional metallic skins themselves) to avoid these results.

On the other hand, a conductive coating has the important advantage of providing an overall electrostatic shield that will virtually eliminate internal electric fields and streamering. Thus, if metallic objects must be located inside the tank in such a manner that streamers are of concern, it may be advisable to consider a conductive coating over the entire nonmetallic skin. This precaution is particularly appropriate if occasional burnoff of part of the coating is acceptable from a maintenance standpoint.

Coating materials which have been tested for such applications include aluminum foils 5 to 15 mm thick, braided and woven aluminum wire meshes of between 40 and 80 wires per centimeter, and various conducting paints. These coatings are discussed more thoroughly in Chapter 7.

6.8 Current Flow Through Structures and Components

In each of the aircraft accidents that were attributed to lightning and that occurred as a result of in-flight explosion of fuel, the exact cause of fuel ignition has remained obscure. Ignition at fuel vent outlets or by meltthrough of integral tank skins has been suspected in several cases, but no conclusive evidence to this effect has ever been found. While it is not correct to eliminate vents and skins from further consideration as the prime factors, it may be possible that minute

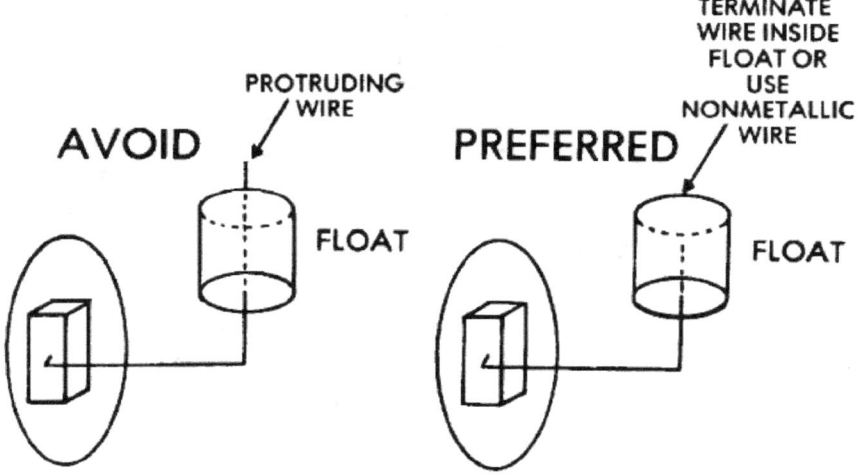

FLOAT-TYPE FUEL TANK SENDER UNIT

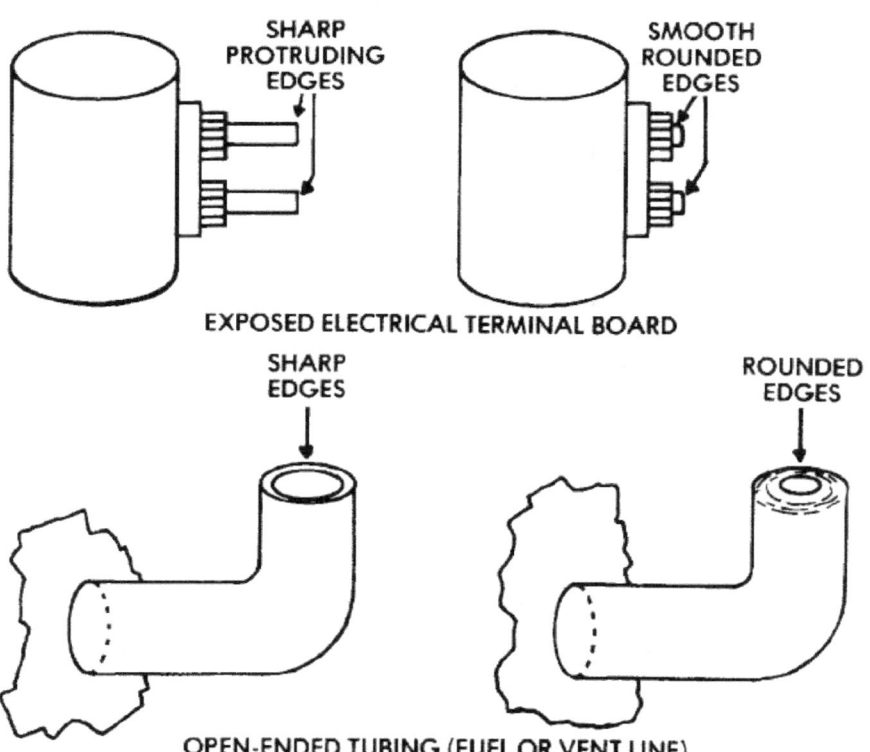

EXPOSED ELECTRICAL TERMINAL BOARD

OPEN-ENDED TUBING (FUEL OR VENT LINE)

Figure 6.29 Avoid sharp edges on metallic parts inside nonmetallic fuel tanks.

sparks resulting from lightning current having been conducted through the fuel tanks caused some of these disasters. Whereas the relatively large amount of thermal energy released at the points of concentrated lightning current entry into the aircraft would be expected to leave tell-tale evidence, the marks left by a several-millijoule spark are very difficult, if not impossible, to detect.

In the past, much attention has been given to keeping lightning currents out of the interior of the aircraft by providing conductive and tightly bonded skins. While most lightning currents, by virtue of their short duration, do not have sufficient time to diffuse completely to interior structural elements such as spars and ribs, it is *not* correct to assume that no current flows in these interior structures. Nor is it correct to assume that none of the lightning current flows in other internal conductors such as fuel and vent lines. Since so little energy is needed for an igniting spark, the behavior of these internal currents is of great importance to fuel system safety. Of course, when the lightning currents which do flow in the skin encounter discontinuities at access doors, filler caps, and the like, a possibility of sparking exists if electrical bonding is inadequate; therefore, the behavior of lightning currents remaining in the skins is important.

Electrical wires entering fuel tanks from other locations in the aircraft pose another potential problem if they are routed in areas which may be reached by high electromagnetic fields when lightning strikes occur.

Figure 6.30 shows the possible lightning current paths in a typical fuel tank and calls attention to these areas of greatest concern. In the following paragraphs we discuss each of these and present guidelines for protection design, referring to related research or testing experience that exists. We also call attention to other areas of concern for which little or no quantitative data is yet available.

6.8.1   Filler Caps and Access Doors

The need to provide a liquid-tight seal around fuel tank filler caps and access doors has led to use of various gaskets and seals between the cap or door and its mating surface in the tank. Most of these have little or no electrical conductivity, leaving the metallic screws or other fasteners as the only conducting paths into the cover. If these fasteners are inadequate or if they present too much inductance, lightning currents may build sufficient voltage along such paths to spark across the nonconducting seals, creating shorter paths. Some of these sparks may occur at the inside surfaces of the joints and be a source of fuel ignition.

Newman, Robb, and Stahmann were among the first to recognize and evaluate this possibility in the laboratory. They demonstrated (Reference 6.54) that direct strikes to filler caps or access doors of (then) conventional design would cause profuse spark showers inside the tank—spark showers of ample energy to ignite fuel. They applied simulated strokes to typical fuel filler caps and access doors. Their test strokes ranged in energy from a very mild 35 kA, 0.006×10^6 A^2-s stroke to a very severe 180 kA, 3×10^6 A^2-s stroke; profuse sparking occurred under all conditions. When an ignitable fuel mixture was

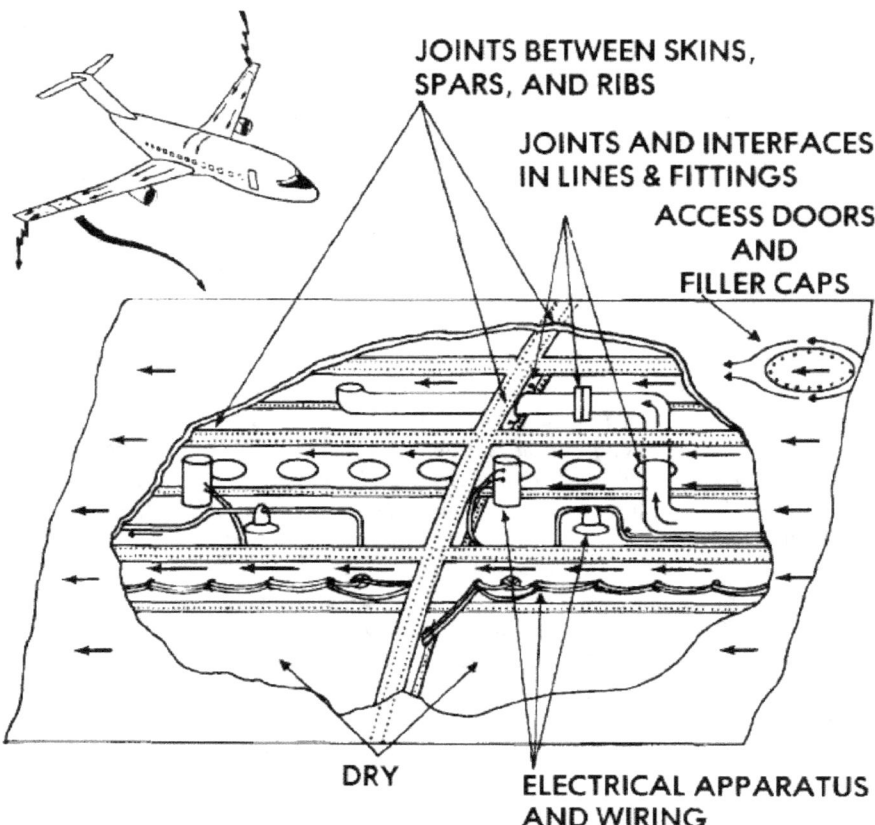

JOINTS BETWEEN SKINS, SPARS, AND RIBS

JOINTS AND INTERFACES IN LINES & FITTINGS

ACCESS DOORS AND FILLER CAPS

DRY

ELECTRICAL APPARATUS AND WIRING

Figure 6.30 Lightning current paths in a fuel tank and potential problem areas.

placed in the tanks, these spark showers readily produced ignition.

Newman and his colleagues did not report sparking or fuel ignition when the same access doors or filler caps were not directly struck but were nevertheless located in a tank skin, which instead was conducting the same simulated lightning currents distributed throughout its surface. Nor has sparking under such a condition been reported elsewhere in the literature. The possibility of such an occurrence, however, has perhaps not been evaluated as thoroughly as it should be.

6.8.1.1 Lightning-Protected Filler Caps

After observing the sparking from direct strikes to the original access doors and filler caps, Newman evaluated several design modifications to prevent this sparking. Figure 6.31 shows the locations of sparking at the inside surfaces of an unprotected filler cap and a modified design intended 'to prevent internal sparking. It incorporates a plastic insert that precludes any sparking at interior

160

faying surfaces and replaces the previous ball chain with a nonconducting plastic strap. Newman successfully tested a mockup of this modified cap with direct strikes (Reference 6.55). Since then lightning-protected fuel tank filler caps embodying this principle have become commercially available and are being used in many aircraft where filler caps must be located in direct or swept-lightning-strike zones (Zones 1A, 1B, 2A, 2B), and at least one specification, MIL-C-38373A (ASG), has been written prescribing a lightning-protected cap.

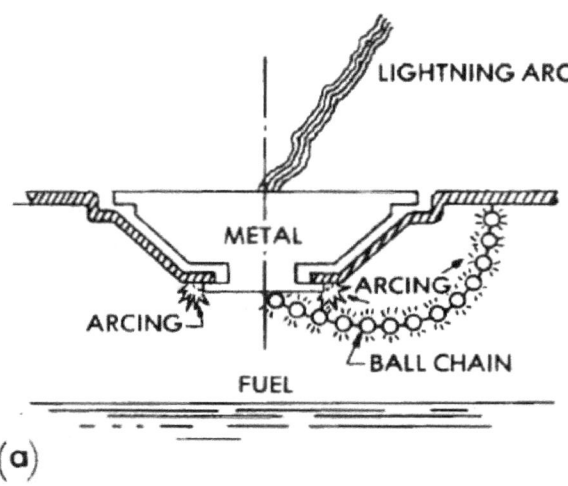

(a)

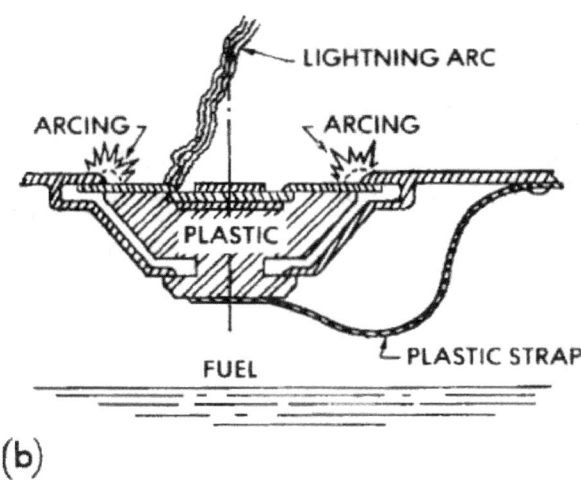

(b)

Figure 6.31 Fuel filler cap designs.
(a) Unprotected
(b) Protected

Lightning-protected caps should always be used if there is *any* possibility that the cap may receive a lightning strike. The consequences of not doing so have been demonstrated by Plumer (Reference 6.56) in tests of a light aircraft nonmetallic wing tip fuel tank equipped with an automotive-type fuel filler cap located in a region where it would be subject to direct and swept strokes. The cap sparked to its grounded adapter even when struck by a very low-amplitude (~5 kA) stroke, igniting a flammable mixture of AVGAS and air. The resulting flame exhausted through a blowout hole that had been cut in the underside of the tank, as shown in Figure 6.32. The same type of cap on the same type of tip tank was also struck during the in-flight incident described in Section 4.5 (Figure 4.16); however, no ignition occurred, presumably because the tank contained sufficient fuel to keep the ullage vapor overly rich, as would be predicted by the AVGAS flammability envelope limits of Figure 6.3.

Figure 6.32 Direct strike to unprotected fuel filler cap resulting in internal sparking and ignition of fuel.

The flammability limits for a particular fuel should not be relied upon, of course, to establish adequate lightning protection because there are factors—such as fuel quantity, misting, and venting—which may cause the actual conditions prevailing in a particular tank to deviate from the limits shown in Figure 6.2 and Figure 6.3. Major emphasis should always be placed upon

162

eliminating possible sources of ignition, regardless of the expected flammability conditions in the tank.

If any doubt about the protection capability of the particular design exists, the cap should be tested as described in the SAE report *Lightning Test Waveforms and Techniques for Aerospace Vehicles and Hardware.* (Reference 6.57).

6.8.1.2 Lightning-Protected Access Doors

Newman also reported sparking at access door interfaces when the mating surfaces had O-ring seals or phenolic or anodized clamp ring surfaces which were electrically insulating. It was found (Reference 6.29) that the application of a *Boeing Service Bulletin* modification (SB-1955) successfully prevented this sparking when the access doors were directly struck. The modification consisted of the following:

(1) Removing phenolic and anodized coatings from the clamp ring
(2) Removing anodized coating from faying surfaces of the access door
(3) Sealing of the door to the surrounding skin adapter with conductive grease consisting of 35% by weight of aluminum powder and 65% by weight of Aeroshell 14 grease.

Other modifications in access door seal design have proved equally effective in preventing sparking. In some cases anodized clamp rings or insulating gaskets have been acceptable when sufficient metal-to-metal conductivity exists via the bolts or fasteners alone. The door requiring the SB-1955 modification described above to resist sparking had only 23 bolts, whereas other doors having over 40 bolts have been found to resist sparking even if insulating finishes or paints such as anodize or zinc chromate remain on the mating surfaces.

A recent survey of different manufacturers' access door designs showed a wide range of different materials and sealing practices in use and relatively little documented lightning test data upon which to judge their adequacy. If access doors are located in Zones 1 or 2, where direct or swept strokes may attach directly to them, they should be tested with direct strokes in accordance with the SAE report *Lightning Test Waveforms and Techniques for Aerospace Vehicles and Hardware* (See Reference 6.57). Even if these doors are not located in direct or swept-stroke zones, they are likely to be involved in conducting some lightning current, since this current passes along the aircraft surface between other attachment points. This report also describes appropriate tests to be performed (for Zone 3) to evaluate this possibility.

The importance of adequate electrical conductivity between access doors and their surroundings can not be overemphasized, especially in cases where the access door is itself a large part of the aircraft skin and where this skin encloses fuel. Such a case is illustrated in the small fighter aircraft of Figure 6.33, in which several large access doors cover a fuselage fuel tank. The doors cover a large area and are located in Zone 2A, a zone that requires them to be designed to safely accept strokes up to 100 kA, 0.25×10^6 A^2-s.

163

FUEL
TANKS

Figure 6.33 Large access doors in Zone 2A over fuselage fuel.
● Swept-lightning strokes will attach to these doors.

Adequate electrical conductivity and protection can not be verified by measuring the dc resistance between the door and the surrounding airframe because this resistance may change under high-current stress, permitting a voltage rise, and also because the *inductance* of current paths through fasteners is equally capable of causing the voltage rises that cause sparking. Structural bonds that do not spark have often had dc resistances below 2.5 mΩ, but there is no universal correlation between this resistance and the degree of sparking. It is, in fact, quite possible that a joint with less than 2.5 mΩ of resistance may nevertheless have enough inductance to cause a spark.

Some guidelines to follow in designing a protected door include the following:

1. Provide as much bare-metal-to-bare-metal contact as possible.
2. If bare-metal-to-bare-metal contact at the mating or faying surfaces is not practical because of corrosion problems, provide as much bare metal contact via screws or fasteners as possible, and make the current paths through these fasteners as short as possible.
3. Seal door with conductive greases or conductive gaskets if possible.

The adequacy of any design should be verified by performing the simulated lightning tests described in the SAE report *Lightning Test Waveforms and Techniques for Aerospace Vehicles and Hardware.*

164

6.8.2 Joints Between Skins, Spars, and Ribs

The large numbers of rivets or fasteners typically employed to attach a metal skin to a metal rib, spar, stiffener, or other internal structural element have usually also provided adequate electrical conductivity for lightning currents. The most common situations found are those where both mating surfaces are painted with an insulating zinc chromate paint and sometimes further insulated with environmental sealers, as shown in Figure 6.34. There are, however, usually enough rivets or fasteners which make bare metal contact with the mating surfaces to provide the necessary electrical conductivity for lightning currents. High-density patterns of rivets or fasteners, such as are commonly utilized to join fuel tank skins to stringers, ribs, and spars, have been found to be capable of conducting 200 kA stroke currents, even when nonconductive primers and sealants are present between the surfaces, as in Figure 6.34 (Reference 6.58). Short, for example, reports 200 kA stroke tests of skin-to-stringer joint samples fastened with double rows of taper-lock fasteners. The sample tested had 16 fasteners on each end of the joint, and no sparking was detected anywhere on the joint (Reference 6.59).

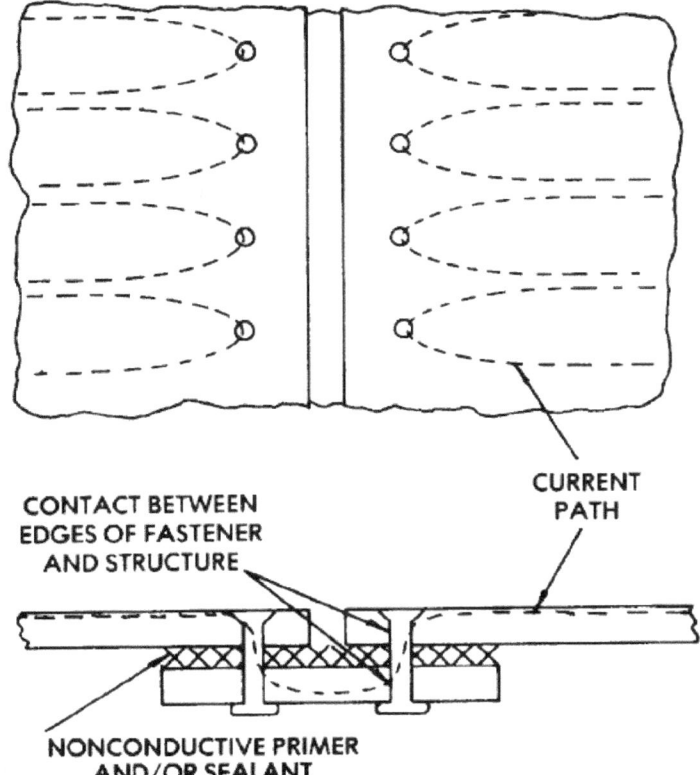

CONTACT BETWEEN
EDGES OF FASTENER
AND STRUCTURE

CURRENT
PATH

NONCONDUCTIVE PRIMER
AND/OR SEALANT

Figure 6.34 Bonding through mechanical fasteners.

165

There is no hard-and-fast rule for the number of fasteners per meter of joint (density) which are necessary to avoid sparking. After all, a single No. 8 AWG copper or aluminum wire is capable of conducting 200 kA, 2×10^6 A^2-s strokes, so a single fastener of this or larger diameter would also, by itself, be capable of conducting this much current. Any sparking usually occurs at the interfaces between the fastener and surrounding metal, and the occurrence of such sparking depends on other physical characteristics, such as skin metal thickness, surface coatings, and fastener tightness. Tests in which simulated lightning currents are conducted through the joint should always be made on samples of joints involving new materials or designs. The type and amount of lightning current to be applied for these tests and diagnostics to be used, should be as described in *Lightning Test Waveforms and Techniques for Aerospace Vehicles and Hardware* for Zone 3.

The foregoing relates to the case where the joint is in Zone 3 and must conduct some portion of the lightning current, but other zones can be, and are likely to be, struck directly by the lightning arc itself. In direct and swept-stroke zones (1A, 2A) and hang-on zones (1B, 2B), it is possible for the lightning arc to attach directly to a single fastener or rivet. If it persists, the hot arc may melt or otherwise damage surrounding skin. Figure 6.35 (Reference 6.60) shows an example of the damage that can be done at such a place. Here, a 1 cm-diameter rivet head and much of its surrounding aluminum skin have melted away. If an arc persists long enough at such a spot, it may melt completely through the skin, beyond the inside member. Such penetration occurred at the point shown in Figure 6.35. The danger involved may require that, if ignitable fuel vapors can exist beneath any joint in a swept-stroke zone (1A, 2A), the skin may have to be thicker than that required on the basis of swept-stroke dwell times only; or some other provision may have to be made to prevent complete meltthrough near a rivet.

Some other guidelines which should be noted in designing integral tank joints are as follows:

1. *Avoid use of anodizing* in joints which may have to carry lightning currents. Anodizing is usually applied to prevent corrosion and results in a hard oxide surface which is nonconductive and difficult to break down at less than about 300 V. Thus, when breakdown eventually does occur, the resulting spark will be more energetic than if a poorer insulator were present. The best coating of all for lightning protection is, of course, *no* coating.

2. *Design short, direct lightning current flow paths* because lightning currents want to flow in straight, direct paths between entry and exit points on the aircraft. Voltages which may cause sparking will build up wherever diversions in these paths exist. The diversions through the fasteners and stringer of Figure 6.34 are acceptable. More extended paths may not be.

3. *Do not depend on resistance measurements* to confirm the adequacy of lightning current conductivity in a joint. The inductance of the lightning current flow path through the joint plays an equally important part, but this inductance is rather small (a few hundredths of a microhenry) and difficult to measure. Resistance measurements (ac or dc) may be useful as a production quality

control tool, but they are not useful for establishing the adequacy of the lightning current path through the joint.

4. *Account for aging and mechanical stress* which may cause reduced electrical conductivity. Continued flexing of structures under flight load conditions may eventually loosen a joint to the point where sparking could occur. To evaluate this possibility, perform simulated lightning tests on joint samples previously given fatigue or environmental tests.

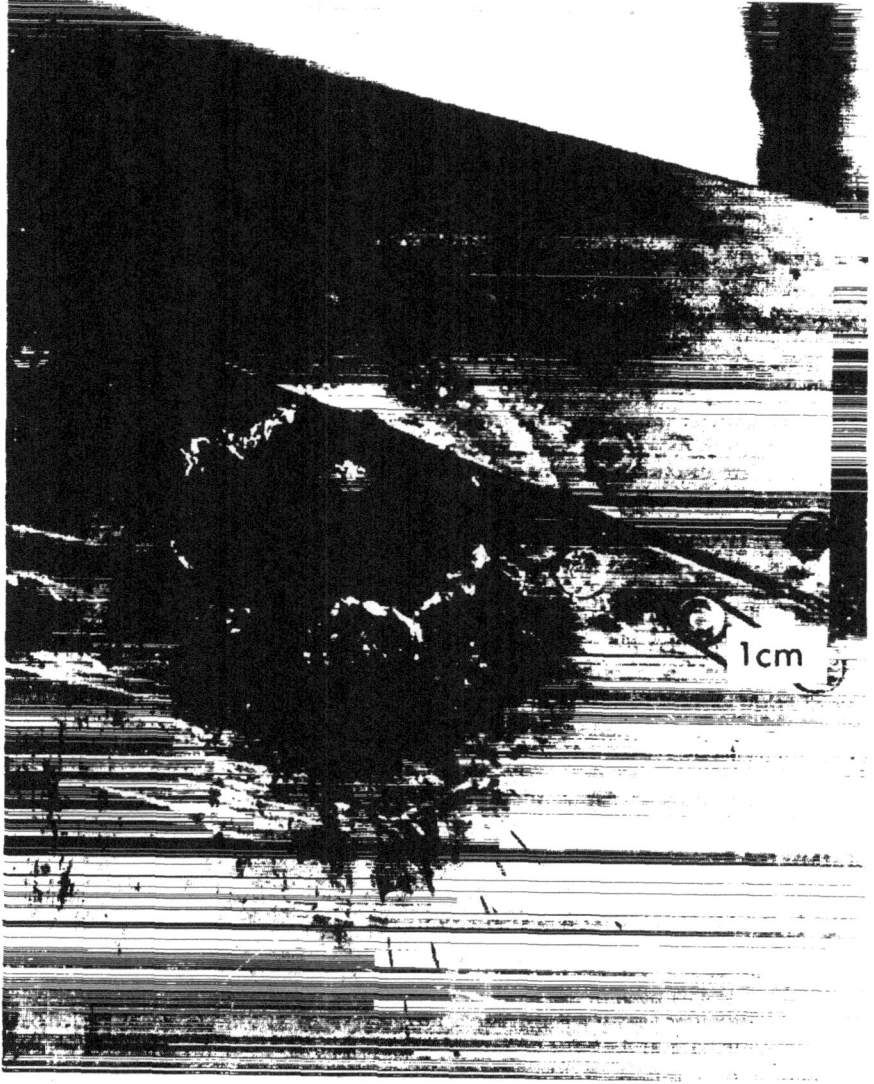

Figure 6.35 Typical lightning damage to a rivet.
● Meltthrough to inside beyond rib surface.

167

6.8.3   Joints and Interfaces in Pipes and Couplings

Because they are metallic and come in contact with structure, the myriad of fuel, hydraulic and vent pipes, and other plumbing inside a fuel tank will have to conduct some, however small, amounts of current when lightning currents are flowing through the structure containing them. This current may cause sparking at pipe joints and couplings where there is intermittent or poor electrical conductivity. Some pipe couplings, for example, are designed to permit relative motion between the mating ends of a pipe so as to relieve mechanical stresses caused by wing flexure and vibration. Also, anodized or other electrically insulating coatings are often applied to the pipe ends and couplings to control corrosion. Relative motion and vibration may wear this insulation away, providing unintentional and intermittent conductive paths. To date, almost no quantitative information has been developed to describe the amounts of current or voltage necessary to cause a spark, nor has this possibility been related to the design or condition of a particular system. Yet this sparking may have been a factor in the cause of the several in-flight explosions associated with lightning strikes cited earlier in this book. This aspect should, therefore, be given particular attention in the design of aircraft fuel systems.

6.8.3.1   Lightning Currents in Plumbing

Because of their relatively short duration, the high-amplitude, short-duration stroke currents will not have time to spread very deeply into interior structural elements or other interior conductors. Most of this current will remain in metallic skins, but this current does begin immediately to diffuse into underlying metallic structure. Appreciable current levels may be reached in spars, stringers, pipes, and other internal conductors if the lightning stroke persists long enough.

The basic penetration, or current *redistribution*, process is described in Chapter 8, and it is shown that penetration times for typical aircraft skins are of the order of several tens of microseconds. If the stroke current were to remain at its peak amplitude for a time longer than this, dc conditions would be approached and the current would be distributed resistively, with a significant portion of it flowing in the internal structure. Fortunately, the stroke current does not remain at its peak level for more than a few microseconds. The amount of this stroke current which, under typical lightning-stroke conditions, can actually flow on internal pipes is not yet known, and determination of the magnitude of this current cannot yet be done confidently by analysis. Laboratory tests in which simulated lightning currents are forced through complete fuel tank structures can be performed, however, and the current in particular pipelines can be measured.

Intermediate and continuing currents do persist for times long enough for the penetration process to be completed, and it is possible to estimate, as follows, the amount of *intermediate* current that may flow in a pipe after dc conditions are reached.

Assume that the leading and trailing edge sections of the wing in Figure 6.30 are nonconductive or sufficiently isolated as to be unavailable for conduction and that the remaining wing box comprised of skins and spars has chordwise dimensions as shown in Figure 6.36. The cross-sectional area of the spars and skins forming this box is 135 cm^2. The tank also contains a 0.05 cm I.D., 10 cm O.D. aluminum vent pipe which is electrically bonded to the structure at each end of the tank. The cross-sectional area of this tube is 1.57 cm^2.

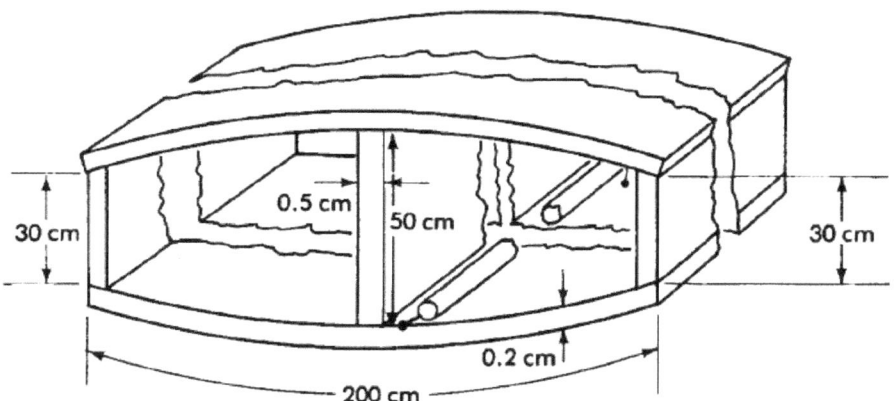

Figure 6.36 Hypothetical wing box with integral fuel tank.

Assuming an intermediate stroke whose average amplitude is 2 000 A for a duration of 5 ms (in accordance with component B of the SAE report *Lightning Test Waveforms and Techniques for Aerospace Vehicles and Hardware*) the current in the pipe can be calculated as follows:

$$I_{tube} \sim \frac{1.57 \text{ cm}}{137 \text{ cm}} = (2\ 000 \text{ A}) = 22.9 \text{ A} \tag{6.6}$$

This current would flow for as long as the lightning intermediate current flows, which is 5 ms if component B is assumed to be representative. The ability of currents of the order of 20 A to cause a spark at pipe interfaces and couplings presently used in aircraft is not known, but the sparking that commonly occurs at similar current levels between other conductors in relative motion (such as battery clamps and motor commutators) raises a distinct possibility of sparking at couplings where relative motion and/or poor electrical conductivity exists.

Electrical bond straps are sometimes installed across poorly conducting pipe couplings, as shown in Figure 6.37.

These bond straps are usually intended to equalize static electrical charges such as those carried into the tank during refueling operations.

169

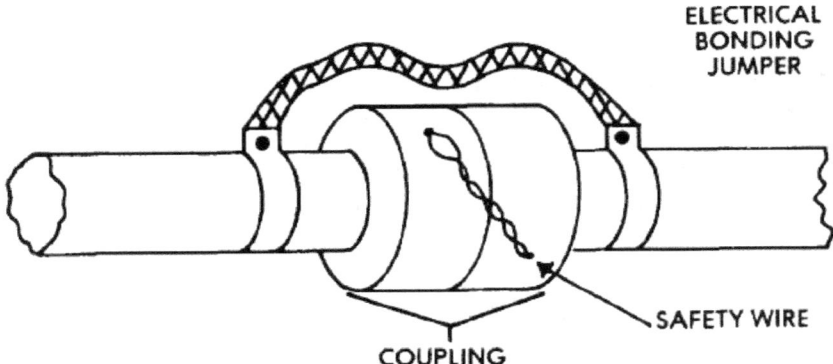

ELECTRICAL
BONDING
JUMPER

SAFETY WIRE

COUPLING

Figure 6.37 Electrical bonding jumper across insulated coupling.

BONDING JUMPERS SUCH AS THOSE SHOWN IN FIGURE 6.37 SHOULD NOT BE RELIED UPON TO PREVENT SPARKING FROM LIGHTNING CURRENTS!

The reason that jumpers like this are unsatisfactory is that, when fast-changing lightning currents begin to flow in these jumpers, voltage rises occur along the inductance of the strap equal to

$$E_{jumper} = L_{jumper} \frac{dI_{jumper}}{dt} \qquad (6.7)$$

where
E = inductance voltage (volts)
L = bond strap inductance (henries)
I = current in tube (amperes)

This voltage appears across all other possible current paths through the coupling. If there exists another possible but unintentional path, such as a safety wire, a retaining chain, or the coupling interface itself, this inductive voltage may be sufficient to force current through these paths and cause sparking if the conductivity is poor or intermittent. For example, if the 20 A portion of the *intermediate current* were to begin to flow in the bonding jumper of Figure 6.37 and if this current were to reach its peak in 100 μs, its rate of rise would be

$$\frac{dI_{jumper}}{dt} = \frac{20 \text{ A}}{1 \times 10^{-4} \text{ s}} = 200\,000 \text{ A/s} \qquad (6.8)$$

If the inductance of the bonding jumper is 0.1 μH, then the inductive voltage developed along the jumper would be

$$E_{jumper} = L_{jumper} \frac{dI_{jumper}}{dt} \qquad (6.9)$$

$$= (0.1 \times 10^{-6} \text{H}) (0.2 \times 10^{6} \text{ A/s}) = 0.02 \text{ V}$$

By itself this voltage is too low to break down coupling insulation such as

170

anodized or zinc chromate primer, but if it is applied across preexisting surfaces where these treatments have been worn through, this voltage may be sufficient to cause current flow and sparking.

As stated earlier, the fraction of *stroke current* which enters a bonding jumper and the inductive voltage *it* produces are not known. It is quite probable that at least some stroke currents would produce even higher rates of current rise in the bonding jumpers than do the intermediate currents.

6.8.3.2 Conductivity through Couplings and Interfaces

Here again is an area where little electrical data is available. The extensive use of anodized coatings to provide noncorrosive mating surfaces in pipe couplings would seem to preclude arcing across the pipe interface, but relative motion between these surfaces can wear through the anodized coating, causing a conductive path. If there happens to be bare metal-to-metal contact within the coupling, this would probably represent the shortest path, and it is altogether possible that no sparking would occur anywhere else in the coupling. A slight change in the relative position of the mating surfaces or in some other aspect, such as the tightness of its threaded assembly or introduction of dirt or residue, may drastically change the electrical conductivity of a coupling. In fact, it is probable that the electrical conductivity of a typical pipe coupling changes many times during a flight as a result of relative motion caused by structural vibrations and flexing.

In the absence of definitive data on the electrical conductivity of pipe couplings under in-service conditions, it is advisable to take the following approach:

1.   Determine the fraction of lightning current expected to flow in a particular pipe, by analyses similar to the analysis of Section 6.8.3.1, or, if available, a more authentic means.

2.   Inject this expected current through a sample of the desired coupling, under simulated in-flight vibration and contamination conditions.

3.   Perform this test in a darkened enclosure to enable inspection for sparks. Repeat the test until a reliable result is established.

If sparking does occur in a coupling, a redesign is likely to be required for assured prevention; however, it *may* be possible to provide adequate conductivity with the addition of two or more short, wide braided jumpers or a braided sleeve, as shown in Figure 6.38(a) and (b).

Alternatively, completely insulated couplings might be designed. If such an insulated coupling were used, a jumper as shown in Figure 6.38(c) could be used to conduct lightning currents and static charges, provided the length of the path through the jumper is not much longer than the path through the air along the surface or through the insulated coupler.

Before using any one of these or any other protective method that has to be added to an existing coupling design, its adequacy should be verified by impulse current tests to check for sparking.

171

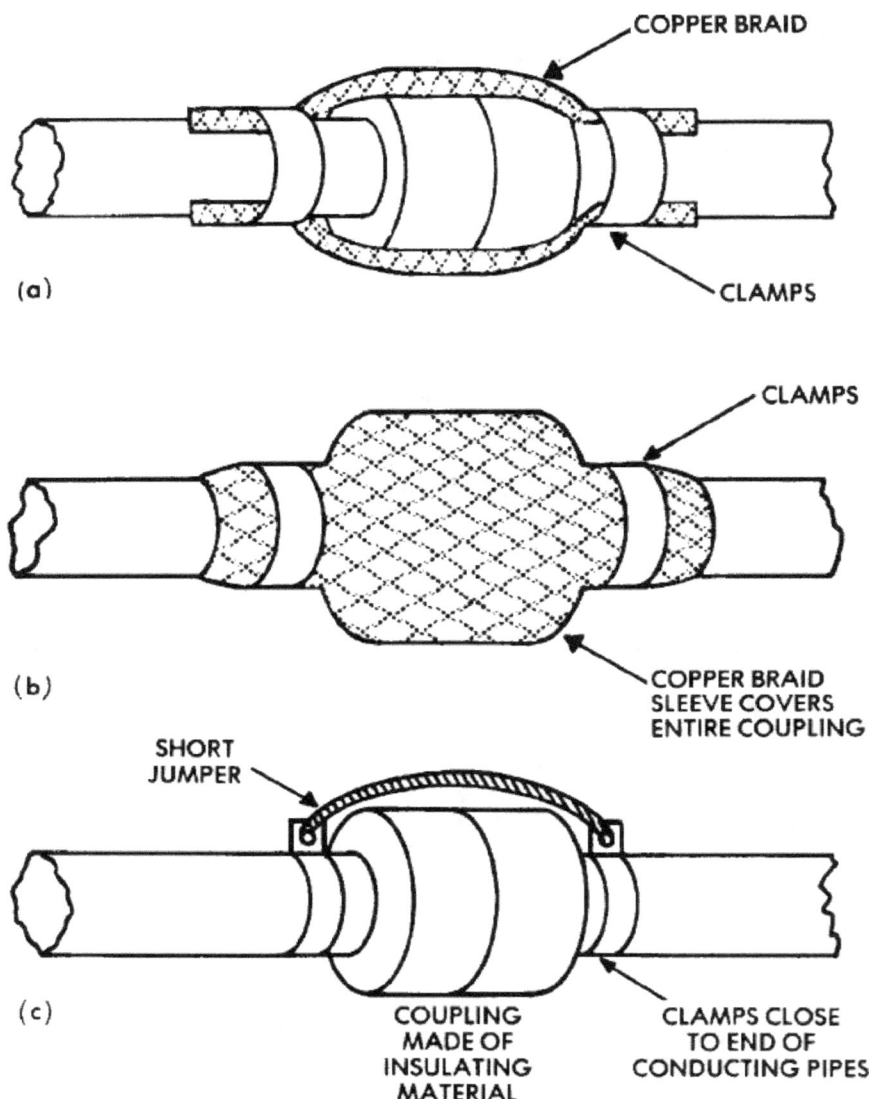

Figure 6.38 Possible methods to reduce sparking at pipe couplings.

6.8.4 Electrical Wiring in Fuel Tanks

Another result of lightning current flow through a structure is that the fields it produces may induce voltages in electrical wiring. If this wiring enters a fuel tank, the possibility of these induced voltages being high enough to cause a spark in the fuel tank is a matter for concern.

Electrical wires found inside fuel tanks are typically those that run to

172

capacitance-type fuel quantity probes or small electric motors which operate pumps or valves. If these wires are totally enclosed by metal skins and ribs or spars, the internal magnetic fields and induced voltages will be relatively low. Electrical devices, such as fuel probes, have been intentionally designed to withstand comparatively higher voltages without sparking. The fuel system designer, however, must be continually alert for structural design or materials changes that may alter this situation and permit excessive induced voltages to appear in fuel tank electrical circuits.

A number of measurements have been made of the voltages which simulated lightning currents may induce in fuel probe wiring found in conventional aircraft. Measurements have also been made of the voltages required to cause a spark between the elements of typical capacitance-type fuel quantity probes. The sparkover voltages have always been found to be much higher than those found to be induced in the wiring. Newman, Robb, and Stahmann (Reference 6.61) found that at least 3 000 V of dc voltage were required to cause a spark between the active and grounded cylinders of a capacitance-type fuel probe found in a KC-135 aircraft wing, and that even higher voltages were needed to spark over the other gaps in this probe. Plumer, (Reference 6.62) ran a similar test using the 1.2 x 50 μs impulse voltage waveform specified as Voltage Waveform C of SAE Report *Lightning Test Waveforms and Techniques for Aerospace Vehicles and Hardware* and found that 12 kV were required to cause a spark between the inner and outer cylinders of the probe. Small gaps like this will withstand more impulse than dc voltage, and the impulse test more realistically represents an induced voltage.

The tests reported above were made at atmospheric pressure on the ground; the same gap would break down at a lower voltage when at the lower air pressures associated with flight altitudes. The amount of reduction to be expected in impulse sparkover voltage at various altitudes can be determined from the curves of Figure 6.39(a) and (b) (Reference 6.63). The product, pd, of pressure (mm Hg) and gap distance (mm) is first determined by obtaining pressure at altitude from the chart of Figure 6.39(c). If it is desired to find the sparkover voltage of a 5 mm gap at 10 000 m altitude, for example, the pressure at this altitude is first determined from the chart to be 198.16 mm Hg. The product, pd, would then be (5 mm) (198.16 mm Hg) = 990.8 mm Hg mm. From Figure 6.39(b), the gap would spark over at about 5500 V.

The sparkover voltages of Figures 6.39(a) and (b) should be considered approximate, since these curves were determined for plane parallel electrodes only. Other electrode configurations, such as a sharp point-to-plane, would spark at voltages perhaps 25% more or less than these. Impulse sparkover tests should be made on particular gap configurations to determine the actual sparkover voltages.

Contaminants on the electrode surfaces may also act to reduce the breakdown voltage; therefore, it is wise in any case to reduce voltage levels measured by laboratory tests by a factor of at least 2 when assigning a withstand voltage to a fuel probe or other electrical gap that might appear in a fuel tank.

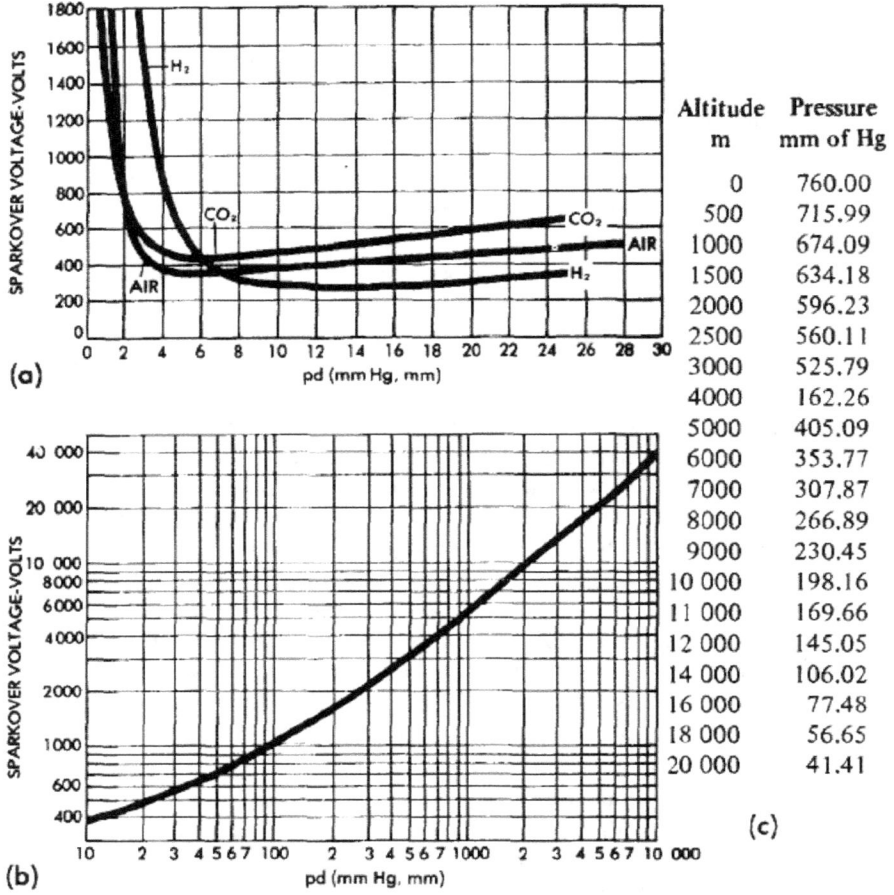

(a)

(b)

Altitude m	Pressure mm of Hg
0	760.00
500	715.99
1000	674.09
1500	634.18
2000	596.23
2500	560.11
3000	525.79
4000	162.26
5000	405.09
6000	353.77
7000	307.87
8000	266.89
9000	230.45
10 000	198.16
11 000	169.66
12 000	145.05
14 000	106.02
16 000	77.48
18 000	56.65
20 000	41.41

(c)

Figure 6.39 Sparkover voltages and altitude pressures.

When determining the breakdown voltages of an object such as a fuel quantity probe, care must be taken to apply test voltage to all of the electrode combinations which may exist (this need not all be done simultaneously). In most cases both the "high" and "low" electrodes in such a device are insulated from "ground" (the airframe), as shown in the simplified drawing of Figure 6.40, but there is always a mounting bracket which brings the probe in proximity to the airframe. Thus, the test voltage should be applied across each of the basic electrode combinations shown in Figure 6.40. Usually, the gaps to the mounting bracket or airframe will be larger than the others and require a higher voltage to cause sparkover. Typical impulse sparkover voltages to be expected across these gaps, at a flight altitude of 3000 m, might be as shown in Table 6.7, where the "high" and "low" electrodes are defined as shown in Figure 6.40.

174

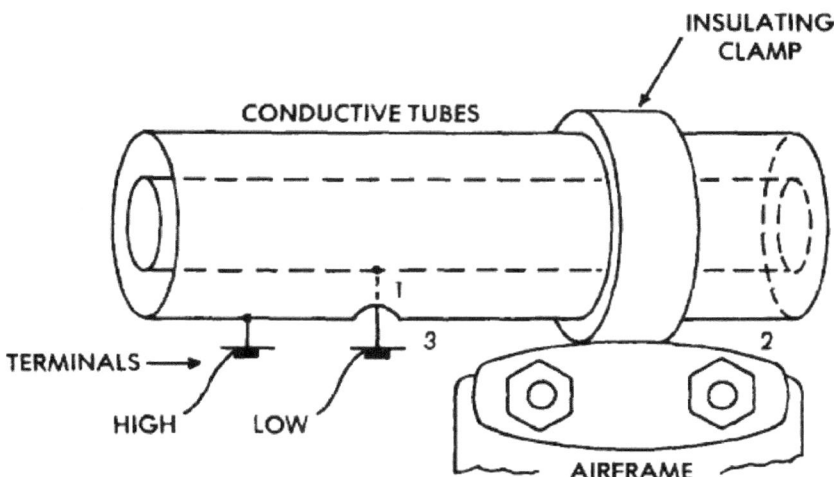

Figure 6.40 Possible breakdown of gaps in capacitance-type probe.
1. High to Low
2. High to Airframe
3. Low to Airframe

**Table 6.7   TYPICAL IMPULSE SPARKOVER VOLTAGES
FOR A CAPACITANCE-TYPE PROBE**

Figure 6.40 Electrode Pair	Sparkover Voltage at 3 000 m Altitude
High to Low	3 kV
High to Airframe	6 kV
Low to Airframe	8 kV

Figure 6.40 illustrates that the proximity of the probe to the airframe structure on which it is mounted is very important in determining breakdown voltages to the airframe. Because of this, design of adequate insulation between the active elements of the probe and the airframe may not be entirely within the probe designer's control.

CARE MUST BE TAKEN BY THE FUEL SYSTEM DESIGNER TO SEE THAT ADEQUATE GAPS ARE MAINTAINED BETWEEN THE ACTIVE ELECTRODES OF THE PROBE AND THE AIRFRAME.

It is particularly important that sufficient insulation be provided between

175

active electrodes and the airframe because the highest induced voltages usually appear between all of the incoming wires and the airframe. The reason is illustrated in Figure 6.41. The wires are most closely referenced to the airframe at the electronics end, which is usually in the fuselage. If one or more of the wires are shielded, usually only the electronics ends of the shields are grounded to the airframe because grounding of both ends permits stray ac fields to induce circulating currents in these shields. These currents are a source of steady state electromagnetic interference (EMI) sufficient to interfere with the operation of the fuel quantity electronics. This EMI has been avoided by leaving the shields *ungrounded* at the fuel tank end.

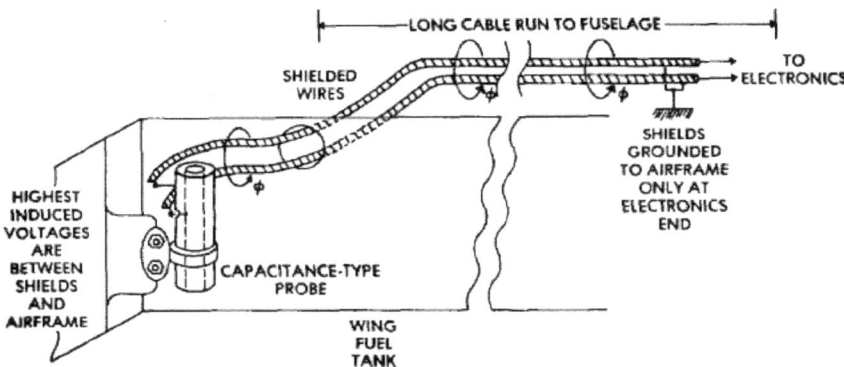

Figure 6.41 Typical fuel probe wiring.
- Most magnetic flux and induced voltage appears between either wire (or shield) and the airframe at probe end.
- Less flux and induced voltage exists between any two wires.
- Even less flux and voltage exists between a wire and its shield.

The performance of shielded circuits within magnetic fields is discussed in Chapter 13, Section 13.2, where it is shown that a shield grounded at only one end actually provides no magnetic shielding of an enclosed wire whose return path (the airframe) is outside of the shield. If a shield is grounded at only one end, the magnetic flux encircled in the loop this shield forms with the airframe will induce a voltage in this loop of

$$E = -\frac{d\phi}{dt} \tag{6.10}$$

where ϕ = magnetic flux

Most of this induced voltage will appear between the open end of the shield and the airframe. The same field will induce a similar voltage between the enclosed wire and the airframe. Only if the shield is grounded to the airframe at both ends will it prevent these voltages from being induced. If the ungrounded end of the shield is in a fuel tank, the highest induced voltages will appear there.

176

Induced voltages will also appear *between* two or more wires or shields at the probe end, but any one of these wire-to-wire voltages will usually be less than the voltage between either conductor and the airframe because the area between wires in a cable bundle is usually less than the area between any one of them and the airframe.

The routing of the fuel probe wires, therefore, has a lot to do with how much induced voltage appears at apparatus inside fuel tanks. Since fuel tanks are often completely enclosed with metallic walls which are electrically bonded and well sealed, the magnetic fields from lightning (or any other source) are likely to be lowest inside the tanks. Thus, wires that must eventually enter the tanks should be routed through them from the start. Newman, Robb, and Stahmann, for example, have measured up to 11 V, or less, in the fuel quantity probe circuits routed inside a KC-135 wing fuel tank (Reference 6.64) through which 50 kA strokes with a 50 kA/μs rate of rise were being conducted. This voltage would extrapolate to a maximum of 44 V at a severe stroke amplitude of 200 kA. Plumer (Reference 6.65) measured up to 4 V in a fuel probe circuit routed inside the tanks of an F-89J wing. The simulated lightning current in this case was 40 kA, with a rate of rise of 8 kA/μs. From these results Plumer predicted that a 200 kA, 100 kA/μs lightning stroke would induce 185 V in the same circuit. At the time these measurements were made, most researchers looked for induced voltages of the same duration as that of the lightning current and paid little attention to other signals, suspected to be "noise," appearing within the first microsecond of an induced voltage trace on an oscilloscope. These "noise" signals, however, often had higher amplitudes than the more familiar, longer duration voltages considered to be authentic. With the advent of better instrumentation, these high-frequency voltages became known to be legitimate, induced perhaps by the higher frequency traveling wave currents that may appear in the airframe as a result of the mechanism described in Chapter 8, Section 8.2. Thus, higher voltages than those reported by Newman and Plumer, may in fact be induced in fuel quantity probe circuits, although the literature shows no recent measurements of induced voltages in fuel tank electrical circuits to establish what the amplitudes of these voltages might actually be.

If it is not possible to route wiring inside the fuel tank, it may be routed outside of a tank as pictured in Figures 6.30 and 6.41; but, if this is done, great care must be taken to minimize the amount of magnetic flux that may link these wires or their shields. This precaution is particularly pertinent for wings where wires are routed within leading and trailing edges, which are often exposed to higher magnetic field levels than elsewhere. Fiberglass has replaced aluminum in some leading edge skins, and, in some aircraft, control surface openings allow magnetic flux to penetrate directly to the aft wing spar, along which electrical cables are often routed.

If fuel system wiring must be routed outside of the tanks, there are several methods that can be used to control induced voltages, often to levels far below those required to cause a spark at a fuel probe. These involve routing improvements, grounding of the shields to the airframe at both ends, and use of an overall shield grounded at both ends to enclose a bundle of individually shielded wires whose shields are left ungrounded at one end for EMI purposes. A relative comparison of six possible wiring methods is shown in Figure 6.42. It is recommended that methods a and b of this figure *not* be used, and that c and d be used only if the (leading or trailing edge) enclosure has an electrically conductive skin or coating that can provide some electromagnetic shielding against the outside world. Methods e and f are preferable because they will provide the lowest induced voltages at the probe. Method f is to be used if the individual shields must be left ungrounded at one end for EMI control purposes. The comparisons shown in Figure 6.42 apply also for wiring to fuel tanks in other parts of the aircraft.

For whatever method is used, assurance should be obtained that the expected induced voltages are lower by a factor of 10 to 1 than the probe sparkover voltage at altitude. An estimate of the induced voltages to be expected in the fuel tank wiring can be obtained by analysis as described in Chapter 12 or by lightning transient analysis tests, as described in Chapter 17.

6.9 Fuel System Checklist

A considerable amount of information on various aspects of fuel systems has been presented in the preceding paragraphs. With reference to these paragraphs, the following checklist is provided to summarize key points and to focus the designer's attention on the important aspects of fuel system lightning protection:

1. Are the fuel tanks located in Zones 1 and 2, where lightning strokes may directly contact their skins? (Chapter 5, Sections 5.2 and 5.3) If so,
    a. What is the expected arc dwell time on these skins? (Section 6.6)
    b. Are the skins thick enough to avoid meltthrough? (Section 6.6)
    c. Can a lightning arc attach directly to an access door or filler cap?; if so, can the lightning current be conducted into the surrounding skin without sparking inside the tank? (Section 6.8)
2. Are there "dry bay" areas into which fuel may leak from adjoining tanks or plumbing? If so, these are also subject to questions 1a, 1b, and 1c.
3. Are fuel vent outlets or jettison pipes located in direct strike zones (Zones 1A or 1B) or swept-stroke zones (Zones 2A or 2B) where the arc may attach or sweep close to the vent outlet? (Chapter 5, Sections 5.2 and 5.3; Sections 6.4 and 6.5)
    a. If so, is an effective flame arrester or surge tank protection (STP) system used?; has its effectiveness been verified by test? (Section 6.4)

178

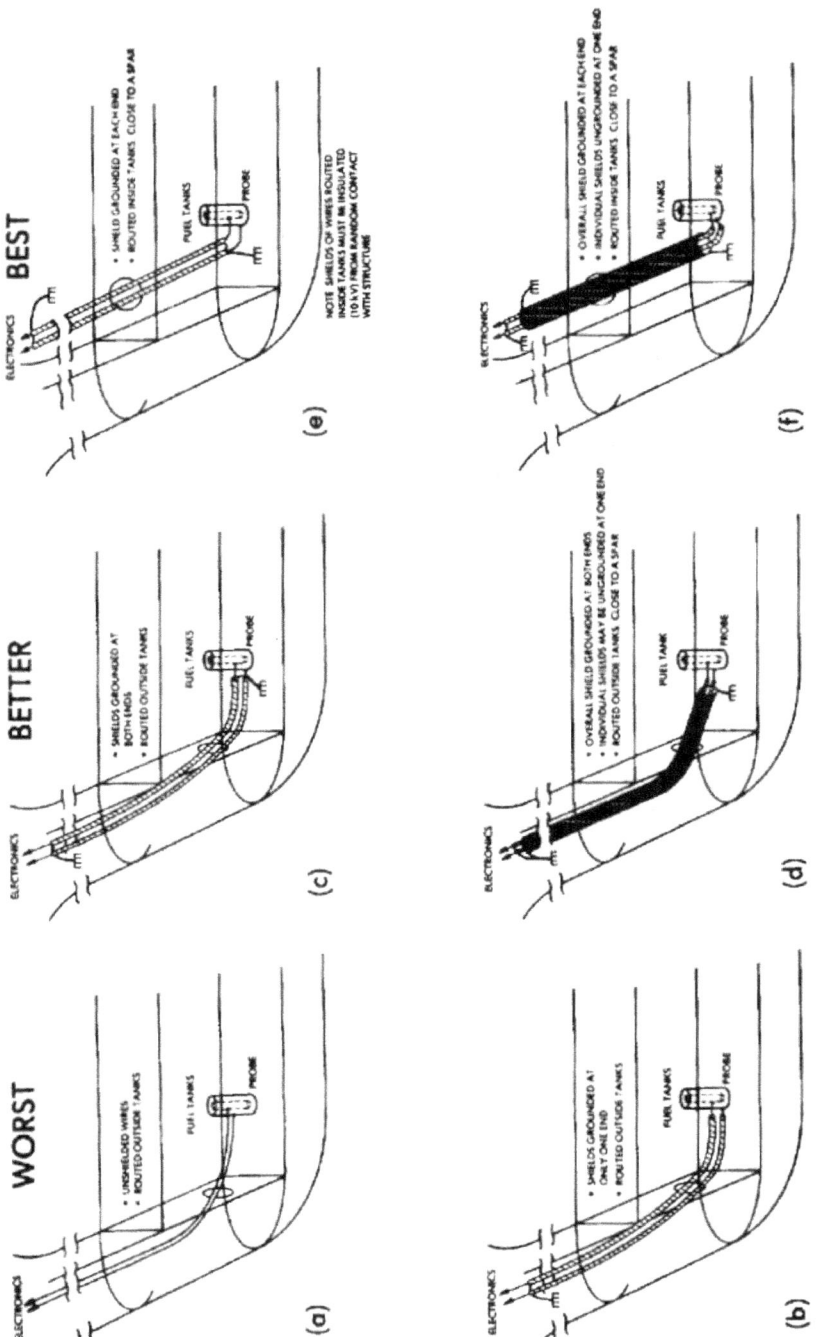

Figure 6.42 Relative comparison of six system wiring methods.

179

b. Is the response time of the STP system shorter than the possible flame propagation time from the STP sensor to the extinguisher? (Section 6.4)

c. Is the STP system protected from false trips resulting from lightning-induced voltages in its electrical wiring or light from nearby flashes?

4. If nonmetallic skins are used, is an adequate diverter or conductive coating system that will prevent skin puncture or internal streamering provided? (Section 6.7)

5. Is the fuel tank structure capable of *conducting* Zone 3 lightning currents even if the tank itself is not located in a direct or swept-stroke zone? Has this been demonstrated by a test in which simulated lightning currents are conducted through a complete tank structure? (Section 6.8)

6. If a simulated lightning test of the complete tank assembly is not feasible, have all of its individual joints, seams, access doors, filler caps, drains, vents, plumbing, and electrical systems been tested for their ability to conduct simulated lightning currents and found to be free of sparking? (Sections 6.5 and 6.8)

7. Are electric circuits entering the fuel tanks protected against high induced voltages? (Section 6.8) Have they been routed away from other wiring, such as navigation lamp circuits, which may be susceptible?

8. Are clearances between exposed electrical parts and each other or between any of them and the airframe sufficient to withstand the induced voltages that may occur without sparking? (Section 6.8)

9. Have the electrical circuits inside nonmetallic tanks been adequately shielded so that excessive induced voltages will not occur? (Section 6.8)

10. Have applicable government airworthiness certification regulations, military standards, or other specifications pertaining to lightning protection been adhered to? (Section 5.4)

REFERENCES

6.1 L. J. Nestor, *Investigation of Turbine Fuel Flammability within Aircraft Fuel Tanks,* Final Report DS-67-7, prepared by the Aeronautical Engine Department, Naval Air Propulsion Test Center, for the Engineering and Safety Division, Aircraft Development Service, Federal Aviation Administration, Department of Transportation, Washington, D.C. (July 1967).

6.2 Adapted from Nestor, *Investigation of Turbine Fuel Flammability,* p. 8.

6.3 Adapted from Nestor, *Investigation of Turbine Fuel Flammability,* pp. 12, 13.

6.4 W. Q. Brookley, "USAF C-141 and C-135 Fuel Tank Nitrogen Inerting Tests," *Report of the Second Conference on Fuel System Fire Safety, 6 and 7 May 1970,* Federal Aviation Administration, Department of Transportation, Washington, D.C. (1970), p. 75.

6.5 Adapted from Nestor, *Investigation of Turbine Fuel Flammability,* p. 23.

6.6 Brookley, "USAF C-141 and C-135 Fuel," pp. 75-82.

6.7 *Investigations of Mechanisms of Potential Aircraft Fuel Tank Vent Fires and Explosions Caused by Atmospheric Electricity,* Final Report under Contract No. NASr-59, prepared by the Lockheed-California Company, for the National Aeronautics and Space Administration, Federal Aviation Administration, Department of Transportation, Washington, D.C. (May 31, 1963).

6.8 Redrawn from *Investigations of Mechanisms of Potential Aircraft Fuel Tank Vent Fires,* p. 84.

6.9 Redrawn from *Investigations of Mechanisms of Potential Aircraft Fuel Tank Vent Fires,* p. 90.

6.10 Redrawn from *Investigations of Mechanisms of Potential Aircraft Fuel Tank Vent Fires,* p. 123.

6.11 *Civil Aeronautics Board Aircraft Accident Report, Boeing 707-121, N709PA, Pan American World Airways, Inc., near Elkton, Maryland, December 8, 1963,* File No. 1-0015, adopted February 25, 1965.

6.12 C. C. Bolta, R. Friedman, G. M. Griner, M. Markels, Jr., M. W. Tobriner, and G. von Elbe, *Lightning Protection Measures for Aircraft Fuel Systems, Phase II,* Technical Report ADS-18, prepared by the Atlantic Research Corporation with the Lightning and Transients Research Institute for the Federal Aviation Agency, U.S. Department of Commerce, Washington, D.C. (May 1964).

6.13 Bolta *et al, Lightning Protection Measures,* pp. 39-43.

6.14 Based on Bolta *et al, Lightning Protection Measures,* pp. 22 and B-3.

6.15 Bolta *et al, Lightning Protection Measures,* pp. 70-79 and C-1 to C-3 (excerpted from Proposal No. PS-139, Fenwal, Inc., Ashland, Massachusetts [January 23, 1964]).

6.16 Bolta *et al, Lightning Protection Measures,* pp. 116-149.

6.17 M. M. Newman, J. R. Stahmann, and J. D. Robb, *Airflow Velocity Effects on Lightning Ignition of Aircraft Fuel Vent Efflux,* FAA Final Report DS-67-9, prepared by the Lightning and Transients Research Institute, with Consultant Staff Cooperation of the Atlantic Research Corporation, for the Federal Aviation Administration, Department of Transportation, Washington, D.C. (July 1967).

6.18 Newman, Stahmann, and Robb, *Airflow Velocity Effects*, pp. 10-20.

6.19 Redrawn from F. L. Kester, M. Gerstein, and J. A. Plumer, *A Study of Aircraft Fire Hazards Related to Natural Electrical Phenomena*, NASA CR-1076, prepared by Dynamic Science for the National Aeronautics and Space Administration, Department of Transportation, Washington, D.C. (June 1968), p. 75.

6.20 J. P. Gillis, *Study of Flame Propagation through Aircraft Vent Systems*, Final Report NA-69-32, prepared by Fenwal, Inc., Walter Kidde and Co., for the National Aviation Facilities Experimental Center, Federal Aviation Administration, Department of Transportation, Atlantic City, New Jersey (May 1969).

6.21 Redrawn from Gillis, *Study of Flame Propagation*, pp. 21, 22.

6.22 Gillis, *Study of Flame Propagation*, p. 23.

6.23 Gillis, *Study of Flame Propagation*, p. 35.

6.24 Federal Aviation Agency, Washington, D.C. (October 6, 1967).

6.25 Society of Automotive Engineers, Warrendale, Pennsylvania, (5 May 1976).

6.26 J. H. Hagenguth, "Lightning Stroke Damage to Aircraft," *AIEE Transactions* 68, American Institute of Electrical Engineers, New York, New York (1949): 1-11.

6.27 Hagenguth, "Lightning Stroke Damage," p. 2.

6.28 Redrawn from Hagenguth, "Lightning Stroke Damage," p. 2.

6.29 J. D. Robb, E. L. Hill, M. M. Newman, and J. R. Stahmann, *Lightning Hazards to Aircraft Fuel Tanks*, Technical Note 4326, National Advisory Committee for Aeronautics, Washington, D.C. (September 1958), pp. 12-14.

6.30 *Lightning Test Waveforms and Techniques for Aerospace Vehicles and Hardware*, Society of Automotive Engineers, Warrendale, Pennsylvania (5 May 1976).

6.31 Kester, Gerstein, and Plumer, *A Study of Aircraft Fire Hazards*, p. 40.

6.32 Robb, *et al*, *Lightning Hazards*, p. 9.

6.33 Society of Automotive Engineers-designated zones.

6.34 Kester, Gerstein, and Plumer, *A Study of Aircraft Fire Hazards*, pp. 51-69.

6.35 R. O. Brick, "A Method for Establishing Lightning-Resistance/Skin-Thickness Requirements for Aircraft," *Lightning and Static Electricity Conference, 3-5 December 1968, Part II: Conference Papers*, AFAL-TR-68-290, Air Force Avionics Laboratory, Air Force Systems Command, Wright-Patterson Air Force Base, Ohio (May 1969), pp. 295-317.

6.36 L. L. Oh and S. D. Schneider, "Lightning Strike Performance of Thin Metal Skin," Session III: Structures and Materials, *Proceedings of the 1975 Conference on Lightning and Static Electricity, 14-17 April 1975, at*

Culham Laboratory, England, the Royal Aeronautical Society of London, (1975).

6.37 Redrawn from Oh and Schneider, *Lightning Strike Performance,* p. 12.

6.38 Kester, Gerstein, and Plumer, *A Study of Aircraft Fire Hazards,* p. 39.

6.39 Redrawn from Oh and Schneider, *Lightning Strike Performance,* p. 12.

6.40 Kester, Gerstein, and Plumer, *A Study of Aircraft Fire Hazards,* pp. 55-68.

6.41 Kester, Gerstein, and Plumer, *A Study of Aircraft Fire Hazards,* pp. 57, 61.

6.42 J. A. Plumer, *Guidelines for Lightning Protection of General Aviation Aircraft,* FAA-RD-73-98, Federal Aviation Administration, Department of Transportation, Washington, D.C. (October 1973), p. 11. Data sources include J. Phillpott, "Simulation of Lightning Currents in Relation to Measured Parameters of Natural Lightning," *Proceedings of the 1975 Conference on Lightning and Static Electricity, 14-17 April 1975, at Culham Laboratory, England,* Session I: Fundamental Aspects and Test Criteria, the Royal Aeronautical Society of London, England (1975); and Brick, "A Method for Establishing Lightning-Resistance/Skin-Thickness Requirements."

6.43 Air Force Cambridge Laboratory Photograph from Rough Rider Program.

6.44 R. O. Brick, L. L. Oh, and S. D. Schneider, "The Effects of Lightning Attachment Phenomena on Aircraft Design," *1970 Lightning and Static Electricity Conference, 9-11 December,* sponsored jointly by the Air Force Avionics Laboratory and the Society of Automotive Engineers, Air Force Systems Command, Wright-Patterson Air Force Base, Ohio (December 1970), pp. 139-156.

6.45 Oh and Schneider, "Lightning Strike Performance," pp. 6-7.

6.46 J. A. Plumer, *Lightning Effects on General Aviation Aircraft,* FAA-RD-73-99, Federal Aviation Administration, Department of Transportation, Washington, D.C. (October 1973), pp. 21-44.

6.47 J. D. Robb, J. R. Stahmann, and M. M. Newman, "Recent Developments in Lightning Protection for Aircraft and Helicopters," *1970 Lightning and Static Electricity Conference, 9-11 December,* sponsored jointly by the Air Force Avionics Laboratory and the Society of Automotive Engineers (December 1970), pp. 25-35.

6.48 K. Berger, "Development and Properties of Positive Lightning Flashes at Mount S. Salvatore with a Short View to the Problem of Aviation Protection," *Proceedings of the 1975 Conference on Lightning and Static Electricity, 14-17 April 1975, at Culham Laboratory, England,* Session I: Fundamental Aspects and Test Criteria, the Royal Aeronautical Society of London (1975), p. 7.

6.49 T. C. Kosvic, N. L. Helgerson, and M. Gerstein, *Ignition of Fuel Vapors Beneath Titanium Aircraft Skins Exposed to Lightning,* NASA CR 120827, National Aeronautics and Space Administration, Lyndon B. Johnson Space Center, Houston, Texas (September 1971), p. 19.

6.50 M. J. Kofoid, *Lightning Discharge Heating of Titanium Aircraft Skins,* D1-82-0752, Plasma Physics Laboratory, Boeing Scientific Research Laboratories, Seattle, Washington (September 1968).

6.51 Redrawn from Kofoid, *Lightning Discharge Heating,* p. 33.

6.52 J. D. Robb, J. R. Stahmann, T. Chen, and C. P. Mudd, "Swept Lightning Stroke Effects on Painted Surfaces and Composites of Helicopters and Fixed Wing Aircraft," *Proceedings of the 1975 Conference on Lightning and Static Electricity, 14-17 April 1975, at Culham Laboratory, England,* Session III: Structures and Materials, the Royal Aeronautical Society of London (1975).

6.53 Plumer, *Guidelines for Lightning Protection of General Aviation Aircraft.*

6.54 M. M. Newman, J. D. Robb, and J. R. Stahmann, *Lightning Protection Measures for Aircraft Fuel Systems--Phase I,* FAA ADS-17, prepared by the Lightning and Transients Institute for the Federal Aviation Agency, Office of Technical Services, U.S. Department of Commerce, Washington, D.C. (May 1964), pp. 22-44.

6.55 Newmann, Robb, and Stahmann, *Lightning Protection Measures,* pp. 47-68.

6.56 Plumer, *Lightning Effects on General Aviation Aircraft,* pp. 11-22.

6.57 pp. 19-21.

6.58 L. E. Short, "Electrical Bonding of Advanced Airplane Structures," *Lightning and Static Electricity Conference, 3-5 December 1968: Part II, Conference Papers,* AFAL-TR-68-290, Air Force Avionics Laboratory, Air Force Systems Command, Wright-Patterson Air Force Base, Ohio (May 1969), pp. 425-441: 433.

6.59 Short, "Electrical Bonding," pp. 435.

6.60 General Electric Company photograph.

6.61 Newman, Robb, and Stahmann, *Lightning Protection Measures,* pp. 17-35.

6.62 J. A. Plumer, "Lightning-Induced Voltages in Electrical Circuits Associated with Aircraft Fuel Systems," *Report of Second Conference on Fuel System Fire Safety, 6 and 7 May 1970,* Flight Standards Service, Engineering and Manufacturing Division, Federal Aviation Administration, Washington, D.C. (1970), pp. 171-191.

6.63 J. D. Cobine, *Gaseous Conductors: Theory and Engineering Application* (New York: Dover Publications, 1958), p. 164.

6.64 Newman, Robb, and Stahmann, *Lightning Protection Measures,* p. 15, 21-44.

6.65 Plumer, *Lightning Induced Voltages,* pp. 176-178, 187-191.

CHAPTER 7
STRUCTURES PROTECTION

7.1 Introduction

It was not easy, however, to escape from Crete, since Minos kept all his ships under military guard, and now offered a large reward for his apprehension. But Daedalus made a pair of wings for himself, and another for Icarus, the quill feathers of which were threaded together, but the smaller ones held in place by wax. Having tied on Icarus' pair for him, he said with tears in his eyes: "My son, be warned! Neither soar too high, lest the sun melt the wax; nor sweep too low, lest the feathers be wetted by the sea!" Then he slipped his arms into his own pair of wings and they flew off. "Follow me closely," he cried, "do not set your own course."

As they sped away from the island in a north-easterly direction, flapping their wings, the fishermen, shepherds, and ploughmen who gazed upward mistook them for gods.

They had left Naxos, Delos, and Paros behind them on the left hand, and were leaving Lebynthos and Colymne behind on the right, when Icarus disobeyed his father's instructions and began soaring towards the sun, rejoicing in the lift of his great sweeping wings. Presently, when Daedalus looked over his shoulder, he could no longer see Icarus; but scattered feathers floated on the waves below. The heat of the sun had melted the wax, and Icarus had fallen into the sea and drowned. Daedalus circled around, until the corpse rose to the surface, and then carried it to the nearby island now called Icaria, where he buried it. Robert Graves, *The Greek Myths I* (Baltimore: Penguin Books, 1955): 312-13.

The materials of which an aircraft is made and the methods used to hold these materials together, forming the aircraft *structure*, are factors as important in protecting a modern aircraft from hazardous natural environments as they were for the mythical Icarus. Conventional aluminum airframes of riveted construction have, by virtue of their excellent electrical conductivity, rarely suffered critical damage from lightning strikes; and these structures have provided excellent protection for more vulnerable systems, as well as for personnel, carried within. But the day of all-metal aircraft is ending. Taking their place will be aircraft constructed partly of new, fiber-reinforced plastics with desirable light-weight and high-strength properties but with poor electrical conductivity. Some of these materials will be fastened together with equally nonconductive adhesives, the strength of which may be lost if the adhesives are exposed to lightning currents. Whereas the designer of the all-metal structure needed to add little, if anything, to his design to achieve adequate lightning protection, the designers of nonmetallic structures must pay particular attention to the lightning environment, taking positive measures to protect against its adverse effects lest safety of flight be endangered.

In this chapter we first review lightning effects on metal structures and discuss how to design protection for the few situations where special protection is necessary. We then review the expected lightning effects on nonmetallic materials such as fiberglass and the advanced composites.

Concern for the potential hazards has prompted a large number of laboratory studies of lightning effects and protection techniques for these materials. We summarize the important results of these studies and discuss those protective techniques and devices which appear most promising. Our discussion emphasizes the electrical performance, leaving to the structures designer (who himself is best equipped to answer them) questions of compatibility with other requirements or manufacturing methods.

7.2 Metallic Structures

Because of their inherently good electrical conductivity, conventional aluminum aircraft structures have rarely suffered serious damage from lightning strikes, and few protective measures have had to be added. The metal skins, ribs, spars, stringers, and other substructures may themselves be utilized to provide the safe conducting path for lightning currents between possible attachment points called for as a basic protection principle in Chapter 5. No special conductor dedicated only for lightning currents need be added to conduct lightning currents through the aircraft, as long as care is taken in the design of basic skins and structures. It would, in fact, be highly impractical, and even dangerous, to design a separate "lightning conductor" to be run through a metal aircraft. No further mention of such a conductor will be made.

Thus, in accord with the principle of providing a safe conductive path for lightning currents, metallic skins and substructures should be designed to provide this conductive path and to minimize, or safely tolerate, the following lightning effects:

- Melting at lightning attachment points
- Resistive temperature rise
- Magnetic force effects
- Arcing across bonds, hinges, and joints
- Ignition of vapors within fuel tanks.

Techniques to assess these effects and design protection, if needed, are discussed in the following paragraphs.

7.2.1 Melting at Lightning Attachment Points

The quantity of metal melted away at a lightning attachment point has been shown to be most closely related to the charge carried into the point by the lightning flash together with the type of metal and its thickness. Most of the charge is delivered by intermediate and continuing current components of the flash, and the amount of this charge that is necessary to initiate a hole (i.e., a pinhole) in a metal skin has been discussed in Section 6.6 of Chapter 6. The charge required to initiate a hole, however, is of little concern except where the skin is over a fuel tank. Of concern from a structural integrity point of view is the eventual quantity of metal that can be melted away. This aspect is of primary concern for skins and structures to which the arc can hang on for prolonged periods, as it may in Zones 1B and 2B.

Plumer (Reference 7.1) has performed laboratory tests to determine the area of holes which may be melted through aluminum or titanium skins of various thicknesses, by varying amounts of continuing current delivered at rates of between 200 A and 800 A. Figure 7.1 shows the hole areas to be expected in typical aluminum and titanium skin metals. The volume of metal melted away may be found by multiplying the area by the thickness of the particular skin material.

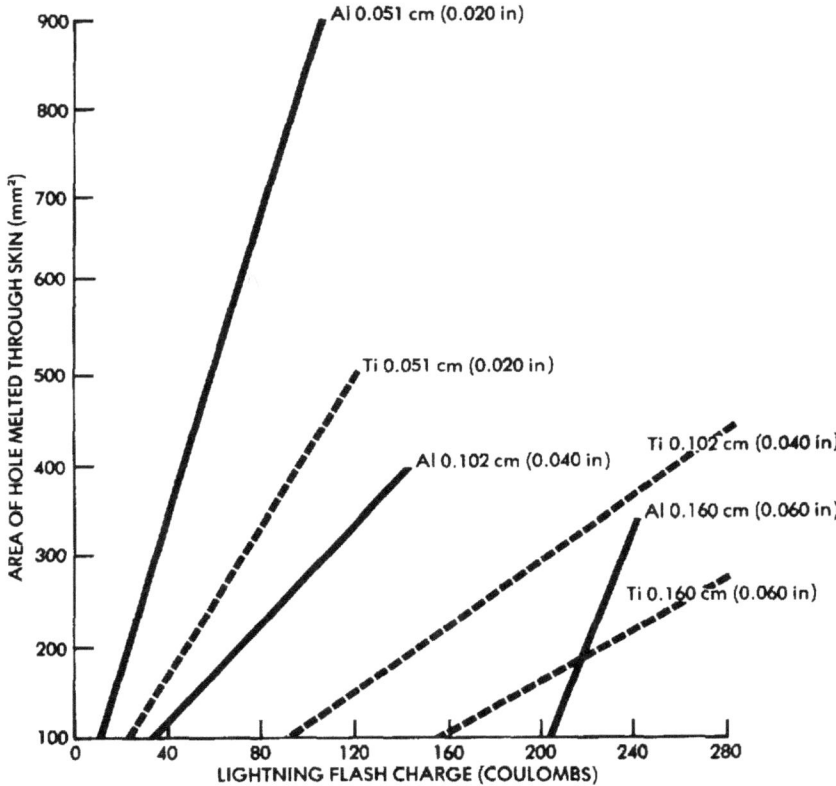

Figure 7.1 Area of holes melted through aluminum and titanium aircraft skins by lightning flash charge.

Normally, holes may be permitted to occur anywhere on the aircraft as long as safety of flight is not impaired. The conditions under which safety of flight might be endangered from hole formation follow.

187

- When the hole is melted through a fuel tank skin or other enclosure of flammable materials
- When the hole is melted through the wall of a pressurized enclosure
- When the hole sufficiently degrades the mechanical strength of a flight-critical component to cause failure

The problem of hole formation in fuel tank skins has been dealt with in Chapter 6. There have been no reports of holes having been melted through the walls of a pressurized container, except for the very small (less than 0.1 cm dia) holes sometimes found at swept-stroke dwell points along the surface of a pressurized fuselage——holes that have been far too small to cause depressurization problems. Since trailing edge areas are usually not pressurized, the larger holes melted by longer duration currents hanging on to trailing edges have also not posed a depressurization problem, although the possibility of this event should be considered in any design that involves pressurization of a trailing edge section.

The structural integrity of a trailing edge closeout member or other load-carrying part may be degraded if a significant portion of metal is melted away by a lightning arc hanging on for a prolonged period of time. For design purposes the worst case should be taken as 200 C, in accordance with the SAE report *Lightning Test Waveforms and Techniques for Aerospace Vehicles and Hardware* (Reference 7.2). Estimates of the amount of metal to be melted away by 200 C of charge entering a hang-on point may be made from Figure 7.1 or by the use of Equations 6.1 and 6.2 of Section 6.6.

7.2.2 Mechanical Property Degradation of Structural Metals

Aside from the burning of holes at attachment points, the flow of lightning current will not degrade the mechanical strength properties of metal skins and other basic structures unless temperature rises approaching the melting temperature occur. Except within a centimeter or two of the points of entry, lightning currents will normally spread out sufficiently so as not to produce a noticeable temperature rise.

A case in which an excessive temperature rise may occur is one in which lightning current remains in a single conductor, such as a bond strap or pipe. In this case, the action integral ($\int I^2 dt$) of the lightning current and the resistance of the conductor may both be high enough to allow sufficient energy to be deposited in the conductor to raise its temperature appreciably. Since the resistance of most metal conductors increases with temperature, an even higher amount of energy will be deposited in the conductor as its temperature rises during current flow, and this, in turn, will increase the temperature even further. If the action integral is high enough, the conductor will be unable to absorb all of the energy and will explode. This will happen for currents whose action integrals are sufficient to raise the conductor temperature above its melting point.

It is therefore important to be able to determine the expected temperature rise which lightning may produce in conductors of small cross-sectional area and to obtain assurance that the melting point will not be reached. The temperature rise in a current-carrying conductor is expressed by the basic equation (units in parenthesis):

$$\wedge T(^\circ C) = \frac{0.2389 \left(\frac{\text{g-cal}}{\text{J}}\right) \int I^2 \, dt \,(A^2 \text{-s}) \text{ Resistivity } (\Omega\text{-cm})}{\text{Specific heat} \left(\frac{\text{g-cal}}{\text{g-}^\circ C}\right) \text{Density} \left(\frac{\text{g}}{\text{cm}^3}\right) \text{ Area}^2 \,(\text{cm}^4)} \qquad (7.1)$$

It is evident that temperature rise is directly proportional to the square of the lightning current and inversely proportional to the square of the cross-sectional area of the conductor. As it stands, however, Equation 7.1 has two shortcomings:

1. It assumes no energy (heat) loss resulting from radiation during the time current is flowing
2. It assumes that resistance does not change with temperature.

Since little energy is lost during the short time duration of lightning strokes, the first shortcoming has a negligible effect on the accuracy of the predicted temperature rise. Resistance does depend on temperature, however, and, since resistance also affects the amount of electrical energy that can be deposited in the conductor, it is important to account for this dependency. This can be done by expressing resistivity as a function of temperature, using the following expression:

$$\rho(\Omega\text{-cm}) = \rho \left(\frac{\Omega\text{-cm}}{\text{at } 20^\circ C}\right) \left[1 + \lambda \left(\frac{1}{^\circ C}\right) \wedge T(^\circ C)\right] \qquad (7.2)$$

where λ is the temperature coefficient of resistivity. Incorporating Equation 7.2 into the expression of Equation 7.1 gives the following equation:

$$\wedge T(^\circ C) = \frac{0.2389 \left(\frac{\text{g-cal}}{\text{J}}\right) \int I^2 \, dt \,(A^2 \text{-s}) \text{ Resistivity } (\Omega\text{-cm}) \left[1 + \lambda \left(\frac{1}{^\circ C}\right) \wedge T(^\circ C)\right]}{\text{Specific heat} \left(\frac{\text{g-cal}}{\text{g-}^\circ C}\right) \text{Density} \left(\frac{\text{g}}{\text{cm}^3}\right) \text{ Area}^2 \,(\text{cm}^4)} \qquad (7.3)$$

Since everything but $\wedge T$ is a material property constant, this equation is most easily solved by combining the conductor dimensions and material properties in a constant, k, in the following expression:

$$\wedge T(^\circ C) = k \left[1 + \lambda \left(\frac{1}{^\circ C}\right) \wedge T(^\circ C)\right] \qquad (7.4)$$

or

$$\triangle T(^\circ C) = \frac{k \,(^\circ C)}{1 - \lambda \left(\frac{1}{^\circ C}\right) k \,(^\circ C)} \qquad (7.5)$$

where k is equal to the right side of Equation 7.1.

189

Physical and electrical properties needed in Equations 7.5 and 7.6 for common structural metals and electrical conductors are provided in Table 7.1. The melting points of these materials are also provided.

Table 7.1 PHYSICAL AND ELECTRICAL PROPERTIES OF COMMON METALS

	Aluminum	Copper	Titanium	Stainless Steel (304)	Magnesium	Silver
Resistivity, (Ω-cm)	2.8×10^{-6}	1.72×10^{-6}	42×10^{-6}	72×10^{-6}	4.45×10^{-6}	1.59×10^{-6}
Temperature coefficient of resistance, λ ($1/^{\circ}C$)	0.00429	0.00393	0.0035	0.001	0.0165	0.0041
Thermal coefficient of linear expansion ($1/^{\circ}C$)	0.254×10^{-4}	0.164×10^{-4}	0.085×10^{-4}	0.120×10^{-4}	0.025×10^{-4}	0.019×10^{-4}
Specific heat (g-cal/g-$^{\circ}C$)	0.215	0.092	0.124	0.120	0.245	0.056
Density (g/cm^3)	2.70	8.89	4.51	7.90	1.74	10.49
Melting point ($^{\circ}C$)	660	1084	1670	1150	650	962

Because of their high amplitudes, the return stroke and restrike currents have the highest action integrals and may be expected to produce higher temperatures than the other components of the lightning flash do when they are conducted through structural metals or other electrical conductors. For design purposes, the action integrals associated with current waveform Components A and/or D of the SAE report *Lightning Test Waveforms and Techniques for Aerospace Vehicles and Hardware* should be used for determining temperature rise in conductors, depending on the lightning-strike zone in which the particular conductor is located.

It should be remembered that if several parallel conductors are available to share the stroke current, the current in each will be divided by the number of conductors but that the *action integral* in each will be reduced by the square of the number of conductors. Consider, for example, a case in which a NAV light is mounted on a plastic vertical fin cap and grounded to the airframe via two bond straps, as shown in Figure 7.2. If this case is considered to be located in Zone 1B, its bond straps must be able to conduct the entire flash current. If it is assumed that the arc will not touch the bond straps, the limiting lightning design criterion for these straps is that together they must be able to carry safely current Components A and D, which have a total of 2.25×10^6 A^2-s. However,

190

the one-half of the total current flowing in each strap will produce only one-fourth of the total action integral, or about 0.51×10^6 A²-s in each strap.

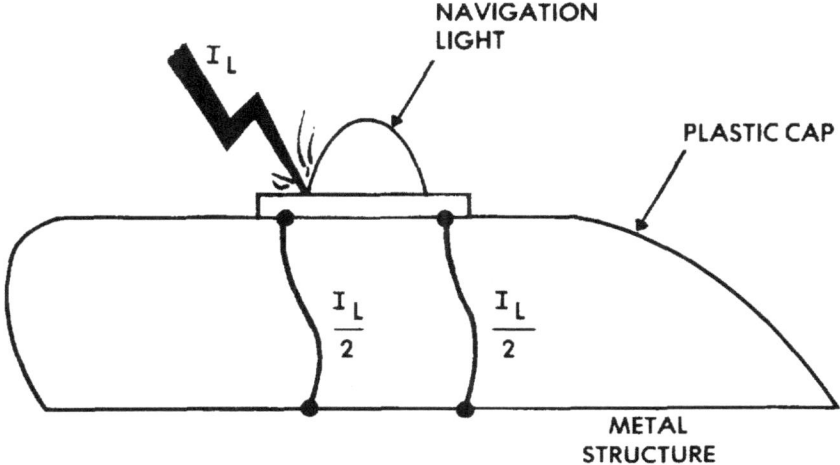

Figure 7.2 Bond straps for NAV light.

If each strap is made of copper with a cross-sectional area of 0.1 cm² (0.015 in²), the temperature rise produced by 0.51×10^6 A²-s in a strap would be determined from Equation 7.4, using the physical and electrical properties for copper given in Table 7.1. Solving for k first

$$k = \frac{\left(0.2389 \frac{\text{g-cal}}{\text{J}}\right)(0.51 \times 10^6 \text{ A}^2\text{-s})(1.72 \times 10^{-6} \text{ }\Omega\text{-cm})}{\left(0.092 \frac{\text{g-cal}}{\text{g-}^\circ\text{C}}\right)\left(8.89 \frac{\text{g}}{\text{cm}^3}\right)(0.1 \text{ cm}^2)^2} \qquad (7.6)$$

$$= 25.6 \,^\circ\text{C}$$

This, of course, would be the actual temperature rise if the resistivity remains constant. But the resistivity actually increases as temperature increases; therefore, Equation 7.5 must be used as follows to calculate the actual rise:

$$\Delta T(^\circ\text{C}) = \frac{(25.6 \,^\circ\text{C})}{[1 - \left(0.00393 \frac{1}{^\circ\text{C}}\right)(25.6 \,^\circ\text{C})]} \qquad (7.7)$$

$$= 28.5 \,^\circ\text{C}$$

A conductor with this cross-sectional area is about the same size as an AWG 15 wire. Since a 28 °C temperature rise would not damage the wire, two bonding straps or wires of this cross-sectional area could, theoretically, be utilized to conduct current Components A and D. If one of them had worked loose, however, all of the current would have to flow through the other one. If the same equations are used in this circumstance, the total action integral of 2.25 x 10^6 A^2-s would cause a calculated temperature rise of over 200 °C, a temperature that should be considered excessive.

In fact, it is prudent to keep conductor temperature rises below 100 °C in order to avoid explosion where flammable materials or vapors may exist and also to avoid mechanical stresses resulting from thermal expansion (See Section 7.2.3). The temperature coefficients of resistivity which are given in Table 7.1 remain reasonably constant only through the range of -50 °C to +100 °C; thus, calculations that are based on these coefficients and which produce temperatures above 100 °C should not be considered accurate.

Consideration of temperature rise effects in conductors carrying lightning currents is most important where all or a large fraction of the stroke currents need to be carried in individual conductors. Examples include radome diverter straps, bonding straps (or *jumpers*), conduits or cable shields, ground wires passing through plastic radomes or wing tips, and certain hydraulic or control lines which may become exposed to lightning-strike currents.

Figure 7.3 presents calculated temperature rises for conductors of the metals of Table 7.1 as a function of cross-sectional area for action integrals of 0.25, 2.0, 2.25, and 3 x 10^6 A^2-s. These curves may be used to determine the expected temperature rise in structural elements, diverter bars, bond straps, or other conductors expected to carry lightning currents. The temperature rises determined from Figure 7.3 should be added to the ambient temperature of the conductor to determine the actual temperature.

If the reader desires to calculate the temperature rise for action integrals other than the four of Figure 7.3, he may do so manually by using Equations 7.1 and 7.5, or he may use the computer program written for the Texas Instruments SR-52 programmable calculator included in Appendix 1. The program requires the action integral, resistivity, specific heat, density, and cross-sectional area of the conductor as inputs from which it calculates temperature rise. The program also calculates the elongation that takes place in the metal as a result of the temperature rise.

A separate SR-52 program is provided in Appendix 2 for calculation of temperature rise and elongation in tubes and pipes. This program accepts dimensions of the outside diameter and wall thickness, thereby saving the user the need to calculate manually the cross-sectional area of the pipe wall.

If several parallel conductors of the same length and cross-sectional area are present, the current may be expected to divide equally among them; the action integral associated with each portion of the current will be

$$\int I^2 \, dt \text{ (each conductor)} = \frac{\int I^2 \, dt \text{ (total)}}{N^2} \qquad (7.8)$$

where N is the number of parallel conductors. This distribution would suggest

192

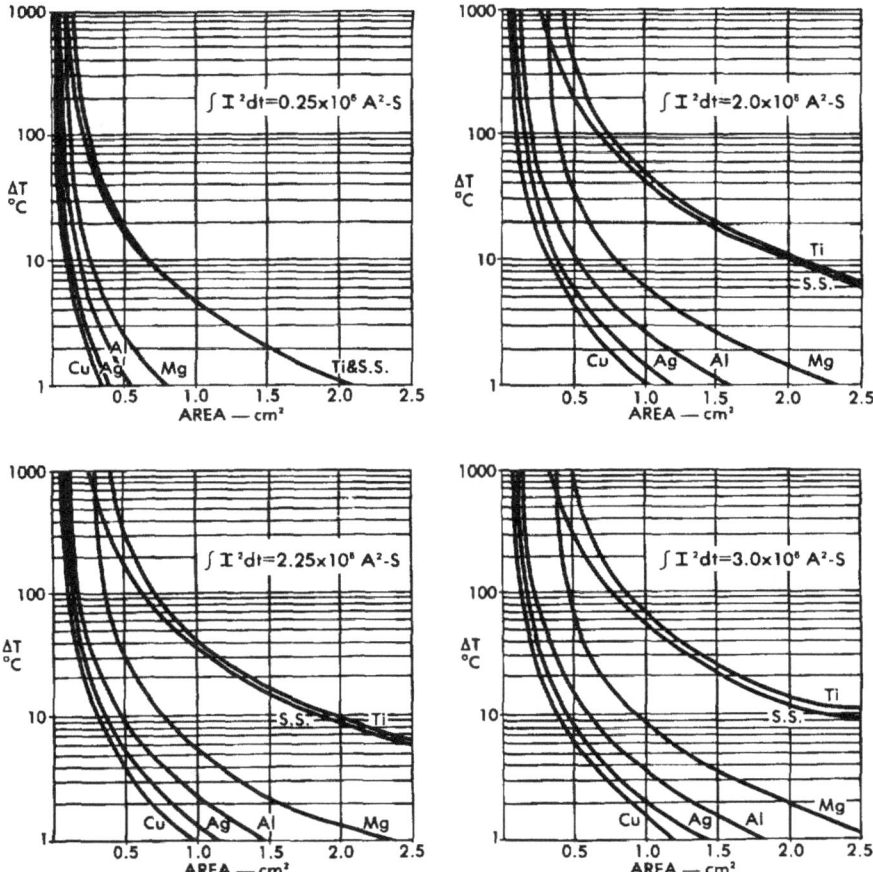

Figure 7.3 Temperature rise as a result of electric current in conductors.

that the use of several conductors of small cross section would be advantageous, since the total amount of metal required for all of them would still be less than that required for one large conductor. This conclusion is sound, but there are two factors that should be given careful consideration before any decision to utilize several thin conductors instead of one thick one to conduct lightning currents is reached:

1. As a result of vibration or corrosion stresses, thin conductors may break sooner, leaving a situation in which the remaining conductor(s) may be greatly overstressed by lightning currents

2. The magnetic forces which act upon parallel current-carrying conductors may cause excessive damage to these conductors.

On the other hand, magnetic forces may sometimes be reduced sufficiently by separating the conductors a sufficient distance from each other (See Section 7.2.4).

193

7.2.3 Thermal Elongation

Most metals expand when their temperatures rise. This phenomenon will not be a problem for flexible conductors which can deflect when expanded, but large stresses may arise in some rigidly held conductors, such as the diverter strap or the pitot probe air pressure tube shown in Figure 7.4. By virtue of its location at the nose of the aircraft, the pitot probe will be in Zone 1A and must therefore be designed to conduct return stroke currents of 2×10^6 A^2-s. The purpose of the air pressure tube is to transmit pitot-static pressure back to the flight instruments, but the tube may also provide a safe path for lightning currents to enter the airframe if the tube is designed to safely withstand the resistive heating, thermal elongation, and magnetic force effects.

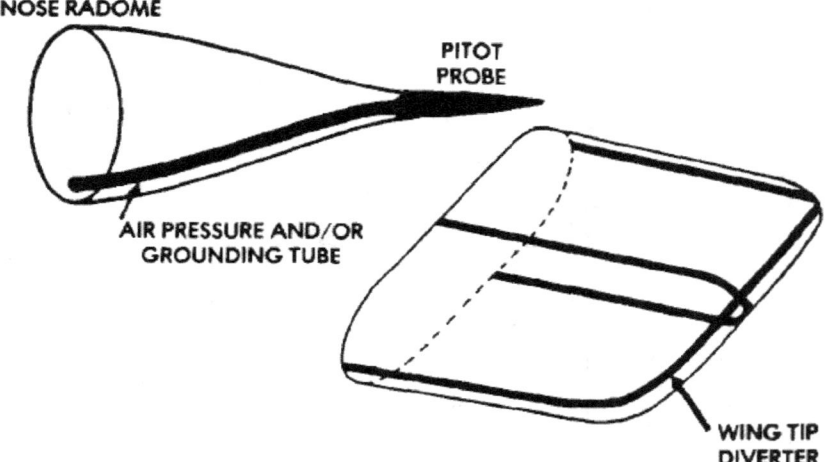

NOSE RADOME

PITOT PROBE

AIR PRESSURE AND/OR GROUNDING TUBE

WING TIP DIVERTER

Figure 7.4 Examples of rigidly held lightning current conductors.
- Cross-sectional area should be sufficient to prevent excessive temperature rise and thermal expansion.

For example, if the air pressure tube in the radome of Figure 7.4 is made of copper and has an outside diameter of 0.476 cm (3/16 in) and a wall thickness of 0.124 cm (0.049 in), it will have a cross-sectional area of 0.138 cm^2 (0.021 in^2). According to the curves of Figure 7.3(c) an action integral of 2×10^6 A^2-s will raise the temperature of a conductor of this cross-sectional area 67 °C.

The amount of thermal expansion to be expected along any dimension of a part is dependent on the temperature rise and the thermal coefficient of linear expansion of the metal, according to the following relation (dimensions in parenthesis):

$$\triangle L = \text{Expansion coefficient} \left(\frac{1}{^\circ C}\right) \wedge T(^\circ C) L(cm) \qquad (7.9)$$

where L is the length of the dimension of concern. Of course all dimensions of a homogeneous material will elongate proportionately, but frequently only one dimension is of much concern. The most significant expansion in a part such as the pressure tube would be along its length.

Based on the coefficient of linear expansion for copper given in Table 7.1 as 0.164×10^{-4} $(1/°C)$ and the assumption that the tube has a length, L, of 2 m when "cold," a 67 °C temperature rise would cause an elongation of

$$\Delta L_{Tube} = \left(0.164 \times 10^{-4} \frac{1}{°C}\right) (67°C)(2m) \qquad (7.10)$$

$$= 2.198 \times 10^{-3} m = 2.2 \text{ mm}$$

If the tube is fastened rigidly, this elongation may cause sufficient stress to break clamps or force the tube to bend. Thus, the tube supports should be designed to accept this elongation without undue stress.

Thermal elongation calculations may be made for other conductors by using Equation 7.9 or the SR-52 programmable calculator programs in Appendices 1 and 2. These programs calculate the elongation in millimeters per centimeter of conductor length for the basic conductors and in inches per foot of conductor length for tubes, since dimensions of tubes are commonly presented in inches.

The temperature rise and elongation expressions described in this and in the previous section have been closely confirmed up to 100 °C by laboratory tests in which simulated lightning currents of known action integral are passed through conductors whose temperature and elongation are measured. Temperature rise is most easily measured by temperature-sensitive tapes or by an optical pyrometer. Elongation can be measured by a mechanical deflection indicator attached to the specimen. Calculated results above 100 °C, however, will begin to diverge from the actual because of physical property changes at higher temperatures.

7.2.4 Magnetic Force Effects

As mentioned earlier, there are situations in which several conductors in parallel may carry lightning currents and, in such cases, each conductor will be acted upon by a magnetic force proportional to the product of the currents in each conductor. One such case is shown in Figure 7.5.

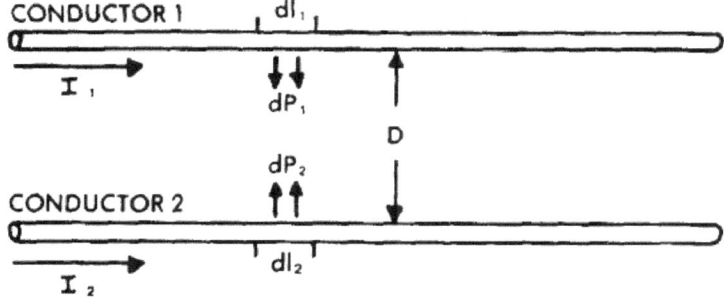

Figure 7.5 Magnetic forces among parallel current-carrying conductors.

Based on Ampere's Law, the force per unit length on wire 2 is

$$\frac{dP_2}{d\ell_2} = \frac{2\mu_a I_1 I_2}{D} \qquad (7.11)$$

where
μ_a = permeability of media between conductors

$= 10^{-7}$ for air

$P_{1,2}$ = force on conductor 1 or conductor 2 (N/m)

$\ell_{1,2}$ = unit of length of conductor 1 or conductor 2 (m)

D = center-to-center distance between conductors (m)

$I_{1,2}$ = currents in conductor 1 or conductor 2

These forces act to pull the conductors together when the currents in them flow in the same direction and to repel the conductors when the currents are in opposite directions. In either case, the magnetic forces are strongest when the conductors are close together. One such case would exist, for example, in a radome-mounted pitot system in which two air pressure tubes are needed to convey both static and dynamic air pressure to the flight instruments. If these tubes are mounted 2 cm apart and half of the lightning current is assumed to flow in each, the peak force in each conductor, as a result of the currents in both, will be

$$\frac{dP_{1,2}}{d\ell_{1,2}} = \frac{2(10^7)\,(100 \times 10^3\ \text{A})\,(100 \times 10^3\ \text{A})}{(2 \times 10^{-2}\ \text{m})} \qquad (7.12)$$

$$= 100\,000\ \text{N/m of tube length}$$

$$= 6\,854\ \text{lb/ft of length}$$

This peak force, of course, would exist only at the instant when the lightning current is at its peak, with lower forces active when the current is lower. During the short duration of the higher amplitude stroke currents, the mass of the conductor may prevent its being deflected appreciably because of its relatively slow mechanical response time, although it can absorb appreciable kinetic energy in that time and continue deflecting afterward. The resulting deflection also depends on the degree of restraint provided by supports, but forces of this magnitude have been shown to cause tubes in one typical installation to slam together, leaving permanent deformations, as shown in Figure 7.6. Deformations like those in Figure 7.6 (Reference 7.3) are sufficient to disrupt the transmission of pitot pressure to flight instruments.

Mathematical calculation of expected deformations from impulsive forces such as these is difficult, and such calculation is further complicated by unknown mechanical factors. James and Phillpott, however, have defined an effective transient force, P_t which, if applied slowly but continuously, results in the same stress as that produced by the peak magnetic force, P_o, calculated by Equation 7.11. This effective force, P_t, is a function of the angular frequency,

196

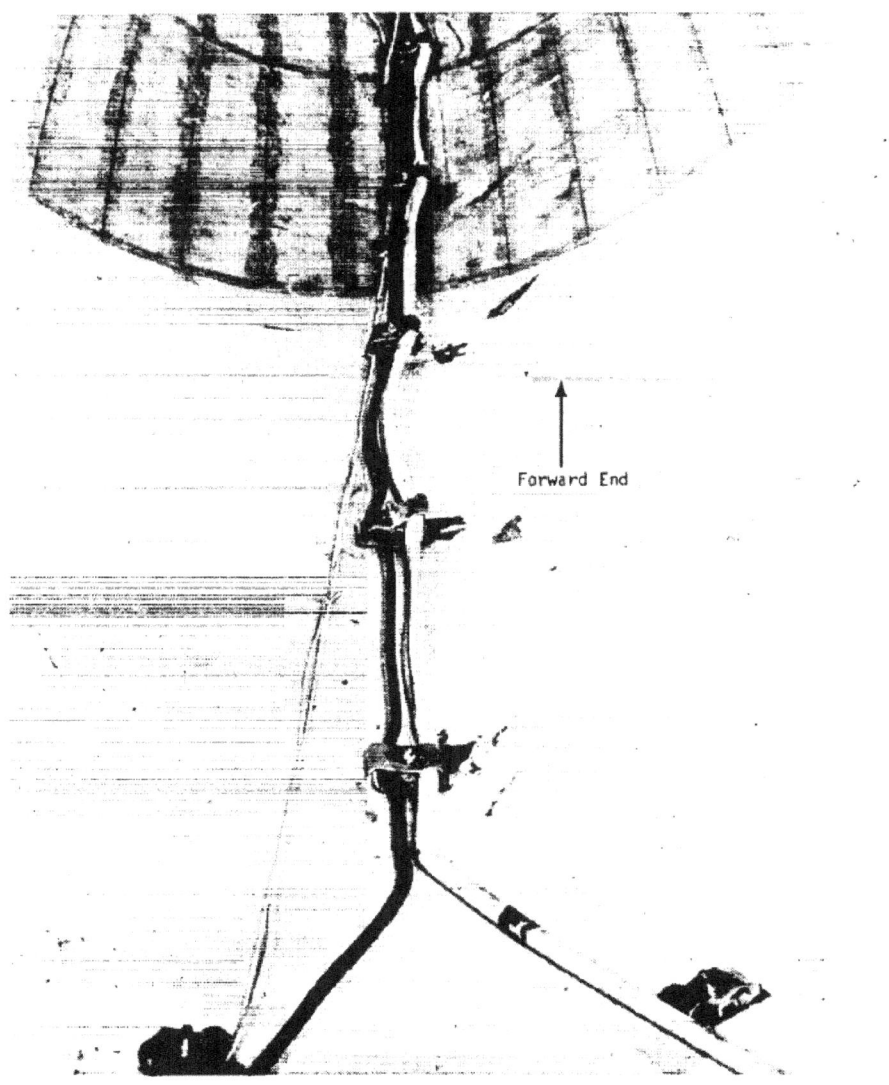

Forward End

Figure 7.6 Air pressure tubes deformed by magnetic forces (after carrying a 120 kA, 2×10^6 A^2-s stroke).

ω, of the mechanical system being acted upon and of the electric current decay time constant, τ. In most practical applications the angular frequency will be between about 10^3 for heavy, flexible systems and 10^4 for light, stiff systems. James and Phillpott give relative values of P_t/P_0 (Table 7.2 [Reference 7.4]) for each of these frequencies and three lightning current waveforms. Their 200 kA and 1 kA currents represent a return stroke and continuing current of the same magnitude as those represented by Components A and C of *Lightning Test*

197

Waveforms and Techniques for Aerospace Vehicles and Hardware. However, the 50 kA intermediate current would deliver much more energy than would the 2 kA intermediate current, represented by Component B of *Lightning Test Waveforms and Techniques.* A 50 kA current decaying in as long as 2 ms would be found only in the severest of positive polarity flashes, which occur very seldom. In any event, Table 7.2 shows that the 200 kA stroke current creates a much higher effective transient force, P_t, than does either of the other currents.

Table 7.2 EFFECTIVE TRANSIENT MAGNETIC FORCE P_t

I (kA)	τ (ms)	Coulombs (C)	$I^2 t$ 10^6 (A²-s)	P_o (Relative Units)	P_t/P_o		P_t (Relative Units)	
					$\omega = 10^3$	$\omega = 10^4$	$\omega = 10^3$	$\omega = 10^4$
200 (Stroke)	0.1	20	2.0	16.0	0.05	0.45	0.8	7.2
50 (Int)	2.0	100	2.5	1.0	0.75	1.73	0.75	1.73
1.0 (Cont)	200	200	0.1	0.0004	2.0	2.0	0.0008	0.0008

An example of how the table is used may be found in the case of the two parallel air pressure tubes 2 cm apart, for which a peak force, P_o, of 100 000 N/m of length was calculated. Assuming that these tubes have an angular frequency of 10^4, Table 7.2 shows that the effective transient force, P_t, would be

$$P_t = 0.45 P_o \tag{7.13}$$

or 45 000 N/m (3084 lb/ft) of length. This is the amount of force which should be used in performing mechanical response calculations or mechanical strength tests.

James and Phillpott also conclude from Table 7.2 that

(a) Continuing current pulses do not give rise to a high effective force because I^2 and therefore the peak magnetic forces are too low.

(b) If $\tau \ll \dfrac{1}{\omega}$ as for relatively heavy flexible systems ($\omega = 10^3$) the effective force is roughly proportional to $I^2 t$. Thus the fast and intermediate components give about the same effective force although the peak forces are very different.

(c) If $\tau \gg \dfrac{1}{\omega}$ as for relatively light stiff systems ($\omega = 10^4$), the effective force equals twice the peak force and is therefore proportional to I^2. For this doubling of the peak force to occur it is assumed that the current rise time is $\ll \dfrac{1}{\omega}$.

(d) If $\tau \approx \dfrac{1}{\omega}$, the effective force is 45% of the peak force for the case of the fast component with $\omega = 10^4$. This alternative also gives the highest effective force because the high value of I^2 more than compensates for the effects of a small value of τ. (Reference 7.5).

These authors conclude that "[t]he effect of magnetic forces on current-carrying components is influenced by a large number of parameters, but the contribution from the continuing current is negligible." Concurring with the authors of this book, they also conclude that future studies of or tests on components likely to be damaged by magnetic forces should be performed with simulated lightning currents having values of I and I^2t similar to Component A of *Lightning Test Waveforms and Techniques.*

Outer portions of the airframe and control surfaces sometimes become badly damaged by magnetic force effects. One example is the wing tip trailing edge shown in Figures 7.7 and 4.4, in which the upper and lower surfaces of the wing were wrinkled, pinched, and pulled forward. The upper and lower surfaces of this wing were made respectively of 0.71 mm (0.028 in) and 0.80 mm (0.031 in) aluminum. Hacker (Reference 7.6) has made a very interesting analysis of the magnetic forces acting upon these surfaces and found, as illustrated in Figure 7.7, that the directions of the calculated forces correlate well with the damage pattern. Hacker calculated that a 100 kA stroke would have created 21 500 N/m (1476 lb/ft) on a 4 cm wide path across the skins. This amounts to 5.4×10^5 N/m^2, or 78 psi, an extremely high compressive pressure for a wing structure reportedly designed for a steady state loading of 1.7×10^3 N/m^3 (0.25 psi). The reader is referred to Hacker's report for a more thorough discussion of the procedure followed in making this analysis.

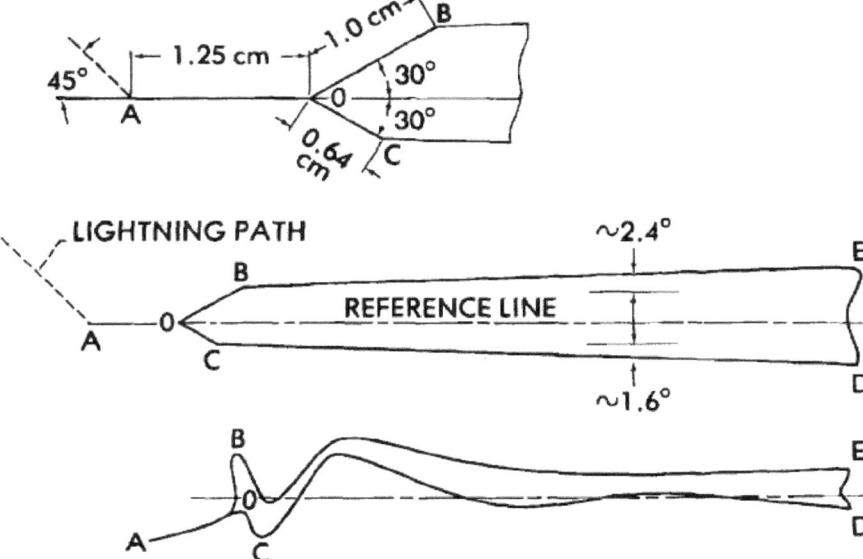

Figure 7.7 Approximate geometry of chordwise cross section of wing tip trailing edge of Figure 4.4.

199

The foregoing examples have illustrated the case in which the magnetic forces are attractive (when parallel currents flow in the same direction). Cases exist, however, where currents in adjacent conductors, or in adjacent legs of the *same* conductor, flow in opposite directions. When this happens, the magnetic force acts to separate the conductors, a fact that may create a problem for the unwary designer. One such case is the folded protective foil splice shown in Figure 7.8.

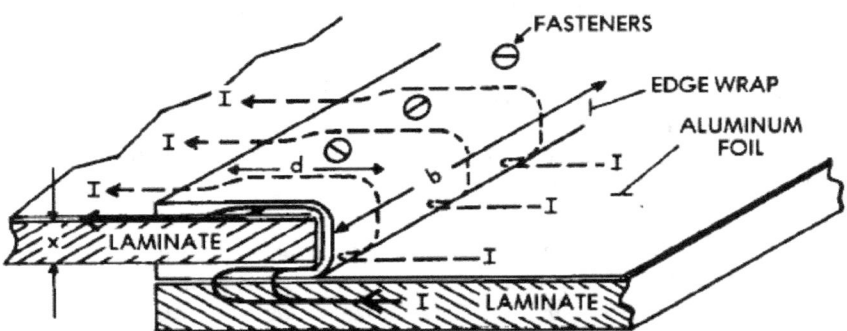

Figure 7.8 Reverse current paths through joint.

A splice of this sort was considered (Reference 7.7) for joining two plastic laminates protected with aluminum foils. An aluminum edge wrap was utilized to provide electrical conductivity between the foils on the lower and upper laminates. This arrangement, however, forced the lightning current to travel in opposite directions through the edge wrap, as shown in Figure 7.8. When a 100 kA simulated lightning current was passed through a sample of this joint, the edge wrap pulled away from the fasteners as shown in Figure 7.9.

Jacobsen (Reference 7.8) gives the force, P, per unit length of current path, d, acting to repel two parallel plates whose separation, χ, is small compared to width, b, and which carry current, I, in opposite directions as follows:

$$P = \mu \frac{I^2}{b} \qquad (7.14)$$

where
$$\mu = 4\pi \times 10^{-7} \text{ H/m} \qquad (7.15)$$

Assuming the aluminum edge wrap of Figure 7.8 was 0.1 m (4 in) wide, the peak force for the 100 kA test current would have been

$$P/\overset{.}{d} = 4\pi \times 10^{-7} \frac{(100 \times 10^3)^2}{(0.1)} \text{ (N/m)} \qquad (7.16)$$

$$= 1.26 \times 10^5 \text{ N/m}$$

For a length of current path, d, of 0.013 m (0.5 in) through the edge wrap

200

$$P = (1.26 \times 10^5 \text{N/m})(0.013\text{m})$$

$$P = 1638 \text{ N} \qquad\qquad (7.17)$$

$$= 368 \text{ lb}$$

This force was evidently adequate to blow off the upper surface of the edge wrap.

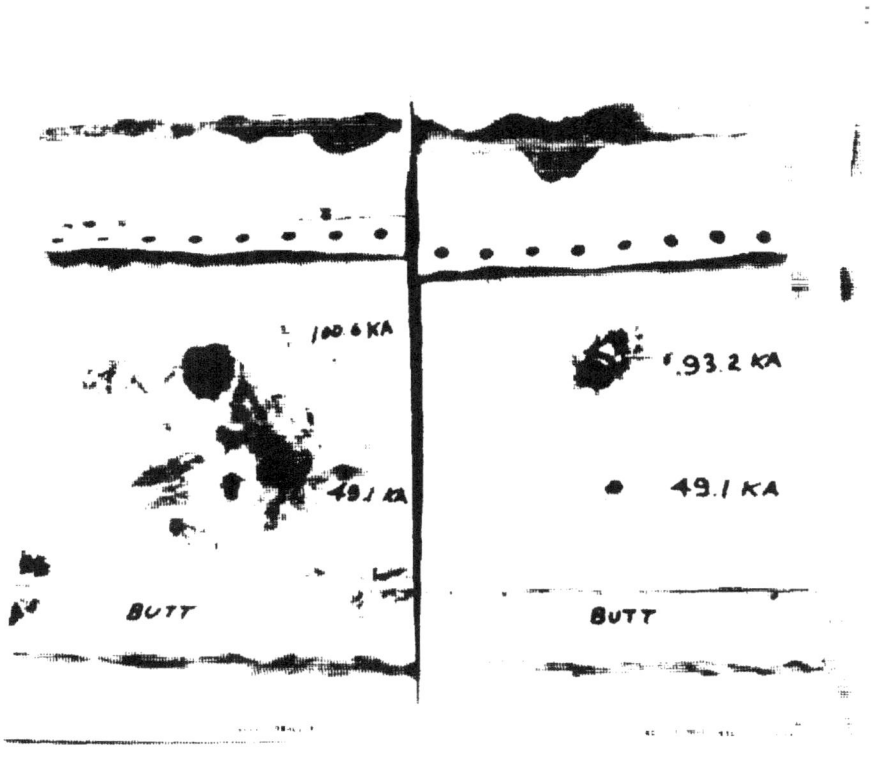

Figure 7.9 Aluminum edge wrap torn away from fasteners by magnetic force.

Another, more common, example of the forces created by currents flowing in opposite directions is the bent bond strap shown in Figure 7.10(a). Even if the strap has a cross-sectional area sufficient to conduct the current, it will frequently break, as shown in Figure 7.10(a), if it forms a bend of more than about 45°. Such straps should be installed, whenever possible, in a straight line, as is the strap in Figure 7.10(b).

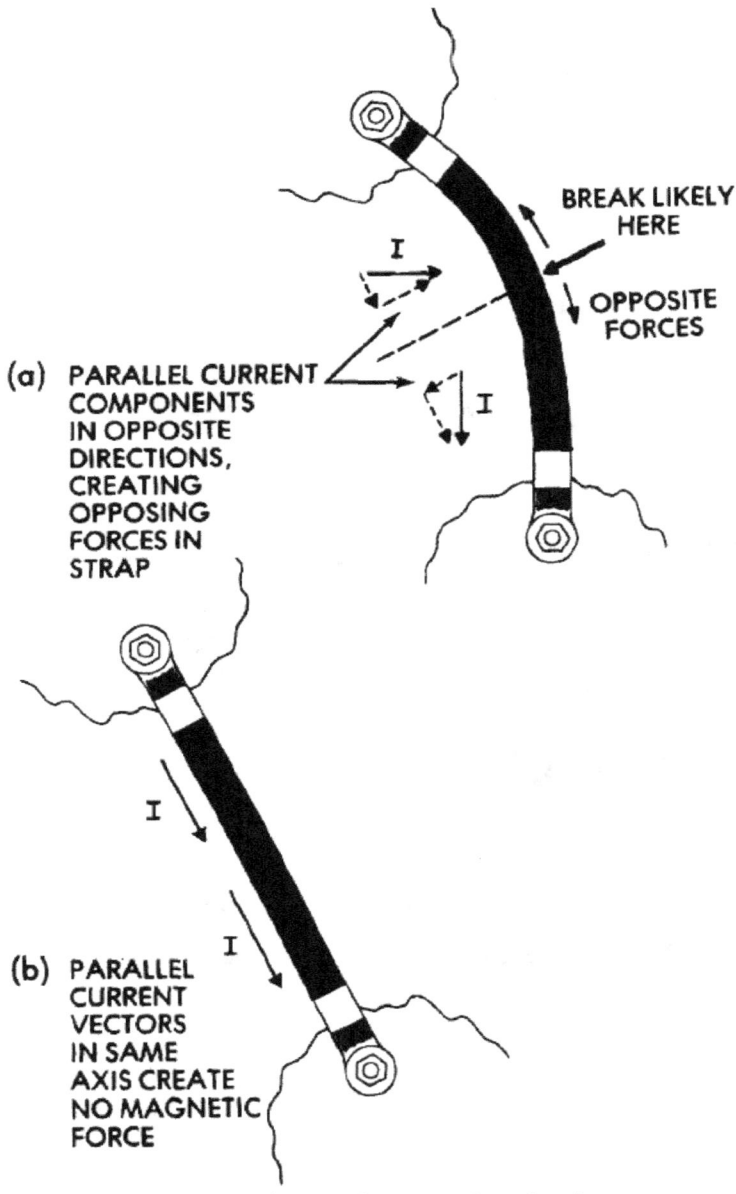

BREAK LIKELY HERE

OPPOSITE FORCES

I

I

(a) PARALLEL CURRENT COMPONENTS IN OPPOSITE DIRECTIONS, CREATING OPPOSING FORCES IN STRAP

I

I

(b) PARALLEL CURRENT VECTORS IN SAME AXIS CREATE NO MAGNETIC FORCE

Figure 7.10 Magnetic force on a bent bond strap.

Another reason for keeping the bond strap straight——and as short as possible——is that the inductive voltage rise occasioned by lightning currents flowing in a longer strap may be sufficient to spark over a direct, shorter path across the air gap. Some other do's and don'ts regarding the design of bond straps intended to carry lightning currents are presented in Figure 7.11. The basic rules to follow are these:

- Use conductors with sufficient cross-sectional area to conduct intended lightning current action integral
- Keep bond straps as short and close to the air gap length as possible
- Avoid bends of more than 45°, or other features that result in reversal of current directions
- Avoid all sharp turns
- If two or more parallel straps are used, separate them sufficiently (usually 30 cm or more) to minimize magnetic force effects.

Of course, the above rules should also be followed not only for bond straps but for any other structural element that must carry lightning currents.

Because of weight limitations, the strength and rigidity of some metallic components, such as wing tips, flaps, and ailerons, may not be sufficient to resist deformation as a result of the magnetic forces from lightning currents concentrated in these extremities. Such deformations will not normally impair safety of flight, but they may require repairs or replacement. Normally, only severe lightning currents cause this deformation; therefore, increased reinforcement of the extremity, if it were deemed economical, could minimize or prevent deformation in many cases.

Since determination of magnetic force effects by mathematical analysis for all but the most elemental geometries is very difficult, laboratory tests may prove the most straightforward and economical way to determine whether or not magnetic force effects are likely to cause deformation of prospective structures. For such tests, simulated lightning strokes in accordance with the SAE criteria of *Lightning Test Waveforms and Techniques* (Reference 7.9) should be passed through the component being evaluated.

7.2.5 Arcing across Bonds, Hinges, and Joints

Basically, the riveted skins and substructures found in conventional aluminum aircraft are adequate to safely conduct lightning currents through the aircraft, and, aside from the special considerations relating to fuel tank areas discussed in Chapter 6, no provisions need be made to improve this conductivity for lightning protection purposes. Even with the profusion of electrically insulating primers and wet sealers now used to coat joined surfaces, the great number of rivets or fasteners needed to meet mechanical strength requirements provides the necessary electrical path between mating surfaces. This generalization has proved to be reliable even when the fasteners and holes themselves have been coated with primers or wet sealants.

If hinges or bearings are located where lightning currents may pass through them (such as on control surfaces in Zone 1B or 2B), they should be able to safely withstand these currents with no impairment of their function. Otherwise suitable means should be provided to divert much of the lightning current around them. In a few cases, passage of lightning currents through some hinges or bearings may roughen or even weld the mating surfaces together, occasionally impairing further movement.

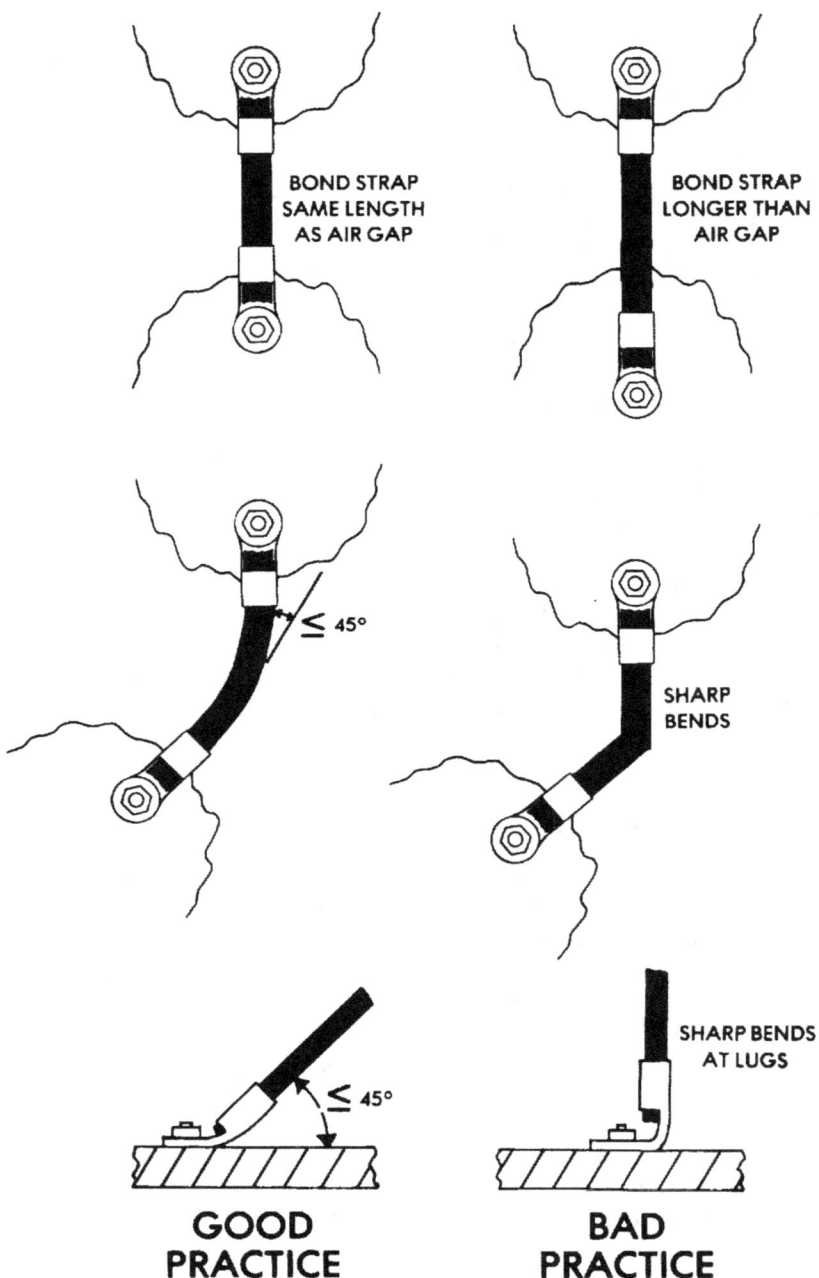

Figure 7.11 Design of bonding straps, or jumpers.

204

Damage such as this occurs only when a hinge or bearing has a single point of contact through which most of the lightning current must pass. In such cases the point(s) of hinge contact may be damaged. Hinges with multiple points of mechanical contact are usually able to safely withstand lightning currents with only minor pitting or erosion. For example, a piano-type hinge such as that shown in Figure 7.12 may be considered as self-bonded provided the dc resistance across the hinge is less than 0.01 Ω (Reference 7.10). In hinges where excessive damage occurs, additional conductivity should be provided to carry a major portion of the lightning current across the hinge to relieve it from excessive lightning current flow. Flexible bonding jumpers of the type shown in Figure 7.11 have often been installed across hinges for this purpose. In most cases, however, these jumpers have provided little or no increase in conductivity for lightning currents. The reason is that they almost always offer a longer and more inductive path than do the paths directly through the hinges, even though the resistances of the latter are high. Because of their fast rates of change, lightning current paths are governed much more by inductance than by resistance. Stahmann (Reference 7.11) has demonstrated that bonding jumpers, in fact, make little or no difference in the amount of superficial pitting that occurs on typical piano-type control surfaces or door hinges, such as those shown in Figure 7.12. Stahmann found the same to be true of ball joints and other hinges utilized on control surfaces or landing gear doors, even when the dc resistance through the hinge was as high as several ohms. No binding or other adverse consequences were found to occur from the small amount of pitting that did occur in Stahmann's tests of piano-type hinges and ball joints.

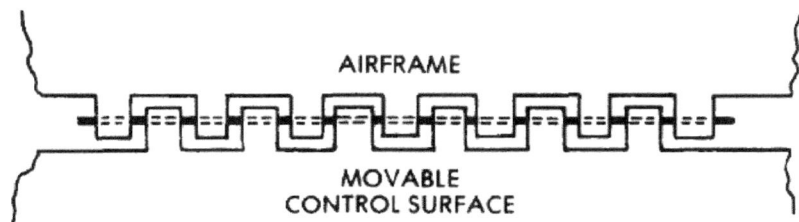

AIRFRAME

MOVABLE
CONTROL SURFACE

Figure 7.12 Piano-type hinge.
• Often capable of conducting lightning stroke currents without need for bonding jumpers.

To be sure of the ability of a particular hinge design to safely conduct lightning currents, a test similar to Stahmann's, in which simulated lightning currents were conducted through a prototype hinge, should be performed. The test waveforms and current described in *Lightning Test Waveforms and Techniques* for the zone in which the hinge is located should be used. If excessive pitting, binding, or welding of the hinge is found to occur, additional conductivity may be necessary. The most satisfactory way of providing this additional conductivity may be to provide additional areas of contact in the hinge itself, or else to provide additional hinges.

Bond straps across hinges *are* sometimes *required* to prevent the electromagnetic interference that arises from precipitation static charges being conducted through hinges with loose or resistive contact. The low currents involved are sometimes unable to follow uninterrupted paths through hinges, with the result that minute sparking occurs. Bond straps are usually able to provide enough conductivity to reduce this sparking. If bond straps are applied for this purpose, the guidelines of Figure 7.11 should be followed.

7.3 Nonmetallic Structures

With the use of nonmetallic materials in aircraft structural and fuel system components, the question of lightning vulnerability becomes important. An all-metallic airframe is in many ways quite compatible with basic lightning protection requirements; often only minor design modifications have been needed to provide adequate lightning protection. Nonmetallic structural components, however, may be considerably more vulnerable to lightning from lack of electrical conductivity. Thus, protective measures must usually be applied. If this is not done, hazardous consequences, such as those illustrated in the example described in Section 4.5 of Chapter 4, may result.

There are three basic types of nonmetallic materials found in aircraft. These are fiberglass-reinforced plastics (commonly called "fiberglass"), boron- or graphite-reinforced plastics (commonly called "advanced composites" or simply "composites"), and polycarbonate resins such as Lexan[1] and Plexiglas[2]. The latter materials are used for windshields or canopies. Fiberglass is used most commonly to achieve light weight in wing tips, fin caps, access doors, and in other applications where high mechanical strength is not required. The advanced composites, on the other hand, are capable of sustaining very high tensile and shear loads; for this reason they are being considered for primary structural applications.

Since the electrical conductivity and high-voltage withstand capability of each of these types of nonmetallic materials vary widely from those of the others, in the following paragraphs we shall discuss the important lightning effects problems and design of protective measures for structures made of each basic material.

7.3.1 Fiberglass Structures

Fiberglass-reinforced plastics have essentially no electrical conductivity. Moreover, they are usually incapable of withstanding the high-voltage stresses created by the approaching lightning leader and frequently will undergo dielectric breakdown, allowing the flash to puncture them and terminate on some conducting object inside. The ensuing return stroke current can blast a

[1] Trademark of the General Electric Company
[2] Trademark of the Rohm and Haas Company

206

large hole in fiberglass skins such as the skin of the nose radome shown in Figure 4.11. Each of the component materials, glass and resin, has relatively good voltage withstand capability; however, the many interfaces which exist between the glass fibers and the plastic resin when the two are put together create voids and seams which will withstand much less voltage stress before ionization and charge mobility (breakdown) begin. Thus, an external conductor, or *diverter*, must usually be placed on the outside surface of a fiberglass skin to provide a place to which lightning flashes can attach. The need for such a diverter and the guidelines for its proper design and location can best be understood by reviewing the lightning attachment mechanism itself.

7.3.1.1 Lightning Attachment to Fiberglass Materials

In Chapter 1, Section 1.2.4, we described the basic lightning breakdown process, and in Chapter 2, Section 2.2, we considered how this process causes attachment to an aircraft in flight. As noted there, when the intense electric field created by the oncoming leader reaches the ionization potential of air in the neighborhood of extremities and sharp edges on the aircraft, ionization occurs and streamers propagate outward from the aircraft in the direction of the oncoming leader. Since fiberglass has no electrical conductivity, the electric field passes directly through it, which phenomenon can cause streamers to originate from objects inside as well as outside of the fiberglass structure. What happens may be viewed as a race between streamers propagating from conducting objects inside and outside of the fiberglass structure, as shown in Figure 7.13.

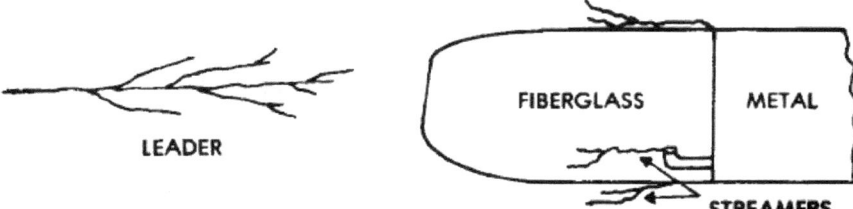

Figure 7.13 Streamers induced by approaching leader.
- Streamers occur from objects inside and outside the fiberglass structure.
- Examples are wing tips, external fuel tanks, and radomes.

Both the voltage withstand capability of the fiberglass and the distances along alternate breakdown paths are important in establishing whether puncture or external flashover will occur. Usually, of course, a higher intensity electric field is necessary to permit the internal streamer to puncture the fiberglass wall and contact the leader than would be necessary merely to draw the external streamer through the air to the leader. As the leader approaches, the field intensity increases until one of the streamers reaches and joins the leader. The electric field about this extremity of the aircraft then collapses (breaks down), and the leader proceeds from another extremity of the aircraft (as described in Chapter 2).

Figure 7.14 (Reference 7.12) shows the results of two simulated lightning attachment tests on a typical fighter aircraft nose radome performed in accordance with SAE Report *Lightning Test Waveforms and Techniques for Aerospace Vehicles and Hardware*. The approaching leader is represented by the striped electrode, to which a high voltage was applied by a Marx-type generator. When the field between the probe and the radome became high enough, streamers began from conducting points both inside and outside the radome and also from the high-voltage electrode. In Figure 7.14(a) the streamer drawn from the external conducting base reached the leader first; thus the base became the attachment point. In 7.14(b), however, an internal streamer, emanating perhaps from the radar antenna, punctured the fiberglass wall and reached the leader before the external streamer from the conducting base. The antenna, therefore, became the attachment point.

Figure 7.14 Simulated lightning attachment point tests of a nose radome.
  (a) Outside streamer intercepts leader.
  (b) Internal streamer punctures fiberglass wall and intercepts leader.

Since these photographs were taken outdoors in daylight, only those streamers that reached the leader and were brightened by the generator discharge current are visible. Figure 7.15 (Reference 7.13) shows similar lightning attachment tests performed on a fiberglass section of an external fuel tank. The tests were made in a darkened area indoors; thus in Figure 7.15(a), in addition to the main discharge, streamers emanating from alternate points can be seen. Streamers also occurred at the same time from objects inside of the tank, and one of these succeeded in puncturing the tank wall, as is evident from Figure 7.15(b). Here, both an external and an internal streamer reached the leader, causing some of the generator discharge current to follow the internal streamer, as evidenced by the reduced brightness of the external flashover. Since the tank had no external diverters, the result illustrates the poor insulating capability of fiberglass alone.

Figure 7.15   Simulated lightning attachment tests of fiberglass fuel tank section.
(a) External flashover
(b) Simultaneous external and internal flashovers

If the test voltage applied between the electrode and the tank were to rise at a slower rate, the electric field might not become intense enough to puncture the fiberglass wall before a streamer coming from an external point reached the leader and collapsed the field. Thus, the rate of voltage rise (dV/dt) is also important in determining whether puncture or external surface flashover will occur.

This important factor can be further explained by comparing the breakdown *timelag* characteristics of solids and air. All insulating materials, whether solids or gases, respond to high voltages and break down according to a

209

timelag curve of the shape shown in Figure 7.16. Timelag effect means simply this: the shorter the time for which a voltage is applied across a given insulation, the higher that voltage must be to cause breakdown; and conversely, the longer the time, the lower the voltage necessary to cause breakdown. There is, of course, a voltage level below which breakdown will not occur at all, even if the voltage is applied for a long time. Most solids show a flatter timelag characteristic than do air or surface flashover paths, as shown in Figure 7.16.

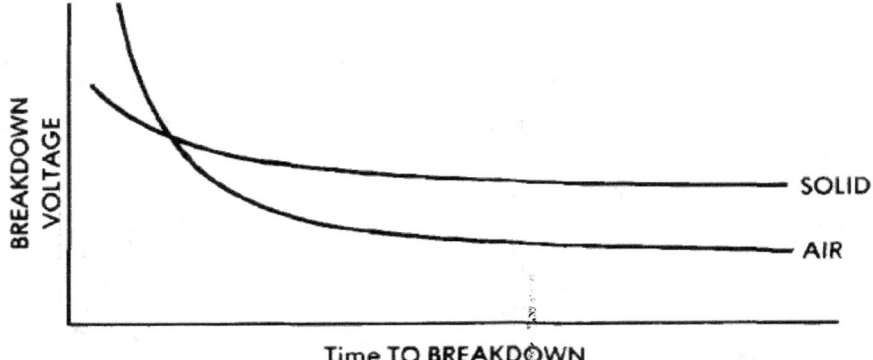

Figure 7.16 Breakdown timelag curves for solids and air.

In general, for fiberglass components, such as the radome of Figure 7.14 and the fuel tank section of Figure 7.15, an oncoming lightning flash has alternate paths to a metallic conductor. One is via a puncture of the nonmetallic skin to an internal metallic component, such as the radar dish. The other is via surface flashover to the nearest exposed metal. While the path via the puncture may be much shorter than that along the outside surface, the added insulation provided by the solid skin often compensates to some degree for this, making both paths viable alternatives. The significance of voltage rate of rise now becomes evident. Because the timelag curve for the alternate paths cross each other, there are voltage waveforms that will intersect either timelag curve, as shown in Figure 7.17, where both a "fast" and a "slow" voltage waveform are superimposed on the breakdown timelag curves of Figure 7.16.

From Figure 7.17 it is evident that the faster rising test voltage is the more severe in terms of increased probability of puncture. Thus, for attachment tests, such as those of Figures 7.14 and 7.15, a test voltage with the fastest rate of rise expected from a natural lightning leader is desirable.

No time-domain measurements have as yet been made of the electric fields surrounding an aircraft during the lightning-strike formation process because such measurements present formidable instrumentation problems. However, in Chapter 1, Section 1.2.3, we state that the leader advances at about 1 to 2 x 10^5 m/s, which would result in an average of 5 to 10 μs for the leader to travel a distance of 1 m. Since about 500 kV are required to break down a 1 m air gap in

5 μs, an appropriate rate of voltage rise to use for attachment tests might be the following:

$$\frac{dV}{dt} = \frac{500\ 000\ V}{5 \times 10^{-6}s} = 100\ kV/\mu s \qquad (7.18)$$

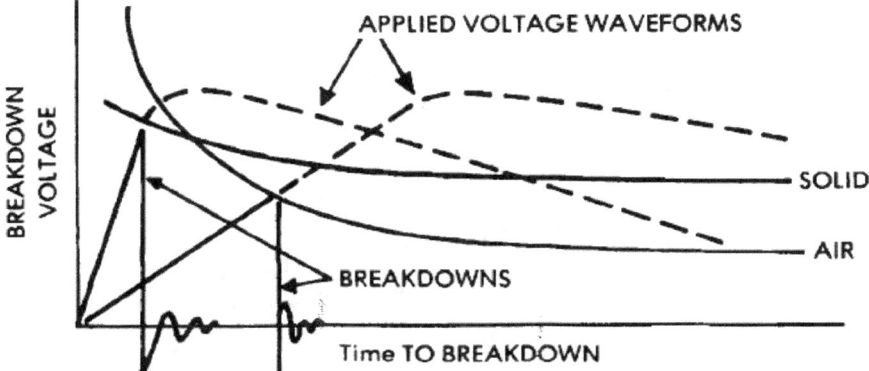

Figure 7.17 Fast and slow applied voltage waveforms superimposed on breakdown timelag curves for solid and air breakdown paths.
- Higher voltage rate of rise intersects solid curve and puncture occurs.
- Lower voltage rate of rise intersects air curve and surface flashover occurs.

On the other hand, if it is remembered that the actual breakdown of a single step of the leader is itself a series of smaller step breakdowns and pauses, it is likely that the rate of voltage rise across segments only a few meters long might be faster (or slower) than the average. Thus, to encompass the worst case, a rate of voltage rise 10 times as fast, or 1000 kV/μs, has been prescribed in SAE Report *Lightning Test Waveforms and Techniques* for attachment tests of aircraft components.

In practice, the timelag curves representing different paths through solid and air insulation are flatter and closer together than those drawn in Figures 7.16 and 7.17, and the point where the solid and air curves cross is less clearly defined. Thus, a relatively wide difference in applied voltage rate of rise exists between waveshapes causing breakdowns 100% of the time along one path as compared with waveforms causing all breakdowns to occur through the other path. This fact is additional support for the selection of 1000 kV/μs as the rate of rise for qualification test purposes. This waveform should create a faster (but not excessively fast) rising electric field than most natural lightning flashes create. In cases where a comparison between the results of laboratory tests using this rate of rise and subsequent in-flight lightning attachments to the same fiberglass component has been made, the laboratory tests have accurately predicted the in-flight attachment points and breakdown paths.

211

Figure 7.18 shows the high voltage applied between the test electrode and the fuel tank of Figure 7.15. In this test the electrode was at negative polarity with respect to the tank; thus the waveform has a negative deflection. Both polarities, of course, should be applied in the course of any complete test program. The rate of rise of this test voltage was 1000 kV/µs; it had reached 1000 kV when complete flashover of the 1 m air gap and fiberglass tank occurred.

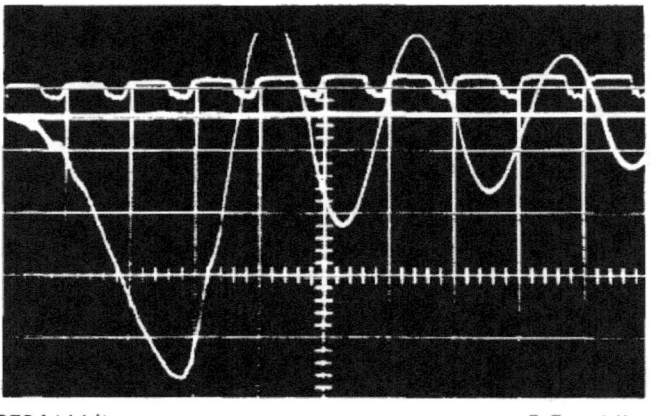

250 kV/div.   0.5 µs/div.

Figure 7.18 Typical high-voltage test waveform for attachment tests.

- $\dfrac{dV}{dt} = 1000 \text{ kV/µs}$
- Flashover on rise at 1000 kV

The 1000 kV reached by this test voltage (Figure 7.18) was of course applied across the total breakdown path between the high-voltage electrode and the metallic parts of the tank, shown in Figure 7.15. That amount of this voltage which must develop across the fiberglass wall for puncture to occur can be determined by applying test voltage directly across the wall. Figure 7.19 shows such a test of the wall of the fuel tank of Figure 7.15, using only 2.54 cm (1 in) of air gap between the electrodes and the wall. Each test voltage was allowed to reach its peak and to decay through the test circuit if breakdown did not occur, and the tests were repeated with successively higher peak voltages until breakdown occurred.

In the tests of Figure 7.19(a), peak impulse voltages of up to 125 kV were withstood by the wall, but partial breakdown occurred when a peak of 152 kV was applied. Complete breakdown occurred at 173 kV. From this test it might be concluded that the path involving puncture of the tank wall would require 173 kV more voltage than would the path along the outside of the tank, and that if the external and internal paths were otherwise the same, the external path would always be fully ionized before the wall was punctured. In practice,

212

however, the weaker segments of a path such as this through different dielectric media break down first and allow the total voltage to become concentrated across the remaining, stronger, segments. Thus, in such situations, where both the internal and external surface flashover paths are long compared with the wall thickness, the presence of the wall makes little difference: punctures occur randomly together with external flashovers.

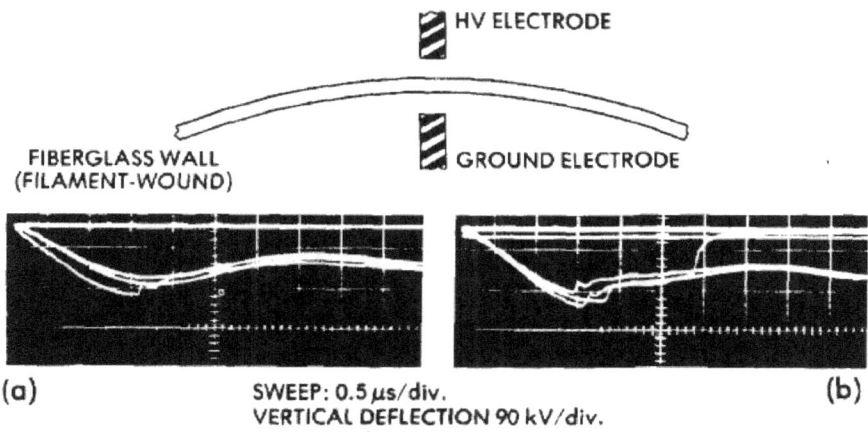

Figure 7.19 Impulse voltage test of typical fiberglass wall.
      (a) Withstands at 110 kV and 125 kV
      (b) Partial breakdown at 141 kV and 157 kV; complete puncture at 173 kV

An important point to remember is that the voltage withstand capability of a path through or along several different insulating materials, including air, is less than the sum of the separate segment withstand voltages. Thus, the breakdown voltage of a complex structure cannot be determined reliably by testing its components individually and adding up the individual withstand voltages. Laboratory tests of the complete structure, such as those of Figures 7.14 and 7.15, offer the best method of determining breakdown voltages and, as a result, the need for protection.

7.3.1.2 Protection for Fiberglass Components

The fiberglass punctures of Figures 7.14 and 7.15 indicate a need for external conductors that will inhibit internal streamer formation and provide external points from which streamers can originate. Such external conductors, called *diverters*, also shorten the surface flashover paths. If properly placed, diverters make puncture of a fiberglass skin unlikely. The following guidelines and tests should be followed to orient and space diverters properly on a fiberglass structure:

213

1. Orient the diverters as nearly as possible in the line of flight so that flashes which originally strike the diverter can reattach farther aft on the same diverter as the aircraft moves forward.

2. While the fore-aft arrangement recommended in Guideline 1 is desirable for reattachment purposes, an adequate path must also be provided for lightning current to be conducted into the airframe. This requirement usually calls for some of the diverters to be oriented perpendicularly to the line of flight. Examples of preferred arrangements incorporating both of these guidelines are shown in Figure 7.20.

3. The surface flashover voltage from any point over the external surface to the nearest diverter strap must be less than the maximum voltage required to puncture the skin and attach to a conducting object beneath. Thus, the diverter must be within a *maximum displacement distance* from a point on the skin directly opposite the enclosed conductor. This relationship is defined as follows:

$$\begin{array}{l}\text{Maximum} \\ \text{displacement} \\ \text{distance (cm)}\end{array} \leqslant \frac{\text{Skin puncture voltage (kV)}}{\text{Surface flashover voltage (kV/cm)}} \qquad (7.19)$$

This criterion should be met for lightning voltage stresses of either polarity applied at up to 1000 kV/μs rate of rise.

4. The maximum voltage drop from the original attachment point to any other point on a lightning arc swept aft directly above a nonconducting surface must not exceed the skin puncture voltage from that point through the skin to any conducting objects inside. The maximum arc voltage drop will occur during a restrike formation in a multiple-stroke flash and can be assumed to be equal to or less than the free air breakdown voltage, or about 500 kV/m of arc length. Since this voltage will be applied from a point directly above the surface, instead of from an approaching leader, this swept-stroke environment may be more severe than one produced by an initial strike.

5. The inductive voltage rise, V_L, along any diverter segment carrying lightning-stroke currents to conducting structure must be less than the skin puncture voltage between the diverter and the nearest conducting object inside the structure. The inductive voltage rise, V_L, may be expressed as follows:

$$V_L = L \frac{dI_L}{dt} \qquad (7.20)$$

where $L =$ diverter segment inductance (henries)
          $I_L =$ lightning current (amperes)
          $t =$ time (seconds)

In practice, L may be assumed to be 1 μH/m for most diverter

214

straps or foils, and dI_L/dt may be assumed to be 100 000 A/μs (1 x 10^{11} A/s).

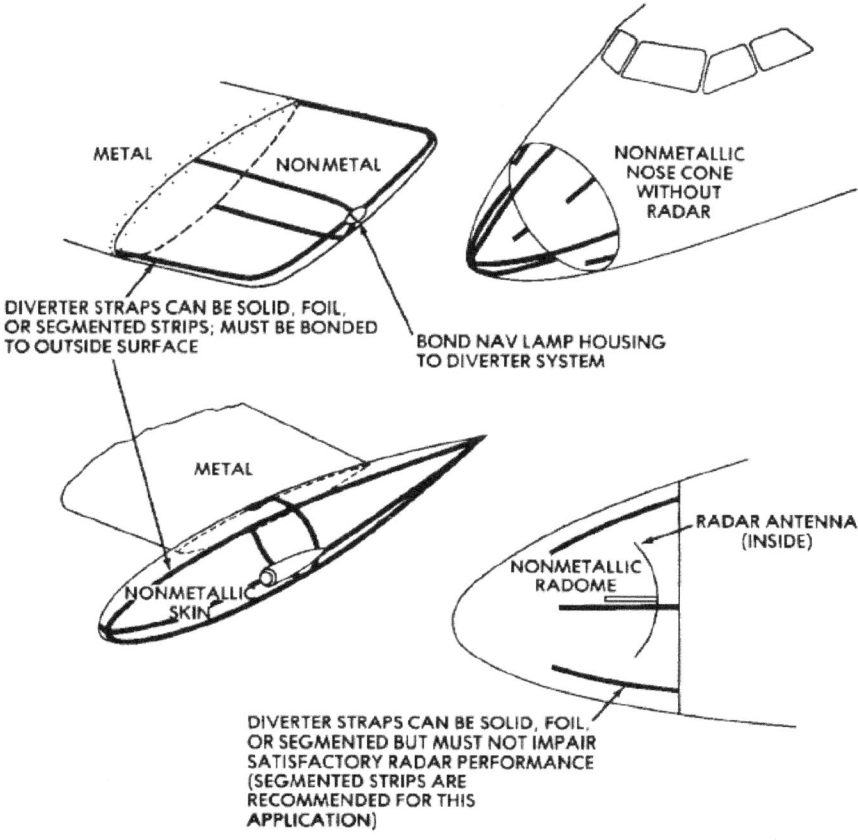

Figure 7.20 Arrangement of diverter straps on nonmetallic structural components.

The intent of Guideline 3 is to assure that sufficient diverters are utilized to prevent punctures resulting from initial strikes. Guideline 4 extends this criterion for swept strokes, and Guideline 5 is aimed at preventing punctures resulting from inductive voltages that arise when lightning currents flow through the diverter. This means that a diverter strap can, in fact, be *too close* to the internal conductor it is supposed to protect. If the maximum inductive voltage along the diverter is greater than the skin-puncture voltage, the diverter will have to be moved a *minimum displacement distance* away from a point on the skin directly above the enclosed conductor. This distance may be written quantitatively in the following way:

If

$$\text{(Diverter inductive voltage)} \leqslant \text{(Skin puncture voltage)}, \qquad (7.21)$$

then

$$\frac{\text{Minimum}}{\text{displacement}} = \frac{\text{Diverter inductive voltage (kV)} - \text{Skin puncture voltage (kV)}}{\text{Surface flashover voltage (kV/cm)}} \qquad (7.22)$$

This distance may also be expressed in terms of the diverter inductance and lightning current rate of rise as follows:

$$\frac{\text{Minimum}}{\text{displacement}} = \frac{L(\mu H)\ dI_L/dt\ (kA/\mu s) - \text{Skin puncture voltage (kV)}}{\text{Surface flashover voltage (kV/cm)}} \qquad (7.23)$$

If a minimum displacement distance applies, then the diverter, to be effective, must be positioned somewhere between the *minimum* and the *maximum* displacement distances from a point directly above the enclosed conductor.

Clearly, the skin-puncture voltages of nonmetallic skins must be known before protective diverter systems can be designed. The skin-puncture voltage for a particular nonmetallic skin construction is obtainable by impulse voltage breakdown tests, and is a function of the material type and thickness, as well as of layup patterns, core fillers, surface treatments, etc. Examples of skin puncture voltages determined from 1000 kV/μs impulse tests are given in Table 7.3.

Table 7.3 IMPULSE BREAKDOWN VOLTAGES OF TYPICAL FIBERGLASS SKIN MATERIALS

Skin Construction	Total Thickness	Breakdown Voltage
One fiberglass sheet	0.163 cm (0.064 in)	21 kV
Two filament-wound fiberglass tape skins enclosing polyimide foam filler	1.27 cm (0.500 in)	150 kV
Two fiberglass skins enclosing foam filler	0.99 cm (0.388 in)	70 kV

Surface flashover voltage levels depend most significantly on the type of surface finish applied. Most rain-erosive paint surfaces will flash over at an average stress of around 10 kV/cm of surface. For example, if a diverter to prevent punctures to the grounding tube located directly beneath the fiberglass skin of the radome of Figure 7.4 was desired, and if this skin was found to have a puncture voltage of 50 kV and surface flashover voltage of 10 kV/cm, then, by

216

Equation 7.19, the maximum displacement distance would necessarily be as follows:

$$\text{Maximum displacement distance (cm)} = \frac{150\,\text{kV}}{10\,\text{kV/cm}} = 15\,\text{cm} \tag{7.24}$$

If this diverter has an inductance of 1.0 μH, then the minimum displacement distance, by Equation 7.23, is as follows:

$$\text{Minimum displacement distance (cm)} = \frac{(1\,\mu\text{H})\,(100\,\text{kA}/\mu\text{s}) - 50\,\text{kV}}{10\,\text{kV/cm}} = 5\,\text{cm} \tag{7.25}$$

Thus, it can be concluded that the diverter strap should be placed between 5.0 and 15 cm away from a point directly over the protected conductor.

The above analysis assumes that the internal conducting objects are at the same potential as the conducting structure to which the diverter is bonded. This assumption is valid because no lightning current flows through the internal conducting objects, and thus there are no potential differences between them and a common conducting structure.

The protection of a nonmetallic structure is most important when the structure contains an integral fuel tank. If this combination is located in Zone 1 or Zone 2, a protective diverter system to prevent punctures and the ignition of fuel should be applied. Of course the number or size of diverters that may be required can be minimized by making as many as possible of the internal fuel tank components of nonconductive materials. By so doing, some objects within the tank which could otherwise be the source of streamers and eventual attachment points will be eliminated. Any conducting parts that must remain inside should then be positioned as far away from the outside skins as possible, and they should be devoid of sharp edges, as discussed in Chapter 6, Section 6.7.

Plumer (Reference 7.14) discusses a particularly challenging situation: an entire wing, including an integral fuel tank, is made of fiberglass; moreover, a metal capacitance probe, vent pipe, end panels, and filler cap are present. The tank is shown in Figure 7.21. The vent line and end plates of this tank could be made of nonconducting materials. The capacitance-type fuel quantity probe, on the other hand, must of necessity be made of conducting material, and the probe should be positioned as far as possible from the top and bottom skins.

Figure 7.21 also shows the protective diverters that were designed for the bottom and top skins of this tank. Diverters will usually be required if there is *any* conducting object within the tank—and sometimes even if there are no conducting objects inside. If the puncture and surface flashover characteristics of the enclosing skin materials are known, the diverter arrangement can be designed using Equations 7.19 to 7.23. Frequently, however, when newer materials are used, puncture and surface flashover information is not available. In this case, a section of the proposed skin material can be given a high-voltage test to determine the required diverter spacings, as illustrated in Figure 7.22. The skin material to be tested should have the same surface treatments and thickness as those proposed for the tank. Conducting objects simulating those in the actual tank should be positioned appropriately beneath the tank skin and together serve as one of the test electrodes. The other electrode should be a rod of at least 1 cm diameter, positioned about 1 cm above the skin, as shown in Figure 7.22. Test voltage at up to 1000 kV/μs rate of rise should be applied between these electrodes.

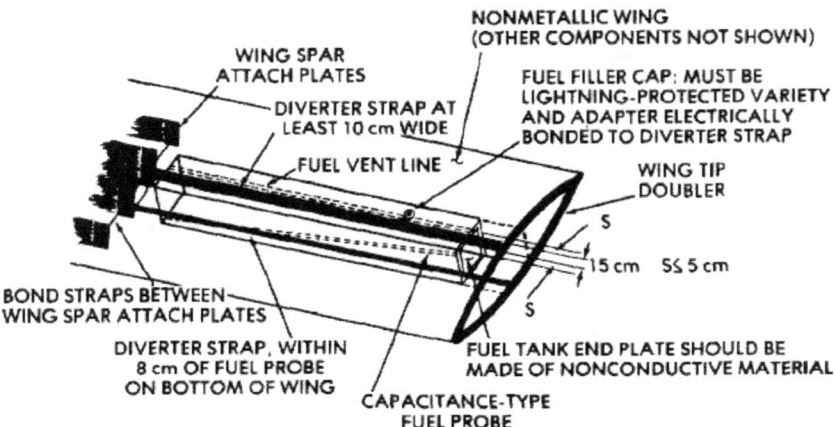

Figure 7.21 Protective measures for integral fuel tank within all-fiberglass wing.

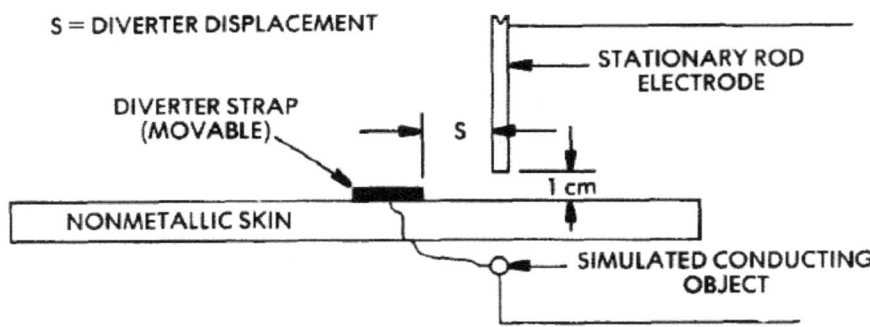

Figure 7.22 Test setup for experimental determination of maximum diverter displacement distance, S.

Proper testing is begun with the diverter strap directly beneath the test electrode. Since all discharges in this position will terminate on the diverter, the diverter can subsequently be moved from a point directly beneath the electrode to a spot where S equals 1 cm, for example. When the test is repeated, arcs will flash down to the nonmetallic surface and along it to the diverter. The test should be repeated with the diverter moved another centimeter away, and so on, until a puncture occurs and the arc terminates on the conducting object beneath the skin instead of on the diverter. When this point has been reached, the diverter is no longer effective, and the largest spacing, S, at which no punctures occurred defines the maximum displacement distance. If more than one diverter is to be used, the diverters should be arranged so that the conducting object is between them. In this case their maximum separation must not exceed 2S, as shown in Figure 7.23.

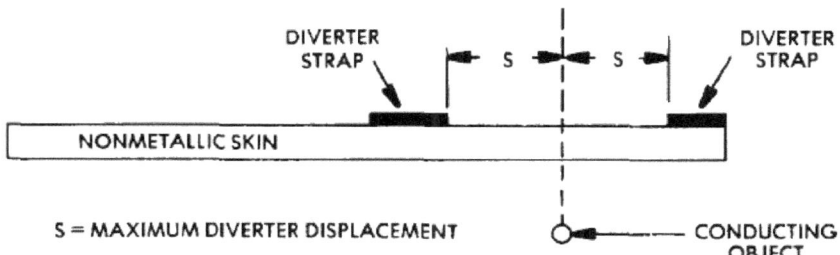

Figure 7.23 Arrangement of multiple diverter straps for maximum separation.

One such test was run by Plumer on a section of the wing tank shown in Figure 7.21, with the aluminum vent line imbedded in the upper tank wall and the capacitance probe along its side wall, as recorded in his report. The diverter was found to be effective within a spacing, S, of 6.7 cm; but a spacing of 9.2 cm caused the arc to puncture the skin and terminate on the vent line beneath. Thus, allowing for a margin of safety, the diverter should be no more than 5 cm away from a line directly above the vent tube.

From the same test performed for the capacitance probe, it was found that the diverters could be over 15 cm away before puncture occurred. The reason for this difference is that the capacitance probe was farther away from the upper skin; hence more voltage was required to puncture the skin and allow termination on the probe. Since the probe was angled, however, and did come closer to the tank surface farther outboard, it can be concluded that the 5 cm criterion determined for protection of the vent tube should also be applied for the capacitance probe. Since in this tank the capacitance probe and vent tubes are about 25 cm apart, two diverter straps (or perhaps a single strap 15 cm wide) would be required to provide adequate protection. Straps would be required on both the top and bottom of the tank for prevention of punctures to the capacitance probe, but only on the top surface to prevent punctures to the vent tube, since the vent tube is located just beneath the top surface. Fabrication of the vent tube from nonconducting material would therefore eliminate the need for a strap as wide as 15 cm.

Each diverter strap utilized should adhere to the outer skin surface and should be electrically bonded to the wing tip doubler or diverter assembly at the outboard end and to those metallic structural components at the inboard end which are capable of conducting lightning current on through the airframe.

Protection against the lightning effects on fuel systems discussed in Chapter 6 should also be accomplished.

A variety of conductors for diverter applications may be considered. Selection of the one to use depends on such factors as weight, obstruction of the airstream, single- or multiple-strike capability, radar antenna pattern interference, and cost. Brief descriptions of typical diverters follow.

External Diverter Strap

The external diverter strap, a simple metal bar, as shown in Figure 7.24, is mounted on an external surface. This type of diverter is designed to receive more than one stroke without sustaining significant damage and is intended, therefore, for permanent applications. External diverter straps are commonly used to protect radomes and wing tips, which receive frequent strikes. The straps are usually made of aluminum, with a rectangular cross section of at least 0.1 cm^2 (20 000 circular mils) to permit conduction of 200 kA, $2 \times 10^6 \text{ A}^2$-s stroke currents without excessive temperature rise or elongation. In order to remain intact by minimizing melting, most straps now in use have areas larger than 0.1 cm^2. Thicknesses of up to 0.64 cm (0.25 in) and widths of up to 1.27 cm (0.5 in) have been used. A common design is 0.32 cm (0.125 in) thick by 0.95 cm (0.375 in) wide.

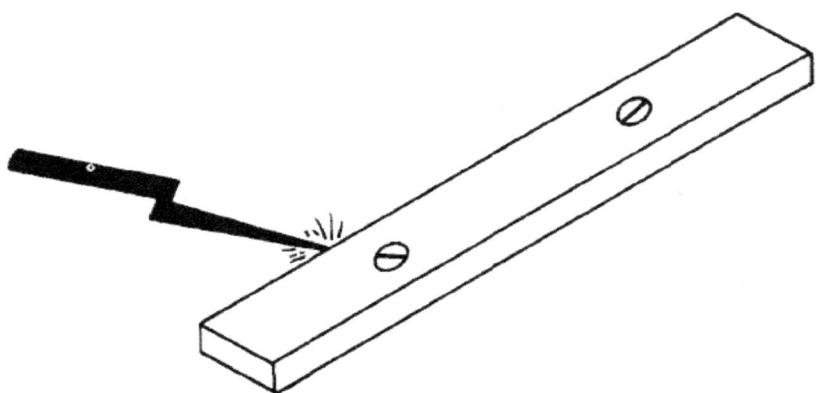

Figure 7.24 External diverter strap.

The straps are usually attached to the wall with steel screws or bolts; the larger dimensions allow holes and countersinks to be made in the strap without reducing to less than 0.1 cm^2 the strap cross section. In nose radomes, the threaded portions of the screws are often surrounded by plastic inserts to prevent sparking from the screws to the radar antenna during lightning encounters.

Since the metal straps are installed on the outside surface, there may be a small drag penalty. This drag can be minimized by orienting the straps parallel to the airstream or by shaping the cross section of the straps aerodynamically.

Internal Strap with Protruding Studs

Mounting of external straps flush within the nonmetallic wall not only creates tooling and manufacturing problems but also degrades the strength of the wall because of stress concentrations along the grooves required for flush mounting. Therefore, metal straps or tubes have sometimes been mounted on the inside surface structure, with metal studs protruding through the wall to serve as lightning attachment points, as shown in Figure 7.25.

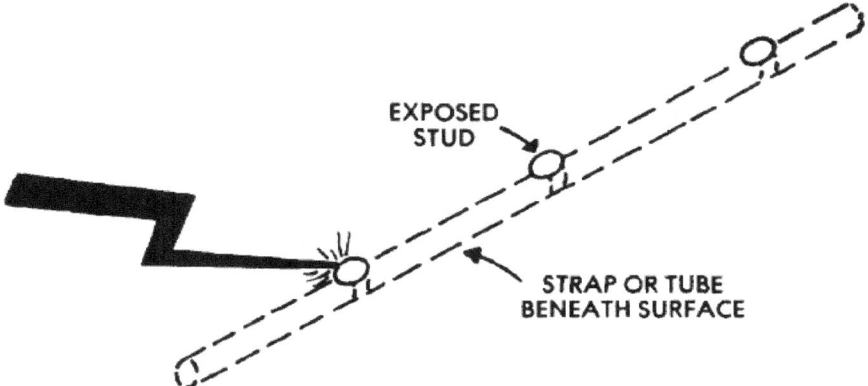

Figure 7.25 Internal strap with protruding studs.

Although this type of diverter installation may reduce the aerodynamic drag created by external mounting, it fails to take advantage of the insulation capability of the nonmetallic wall, and, in fact, may even intensify the electric field stress through the wall, promoting punctures if the protruding studs are too far apart. In general, the thicker the wall, the farther apart may be the studs. In many cases, 30 cm (1 ft) is close enough, but individual cases should be evaluated by high-voltage tests similar to that of Figure 7.22, or by tests of the complete structure similar to those of Figures 7.14 and 7.15. Appropriate test conditions for the latter are described in *Lightning Test Waveforms for Aerospace Vehicles and Hardware.* The possibility of puncture to an internal strap or tube can be minimized if the portion of it between the protruding studs is covered with an insulating sleeve.

When lightning currents travel through any internal conductor, of course, there is also the possibility of the inductive voltage rise causing sparking to nearby conductors which are not carrying current. Covering of the lightning conductor with an insulating sleeve in the vicinity of nearby conductors will reduce the possibility; the internal strap or tube can be used without such a sleeve if the spacing between the strap and adjacent conductors is great enough to prevent sparking. For purposes of estimating what this spacing should be, assume that the diverter strap has an inductance, L, of 1.0 μH/m of length. Assume a maximum lightning current rate of rise of 100 kA/μs, and determine the air-gap impulse sparkover voltage from Figure 6.39 or from impulse voltage tests of the particular configuration. Then make sure that the inductive voltage rise

$$V_L = L \frac{di_L}{dt}$$

is less than the gap sparkover voltage. It is prudent to add a safety margin of 50% or more to the calculated inductive voltage when making this comparison.

Thin Foil Strips

Thin foil strips, usually made of aluminum 0.008 to 0.040 cm (0.003 to 0.015 in) thick, are capable of conducting moderate strokes or of melting and vaporizing on severe strokes. In the latter case they leave an ionized channel through which subsequent currents in the flash may travel. Because they are thin, foil strips have good aerodynamic characteristics, but they usually provide protection against only one severe flash. The thin foil strips are generally used when aerodynamic drag requirements prevent thicker straps from being installed on external surfaces. A typical strip is shown in Figure 7.26.

Figure 7.26 Thin foil strip.

The skin of the structure must be thick enough to withstand safely the blast pressure that occurs if the strip vaporizes. Thick foil strips may also be used; however, they release more energy during vaporization than the thinner foil strips and, therefore, require a thicker radome outer skin. Wide foil strips are preferred because they distribute the explosive forces over a larger surface area.

A 0.008 cm (0.003 in) thick by 0.64 cm (0.375 in) wide aluminum strip will protect a 0.1 cm (0.040 in) thick fiberglass wall from one 200 kA, 2 x 10^6 A^2-s stroke and associated intermediate or continuing current, but the strip will have been vaporized and must be replaced.

Thus, the foil strips should not be used for applications where more than one strike might occur in the same flight. This caution means that they should not be used to protect fiberglass fuel tanks or other structures in which a puncture might result in a flight safety hazard, for, contrary to the old adage, lightning *does* strike the same place twice, even in the same flight! Thin foil strips are appropriate only for inexpensive protection of fiberglass fairings, wing tips, fin caps, or other non-flight-critical components.

Radio frequency interference may be created if the strips are improperly grounded to metallic structures, or if fatigue cracks occur in the strips, which may result in a broken electrical connection. Fatigue cracks are often caused by adhesive bonding of a single strip to a surface with several panels or segments that may move with respect to one another. Cracks may cause severe damage when lightning currents flow across them, as illustrated in Figure 7.27. This figure shows a 2.54 cm (1 in) wide by 0.018 cm (0.008 in) thick strip of pressure-sensitive aluminum foil tape after a test in which a (moderate) 54 kA stroke, followed by a 90 A continuing current, was discharged into the strip at Point b. The continuing current conveyed 18 C of charge into the strip. A razor-thin cut had been made in the strip at Point c; the damage which resulted as the current arced across this confined cut was more extensive than that produced by the same current when it entered the strip at Point b. Unblemished portions of the strip itself, however, were able to conduct these currents without damage. In fact, a previous test had delivered a 54 kA stroke and 187 C of charge to Point a, causing only the erosion shown at this point. The cross-sectional area of foil necessary to conduct more severe stroke currents can be determined from the curves of Figure 7.3.

Figure 7.27 Damage produced by moderate stroke and continuing current in 0.018 (0.008 in) thick, 2.54 cm (1 in) wide aluminum foil with crack.
- Point a – erosion resulting from previous stroke
- Point b – arc entry
- Point c – location of crack

223

Metallic Coatings

Coatings are thin metallic foils, paints, or plasma flame sprays that completely cover the nonmetallic structure to make the surface entirely conductive and enable it to accept strikes anywhere. Such coatings, however, may suffer considerable vaporization unless they are prohibitively thick. They are recommended only in cases where diverters are not practical.

Woven Wire Fabrics

Flexible metallic fabrics woven of fine aluminum, copper, or stainless steel wires about 80 to the centimeter, as shown in Figure 7.28, are usually more effective in resisting damage than are coatings because woven wire fabrics permit gases from arcing to be vented through the holes in the mesh instead of containing them, as do continuous foil and paint coating. Development of these fabrics came about in response to a need for the effective protection of advanced composites. Their use is discussed more fully in that context in Section 7.3.2.5.

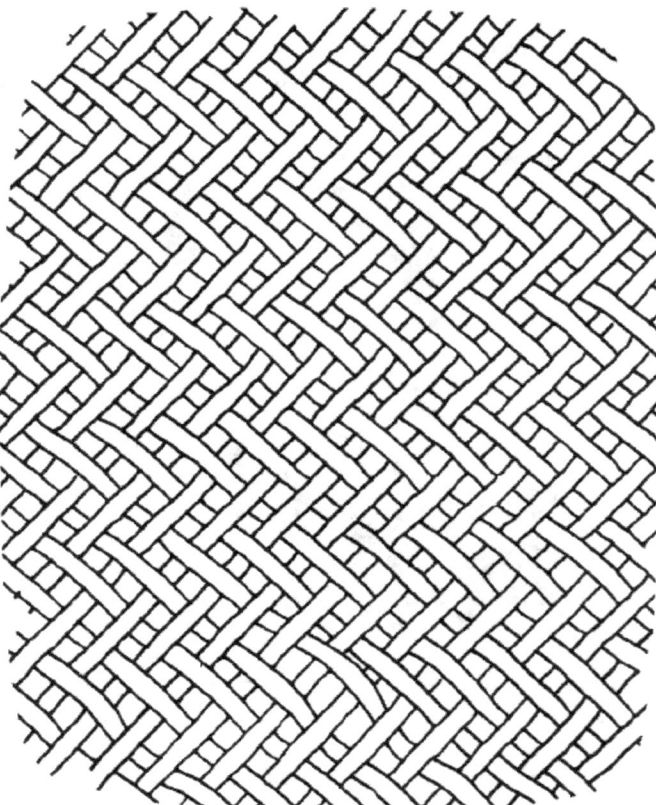

Figure 7.28 Woven wire fabric.

224

Segmented Strips

Segmented strips, sometimes called "button strips," are one of several developments in lightning protection technology. These segmented strips, developed by Amason and Cassell of Douglas Aircraft Company (Reference 7.15), consist of a series of thin, conductive segments, or "buttons," fastened to a strap of resistive material, which is then bonded to the surface to be protected, as shown in Figure 7.29. Instead of attempting to conduct lightning currents, these segmented strips provide many small air gaps that ionize and spark when an intense electric field is applied along the strip, as is the case when a lightning leader approaches. Since the small gaps are close together, the resulting ionization is nearly continuous. Thus the segmented strips provide a conductive path, initially for streamers to follow outward from the aircraft and, after these reach the leader, for return stroke and other currents to flow into the airframe. The construction of the strip and breakdown process are pictured in Figure 7.30.

Figure 7.29 Segmented strips.

Laboratory tests such as the one illustrated in Figure 7.31 show that the segmented diverters are nearly as effective in intercepting strikes and preventing punctures of fiberglass walls as are the solid conductors. Since ionization of the strips requires that an electric field be able to exist in the direction of their length, the segmented strips will perform best if oriented normal to conductive portions of the airframe rather than parallel to such structure. Also, the strips should not be positioned parallel to a conducting object lying beneath the nonmetallic surface. For example, the grounding tube within the radome of Figure 7.4 will reduce the electric field gradient along its length to nearly zero, so that a segmented diverter positioned near the tube may not ionize and an approaching flash may puncture the radome wall and attach, instead, to the tube itself. Thus, segmented diverters should be positioned on either side of the grounding tube but not directly over it. This positioning is also necessary to prevent inductive voltage along an ionized strip from puncturing the radome wall, as discussed in Section 7.3.1.2.

The maximum spacing between segmented diverters——or the minimum permissible spacing between them and underlying conductors——is dependent, among other things, upon the amount of voltage required to ionize the

225

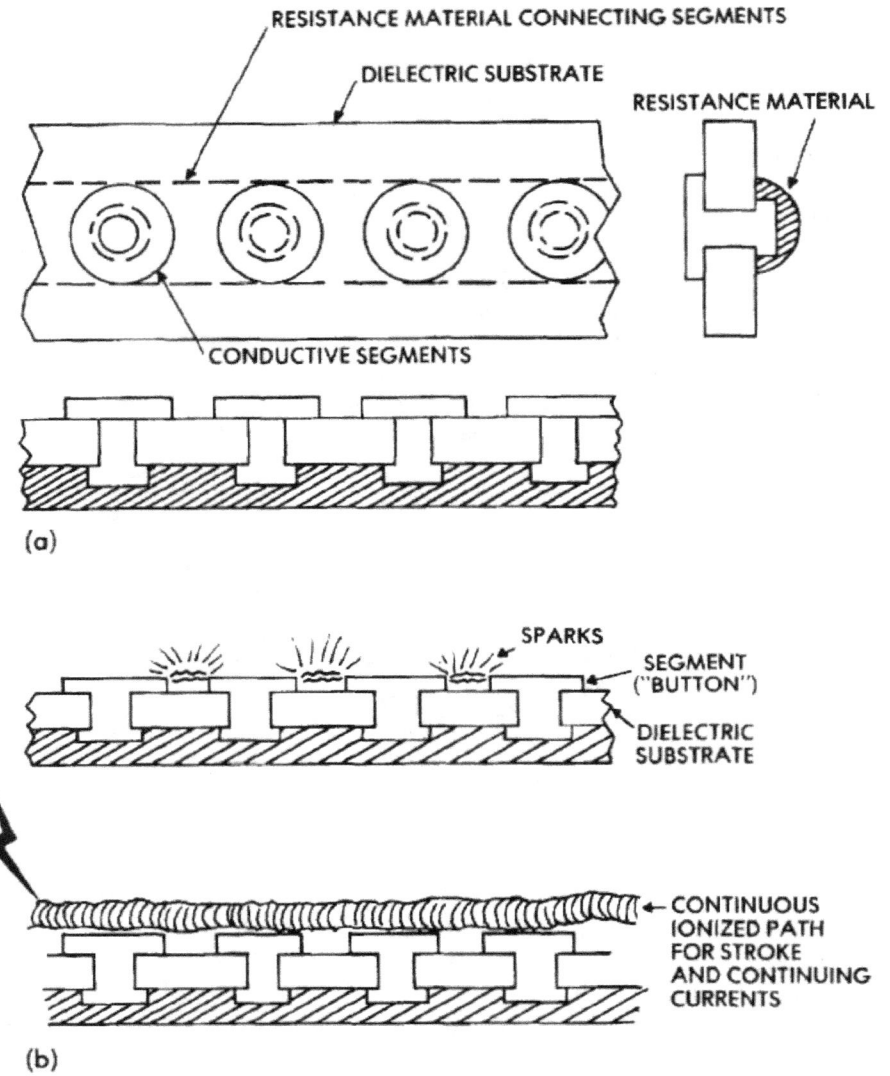

Figure 7.30 Segmented strip design and breakdown process.
    (a) Segmented strip design
    (b) Segmented strip breakdown and conduction process

segmented strips. Ideally, a very low voltage would be desirable——a voltage even lower than that required to ionize a path along the bare surface itself and much lower than that required to puncture the surface. While the developers report (Reference 7.16) that the strips ionize at about 10 kV/30 cm, they do not state what rate of voltage rise this applies to, nor have any oscillographic measurements showing test voltage rate of rise and strip ionization voltage yet

226

been reported in the literature. The authors, meanwhile, have found the strips to ionize at voltages of up to 70 kV/30 cm (1 ft) when tested with a voltage rising at 200 kV/μs. The conductivity of the surrounding surface and the proximity of underlying conductors undoubtedly play a role in the function of the strips, but no laboratory investigations of these factors have yet been reported either.

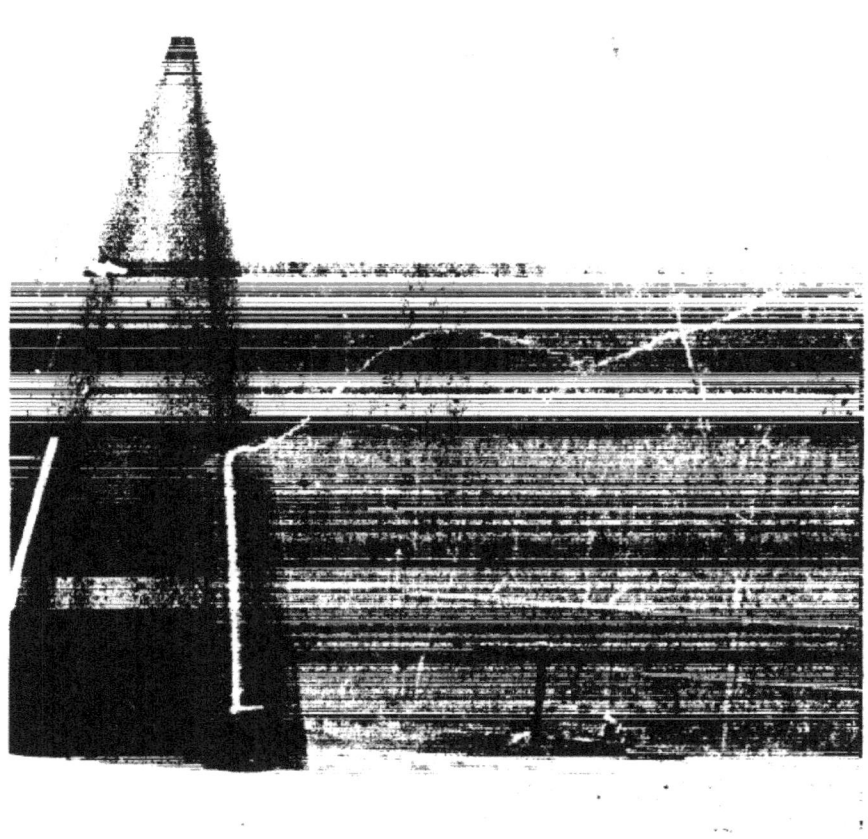

Figure 7.31   Simulated lightning strike to a segmented diverter.
  ● Streamers propagate up diverter and outward to join oncoming leader.

227

The developers do report (Reference 7.17) that the segmented strips remain undamaged after passage of a 170 kA stroke and 200 C continuing currents. This capability has been confirmed by the authors, who have found the strips capable of sustaining two strokes of 180 kA and 2×10^6 A^2-s (Component A) before destroying their ability to ionize another time. The authors have also found the segmented strips to be capable of sustaining a much larger number of the more common 50 kA strokes. Other investigators, including Conti and Cary (Reference 7.18), have reported the strips to be severely damaged after one 200 kA discharge or one "high coulomb" continuing current discharge has been delivered, a finding that has aroused controversy over the effectiveness of the strips. The continuing current charge (500 C) delivered by Conti and Cary, however, is more charge than that conveyed by any natural lightning discharge yet measured.

The segmented strip design might be modified to lower the ionization voltage or to improve durability; however, the basic principle is sound, and there are many applications where the presently available strips will afford lightning protection improved over other methods, while not interfering with radar performance. The present lack of strip performance data simply means that the proper location of the strips can not yet be determined analytically but, instead, must be accomplished with the aid of laboratory tests of candidate arrangements on the particular structure being protected.

The selection of protective devices and the arrangement of them on a fiberglass structure comprise a major part of the protection task. The best design can be completely undone, however, if other, seemingly innocuous, features of the structure are left unchanged. Figure 7.32 illustrates some particularly troublesome situations. They usually arise when practices perfectly acceptable in metallic structures are continued without modification in nonmetallic structures.

The several ways of providing protection for the fiberglass wing tip of Figure 7.30 should include the following:

- Remove paint from the diverters
- Electrically bond the lamp housing directly to the diverters
- Enclose the lamp power wires in a conduit or braided shield inboard to the metal rib
- Electrically bond the diverters to metallic structure.

7.3.2 Advanced Composites

Advanced composites, consisting of boron- or graphite-reinforced plastics, are being used increasingly often as replacements for aluminum load-bearing applications in aircraft. The high strength-to-weight characteristics of these materials make them attractive for structural applications, but because they are nonmetallic, they are inherently more vulnerable to lightning effects.

In contrast to aluminum, which by virtue of its electrical and thermal conductivity suffers few direct effects of lightning and which provides excellent electromagnetic shielding, the advanced composites are neither good conductors, nor good insulators, nor good providers of shielding against electromagnetic fields. If these materials are used in lightning-strike zones, it is possible that they will require some form of protection. The questions which must be answered are these: Just how vulnerable are particular advanced composites to lightning? and What protection, if any, may be necessary?

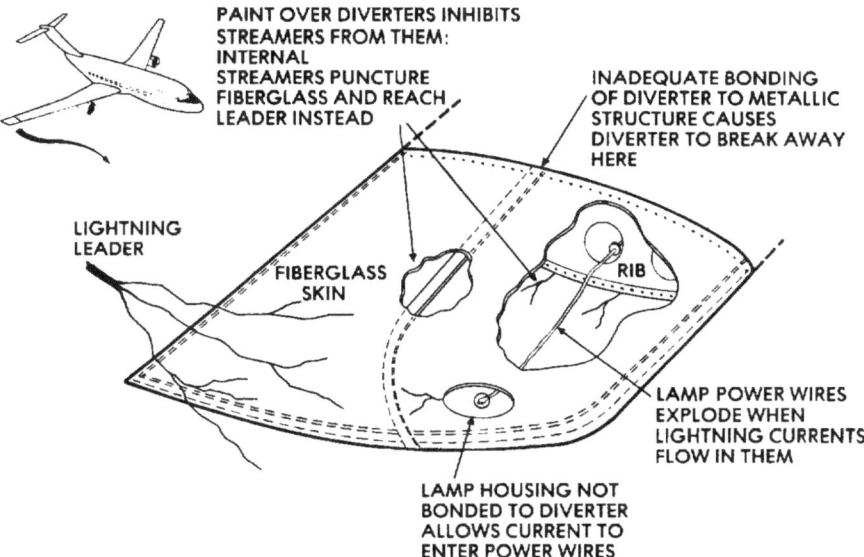

Figure 7.32 Inadvertent causes of disaster.

Unlike fiberglass-reinforced plastics, which have no electrical conductivity, the boron filaments or graphite fibers which are imbedded in the matrix are resistive conductors and will conduct some lightning currents. However, the fact that, like any resistor, they heat up when electric current flows in them causes a serious problem. For example, typical boron and graphite composites have resistivities of 4 to 9 x 10^{-3} Ω-cm and 0.9 to 1.1 x 10^{-4} Ω-cm, respectively, as compared to 2.8 x 10^{-6} Ω-cm for aluminum. This means that the composites will try to absorb between 100 and 1000 times more energy from the lightning current than will aluminum. Moreover, the materials may not be able to dissipate this much heat without some change in or destruction of physical properties. Simulated lightning-strike tests of typical composite laminates have demonstrated this problem, with results similar to those shown in Figures 4.12 and 4.13.

A number of careful studies of the electric current conduction and damage mechanisms within advanced composites have been made in order to learn the quantitative relationships between current intensity and the amount of mechanical strength degradation. It has been confirmed that damage results primarily from electric current flow through the boron filaments or graphite

229

fibers, but the mechanisms differ for these two types of composites. The current flow processes and degradation mechanisms for each type are reviewed in the following paragraphs.

7.3.2.1 Boron Composites

Because of initial interest in boron composites for aircraft structural applications and because of their higher resistivities, the lightning current degradation mechanisms for boron were studied first. Fassell, Penton, and Plumer (Reference 7.19) began with a study of the degradation process in a single boron filament. These filaments were formed by deposition of boron on a 0.13 mm (0.0005 in) diameter tungsten filament called the *substrate*. The boron sheath, which provides the filament's strength, was about 1.3 mm (0.005 in) in diameter. The tungsten substrate is a resistive conductor, but the resistance of the boron is much greater by several orders of magnitude. Thus, electric currents entering a boron composite will want to flow in the tungsten substrates instead of in the boron. Tungsten has a positive temperature coefficient of resistivity, which means that, as its temperature increases, its resistivity increases as well, and it becomes even hotter as current flows. When temperature rise is calculated as a function of current action integral by Equation 7.5, the temperature rise curve of Figure 7.33 results. The quantity $\int I^2 dt$ is, of course, equal to the unit energy dissipation in the substrate (per ohm of resistance) and is dependent solely upon the lightning current amplitude and waveform.

Figure 7.33 shows that very little temperature rise occurs as $\int I^2 dt$ is increased to 20×10^{-6} A^2-s. However, increases beyond this value cause moderate temperature increases, and increases beyond 100×10^{-6} A^2-s result in extreme temperature rise. Fassell, Penton, and Plumer injected currents beyond 100×10^{-6} A^2-s into single filaments and found them to become severely cracked and to lose all mechanical strength. Correspondingly, filaments exposed to pulses with $\int I^2 dt$ values of 100×10^{-6} A^2-s or less exhibited little or no evidence of degradation. The mechanism of filament failure was found to be transverse and radial cracking of the boron sheath, presumably caused by excessive thermal expansion of the tungsten substrate. (Examples of cracked filaments are shown in Figure 4.12.) Thus, the calculated temperature rise is a significant indication of filament failure onset.

In comparison to natural lightning currents, it is noted that when distributed over several thousand filaments, a stroke of several thousand amperes may provide only 1 A of current per filament. If this current exists for 100 μs, it will indeed provide an $\int I^2 dt$ of 100×10^{-6} A^2-s in a single filament; if this current exists for 20 μs, it will provide an $\int I^2 dt$ of 20×10^{-6} A^2-s in a single filament, etc.

While the previous example illustrates the manner in which lightning currents may cause mechanical property degradation of composite materials, it also illustrates the relatively flat region in which increasing values of $\int I^2 dt$ do not result in corresponding increases in temperature rise. For protection design, this fact emphasizes the desirability of distributing lightning currents throughout as many filaments as possible.

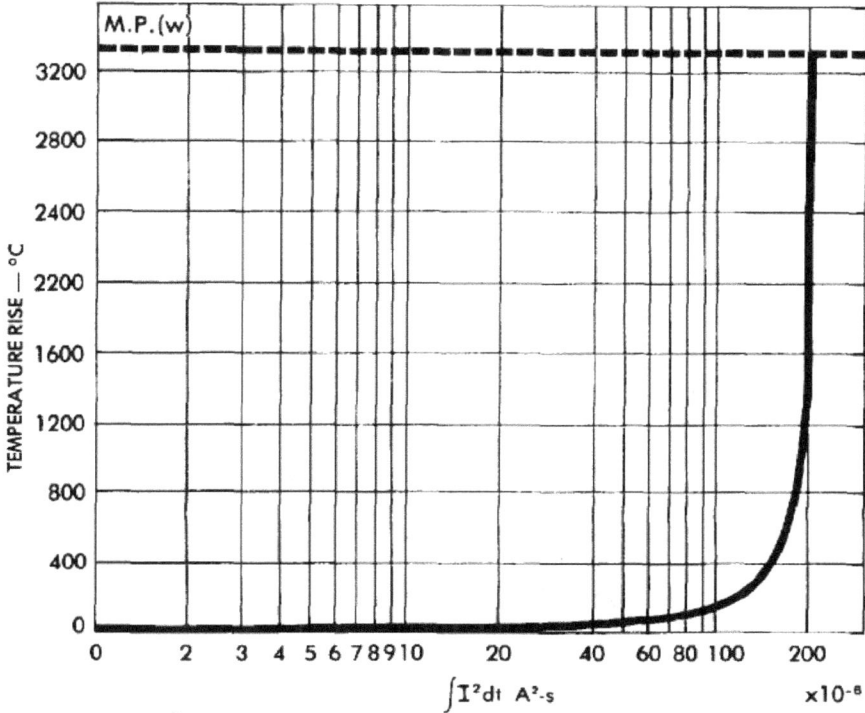

Figure 7.33 $\int I^2 dt$ vs temperature rise in the tungsten substrate of a boron filament.

To explain the process of current flow into the many boron filaments in a composite, Penton, Perry, and Lloyd (Reference 7.20) formulated an electrical model of two boron filaments in a resin matrix, as shown in Figure 7.34. Each filament is made up of incremental portions of the tungsten substrate resistance. These researchers found that the filament-to-filament breakdown voltage is about 300 V and that the insulation breakdown at this level is represented by the back-to-back Zener diodes, which have a reverse conduction (Zener) voltage of 300 V.

A lightning current flowing in the outer filament, they found, would remain in this filament until the resistive voltage rise along it exceeds the boron and resin breakdown voltage represented by the Zener diodes, at which point the diodes in the electrical model conduct and permit current to enter the neighboring filament. In most cases this would be a filament in the next laminate. While some material strength degradation undoubtedly occurs from the breakdown of the resin and boron sheaths, the major loss of strength was

231

found to be a result of temperature rise in the filaments and cracking of the boron sheaths along the entire length of current flow. The appearance of a filament so damaged is shown in Figure 4.12.

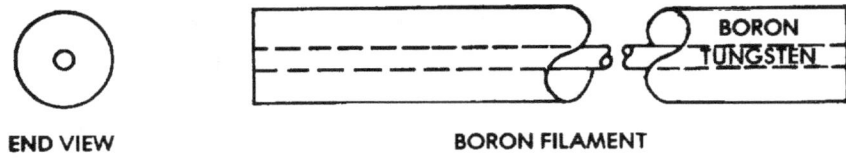

END VIEW   **BORON FILAMENT**

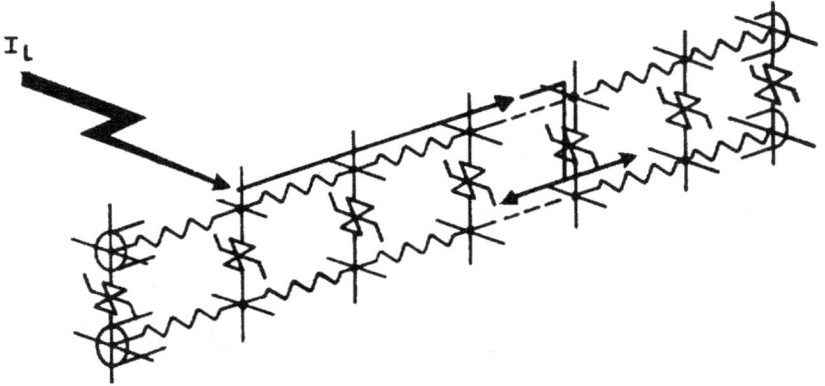

Figure 7.34 Electrical model of two boron filaments within a laminate.

When the filaments in an outer layer fail by the transverse, or axial, cracking, the resistances of their tungsten substrates also increase (by cracking) and so does the voltage rise along this layer, until the breakdown level of the boron sheaths and resin between it and the next layer of filaments is reached. When this happens, the current enters this next layer of filaments, and the process continues. The depth of penetration, of course, depends on both the amplitude and duration of the lightning current and the number of filaments into which this current can distribute.

In addition to the current conduction effects just described, the high temperature and blast forces from the arc have been found to cause substantial damage to a boron composite at the point of attachment; this, in fact, may be the only damage which is visible. Yet, because loss of strength also results from cracking of the embedded filaments, there may be little or no visible change in the outer appearance of a boron composite away from the arc attachment point. An example of this is the 16-ply boron composite panel (pictured in Figure 7.35 [Reference 7.21]) after receiving an 88 kA simulated stroke. The action integral of this stroke was 0.2×10^6 A^2-s, one that might be expected from a restrike to

232

a wing or fuselage surface in Zone 2A. A damaged area approximately 7 cm wide and 10 cm long is visible in Figure 7.35; yet the photomicrograph of Figure 7.36 (Reference 7.22) shows that nearly all the filaments in a cross section cut from the visibly undamaged area near the base of this panel are cracked. This cracking resulted in a 50% reduction in the shear strength of this region.

A hole was blown completely through the panel of Figure 7.35 at the arc attachment point by this 88 kA stroke. Strokes of somewhat less amplitude and action integral would not puncture a hole; strokes of higher energy would produce larger holes in this 16-ply boron panel.

The amount of damage to expect in unprotected panels of other thicknesses depends primarily upon the action integral of the lightning current and the number and orientation of the plies in the composite. An estimate of the *visible* damage to typical boron laminates as a function of lightning current amplitude and number of plies is presented in Figure 7.37.

The area throughout which *invisible* damage will occur in a boron composite depends on the cross-sectional area available for conduction of current away from the arc attachment point. This area becomes greater with distance away from this point, and can be estimated from Table 7.4 in the next section.

7.3.2.2 Graphite Composites

Graphite yarns provide the reinforcement in graphite composites. These yarns are made up of a great many graphite fibers, each much thinner than a boron filament. Since the melting point of graphite (3735 °C) is much higher than that required to cause melting and pyrolysis of the surrounding resin (315 °C), damage to graphite composites occurs by pyrolysis and, usually, ignition of the resin, which in turn causes gross delamination. This result is in contrast to the initial destruction of the reinforcing fibers which occurs in boron composites. The burning of the resin then leaves the entire composite in disarray, as shown in Figure 4.13.

The spectacular nature of the graphite composite damage mechanism and the gross disarray in which the composite is left might lead one to worry more about exposing graphite composites to lightning than boron composites. The reverse is actually the case, however, because, for the same cross-sectional area, the graphite composites have a much lower resistivity than has boron and therefore can conduct much more lightning current before dangerous temperatures are reached.

Arc-entry damage to graphite is somewhat similar, though less extensive than that produced by comparable arcs striking boron composites. Figure 7.38 (Reference 7.23) shows a typical 12-ply graphite composite panel after a 98 kA, 0.25×10^6 A^2-s stroke. Photomicrographic analysis and mechanical strength tests showed no damage other than that which is visible at the attachment point. Figure 7.39 presents an estimate of the amount of damage to expect in unprotected graphite laminates under other conditions.

233

Figure 7.35 Visible damage to 16-ply unprotected boron composite panel.
- 88 kA, 0.2×10^6 A^2-s
- Arc attachment to middle of panel
- Current exit from front and back plies at bottom

The literature abounds with simulated lightning test data for both boron and graphite, but since most of these tests were conducted before the

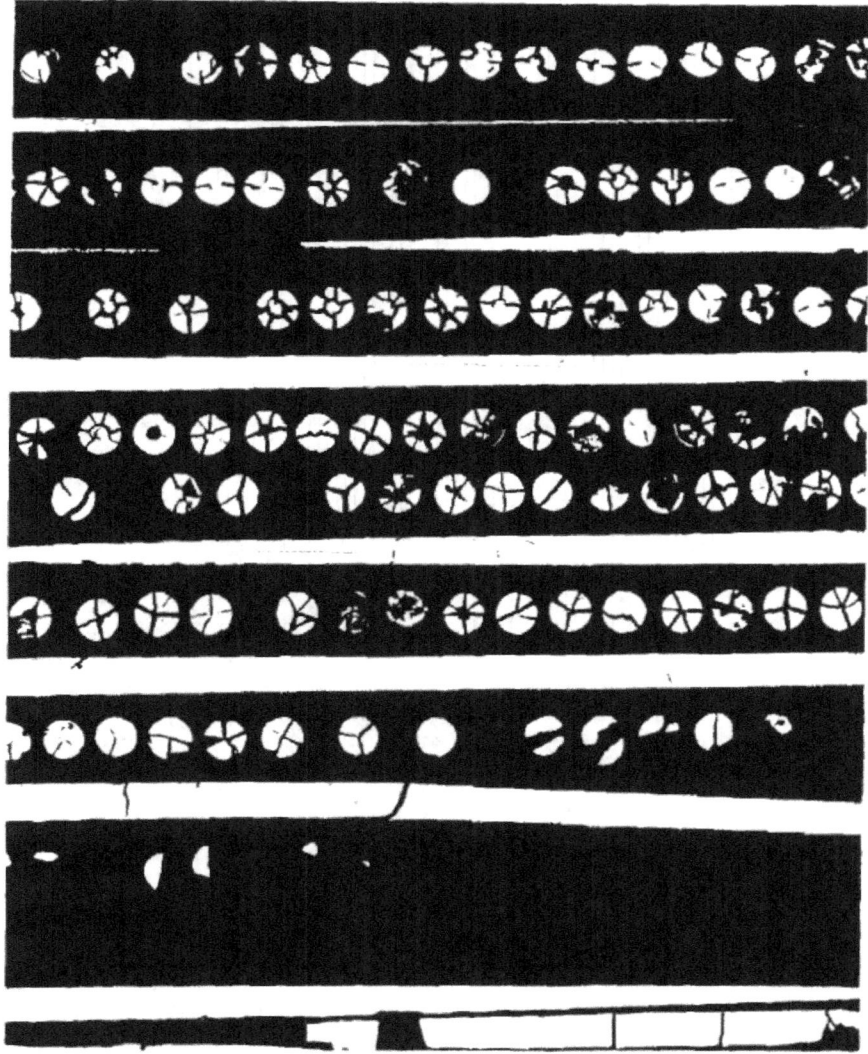

Figure 7.36 Boron filaments in visibly undamaged region at base of panel in Figure 7.35.
- Filaments are extensively cracked.
- Current has entered all 16 plies.
- Front of panel is at top.

standardization of test conditions was attempted, the tests have produced results which, while valid for the particular cases of interest, are too voluminous and diverse to normalize and reproduce here. From tests in our own laboratory, however, we have learned that composite strength degradation is closely related

235

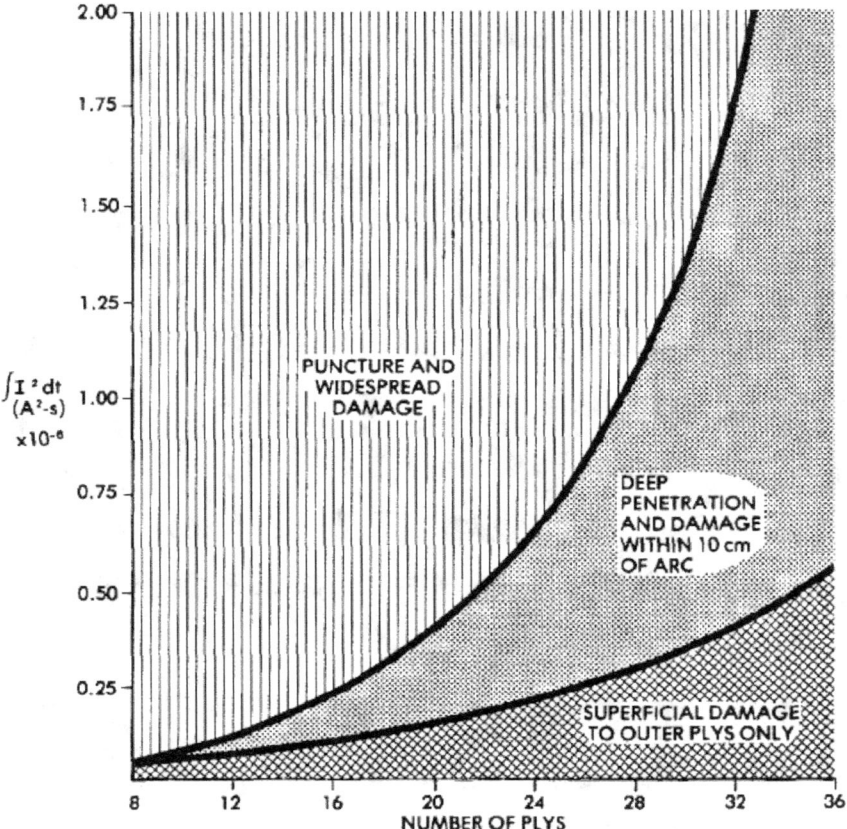

Figure 7.37 Estimated visible damage to unprotected boron composites.

to the lightning current action integral, even when this quantity is delivered over a period of up to one second, as it is in continuing currents. Table 7.4, which is based on these tests, compares the capability of boron and graphite composites to *conduct* lightning currents with and without degradation of their mechanical strength properties.

Since this table has been prepared from a variety of tests, the numbers in it must be considered approximate. When results were compiled, a range of about one order of magnitude of action integral was found to separate samples which experienced no discernible loss of strength from those which lost 50% or more of their former tensile or shear strength. From Table 7.4 it is evident that a cross-sectional area of about 16 cm^2 of boron composite is necessary to conduct a severe stroke of 2 x 10^6 A^2-s; yet only 2 or 3 cm^2 of graphite composite would be needed to conduct the same current.

In many cases it may not be practical or necessary to prevent localized damage at the arc attachment point, but it is important to assure that the

Figure 7.38 Visible damage to 12-ply unprotected graphite composite panel.
- 98 kA, 0.25 x 10^6 A^2-s
- Arc attachment to middle of panel
- Current exit from front and back plies at bottom

surrounding composite can conduct away the lightning current without being damaged in the process. The information of Table 7.4 can be used for this purpose as follows:

237

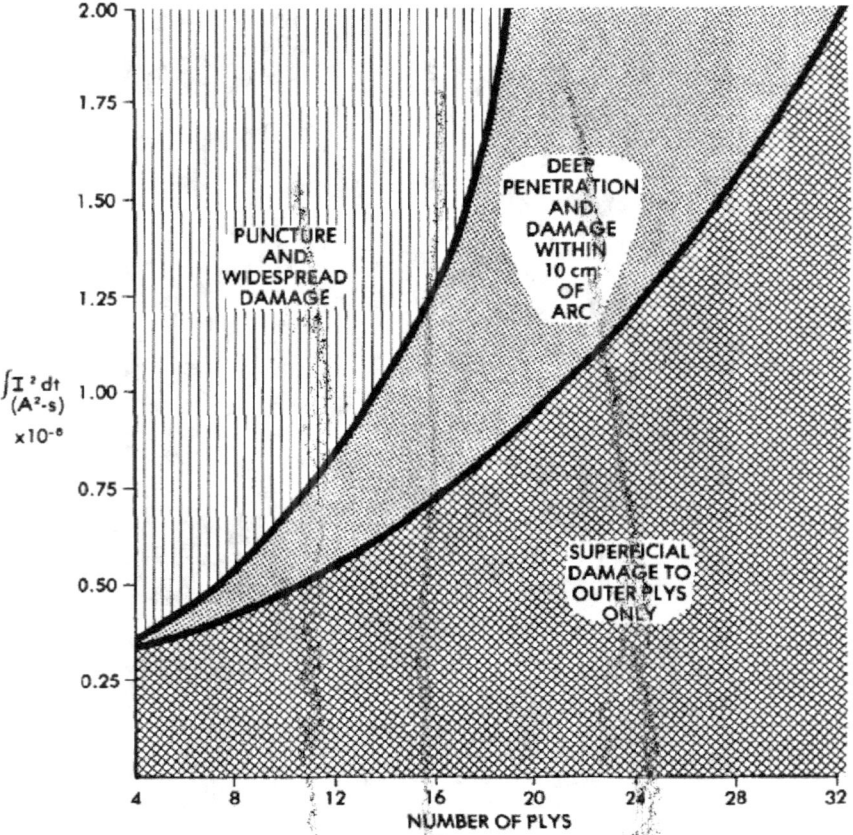

Figure 7.39 Estimated visible damage to unprotected graphite composites.
  • For graphite composites, the damage extends only throughout the visibly damaged region.

Table 7.4 COMPOSITE $\int I^2 dt$ CONDUCTION CAPABILITY

Composite Cross-Sectional Area		$\int I^2 dt$ (A²-s)			
		Boron		Graphite	
cm²	in²	Non-degraded	Degraded	Non-degraded	Degraded
0.016	(0.0025)	1	10		
0.161	(0.025)	10^2	10^3	2×10^3	2×10^4
1.61	(0.25)	10^4	10^5	2×10^5	2×10^6
16.1	(2.50)	10^6	10^7	2×10^7	2×10^8

238

1. Assume that only half of the surrounding composite is available to conduct current, as shown in Figure 7.40. The reason for this is that the lightning current will want to flow in one predominant direction——toward its eventual exit point somewhere else on the aircraft.

2. In this case the relationships among damage, radius, cross-sectional area, and composite thickness are given by the following:

$$d \geqslant \frac{A}{\pi r} \qquad (7.26)$$

where d = the necessary skin thickness (cm)

A = cross-sectional area required to conduct the required $\int I^2 dt$, from Table 7.4 (cm^2)

r = radius of permissible damage (cm)

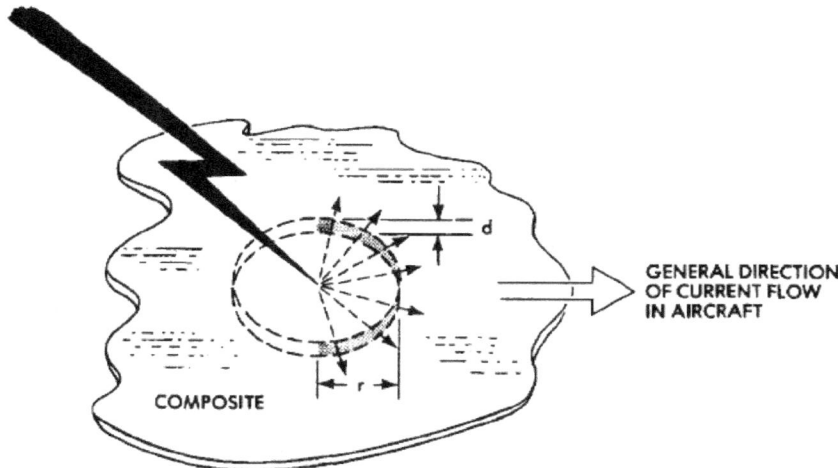

Figure 7.40 Composite thickness vs area of damage from conducted currents.

As an example, consider fuselage skin in Zone 2A to be made of graphite composite. It is desired that the area of strength loss extend no more than 5 cm outward from an arc attachment point. What thickness must the composite be to meet this requirement?

The SAE report *Lightning Test Waveforms and Techniques for Aerospace Vehicles and Hardware* defines the lightning current action integral for Zone 2A as 0.25×10^6 A^2-s. Table 7.4 shows that 1.61 cm of graphite composite will *conduct* this action integral without being structurally degraded. Thus, from Equation 7.26, the necessary composite thickness needed to limit the degraded region to within a 5 cm radius from the arc attachment point may be calculated as follows:

$$d = \frac{A}{\pi r} = \frac{1.61 \text{ cm}}{\pi \,(5 \text{ cm})} = 0.1 \text{ cm} \qquad (7.27)$$

If a boron composite were to be used, its thickness would have to be at least ten times greater to restrict the damage radius to the same 5 cm. The thicknesses of either composite would have to be even greater, of course, if the composites were located in Zone 1, where an action integral of $2 \times 10^6 \text{ A}^2$-s is to be expected. In such a case, it is likely that either the area of damage or the composite thickness required to reduce this damage would be intolerable. And in such a case the addition of some protective device(s) capable of keeping some or all of the lightning current out of the composite should be considered. The coatings and diverter strips for fiberglass described in Section 7.3.1.2 may also be applicable for advanced composites; and several coatings were developed specifically for use on composites. The most promising of these are discussed in Section 7.3.2.5.

7.3.2.3 Problems at Joints

In addition to lightning currents causing direct effects by entering or flowing in composite materials, these same currents may also cause damage at the places where they *leave* the composite—for example, a place where the original panel is joined to another one or to substructure such as ribs and spars. Two basic methods of joining composites are in use. One of these is *bonding*, in which the composite is joined (glued) with an adhesive to another member. The other method is with *bolted* joints, in which bolts, rivets, or other mechanical fasteners are employed to hold two members together.

Unfortunately, most adhesives used in bonding are highly resistive and by themselves are not able to conduct lightning currents. Instead, arcing occurs within the adhesive, releasing sufficient gas pressure to make the structure literally "come unglued." Either the adhesive must be made electrically conductive or a separate conducting path must be provided to carry lightning currents across the joint. A conductive paint or foil may be able to accomplish this if it is necessary only to transfer lightning current from one composite panel to an adjacent one. But if the composite is also fastened to substructure, electrical contact must be made to this substructure as well. A metal foil wrapped around the edge of the composite panel may suffice; however, such a foil must be held securely to the composite surface so that the magnetic force (a problem of Figure 7.8) does not cause the foil to pull away. If bolts or fasteners are used, they may provide the necessary electrical path, whether intentional or not, but the concentration of current near them may degrade the composite's strength in this important region.

Relatively little work has yet been accomplished to evaluate the degree of strength loss caused by lightning currents in composite joints; however, further research is now in progress. Schneider, Hendricks, and Takashima (Reference 7.24) are studying the effects of lightning currents on typical bonded and bolted joints of graphite composites, and their preliminary findings show that bonded joints suffer more damage from lightning currents than do bolted joints. Other tests (Reference 7.25) have also been made on various joints, with similar results. The reader is referred to these sources for further information on the specific joints researchers have evaluated and the results they have obtained.

7.3.2.4 Electromagnetic Field Penetration

The importance of keeping the electromagnetic fields produced by lightning currents away from the aircraft's electric wiring is by now clear. In conventional aircraft this is conveniently and successfully accomplished by the aluminum skins, and the electrical system designer benefits greatly from this assistance heretofore given him by the structures designer. Substitution of composites for aluminum, however, poses an additional threat to the electrical systems because composites do not possess the excellent shielding property of aluminum, with the result that the electrical system designer may now have to provide his own protection against these fields.

The relative shielding effectiveness of composites as compared with various metals has been measured for magnetic (H) fields by Fisher and Fassell (Reference 7.26) by establishing an H-field of known intensity on one side of a composite surface and measuring the amount of it reaching the other side. Continuous sine-wave fields at several different frequencies were applied to the top side of a test sample in the arrangement of Figure 7.41, and the shielding capability of the sample was determined by measuring the voltage induced in a receiving coil on the other side by that portion of the field which penetrated it. The current, I, in the transmitting coil was kept constant during the tests to make sure that reflected fields did not change the level of field radiated in the first place. The amount of H-field attenuation was then determined by the usual expression:

$$\text{Attenuation} = 20 \text{ Log} \frac{V_1}{V_2} \tag{7.28}$$

where V_1 is the voltage induced in the receiving coil when no test sample is present and V_2 is the voltage when a sample of composite or other material is placed between the coils. The metal enclosure was used to avoid fringing effects. Figure 7.42 shows the H-field attenuations for boron and graphite composites by themselves, and Figure 7.43 shows the improvements in shielding effectiveness gained by application of conductive coatings (Reference 7.27).

The measurements of the shielding effectiveness of panels covered with flame-sprayed aluminum and aluminum foil indicated that most of the shielding came from the aluminum. Values of 16 dB and 20 dB were measured at 0.5 MHz and 1.0 MHz for 2.1 mm (0.083 in) boron with a coating of flame-sprayed aluminum. The thickness of the aluminum was about 0.4 mm.

Many investigations into the shielding effectiveness of conductive coatings of enclosures have, of course, been made for other purposes. One is that of Uhlig (Reference 7.28), who evaluated several conductive coatings applied to 1.2 m

cubic boxes made of wood. The coatings tested were the following:
1. Flame-sprayed aluminum approximately 0.18 mm (0.007 in) thick
2. Aluminum foil 0.025 mm (0.001 in) thick
3. Two coats of conducting paint (300-600 Ω/square)

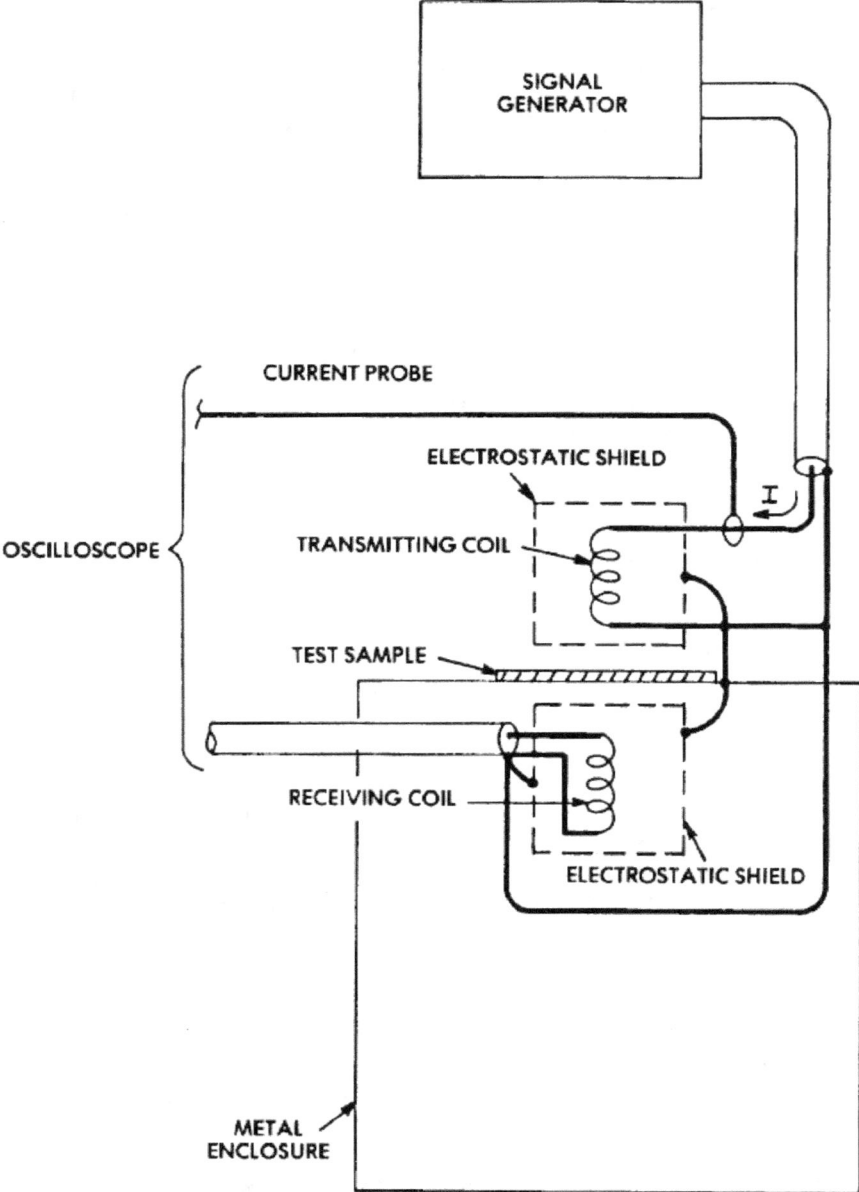

Figure 7.41 Arrangement for shielding effectiveness tests.

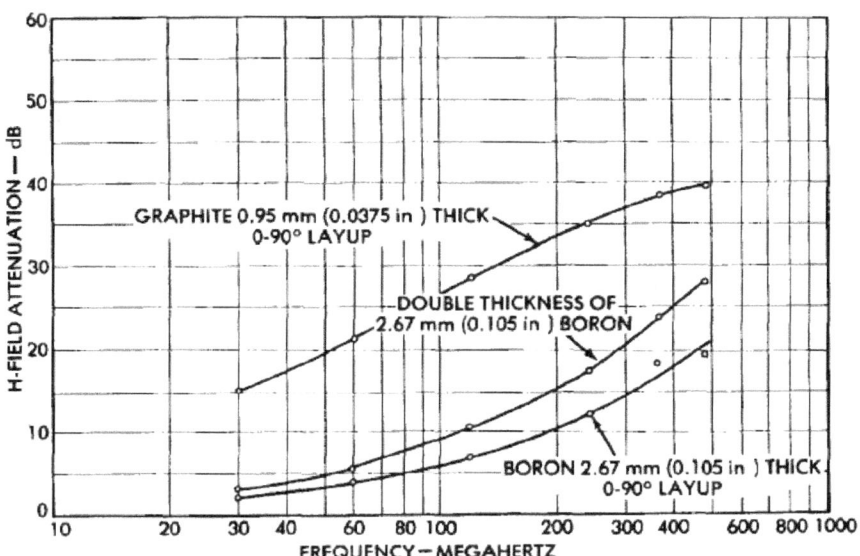

Figure 7.42 H-field shielding effectiveness of boron and graphite composite materials.

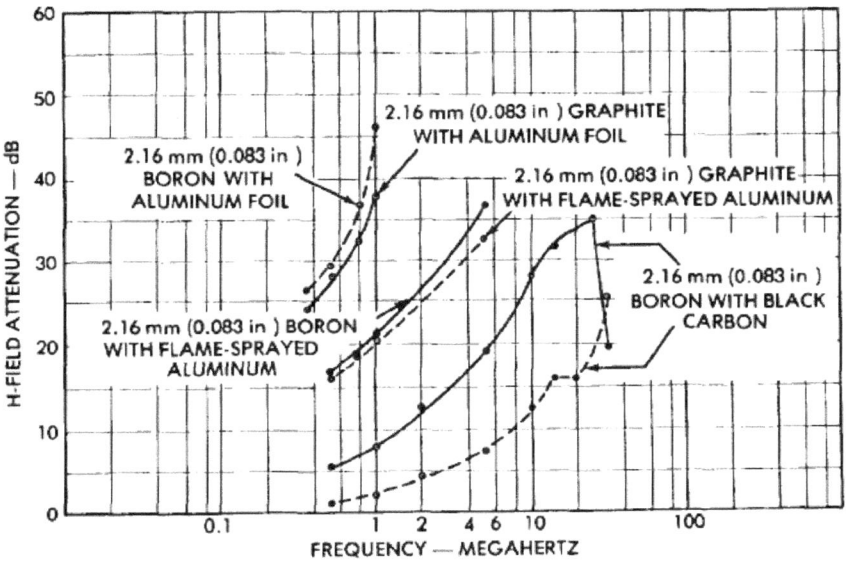

Figure 7.43 H-field shielding effectiveness of boron and graphite composites with protective coatings.

243

The box was placed in an impulse-type H-field with amplitude and waveform as shown in Figure 7.44(a) (Reference 7.29). This field has a rise time of approximately 2 μs and a decay time of approximately 60 μs, such as would be produced by a lightning-stroke current with the same rise and fall times. (Component F of the SAE report). Its amplitude of 150 A/m is of the order produced by a typical lightning-stroke current (discussed in Chapter 9).

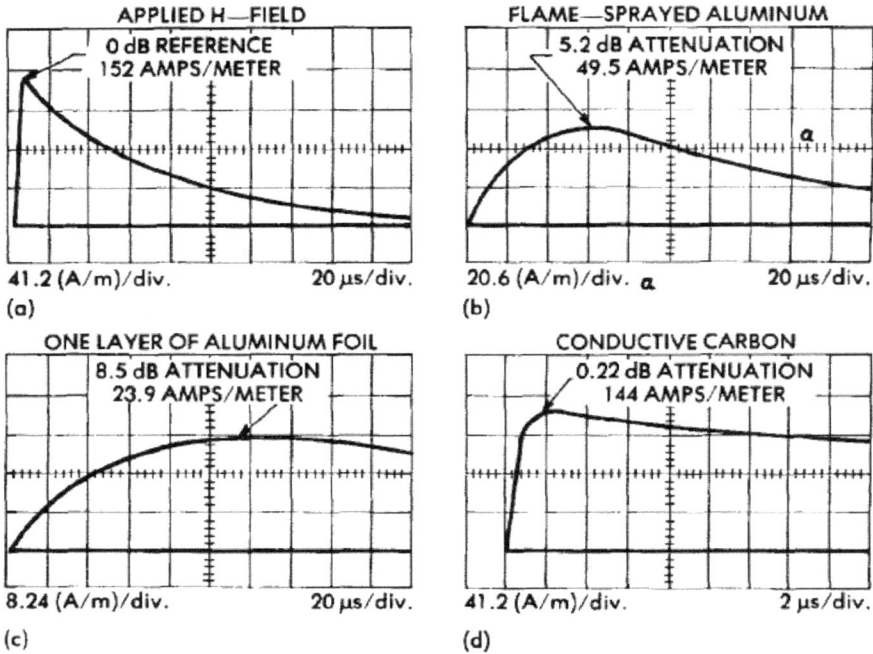

Figure 7.44 H-field shielding effectiveness of various conductive coatings on a 1.2 m cube.

Figure 7.44(b), (c), and (d) shows the actual field strength within the center of the box when coated with each of the three materials. The 0.18 mm flame-sprayed aluminum gave an H-field attenuation of 5.2 dB produced by a 0.025 mm thickness of aluminum foil. Conductive paint gave practically no attenuation.

The numerical data presented in Figures 7.42, 7.43, and 7.44 are of course applicable only for the coil spacing used in these tests, since the coil spacing affects the impedance of the magnetic field and the shielding is determined in part by the impedance mismatch between the field and the shielding material. However, implications of the data are quite clear:

1. The shielding effectiveness of both graphite-epoxy and boron-epoxy materials is much less than an equivalent thickness of aluminum.
2. A thin piece of aluminum foil has much greater shielding effectiveness than has even a 3 mm panel of graphite-epoxy.

3. Graphite composites, by virtue of their higher conductivity, provide more effective shielding than do boron composites.
4. When conductive coatings are present for control of the direct effects of lightning, they will have more effectiveness as electro-magnetic shields than the composites they protect.
5. When conductive coatings are not needed for direct effects protection, electrical wiring enclosed within these composites may need to be placed in conduits or otherwise shielded from the intense fields which will pass through the composites.

Electric fields (E-fields) are also associated with lightning strikes and may cause as much interference in some electrical systems as do the magnetic fields, or more. No E-field shielding effectiveness data have as yet been reported, however.

Furthermore, no measurements have yet been made of the voltages which lightning may actually induce in the electrical circuits inside of composite aircraft structures because there have been few such structures yet available. Nor are there analytical techniques yet available for prediction of field levels within composite structures. The shielding effectiveness measurements reported by Fisher and others have always involved small panels usually clamped to a simple test receptacle, like that of Figure 7.41. However, when panels are employed in a structure, with provisions made for conducting currents out of edges and joints, it is likely that some improvement in shielding effectiveness will result, and the situation will not be as bleak as it is indicated to be by the small panel data now available.

7.3.2.5 Protection for Advanced Composites

That it is necessary to supply lightning protection for all composites wherever they are employed on the aircraft is certainly not a foregone conclusion. Graphite composites greater than 4 plies thick and located in Zone 2A, for example, will not be completely punctured and will suffer only localized surface damage, as indicated by Figure 7.39. Some lightning protection may be necessary, however, for Zone 1 applications of graphite and for virtually *any* lightning-stroke zone application of boron.

A number of programs have been undertaken to develop suitable protective coatings for composites. The most important of these was conducted by Quinlivan, Kuo, and Brick (Reference 7.30), who developed and tested a wide variety of conductive coatings, and Brick, King, and Quinlivan have continued this work (Reference 7.31). Coatings evaluations have also been carried out by Fisher and Fassell (Reference 7.32) and others; the reader is urged to study these references to obtain the performance data necessary for the selection of a coating for a particular application. Coatings that have performed especially well include the five discussed below.

Wire Fabrics

Wire fabrics are flexible meshes woven or knitted from very fine aluminum or copper wires, as shown in Figure 7.28. These perform well as lightning protective coatings because of their use of the skin effects for electrical conduction and because their transparency allows hot gasses to escape without damaging the composite beneath. For example, the skin area of an 80 by 80 mesh (80 wires/cm) fabric of 0.05 mm wire is over 200 times that of the composite surface area actually coated by the fabric. Consequently, the fabric is a highly efficient conductor of current away from the arc contact point. This fabric has been found capable of withstanding successive 100 kA discharges at the same location with little visible damage to the fabric and no reduction in the mechanical strength of the coated laminate. At the 200 kA, 2×10^6 A^2-s level, the only significant damage to the coated boron laminate is that directly under the arc contact region, and, even in this region, the tensile strength is reduced only about 20%.

One can expect paint coatings to be used on aircraft structures for various reasons such as camouflage, environmental protection, and marking. Because paints over 80 by 80 mesh wire fabric reduce the area of arc entry, damage to the coating increases at the point of arc entry, allowing some of the wire to be melted or vaporized. However, there is no additional damage to the composite. If mechanical conditions warrant its use, 50 by 50 wire mesh may also be used; however, this mesh offers somewhat less protection because the wire surface area is less.

Aluminum fabrics were found by Quinlivan, Kuo, and Brick to provide the best combination of lightning protection, light weight, environmental resistance, and ease of application. For 80 by 80 mesh woven aluminum wire fabric, the area density is 0.019 lb/ft^2; for a 50 by 50 mesh fabric the area density is 0.042 lb/ft^2. These weights are increased to 0.036 and 0.072 lb/ft^2, respectively, if the resin required for encapsulation of the fabric is accounted for. Employing calendered wire cloth permits some weight saving. Calendering reduces the thickness of the cloth by flattening the intersections of the wires; weight savings occur because less resin is required to encapsulate the flattened mesh. As a point of reference, 6-mil-thick aluminum foil has an area weight of 0.084 lb/ft^2. The weights of environmentally protective topcoats or adhesive required for bonding the foil (or fabric) must be added to these figures.

Wire fabrics possess the hand and drape necessary for use as an overlay on complex contoured parts, and the composite matrix which fully encapsulates the fabric has been found to protect it from environments such as humidity, salt spray, and jet fuel.

Aluminum Foils

Aluminum foils from 0.05 mm to 0.2 mm thick will provide protection similar to that of wire fabrics, but the mechanical properties of the foil, its weight, and the need for an adhesive between the foil and the composite may make aluminum foil less attractive as a protective device.

Flame-Sprayed Aluminum

An equal weight of flame-sprayed aluminum offers approximately the same protection as that of aluminum foil. It has sometimes been difficult to control the thickness of flame-sprayed coatings, a phenomenon that could create inconsistencies in the protection afforded by this type of coating, but flame-sprays may be easier to apply to some contoured surfaces than are foils.

Diverter Straps and Foil Strips

One might wonder if the metal diverter straps or foil strips recommended for protection of fiberglass in Section 7.3.1.2 are also applicable for advanced composites. As discussed in that section, these devices provide an external conductor from which a streamer may originate, reaching an approaching leader before a streamer from an internal conductor has a chance to puncture the fiberglass wall. The electrical conductivity of the strap, in contrast to the surrounding nonconductive fiberglass, is depended upon to assure that this streamer will come from the strip and not from some other place.

Advanced composites, of course, have *some* electrical conductivity —— enough to supply the few milliamperes of current needed for a streamer. Thus, a metal diverter on a poorly conducting surface is not nearly as effective as it is on a nonconducting surface. This fact has been demonstrated repeatedly on nose radomes protected by diverters but with partially conductive antistatic paints also applied to their outside surfaces to reduce EMI from P-static charge buildups on the radome surface. Streamers have occasionally initiated from the antistatic paint instead of from a diverter, causing the ensuing stroke to attach to the painted surface and destroy a large part of the radome while finding its own path to metal structure.

Lubin and Dastin (Reference 7.33) report that, in laboratory tests, closely spaced aluminum foil strips did keep strikes from attaching to the boron composite surface of a horizontal stabilizer. Their tests —— which are the most extensive yet reported of conducting strips on a composite —— showed that 2.54 cm (1 in) wide, 0.2 mm (0.008 in) thick aluminum foil strips had to be spaced within 6.35 cm (2.5 in) of each other to keep strikes from hitting the boron composite surface between them. A similar spacing was necessary for 6.35 cm (2 in) wide strips that were half as thick. Wider spacings permitted some strikes to terminate on the boron composite surface.

It is likely that such strips would not work as well as this over a graphite composite, which is more conductive than boron. The performance of strips on either composite, of course, would be enhanced if the composite surface were covered with a dielectric (insulating) material that would inhibit streamer formation. A system in which a dielectric film is first applied to the entire composite surface, with the conducting strips then applied over the dielectric, is likely to work best, since this would also help to keep lightning currents in the foil strips from entering the composite. No thorough tests of such a system have been reported, however.

247

Dielectric Coatings

Dielectric coatings such as polyurethanes or Kapton[1] film have been proposed by Kung, Amason (Reference 7.34), and others, as a means of protecting composite surfaces in swept-stroke zones (Zone 2A). These researchers acknowledge that some external conductors must still be used, but they suggest that the distance between such conductors could be extended by applying a dielectric coating to the composite surface, as shown in Figure 7.45. In tests of a 0.076 mm (3-mil) polyurethane coating over a graphite composite, Kung and Amason report that a 78 cm (31 in) separation is possible between conducting strips. In their tests a strike to the first conductor is initiated and caused to sweep aft in a wind stream across the dielectric coating until it reattaches to the second conductor, in the manner of Figure 6.21. As discussed in Section 6.6.2 of Chapter 6, the distance this arc will sweep depends on the voltage withstand capability of the surface paint versus the amount of voltage which can be developed along the arc. The greater the insulation capability of the coating, the longer the sweep to the next attachment point; the higher the arc voltage, the shorter the sweep. The principle of the dielectric coating concept is simply to make the coating from a good dielectric able to withstand high voltages. But there is a pitfall: if a puncture of the dielectric *should* occur, the arc will remain attached to this point much longer than it would have had no dielectric coating been present, and in so doing the arc will inflict more damage to the composite than if the arc had been uninhibited by the dielectric coating and allowed to move quickly from one place to another.

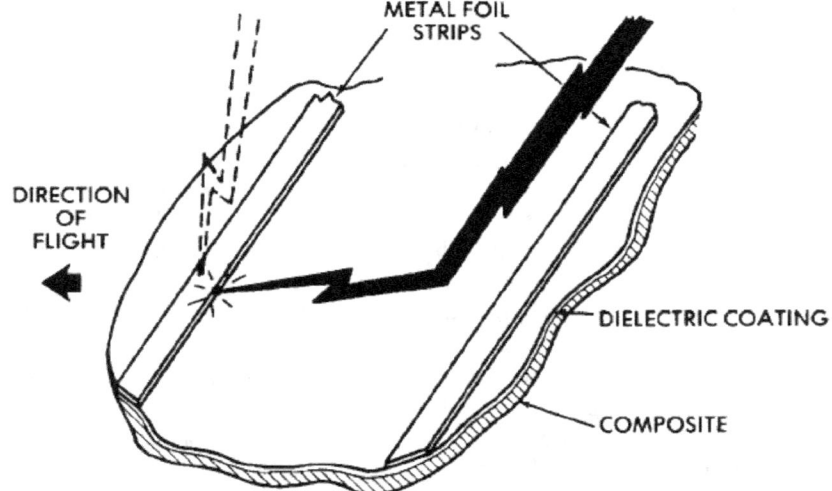

Figure 7.45 Dielectric coating protection system.

[1] Trademark of the E. I. duPont de Nemours Company, Inc.

The voltage withstand capability of polyurethanes and similar materials considered for this application is very high, a characteristic that may permit large spans of the materials to exist between conductors in laboratory tests, such as those of Figure 7.45. The withstand capability of these dielectrics is drastically reduced, however, if their surfaces become scratched or cracked, as might occur in service. Thus, if a dielectric protection scheme is considered, it should be given the lightning tests described in *Lightning Test Waveforms and Techniques* for nonmetallic materials in Zone 2A—*after* the test sample has been exposed to vibration, temperature cycling, abrasion, and any other environment that is likely to degrade the condition of the dielectric coating. Even then, such a protection system may become damaged by unexpected contact with a foreign object and its degraded status go unnoticed until a lightning strike occurred. These coatings contrast with conductive coatings, which may sustain many abrasions or cracks with much less, if any, degradation in their protection capability.

Dielectric coatings, at their present stage of development at least, should not be considered, therefore, for protection of flight-critical composite structures except where a puncture of the coating would not create an unacceptable hazard. Rather, dielectrics are likely to be most beneficial when they are used to supplement the performance of a primarily conductive protection system, such as the closely spaced foil strips discussed previously.

In addition to protecting composite surfaces, it is necessary to provide a means to safely conduct lightning currents through joints, as discussed in Section 7.3.2.3. If the composite is protected by a conductive coating and bolted joints are used, the coating should be extended into tapered bolt holes (to make good electrical contact with the bolts) and the bolts used to transfer current through the joint. If adhesive bonding joints are used instead, a conducting additive should be added to this adhesive; otherwise a different conducting path should be provided.

If a composite is to be used in a flight-critical application, in every case the performance of its lightning protection system should be verified by the simulated lightning tests described in SAE Report *Lightning Test Waveforms and Techniques for Aerospace Vehicles and Hardware.*

249

REFERENCES

7.1 J. A. Plumer, *Guidelines for Lightning Protection of General Aviation Aircraft*, FAA-RD-73-98, Federal Aviation Administration, Department of Transportation, Washington, D.C. (October 1973), p. 8.

7.2 *Lightning Test Waveforms and Techniques for Aerospace Vehicles and Hardware*, Society of Automotive Engineers, Warrendale, Pennsylvania (5 May 1976), pp. 9-10.

7.3 J. A. Plumer, *Lightning Protection of the F-111 Radome Pitot System*, 76CRD018, Corporate Research and Development, General Electric Company, Schenectady, New York (February 1976), p. 19.

7.4 General Electric Company photograph.

7.5 James and Phillpott, *Simulation of Lightning Strikes*, p. 12.

7.6 Paul T. Hacker, *Lightning Damage to a General Aviation Aircraft—— Description and Analysis*, NASA TN D-7775, National Aeronautics and Space Administration, Lewis Research Center, Cleveland, Ohio (September 1974), p. 49.

7.7 *Advanced Development of Conceptual Hardware for the Lightweight Fighter*, prepared by General Dynamics under Contract F 33615-73-C-5130 for the Advanced Development Division, Air Force Materials Laboratory, Wright-Patterson Air Force Base, Ohio (15 February 1976), p. 39.

7.8 R. S. Jacobsen, "Magnetic Acceleration of Flyer Plates for Shock Wave Testing of Materials," *Pulsed Electric Power Flyer Plate Notes*, AFWL PEP 2-1, Vol. I, Air Force Weapons Laboratory, Kirtland Air Force Base, New Mexico (July 1971): 6-5, 6.

7.9 (5 May 1976), pp. 14-21.

7.10 MIL-B-5087B (ASG), *Military Specification: Bonding, Electrical, and Lightning Protection for Aerospace Systems* (15 October 1964).

7.11 J. R. Stahmann, "Control Surface and Door Hinge Bonding Effectiveness in Modern Aircraft," *Proceedings of the 1972 Lightning and Static Electricity Conference, 12-15 December 1972*, AFAL-TR-72-325, in cooperation with SAE Committee AE-4 on Electromagnetic Compatibility and the Air Force Avionics Laboratory, Air Force Systems Command, Wright-Patterson Air Force Base, Ohio (1972), pp. 527-34.

7.12 General Electric Company photographs.

7.13 General Electric Company photographs.

7.14 Plumer, *Guidelines for Lightning Protection of General Aviation Aircraft*.

7.15 M. P. Amason, G. J. Cassell, J. T. Kung, "Aircraft Applications of Segmented-Strip Lightning Protection Systems," *Proceedings of the 1975 Conference on Lightning and Static Electricity, 14-17 April 1975, at Culham Laboratory, England,* Session IV: Aircraft Applications, the Royal Aeronautical Society of London (1975).

7.16 S. A. Moorefield, J. B. Styron, and L. C. Hoots, *Manufacturing Methods for Advanced Radome Production*, Interim Technical Report 6, prepared by the Brunswick Corporation for the Air Force Materials Laboratory, Air

Force Systems Command, United States Air Force, Wright-Patterson Air Force Base, Ohio (November 1975), pp. 32-40.

7.17 Amason, Cassell, and Kung, "Aircraft Applications of Segmented-Strip Lightning Protection," p. 9.

7.18 D. A. Conti and R. H. J. Cary, "Radome Protection Techniques," *Proceedings of the 1975 Conference on Lightning and Static Electricity, 14-17 April 1975, at Culham Laboratory, England,* Session IV: Aircraft Applications, the Royal Aeronautical Society of London (1975).

7.19 W. M. Fassell, A. P. Penton, and J. A. Plumer, "The Susceptibility of Advanced Filament Organic Matrix Composites to Damage by Simulated Lightning Strikes," *Lightning and Static Electricity Conference, 3-5 December 1968, Part II: Conference Papers,* AFAL-TR-68-290, Air Force Avionics Laboratory, Air Force Systems Command, Wright-Patterson Air Force Base, Ohio (May 1969), pp. 530-69.

7.20 A. P. Penton, J. L. Perry, and K. J. Lloyd, *The Effects of High Intensity Electrical Currents on Advanced Composite Materials,* U-4866, prepared by the General Electric Company for the Naval Air Systems Command, Department of the Navy, Washington, D.C. (15 September 1970), pp. 2-30, 2-32.

7.21 F. A. Fisher and W. M. Fassell, *Lightning Effects Relating to Aircraft, Part I Lightning Effects on and Electromagnetic Shielding Properties of Boron and Graphite Reinforced Composite Materials,* AFAL-TR-72-5, Air Force Avionics Laboratory, Air Force Systems Command, Wright-Patterson Air Force Base, Ohio (January 1972), p. 55.

7.22 Fisher and Fassell, *Lightning Effects, Part I,* p. 79.

7.23 Fisher and Fassell, *Lightning Effects, Part I,* p. 29.

7.24 S. D. Schneider, C. L. Hendricks, and S. Takashima, *Vulnerability/Survivability of Composite Structures - -Lightning Strike,* D6-42673-3, prepared by the Boeing Commercial Airplane Company for the Air Force Flight Dynamics Laboratory, Aeronautical Systems Division, Wright-Patterson Air Force Base, Ohio (April 1976).

7.25 *Advanced Development of Conceptual Hardware for the Lightweight Fighter.*

7.26 Fisher and Fassell, *Lightning Effects, Part I,* pp. 178-204.

7.27 Fisher and Fassell, *Lightning Effects, Part I,* pp. 187, 201, 198.

7.28 E. R. Uhlig, *Investigations of H-Field Shielding Effectiveness of Thin Conductive Coatings and Reduction of Induced Currents on Penetrations,* submitted by author, General Electric Company, under Contract DA-49-129-ENG-543 and placed by Corps of Engineers, Department of the Army, Huntsville, Alabama.

7.29 Uhlig, *Investigations of H-Field Shielding Effectiveness,* pp. 11, 13, 15.

7.30 J. T. Quinlivan, C. J. Kuo, and R. O. Brick, *Coatings for Lightning Protection of Structural Reinforced Plastics,* AFML-TR-70-303 Pt. I, Air Force Materials Laboratory, Nonmetallic Materials Division, Air Force Systems Command, Wright-Patterson Air Force Base, Ohio (March 1971).

7.31 R. O. Brick, C. H. King, and J. T. Quinlivan, *Coatings for Protection of Structural Reinforced Plastics,* AFML-TR-70-303 Pt. II, Air Force Materials Laboratory, Nonmetallic Materials Division, Elastomers and Coatings Branch, AFML/LNE, Wright-Patterson Air Force Base, Ohio (February 1972).

7.32 Fisher and Fassell, *Lightning Effects, Part I*, pp. 3-177.

7.33 G. Lubin and S. Dastin, "Lightning Protection for Aircraft Sandwich Structures with Boron/Epoxy Composite Skins," *Proceedings of the 1972 Lightning and Static Electricity Conference, 12-15 December 1972,* AFAL-TR-72-325, in cooperation with the SAE Committee AE-4 on Electromagnetic Compatibility and the Air Force Avionics Laboratory, Air Force Systems Command, Wright-Patterson Air Force Base, Ohio (1972), pp. 359-411.

7.34 J. T. Kung and M. P. Amason, "Dielectric Shielding Lightning Protection for Composite Aircraft Structures," *Proceedings of the 1972 Lightning and Static Electricity Conference, 12-15 December 1972,* AFAL-TR-72-325, in cooperation with the SAE Committee AE-4 on Electromagnetic Compatibility and the Air Force Avionics Laboratory, Air Force Systems Command, Wright-Patterson Air Force Base, Ohio (1972), pp. 337-58.

CHAPTER 8
VOLTAGES AND CURRENTS INDUCED BY LIGHTNING

8.1 Definition and Elementary Considerations

The term *indirect effects* of lightning refers to the damage to or malfunction of electrical equipment that results from lightning flashes. These effects may range from tripped circuit breakers to computer upset, to physical damage to input or output circuits of electronic equipment. There may be other indirect effects of lightning flashes that pertain to aircraft safety, such as flash blindness of the crew or acoustic shock waves. These effects, however, are not treated in this chapter. Included in this definition and discussed here are the voltages and currents induced by lightning on the electrical wiring of the aircraft, regardless of whether or not such voltages and currents cause damage or upset of electrical equipment.

The demarkation line between direct and indirect effects may be somewhat arbitrary in some instances. An example might involve a lightning flash terminating on a wing tip navigation light. In this case, the burning or blasting damage to the light fixture would be considered a direct effect of the lightning. If the lightning flash were to contact the filament of the bulb and inject current into the wiring, there would be produced electrical effects ranging from overvoltage breakdown of the insulation at the socket to tripping of remote circuit breakers or upset of equipment, effects of the resulting surge voltages on the power system. All of these effects will be considered indirect effects in this discussion, even though the initiating event was the direct injection of current into the filament. Another hypothetical possibility would involve a lightning flash that passed close to but did not contact the aircraft. The changing electromagnetic field produced by that flash might upset electronic equipment. The circuit upsets produced by such a flash would clearly be of an indirect nature, since there was no direct involvement of the aircraft with the lightning flash. Usually, however, the indirect effects of concern are produced by a flash that contacts the aircraft.

In this section as a whole we shall attempt to explain some of the electrical phenomena involved and to discuss some of the analytical techniques through which indirect effects may be analyzed, some of the measurements that have been made on aircraft to determine the nature of the indirect effects, and some of the measures one may take to control the indirect effects of lightning.

The arts of analysis and control of indirect effects are embryonic at present. Accordingly, the analytical procedures to be described are not as well developed as might be desired. Neither can very many examples of successful protective measures be given, since protective measures against indirect effects are as yet seldom included in aircraft design.

The problem of analysis of indirect effects is a complex one, but, fortunately, analysis can be divided into several stages. These stages are illustrated in Figure 8.1. The overall task is to determine how the lightning current, I_L, leads to voltages and currents on the internal wiring. The individual tasks are as follows:

(a) To determine the amplitude and waveform of the current on the exterior of the aircraft. This external current may or may not be the same as the undisturbed lightning current.

(b) To relate the internal response of the aircraft structure to the current flowing on its exterior. This internal response may be expressed either in terms of an internal electric field or an internal magnetic field.

(c) To determine how the internal response of the structure affects the response of the aircraft's wiring. The response of the wiring may be expressed either in terms of voltages or currents induced on active electrical conductors or in terms of currents induced on shields.

(d) To determine how these currents and voltages affect the electrical equipment in the aircraft.

8.2 The Exterior Response of the Aircraft Structure

The problem of how the aircraft as a whole responds to the passage of lightning current is illustrated in Figure 8.2. There are two parts of this problem, one involving a direct flash to the aircraft and the other involving a nearby flash. While it is generally true that the indirect effects produced by a direct flash are more severe than those produced by a nearby flash, they are not necessarily so, since the response of the aircraft external structure is different in the two cases.

Consider first a direct hit on the aircraft. The return stroke current enters the aircraft at one extremity, flows through it, and exits from the other end. As a result of the different impedances of the lightning channel and the aircraft, the current flowing through the aircraft may undergo some distortion in waveform. The result is that the waveform of the current leaving the aircraft may be different from that entering the aircraft. Typically, the exit current rises to crest over a longer time than does input current.

The phenomena involved are shown in Figure 8.3. A waveform is shown traveling along the conductor having surge impedance Z_1. Surge impedance is defined as

$$Z = \sqrt{L/C} \qquad (8.1)$$

where
Z = surge impedance (ohms)
L = inductance (henries) per unit length
C = capacitance (farads) per unit length

at the point under consideration. At some point the current encounters a *transition point between the conductor having surge impedance Z_1 and a different conductor of surge impedance Z_2.* At this transition point, part of the incident current is transmitted onto the second conductor, but part of it is reflected back in the direction from which it came. The magnitudes of the transmitted and reflected components of voltage or current at this transition

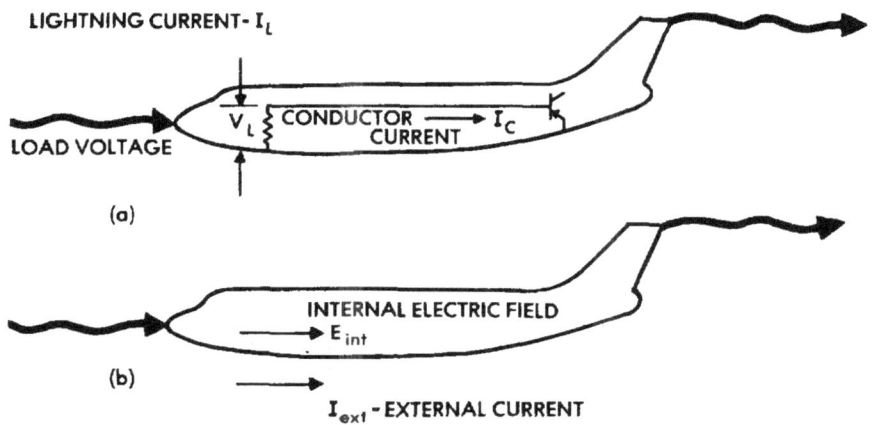

LIGHTNING CURRENT- I_L

LOAD VOLTAGE

V_L { CONDUCTOR → I_C CURRENT

(a)

INTERNAL ELECTRIC FIELD
→ E_{int}

(b)

I_{ext} - EXTERNAL CURRENT

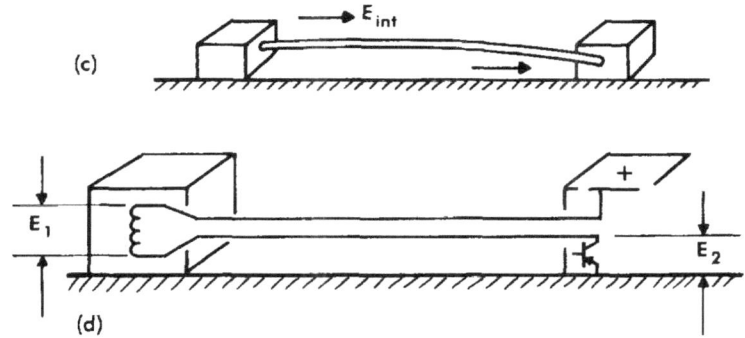

E_{int}

(c)

E_1

E_2

+

(d)

Figure 8.1 Indirect effects.

   (a) The total system
   (b) The internal response of the structure
   (c) The response of cables
   (d) The response of piece parts

point are given by the values of transmission and reflection operators, a and β for voltage and λ and δ for current, shown in Figure 8.3(a).

    If there are two discontinuities, as in the case of Figure 8.3 (b), there will be reflections at each discontinuity with currents traveling back and forth, and thus oscillating, in the intermediate conductor. The amplitudes of these oscillatory currents will diminish as energy is transmitted from the intermediate conductor to the conductors on either end.

    The geometry of Figure 8.3(b) approximates that of a lightning current entering an aircraft, since the surge impedance of the lightning channel is higher than the surge impedance of the aircraft. When account is taken of all the reflections and transmissions at the various transition points, the result is that

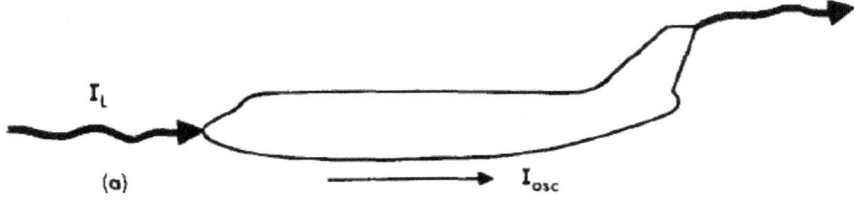

(a)

I_L I_{osc}

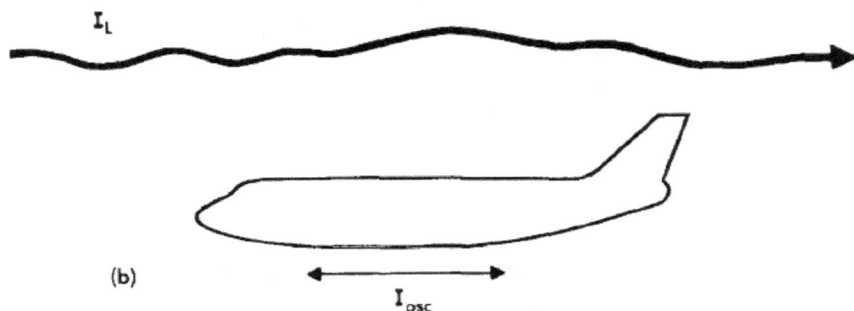

(b)

I_L I_{osc}

Figure 8.2   Response of the structure to the lightning current.
     (a) To a direct flash
     (b) To a nearby flash

the current waveforms will appear as shown in Figure 8.3(c). If the entering lightning current has a fast front and a slower decay, the current at the entrance of the low-impedance section (in the aircraft) will exhibit an overshoot. The current at the center of the aircraft will have an oscillatory component superimposed upon a waveform fundamentally like that of the input current, and the current leaving the aircraft will have a slower rise time than that of the incident lightning current.

    A way in which one can tally the various reflected and transmitted currents is through the use of the lattice diagram shown in Figure 8.4. A current wave of unit amplitude enters from the left, sees the transition point, and is partly transmitted and partly reflected. Since Z_2 is less than Z_1, the exiting (leaving) current is of higher amplitude than the entering current. In the limit, a current wave traveling on a conductor and encountering a short circuit ($Z_2 = 0$) would double at that discontinuity. The transmitted current then passes along the low-impedance conductor and after a time delay of Δt encounters the discontinuity at the exit end. Here part of the current on conductor 2 is again reflected. Since the impedance of conductor 3 is greater than the impedance of conductor 2, the transmitted current is less than the incident current. A limiting condition of this would consist of a current wave traveling on a conductor and then encountering an open circuit. The current transmitted into the open circuit is of course zero.

256

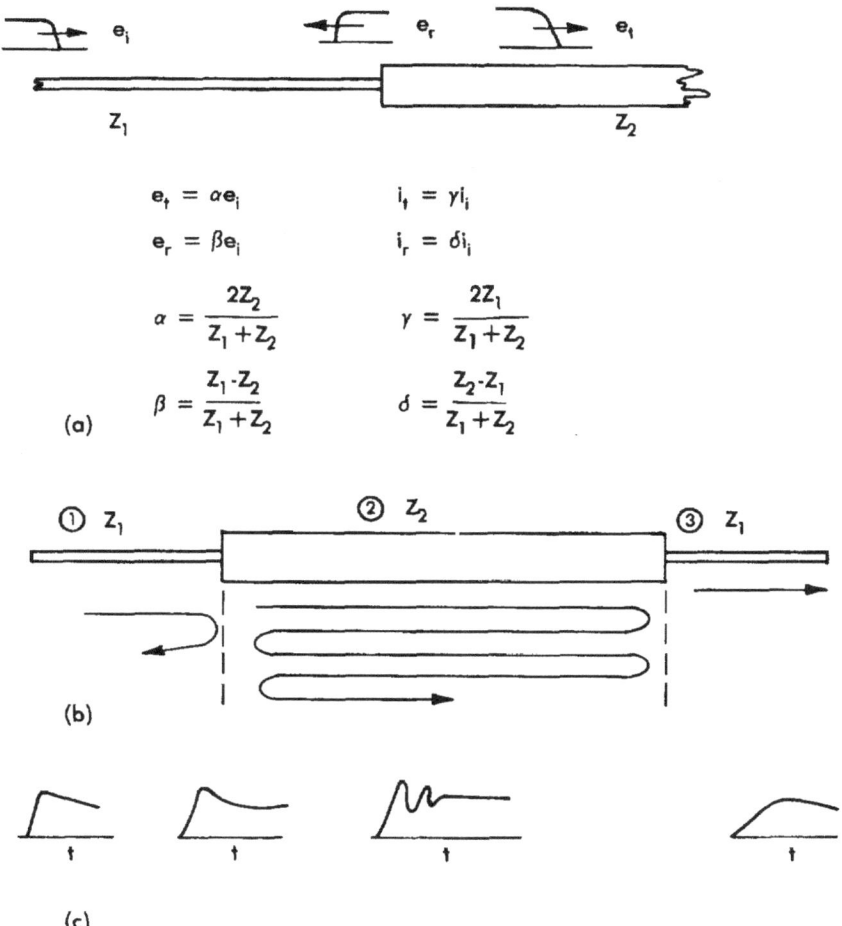

$$e_t = \alpha e_i \qquad i_t = \gamma i_i$$

$$e_r = \beta e_i \qquad i_r = \delta i_i$$

$$\alpha = \frac{2Z_2}{Z_1 + Z_2} \qquad \gamma = \frac{2Z_1}{Z_1 + Z_2}$$

$$\beta = \frac{Z_1 - Z_2}{Z_1 + Z_2} \qquad \delta = \frac{Z_2 - Z_1}{Z_1 + Z_2}$$

(a)

(b)

(c)

Figure 8.3 Surge propagation at transition points.
      (a) A single transition point
      (b) Two transition points
      (c) Waveshapes

The current amplitude at any point is then the algebraic sum of the reflected and transmitted components at that point. For example, the current leaving the intermediate conductor is 0.556 at $1\Delta t$ after the current enters the left-hand portion of it, 0.803 after $2\Delta t$, and 0.913 after $3\Delta t$, etc. The exit current thus rises more slowly than does the incident current. At the center of the low-impedance intermediate conductor, the current is oscillatory.

If instead of a step function a current wave of finite rise time encounters the transition, there will be a less pronounced oscillation of current in the intermediate conductor and less difference between the waveforms of the entering and exiting currents.

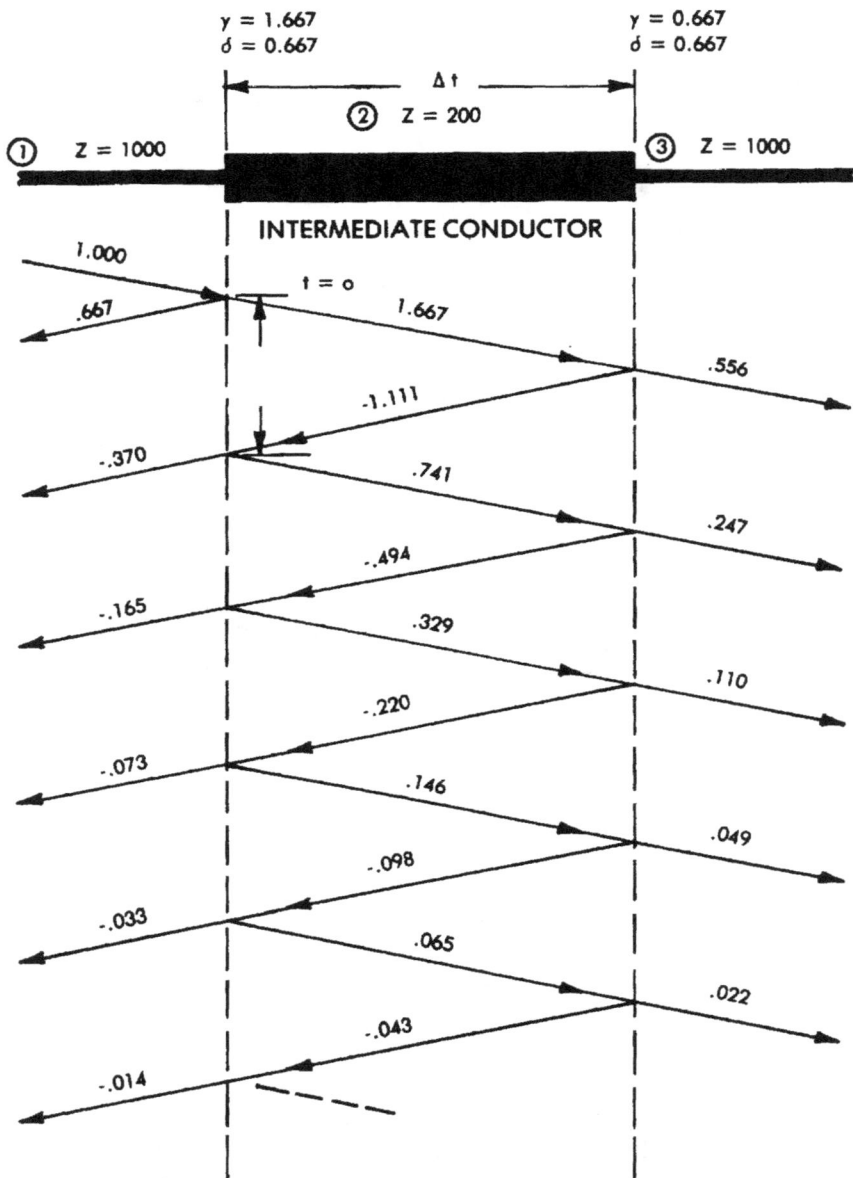

$\gamma = 1.667$
$\delta = 0.667$

$\gamma = 0.667$
$\delta = 0.667$

Δt

② $Z = 200$

① $Z = 1000$ ③ $Z = 1000$

INTERMEDIATE CONDUCTOR

1.000

.667

$t = 0$

1.667

.556

-1.111

-.370

.741

.247

-.494

-.165

.329

.110

-.220

-.073

.146

.049

-.098

-.033

.065

.022

-.043

-.014

Figure 8.4 Lattice diagram showing amplitude of reflections

An example of this effect was demonstrated on a simple model of the *Space Shuttle* (Reference 8.1). A two-dimensional outline of the *Shuttle* was cut from metal foil and the outline split in the center so that the two halves could be slightly separated. The model was suspended in the air and connected to a fine

258

wire; a pulse current was passed along the wire, through the model, and along the rest of the fine conductor wire to a termination resistor. The fine wire represented the channel of the lightning arc. The input and output currents and the current at the center of the model were measured with small transformers.

Typical results are shown in Figure 8.5. One significant point about the results was that fast-rising incident currents excited a higher degree of oscillation of the aircraft than did slower rising currents. The second significant point demonstrated was that in most cases the waveform of the lightning current passing through the center of the aircraft was sufficiently similar to the waveform of the basic lightning current that the oscillatory component superimposed as a result of the change in impedance between the arc channel and the aircraft was not significantly large.

In the case of a nearby lightning flash, the electric field from the flash will excite a dipole oscillation of the aircraft. In terms of Figure 8.4, the effect could be viewed as one in which the impedance of conductors 1 and 3 was infinite, leading to complete reflection and no transmission at the entry and exit points on a low-impedance aircraft. A dipole kind of oscillation excited on an aircraft structure would be the typical response considered if one were evaluating the effect on aircraft of the electromagnetic fields produced by nuclear explosions. The period of oscillation would be proportional to the length of the aircraft; thus, large aircraft would tend to ring at lower frequencies and higher amplitudes than would shorter aircraft. While the response of aircraft and missile systems to the rapidly changing electromagnetic fields produced by nuclear explosions has been extensively studied, there have not been any corresponding studies of the response of aircraft to the electric fields associated with the passage of a nearby lightning flash. Accordingly, in this record we will not treat further the case of a nearby lightning flash, except to add the cautionary statement that it has not been proven finally that the indirect effects associated with a nearby lightning flash are necessarily lower than those associated with a lightning flash that contacts the aircraft. In the case of a nearby lightning flash it would appear that the aircraft structure would be subjected to currents of much lower amplitude but currents much more oscillatory in nature. These lower amplitude—higher frequency currents might lead to more upsets of sensitive electronic circuits than would currents of higher amplitude and lower frequency.

8.3 The Internal Response of the Aircraft Structure

A metallic aircraft is often viewed as a Faraday Screen, a concept from electrostatics which implies that the electrical environment inside the aircraft is separate and distinct from the environment outside. To some extent this is true for the electrical environment inside the structure: the environment is not nearly as harsh as is the external environment. There are, however, some important mechanisms by which electrical energy couples to the interior of the aircraft.

The basic coupling mechanisms are shown on Figure 8.6. The first of these relates to the electric field produced along the inner surface of the aircraft. This

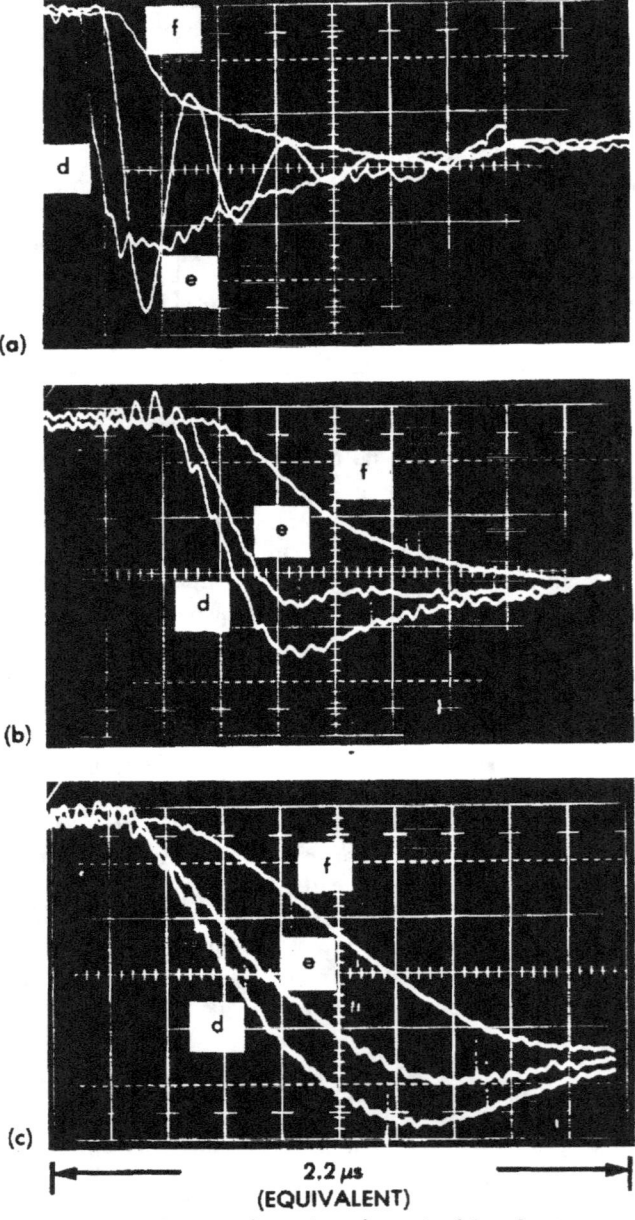

Figure 8.5 Dipole oscillation of an aircraft excited by the passage of lightning current.

(a) Fast rise of current

(b) Average rise of current

(c) Slow rise of current

On each oscillogram

d is the current entering the aircraft

e is the current at the center of the aircraft

f is the current leaving the aircraft

260

coupling mechanism might be defined as a resistive voltage drop. In some cases a definite resistance will be involved, though frequently the resistance will be of a distributed nature and probably frequency- or time-dependent.

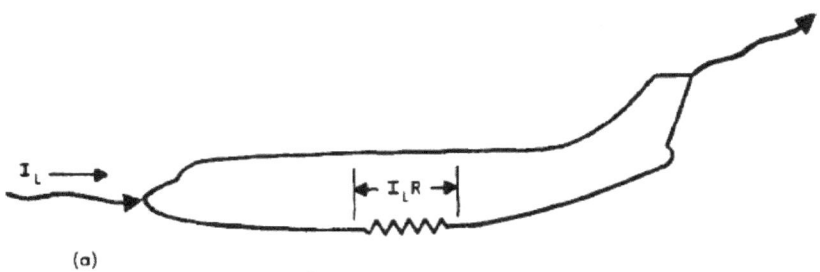

(a)

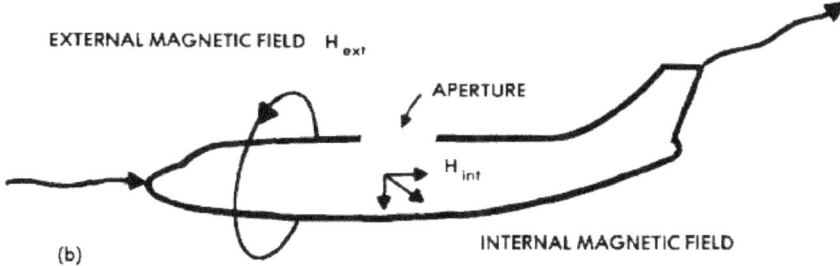

(b)

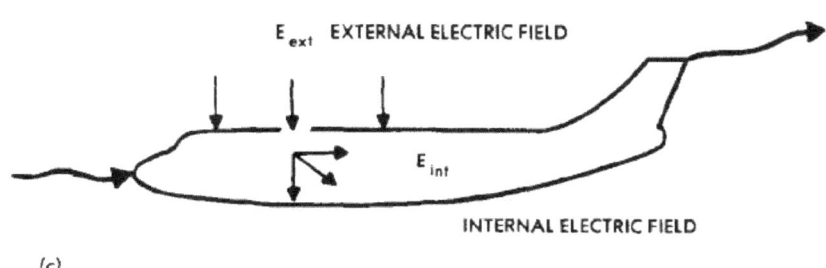

(c)

Figure 8.6 Coupling mechanisms.
(a) Resistive
(b) Magnetic fields
(c) Electric fields

The second coupling mechanism involves magnetic fields in the interior volume of the aircraft. The most common and important type of magnetic field is that drawn through apertures from the outside of the aircraft to the interior,

261

as shown in Figures 8.6 and 8.7. This is frequently called the *aperture field*. There will also be magnetic fields produced by the diffusion of lightning currents to the inside surfaces of the aircraft skins. These are referred to loosely as the *diffusion fields*. The diffusion fields are also related to the frequency-dependent properties of the resistively generated electric field. Because some of the concepts involved in the study of the diffusion fields are central to an understanding of other effects, particularly with respect to the response of shielded wires, they will be discussed in detail before fields of other origins are considered.

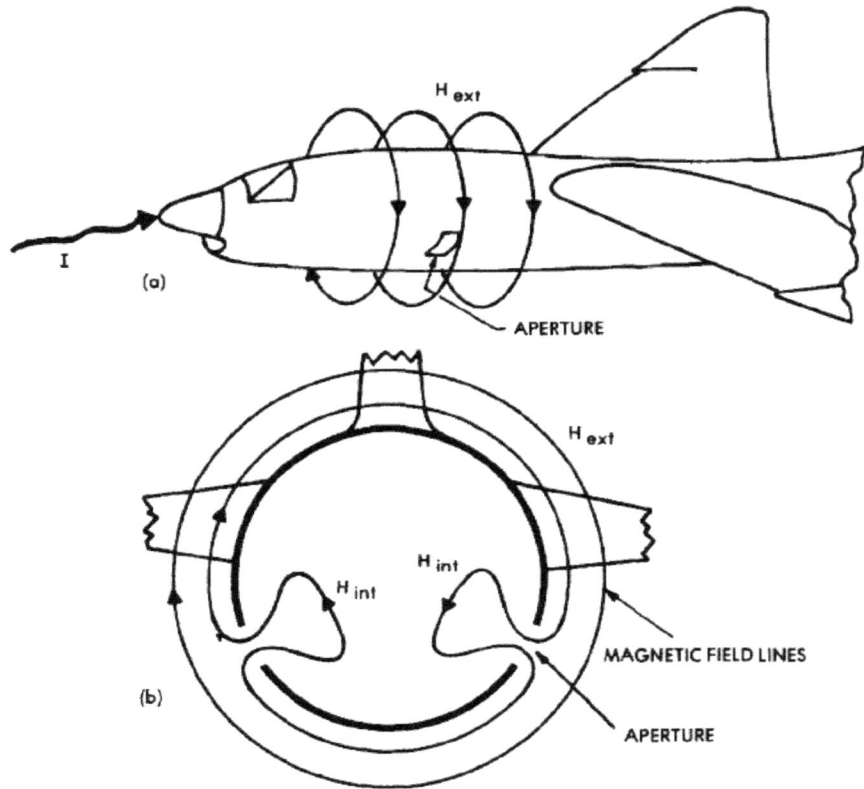

Figure 8.7 Aperture-type magnetic field coupling.
(a) External field patterns
(b) Internal field patterns

The third type of coupling involves electric fields passing directly through apertures, such as windows or canopies, to the interior of the aircraft. In metal aircraft this coupling is entirely through apertures, since virtually any thickness of metal provides comparatively good shielding. Aperture-type electric field coupling is shown in Figures 8.6(c) and 8.8.

262

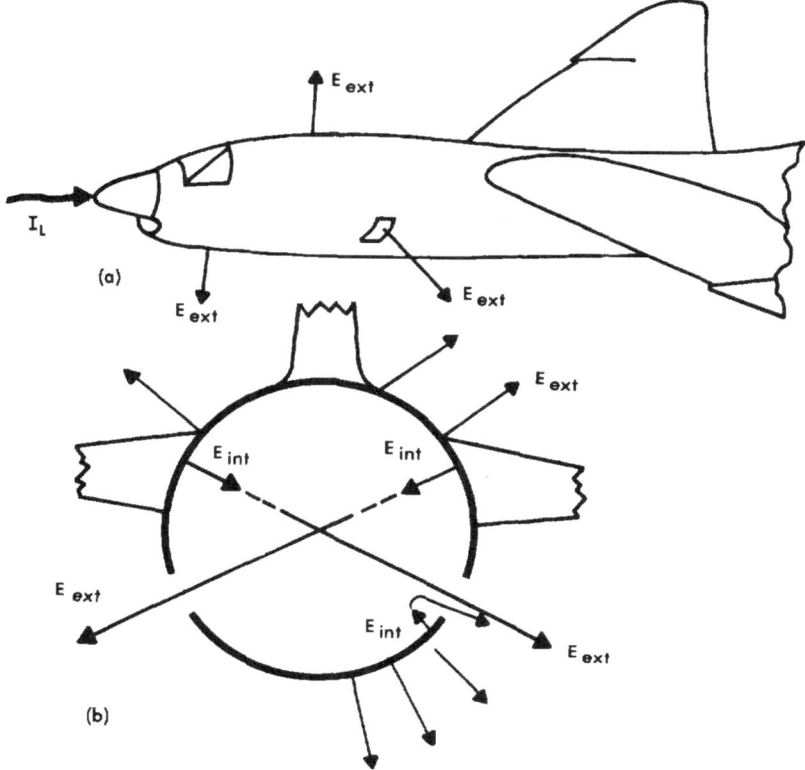

Figure 8.8 Aperture-type electric field coupling.
(a) External field patterns
(b) Internal field patterns

The most easily understood mechanism by which the passage of lightning current gives rise to voltages on aircraft electrical circuits is that in which the current, flowing through joint resistances, produces a voltage by the elementary IR voltage drop. Such a case is shown in Figure 8.9. Here lightning is shown contacting a wing tip navigation light. The lightning current flowing through the resistance of the mechanical mounting structure of the lamp housing produces a voltage across that resistance. The voltage drop across the resistance will have the same waveform as that of the lightning current. The voltage at some remote point, however, may not have the same waveform, since the distributed inductance and capacitance of the wire supplying power to the filament of the light will be set into oscillation by the suddenly developed voltage. The result is that at remote points there probably will be an oscillation superimposed upon the basic IR voltage.

Figure 8.10 shows two other examples of cases in which resistive voltages might be encountered. The first would be at the pylons for mounting external

263

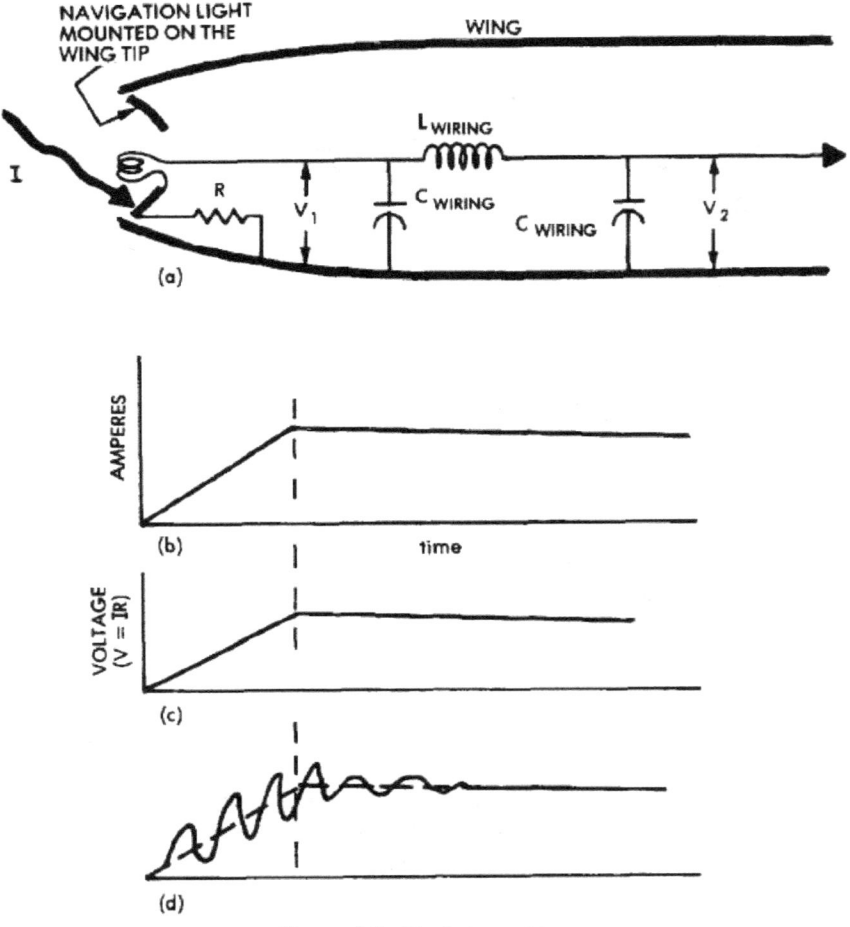

Figure 8.9 Resistive voltages.

    (a) Physical structure

    (b) Current waveshape

    (c) V_1

    (d) V_2

stores, shown in 8.10(a). If lightning current were to contact such external stores, it would have to flow through the pylons to enter the aircraft. The pylons, not generally designed as current-carrying members and being points where the lightning current would be concentrated, might have a high voltage developed across them. Another example might be the structural bolts attaching a large segment of the airframe, such as the vertical stabilizer shown in Figure 8.10(b).

    The effects of joint resistance on circuits are strongly influenced by the manner in which circuits are grounded, as shown in Figure 8.11. Current flowing

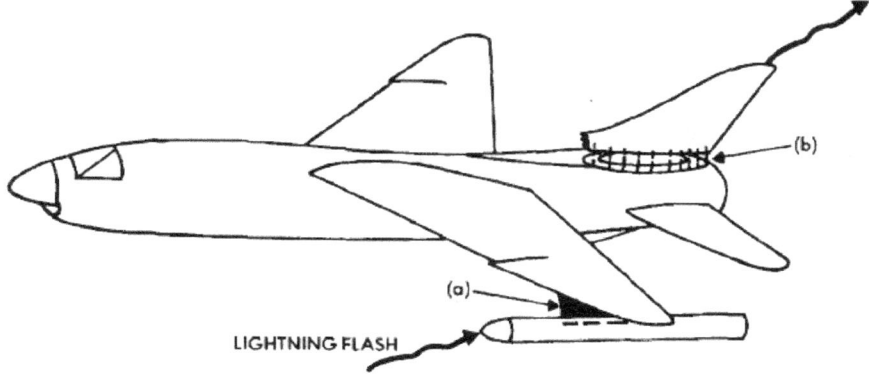

LIGHTNING FLASH

Figure 8.10 Other examples of resistance.
(a) The pylons for external stores
(b) Joints in structural members

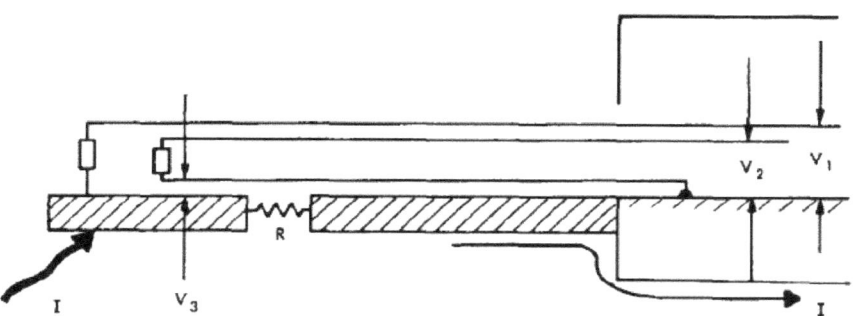

Figure 8.11 Effects of grounding.
(a) Structural return, $V_1 = IR$
(b) Single-point ground, $V_2 =$ low
(c) Single-point ground, $V_3 = IR$

across the joint resistance, R, produces a driving voltage: $V = IR$. Since the circuit across which V_1 is measured employs the structure as a ground-return path, the circuit couples all of this voltage; thus V_1 would be high. A circuit employing a single-point ground does not include this resistive drop; hence V_2 would be low. The use of a single-point ground, however, does not eliminate the voltage, since in this latter case the voltage at the source end of the circuit, V_3, would be high.

These elementary descriptions of joint resistance should not be relied upon to predict coupling into circuits extending throughout the entire aircraft. The more massive the joint and the lower the dc resistance, the greater will be the dependence of resistance on the waveform and frequency content of the lightning current, and the greater will be the proportionate effects of changing magnetic fields. While these effects will be discussed in more detail in other sections, one common oversimplification, shown in Figure 8.12, should be pointed out here. If the total end-to-end resistance of the aircraft were 2.5 mΩ and a lightning current of 200 000 A were flowing through the aircraft, the end-to-end voltage on any circuit could not be depended upon to be less than 500 V, the product of the lightning current and the dc resistance.

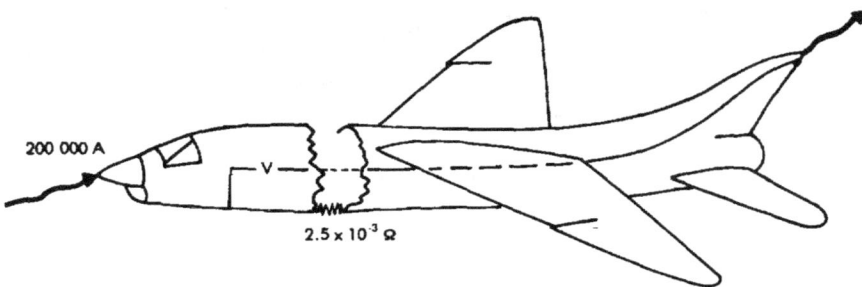

Figure 8.12 An oversimplified model.
● Maximum voltage is not determined only by total end-to-end resistance.

REFERENCE

8.1   F. A. Fisher, *Analysis of Lightning Current Waveforms through the Space Shuttle,* Aircraft Lightning Protection Note 75-1, National Aeronautics and Space Administration, Lyndon B. Johnson Space Center, Houston, Texas (January 17, 1975).

CHAPTER 9
THE EXTERNAL MAGNETIC FIELD ENVIRONMENT

9.1 Elementary Considerations

If a long conductor is carrying a current, I, and the return path is far removed, the average field intensity, H_{av}, at a distance, r, from the conductor, as shown in Figure 9.1(a), is

$$H = \frac{I}{2\pi r} \tag{9.1}$$

If instead of a solid wire the current is carried on a hollow tube of radius r, as shown in Figure 9.1(b), the field intensity, H, at radius r is again

$$H = \frac{I}{2\pi r} \tag{9.2}$$

and at the surface of the tube, where $r = r_0$, the field intensity is

$$H = \frac{I}{2\pi r_0} \tag{9.3}$$

In the interior of the tube, $r < r_0$, the field intensity is zero, a concept that will be treated in more detail later.

Since the circumference, P, of the tube is

$$P = 2\pi r_0 \tag{9.4}$$

it follows that the field intensity at the surface of the tube is also equal to the total current divided by the circumference.

$$H = \frac{I}{P} \tag{9.5}$$

In all the cases the units of field intensity are amperes per meter if the radii are measured in meters.

If the conductor is not cylindrical, as shown in Figure 9.1(c), the field intensity at different points on the surface will be different. However, the average field intensity will still be equal to the total current divided by the circumference:

$$H_{av} = \frac{I}{P} \tag{9.6}$$

The actual field intensity will be greater than average at points where the radius of curvature is less than average; it will be less than average at points where the radius of curvature is greater than average, as shown in Figure 9.2. For example, the circumference of the fuselage of a typical fighter aircraft just forward of its wing is about 5.5 m. Assuming a lightning stroke current of 30 000 A to flow through the fuselage, the average field intensity at the surface would be

$$H = \frac{I}{2\pi r} = \frac{I}{P} = \frac{30\,000}{5.5} = 5455 \text{ A/m} \qquad (9.7)$$

Since there are no points of very sharp radius, the field intensity around the fuselage would probably not vary greatly from the average value.

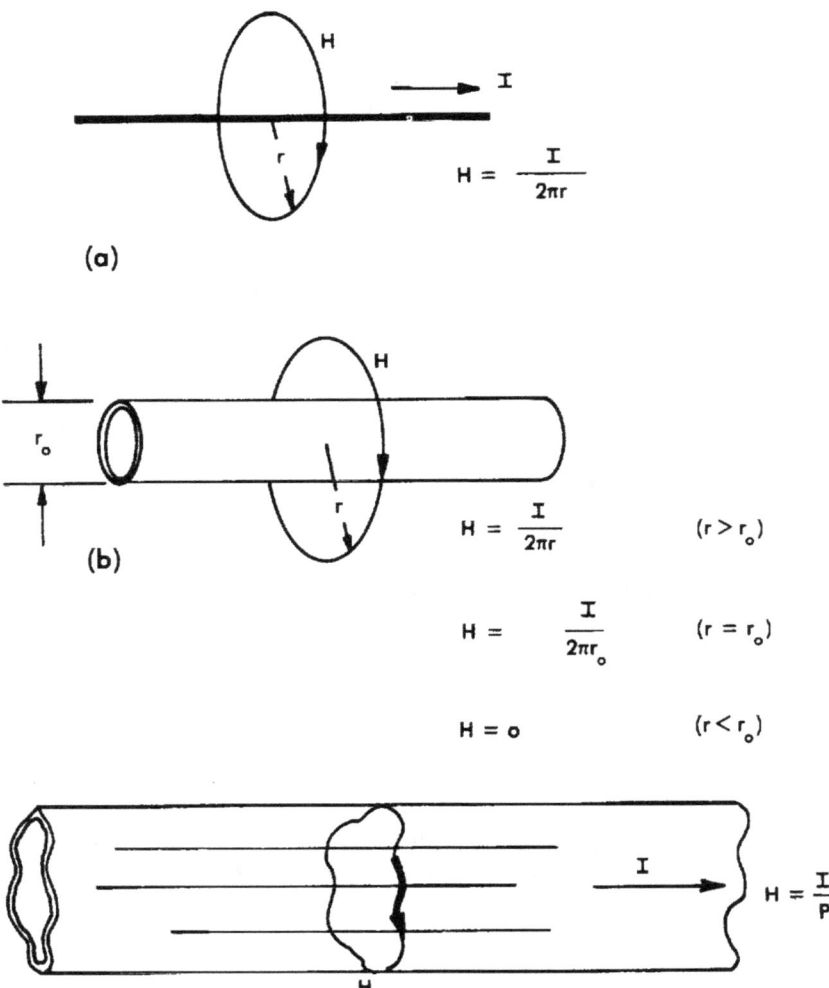

$$H = \frac{I}{2\pi r}$$

(a)

$$H = \frac{I}{2\pi r} \qquad (r > r_o)$$

$$H = \frac{I}{2\pi r_o} \qquad (r = r_o)$$

$$H = 0 \qquad (r < r_o)$$

(b)

$$H = \frac{I}{P}$$

(c)

Figure 9.1 Magnetic fields around current-carrying conductors.
   (a) Current-carrying filament
   (b) Tubular conductor
   (c) Irregular conductor

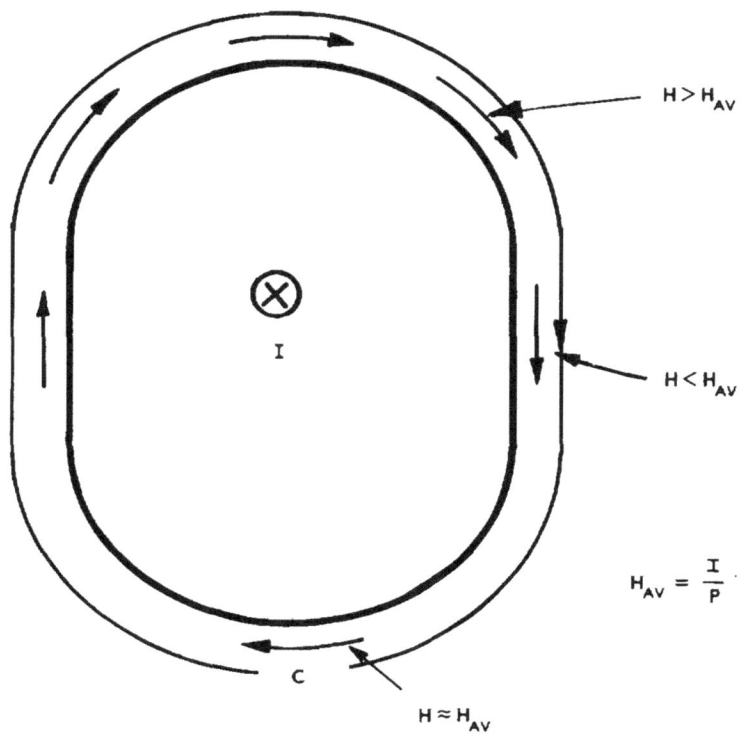

Figure 9.2 Field intensity vs radius of curvature.

The situation along a wing carrying lightning current is considerably different in that the leading and trailing edges have radii of curvature much less than the average. Field intensity along the leading and trailing edges would be quite high compared to the field intensity along the top and bottom surfaces, for example.

Figure 9.3 shows, in general, how the magnetic field strengths would vary with position on an aircraft if a lightning flash enters through the nose pitot boom and leaves through the vertical stabilizer. The field intensity would be highest around the pitot boom, lowest around the midsection of the fuselage, and high again around the vertical stabilizer. In the vicinity of the nose equipment bays, the field would be of greater than average intensity. Since the field intensity is inversely proportional to the radius of curvature, it then follows that the field intensity outside the fuselage of a large transport aircraft would be considerably less than that outside the small fighter aircraft shown on Figure 9.3.

Since both the average current density, J_{av}, and average field intensity, H_{av}, are equal to the total current divided by the circumference, or

271

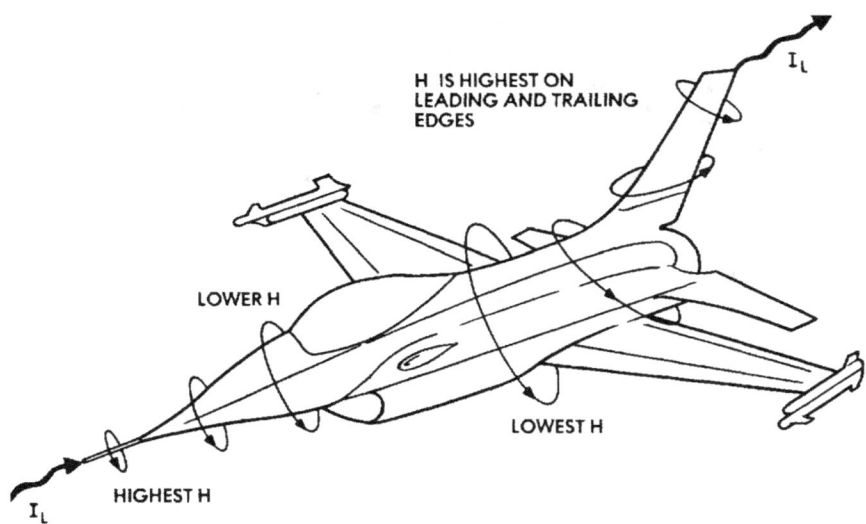

Figure 9.3 Variation of magnetic field strength with aircraft radius of curvature.

$$J_{av} = H_{av} = \frac{I}{P} \tag{9.8}$$

it follows that the tangential field intensity at the surface of a conducting object is equal to the current density at that point. This is in fact true, at least for transient currents. The relation is not true for dc currents or transients sufficiently slow that appreciable magnetic fields penetrate the skin. The orientation of the H field vector is always at right angles to the direction of the current vector. While small gaps in the structure (Figure 9.4) direct the current around the gap, the magnetic field is virtually unaffected, except directly on the surface and on a length scale that is small compared to the dimensions of the gap interrupting the current flow.

9.2 Determination of the External Current and Magnetic Flux Density

Both the resistive voltage drops inside a structure and the amounts of magnetic or electric field which penetrate through apertures depend upon the external field distribution, most importantly upon the magnetic field distribution. This section will discuss some of the ways in which this field distribution might be determined.

The current density at the surface of a conductor is seldom determined by the dc resistance of the conductor; rather, it is primarily controlled by magnetic effects. The magnetic distribution of current density can be calculated in simple geometries. Around the periphery of a cylinder, for example, the current density is uniform, assuming the return path for that current is far removed from the cylinder: greater than ten times the diameter of the cylinder.

272

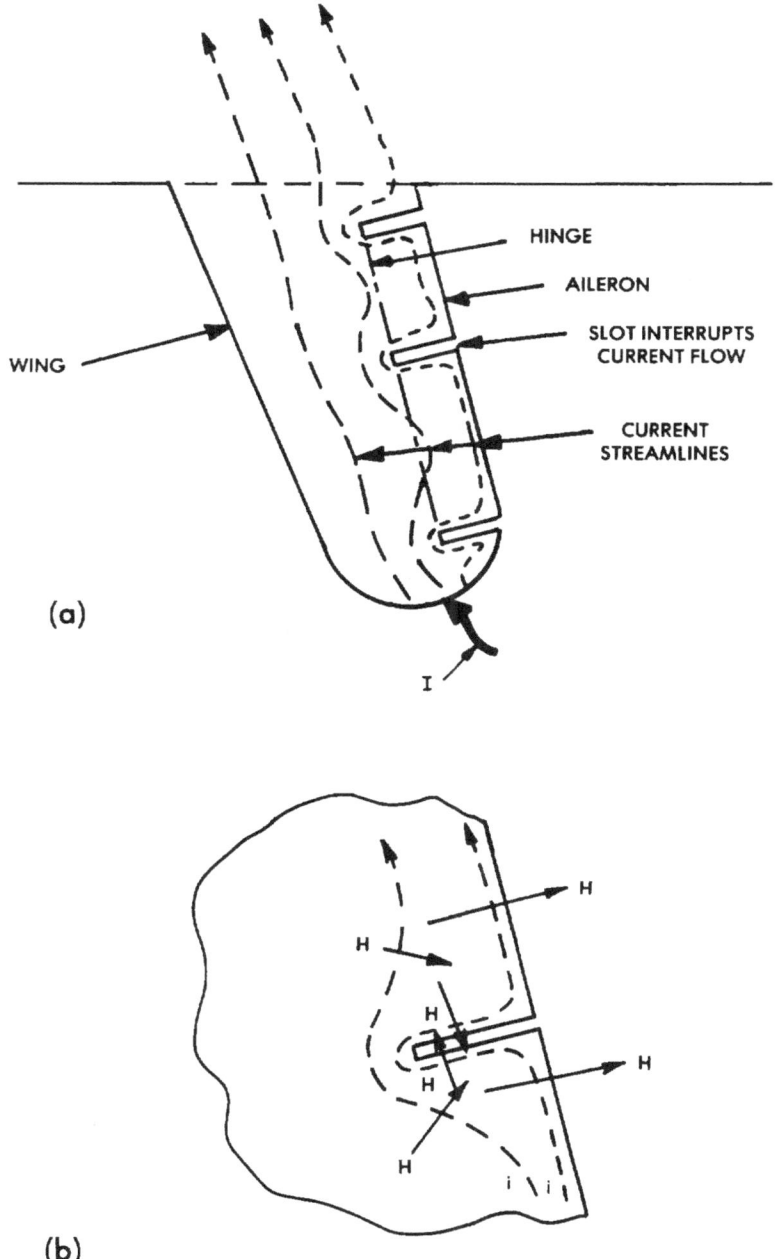

Figure 9.4 Current flow and magnetic field around structural gaps.
    (a)   Current entering typical wing
    (b)   Resultant magnetic field virtually unaffected
           by slot interrupting flow of current

9.2.1 Elliptical Conductors

Another geometry of considerable practical importance is the ellipse, shown in Figure 9.5. At the center $(X = 0, Y = \pm \frac{d}{2})$ the magnetic field intensity is given by

$$H = \frac{I}{\pi b} \tag{9.9}$$

and at the edge $(X = \pm \frac{b}{2}, Y = 0)$

$$H = \frac{I}{\pi d} \tag{9.10}$$

At intermediate points the magnetic field intensity or current density is given by the expression (References 9.1 and 9.2)

$$H \text{ surface} = \frac{I}{\pi} \frac{1}{\sqrt{b^2 - (2x)^2 \left[1 - \frac{d^2}{b^2}\right]}} \tag{9.11}$$

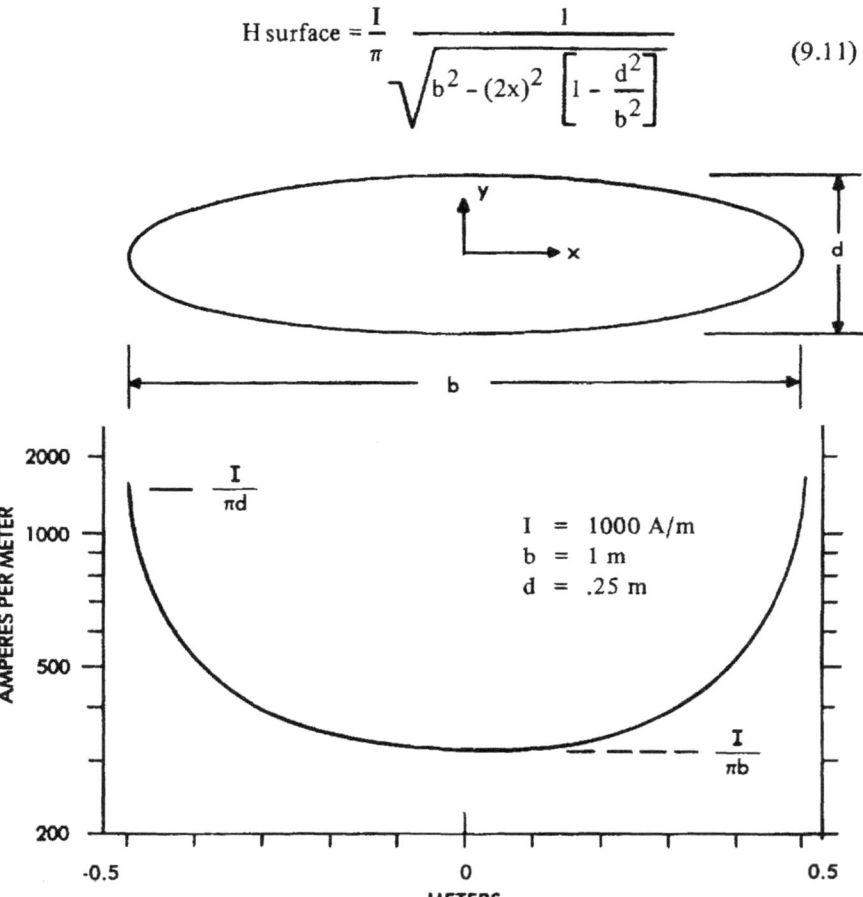

$I = 1000 \text{ A/m}$
$b = 1 \text{ m}$
$d = .25 \text{ m}$

Figure 9.5 Magnetic field intensity at the surface of an elliptical conductor.

274

This current distribution does not hold for dc currents where the current density over the surface is determined by the dc resistance. If the conductor were an elliptical cylinder of uniform thickness, the current density would be uniform. The region over which a transition takes place between the uniform distribution of current governed by resistance and the nonuniform distribution governed by magnetic fields (Equation 9.11) is indicated as follows:

$$\sqrt{\frac{bd}{\delta}} \leqslant 1\text{-}10 \qquad\qquad (9.12)$$

where
b = width
d = depth

The classical skin depth, δ, is

$$\delta = \sqrt{\frac{2\rho}{\mu\omega}} \qquad\qquad (9.13)$$

where
ρ = resistivity
μ = permeability
ω = angular frequency

As an example of the frequency range over which the transition takes place, consider an elliptic cylinder made from aluminum ($\rho = 2.69 \times 10^{-8}\ \Omega$m) with width (b) 1 m, depth (d) 0.25 m, and a wall thickness of 1 mm. The skin depth and value of the term

$$\sqrt{\frac{bd}{\delta}}$$

is given in Table 9.1.

Table 9.1 SKIN DEPTH AS A FUNCTION OF FREQUENCY

Frequency f	Skin depth δ	$\sqrt{\dfrac{bd}{\delta}}$
1 Hz	8.25×10^{-2}m	6.06×10^{0}
10 Hz	2.61×10^{-2}m	1.92×10^{1}
100 Hz	8.25×10^{-3}m	6.06×10^{1}
1000 Hz	2.61×10^{-3}m	1.92×10^{2}
10 000 Hz	8.25×10^{-4}m	6.06×10^{2}

Even at 10 Hz the current crowds to the edges of the ellipse and has a current density given by Equation 9.11. At 1 kHz the skin depth is about 0.25 cm, and at 10 kHz it is less than 1 mm.

Thus, the region over which the current density changes from its uniform dc value to the limiting ac distribution is 1 to 10 Hz. The current density remains at that value up to indefinitely high frequencies, probably until the width of the cylinder becomes on the order of a tenth of a wavelength. The current density through the wall thickness of the cylinder remains constant over the range 0 to 3 kHz (approx.).

In many cases the external current density may be determined with accuracy sufficient for practical purposes by approximating the surface under consideration by an ellipse or ellipses. If such an approximation does not give sufficient accuracy, there are other techniques that may be used.

9.2.2 Numerical Solution of Fields

Analytically the solution of the field around a current-carrying conductor may be determined by a solution of Laplace's equation:

$$\nabla^2 \phi = 0 \qquad (9.14)$$

where ϕ = the potential.

In rectangular coordinates Laplace's equation becomes

$$\frac{\partial^2 \phi}{\partial x^2} + \frac{\partial^2 \phi}{\partial y^2} + \frac{\partial^2 \phi}{\partial z^2} = 0 \qquad (9.15)$$

For some geometries Laplace's equation can be solved analytically: an elliptical geometry is one of those cases. The equations defining the field distribution will not be presented here, but Figure 9.6 shows an example of the field around an elliptical conductor. This field may be viewed as the electric field around an isolated ellipsoid, the figure showing the field in one plane passing through the major axis of the ellipsoid or the field around an elliptic cylinder of infinite length. Alternatively, it may be viewed as the magnetic field around an infinitely long elliptic cylinder carrying current into the plane of the figure. If the field is viewed as an electric field, the flow lines represent the path along which displacement currents flow, and if it is viewed as a magnetic field, the equipotential lines define the paths along which magnetic flux lines flow. The return path for either the voltage maintaining the charge or the return path for the current is assumed to be sufficiently far away from the conductor that it does not influence the field around the indicated region.

The indicated flow lines divide the region into 44 sectors. At the surface of the conductor, the magnetic field strength is inversely proportional to the spacing between the flow lines. Since the average field strength around the surface is

$$H = \frac{I}{P} \qquad (9.16)$$

it follows that the field strength at the surface between any two flow lines is

$$H_{surface} = \frac{I}{44\Delta S} \qquad (9.17)$$

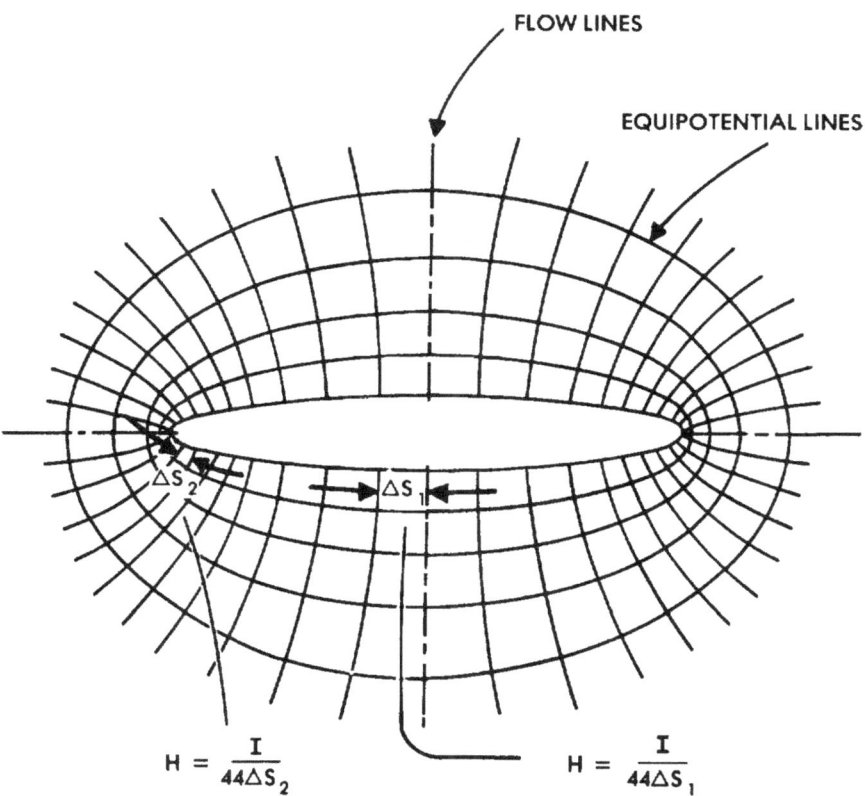

FLOW LINES

EQUIPOTENTIAL LINES

$$\Delta S_2$$

$$\Delta S_1$$

$$H = \frac{I}{44\Delta S_2} \qquad\qquad H = \frac{I}{44\Delta S_1}$$

Figure 9.6 The field around an elliptical conductor.

In only the simplest geometries is it possible to calculate the field analytically. Usually one must resort to some numerical or graphical method of determining the field. Numerically, the field may be determined by a numerical solution of Laplace's equation. Figure 9.7 shows a conductor at potential P surrounded by a return conductor, a circle in this case, at potential 0. To this geometry is fitted a rectangular grid, shown here as a very coarse grid.

Initially, all the grid points that lie on the conductor would be assigned a

277

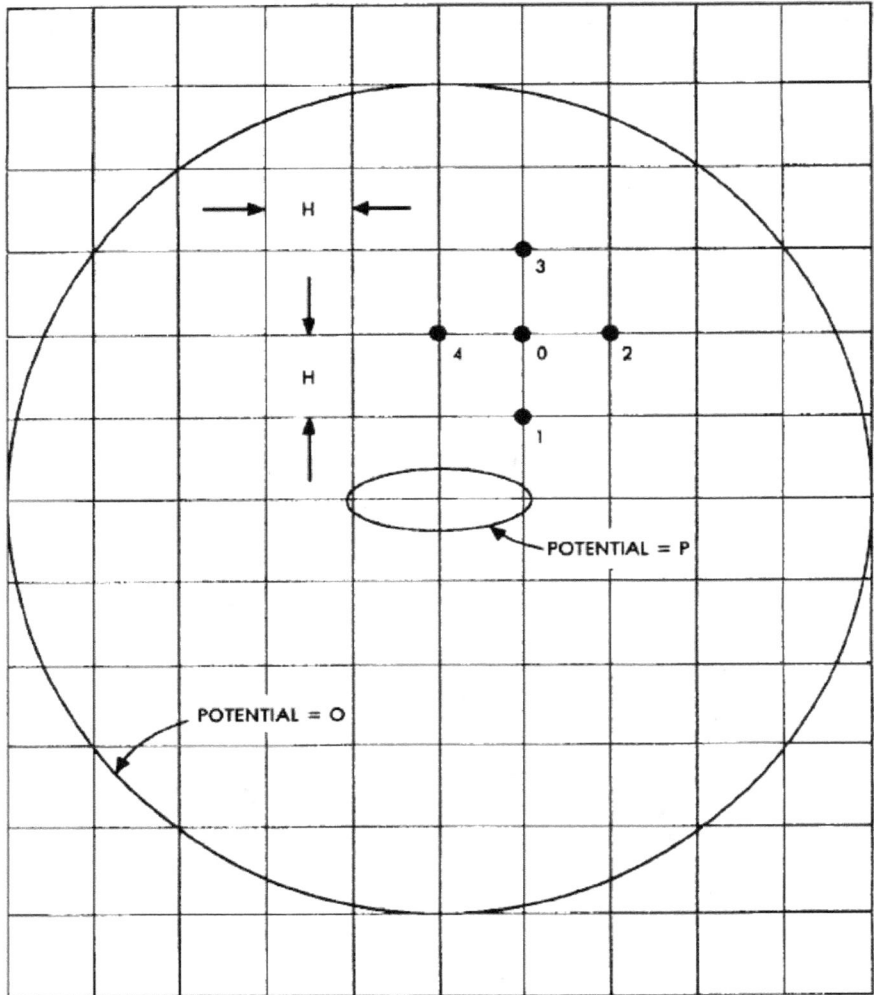

Figure 9.7 A rectangular grid for evaluation of Laplace's equation.

potential P, and all the grid points that lie on the return path would be assigned a potential zero. Laplace's equation in two dimensions can be shown to be approximately

$$\frac{\partial^2\phi}{\partial x^2} + \frac{\partial^2\phi}{\partial y^2} \cong \frac{1}{k^2}\left[(\phi_1 + \phi_2 + \phi_3 + \phi_4 - 4\phi_0)\right] \qquad (9.18)$$

From this it follows that Laplace's equation is satisfied if the numeric values at four points surrounding a central point have values that satisfy the equation

$$\phi_1 + \phi_2 + \phi_3 + \phi_4 - 4\phi_0 = 0 \qquad (9.19)$$

278

A determination of the field around the conductor then involves assigning field values at all of the points between the conductor and its return path, and adjusting the value of these points until Equation 9.19 is satisfied everywhere within the grid. The literature (Reference 9.3 and 9.4) indicates a number of the numerical techniques by which the potentials at the points may be adjusted to their final values. While the process is tedious, it is not completely impractical to do by hand. Usually the process is done by computer routines that solve the field equations. In addition to tabulating the numerical values of the field at the grid points, frequently such computer routines allow one to plot the flow and equipotential lines.

Any solution of Laplace's equation will also allow the pattern of coupling through an aperture, as well as the external field pattern, to be calculated. Most computer routines deal either with two-dimensional problems or problems with rotational symmetry. In a two-dimensional case, the aperture would run the length of the conductor being studied. This is not necessarily a disadvantage, since some important geometries and apertures are basically two-dimensional. An important aperture is that which may exist on the trailing edge of a wing when the flaps have been extended. Since it is a convenient region to reach, wiring is often placed in this region. Electrically it is a poor place, since the aperture is near a region where the magnetic fields external to the wing are high. Another important set of apertures that may be approximated as a continual opening is that formed by the windows in the fuselage of a transport aircraft.

In principle, the field around three-dimensional objects may also be solved numerically by extending Equation 9.18 to include 8 points on a cube surrounding a central point. Solutions of the fields around three-dimensional objects not involving rotational symmetry are seldom attempted because of the great increase in the amount of computer storage locations required to define the field satisfactorily in three dimensions.

9.2.3 Hand Plotting of Fields

The fields around any geometry may also be determined graphically by a cut-and-try process in which flow lines and equipotential lines are drawn on the geometry and adjusted until repeated subdivision always yields small squares and the flow and equipotential lines intersect at all points at right angles.

Cut-and-try field plotting, of which Figure 9.8 is an imperfect example, may with care, patience, and liberal use of a soft pencil and eraser yield a field pattern of any desired accuracy.

9.2.4 Calculation Using Wire Grids

Another approach to determining current density, shown in Figure 9.9, is based on the premise that a two-dimensional geometry can be represented as an array of parallel wires. If the current in each wire is known, the average current density along the surface defined by any two wires will be the average of the current on the two wires divided by the spacing between the wires.

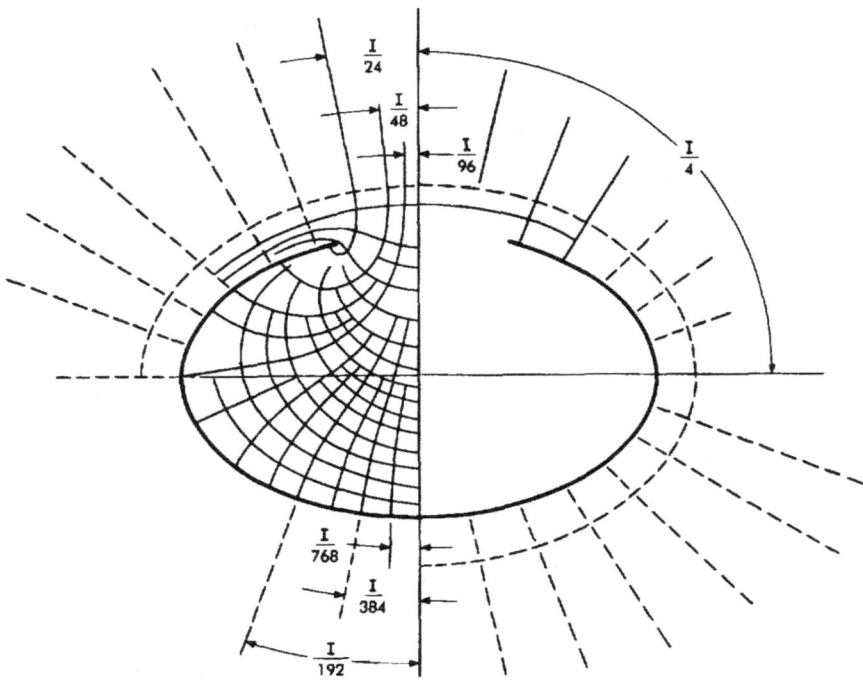

Figure 9.8 Cut-and-try field plotting

If the wires are all of infinite length, so that no end effects need be considered, and are all connected together at their ends, the manner in which current divides among the wires may be calculated with the aid of the simple computer program MAGFLD, the elements of which are discussed below, and the program given in Appendix 1. The equations governing the current distribution are given below.

Let the location of the wires be defined in terms of the rectangular coordinates x_i and y_i, and let the radius of the conductors be r_1. The self-inductance per unit length of each wire is

$$L = 2 \times 10^{-7} \log_\epsilon \frac{R}{r_1} \qquad (9.20)$$

henries per meter. In this equation R is defined as the distance from the conductor, or from the group of conductors, to an arbitrary return path. The numerical accuracy of the current distribution to be calculated does not depend critically upon the value assigned to R, but it should be of the order of 10 to 20 times the greatest dimension of the structure being modeled.

Between any two conductors, ij, there will be a mutual inductance:

$$M_{ij} = 2 \times 10^{-7} \log_\epsilon \frac{R}{r_{ij}} \qquad (9.21)$$

henries per meter.

280

The spacing between conductors, r_{ij}, can be determined from the coordinates of the conductors. If a group of conductors, each carrying a current, (Figure 9.10) is considered, and if the self- and mutual inductances of and between each conductor are known, the voltages across the self-inductance of each conductor will be as follows:

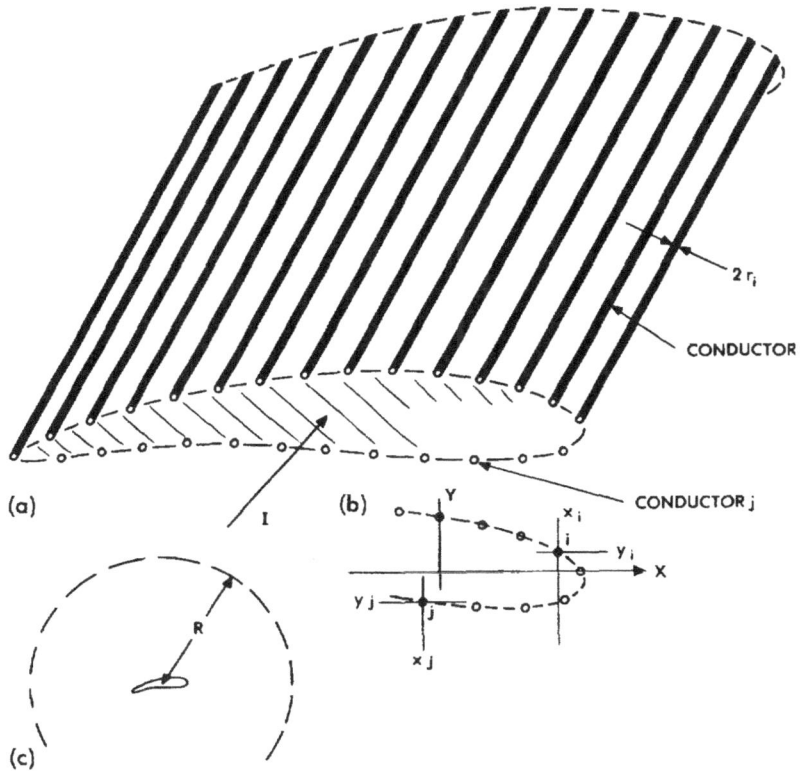

Figure 9.9 A structure defined as an array of wires.
    (a)   The array
    (b)   Coordinates defining location
    (c)   Definition of the return path for current

$$V_1/\omega = L_1 i_1 - M_{12} i_2 - M_{13} i_3 \ldots -M_{1n} i_n \qquad (9.22)$$

$$V_2/\omega = -M_{21} i_1 + L_2 i_2 - M_{23} i_3 \ldots -M_{2n} i_n \qquad (9.23)$$

$$V_3/\omega = -M_{31} i_1 - M_{32} i_2 - L_3 i_3 \ldots -M_{3n} i_n \qquad (9.24)$$

$$V_\eta/\omega = -M_{n1} i_1 - M_{n2} i_2 - M_{n3} i_3 \ldots L_n i_n \qquad (9.25)$$

Equations 9.22 through 9.25 may be placed in matrix notation as follows:

281

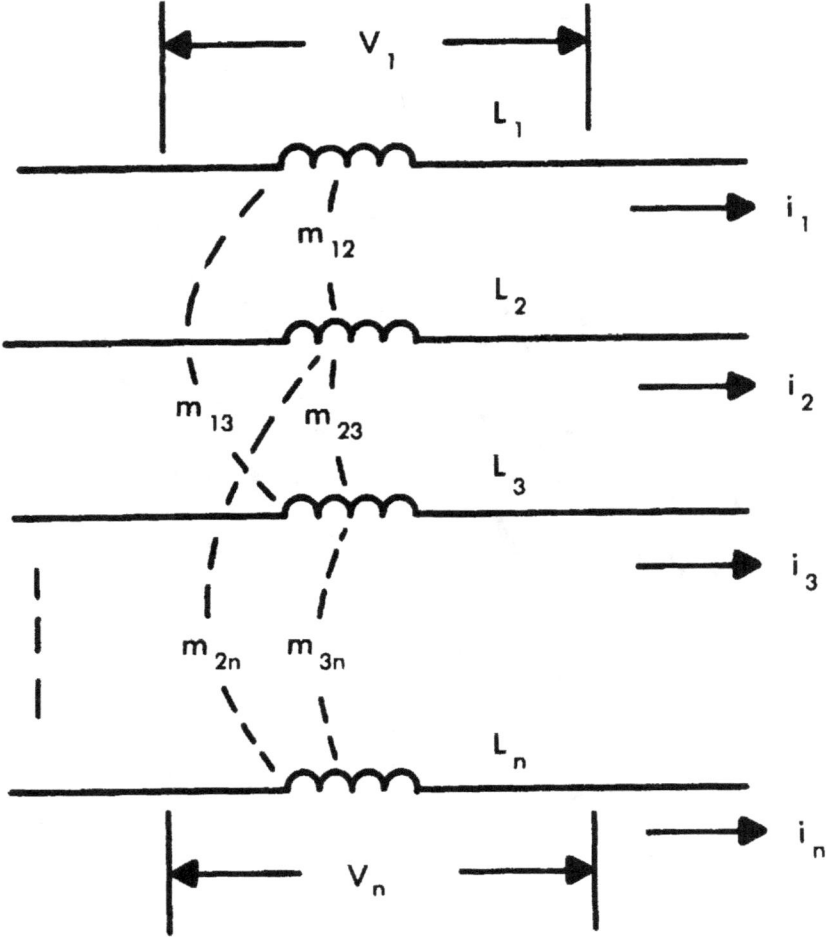

Figure 9.10 Mutually coupled inductances.

$$
\begin{vmatrix} V_1 \\ V_2 \\ V_3 \\ \cdot \\ \cdot \\ \cdot \\ V_n \end{vmatrix} = \begin{vmatrix} L_1 & -M_{12} & -M_{13} & \cdots & -M_{1n} \\ -M_{21} & +L_2 & -M_{23} & \cdots & -M_{2n} \\ -M_{31} & -M_{32} & -L_3 & \cdots & -M_{3n} \\ \cdot & & & & \\ \cdot & & & & \\ \cdot & & & & \\ -M_{n1} & -M_{n2} & -M_{n3} & \cdots & -L_{nn} \end{vmatrix} \times \begin{vmatrix} i_1 \\ i_2 \\ i_3 \\ \cdot \\ \cdot \\ \cdot \\ i_n \end{vmatrix} \qquad (9.26)
$$

or, in more compact notation

$$| V | = | M | \times | i |$$ (9.27)

In Equations 9.26 and 9.27 the normalized angular frequency ω has been set equal to unity.

Multiplying by the inverse of the M matrix, $| M |^{-1}$, gives the following:

$$| M |^{-1} \times | V | = | M |^{-1} \times | M | \times | i |$$ (9.28)

or

$$| i | = | M |^{-1} \times | V |$$ (9.29)

$$
\begin{vmatrix} i_1 \\ i_2 \\ i_3 \\ \cdot \\ \cdot \\ \cdot \\ i_n \end{vmatrix} = \begin{vmatrix} m_{11}m_{12}m_{13} \cdots m_{1n} \\ m_{21}m_{22}m_{23} \cdots m_{2n} \\ m_{31}m_{32}m_{33} \cdots m_{3n} \\ \cdot \\ \cdot \\ \cdot \\ m_{n1}m_{n2}m_{n3} \cdots m_{nn} \end{vmatrix} \times \begin{vmatrix} V_1 \\ V_2 \\ V_3 \\ \cdot \\ \cdot \\ \cdot \\ V_n \end{vmatrix}
$$ (9.30)

where m_{11}, m_{12}, m_{13} are the elements of the inverse of the M matrix.

If all of the voltages are the same and equal to V, as is the case if all of the inductances are connected in parallel, the absolute current in each element is

$$i_1 = (m_{11} + m_{12} + m_{13} \cdots + m_{1n}) V$$ (9.31)

$$i_2 = (m_{12} + m_{21} + m_{23} \cdots + m_{2n}) V$$ (9.32)

$$i_3 = (m_{31} + m_{32} + m_{33} \cdots + m_{3n}) V$$ (9.33)

$$\cdot$$
$$\cdot$$
$$\cdot$$

$$i_n = (m_{n1} + m_{n2} + m_{n3} \cdots + m_{nn}) V$$ (9.34)

The total current that flows, which is proportional to the impressed voltage, is

$$i_r = (i_1 + i_2 + i_3 + \cdots i_n) V$$ (9.35)

The fraction of the total current that flows in each circuit is

$$I_1 = \frac{i_1}{i_r}$$ (9.36)

283

$$I_2 = \frac{i_2}{i_r} \qquad\qquad (9.37)$$

$$I_3 = \frac{i_3}{i_r} \qquad\qquad (9.38)$$

.
.
.

$$I_n = \frac{i_n}{i_r} \qquad\qquad (9.39)$$

9.2.5 Measurements Using Wire Grids

An alternate experimental method of determining current distribution is shown in Figure 9.11. It would be entirely feasible to model the surface of an aircraft by a grid of wires spaced over a supporting framework or placed on the surface of a complete model made from nonconductive materials. If a high-frequency current were passed through the wire grid in a path similar to that expected for a lightning flash, it would be possible to measure the current in the individual wires by a clamp-on current probe.

To the author's knowledge such a wire grid model has not yet been used to study current distribution resulting from lightning flashes striking aircraft, but somewhat similar models have been used to measure the current distribution in ground-based launching facilities (References 9.5 and 9.6), for studies of the interaction of aircraft and missile systems with the electromagnetic fields generated by the explosions of nuclear weapons, and for the study of magnetic fields in and around the towers used to support high-voltage transmission lines. In principle, the wire grids should follow as closely as possible the expected current distribution paths and should be interconnected frequently by other wires circling the structural member being modeled. It would probably be best if the interconnecting wires tended to form orthogonal squares with the wires running along the flow lines, in a manner similar to the orthogonal squares formed by flux and equipotential lines shown on Figure 9.6.

In the studies mentioned for current distribution on ground-based launching facilities, current with a frequency on the order of 50 to 100 kHz proved satisfactory. Clamp-on current transformers of the requisite size and sensitivity are commercially available.

It may well be that the cost of building such a model and making actual measurements of current distribution might be cheaper than the cost of trying to study current distribution by field plotting methods, either manual or computer aided, or by analytical determination of the distribution of currents in parallel filaments.

284

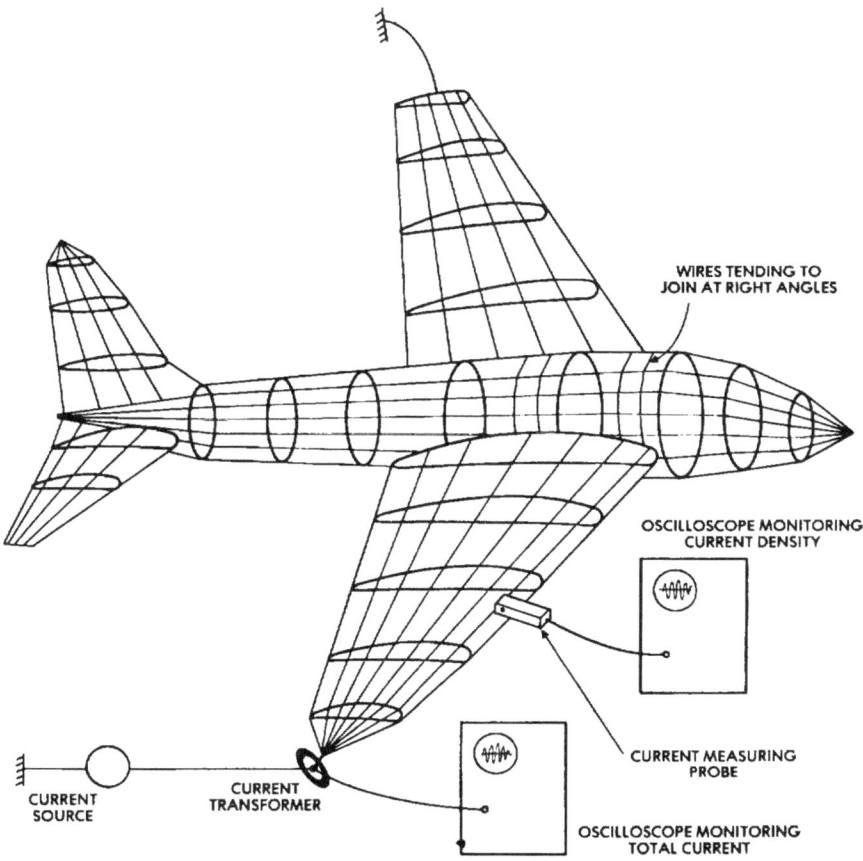

WIRES TENDING TO
JOIN AT RIGHT ANGLES

OSCILLOSCOPE MONITORING
CURRENT DENSITY

CURRENT
SOURCE

CURRENT
TRANSFORMER

CURRENT MEASURING
PROBE

OSCILLOSCOPE MONITORING
TOTAL CURRENT

Figure 9.11   The wire grid model.

9.3 Examples of External Magnetic Fields

The magnetic field within and around the cluster of wires can be determined by taking the proper summation of the magnetic field produced by each individual wire. The computer program MAGFLD provides such a summation.

Some examples of calculations performed with MAGFLD will now be described. They will serve to illustrate some of the points discussed in previous sections. The geometry chosen for analysis is the fuselage of the hypothetical aircraft shown in Figure 9.12. The aircraft, whose airworthiness is not under discussion, has a fuselage of elliptical cross section, two meters along the major axis and one meter along the minor axis. The fuselage is considered long enough that no end effects need to be considered. A lightning current of 1000 A is assumed to enter the nose and to exit through the rear of the fuselage. This elliptical fuselage is represented by 48 parallel conductors distributed in the manner shown in Figure 9.13. Figure 9.13 also shows the current density, or magnetic field strength, at the surface of the cylinder, both as determined from the program MAGFLD and analytically according to Equation 9.11.

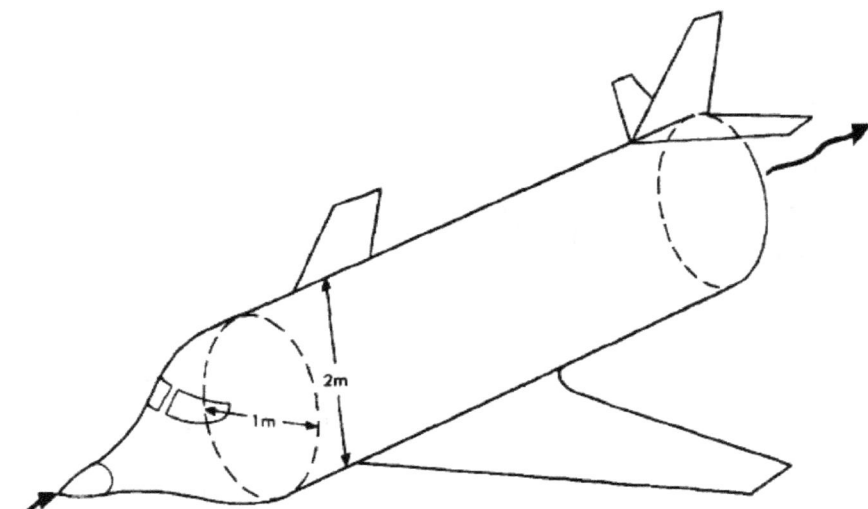

Figure 9.12 A hypothetical fuselage to be modeled as a wire grid.

One quadrant of the elliptic cylinder is shown on Figure 9.14. This figure shows the magnetic field strength both within and around the wire grid as calculated by MAGFLD. The orientation of the magnetic field is shown by the direction of the arrows, and the strength both by the length of the arrows and by the indicated contour lines. The magnetic field strength inside the grid is much smaller than that outside, since on the inside the fields from each of the filaments tend to cancel, whereas outside the grid they tend to add. In Figure 9.14 the fields inside the grid are largely the result of the finite number of wires defining the elliptic cylinder. If the cylinder were defined by more conductors, the fields inside would be smaller, presumably vanishing if the number of wires were to become infinite. Figure 9.15 shows a similar plot with one important difference: the current was forced to be equal in each wire. The magnetic field pattern produced here would be that determined by the resistive current distribution, the pattern that represents the final stage after currents and current density have become uniform. The orientation of the field external to the grid shows only relatively slight differences from that in Figure 9.14, but those differences are sufficient to account for the increased field intensity in the interior of the grid. A third example of field distribution, shown in Figure 9.16, assumes that on each side of the fuselage there is a recessed cavity. Such a cavity would be an approximation of the equipment bays for electronic equipment frequently found on military aircraft. Figure 9.16 again assumes the current in

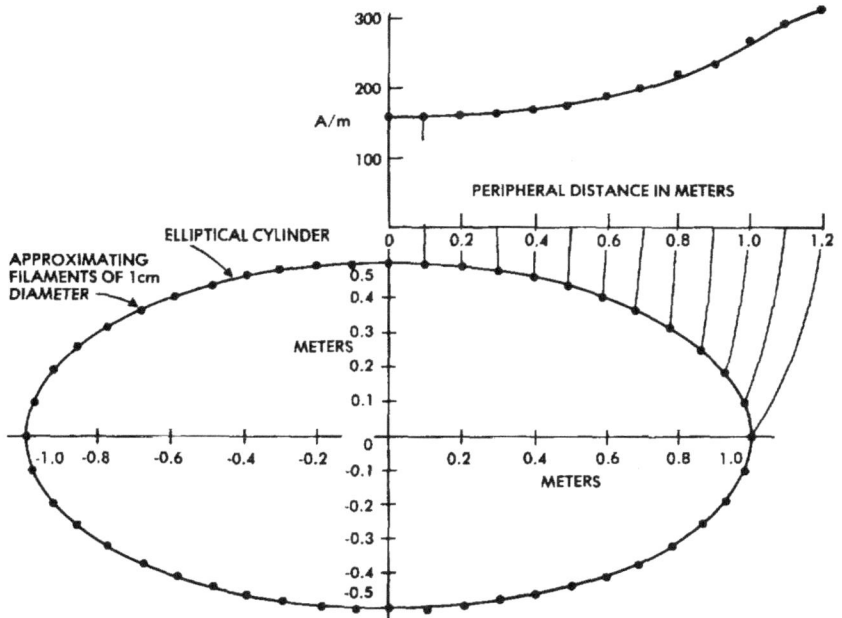

Figure 9.13 Wire grid approximation of the elliptical fuselage.

each filament to be controlled by the magnetic distribution. The field patterns clearly show the field in the recessed cavity to be less than the field at other exterior points on the fuselage.

The degree to which the MAGFLD program can calculate the magnetic field external to the aircraft fuselage, while interesting, may not be of great use for aircraft studies, since it is usually only the field intensity at the surface of the structure that is of interest. The program, however, is capable of calculating the field distribution interior to these simple geometries with sufficient accuracy for many purposes. One such purpose is to help determine the effective area through which circulating currents flow. For such calculations, as shown on Figure 9.15, the number of filaments defining the structure can be greatly increased, since each of them will carry a uniform current. Determining the current distribution on many more filaments when that current is determined magnetically might lead to numeric problems, since the order of the matrix that must be inverted is equal to the number of conductors involved. The greatest utility of the MAGFLD program would be to calculate the current density or magnetic field density at the surface of a structure. Figure 9.17 shows the current distribution on the filaments for the geometry previously shown on Figure 9.16. The current density here is determined magnetically.

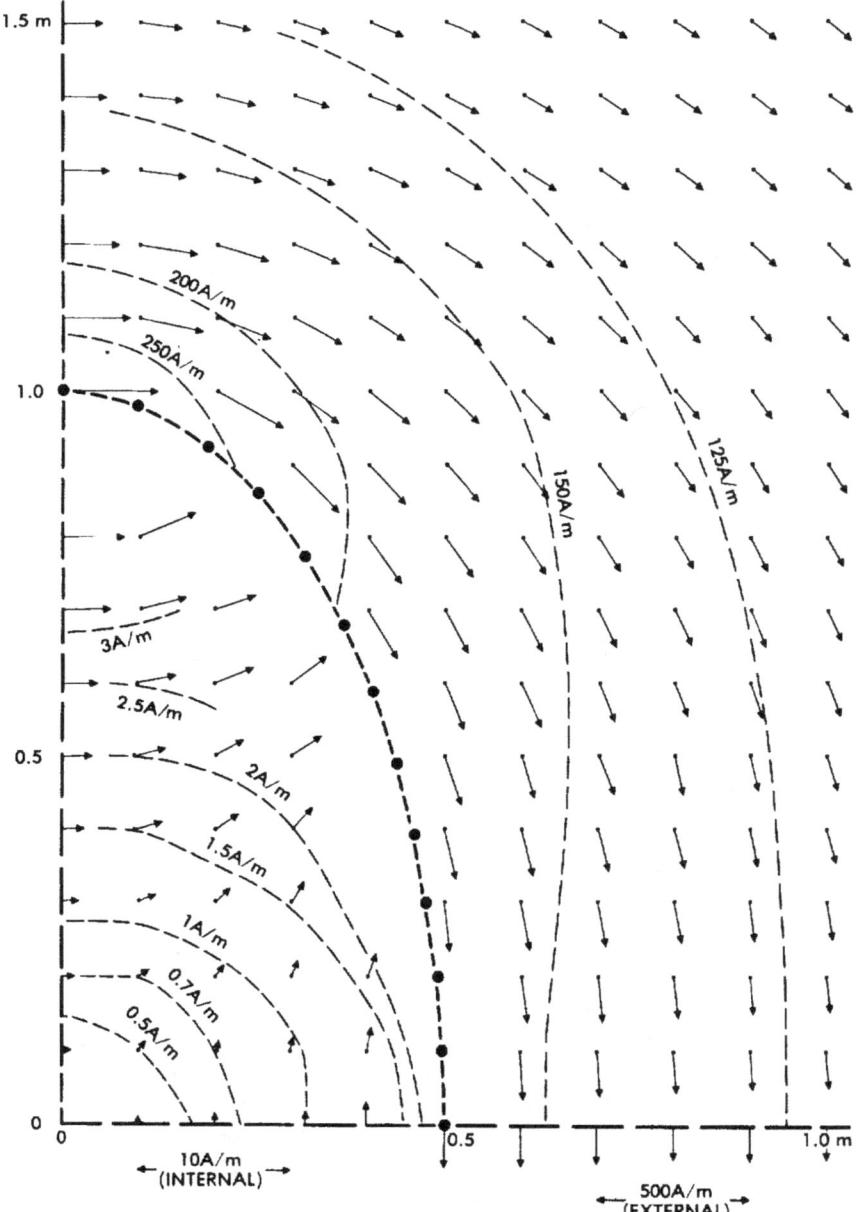

Figure 9.14 Field with unequal current distribution.

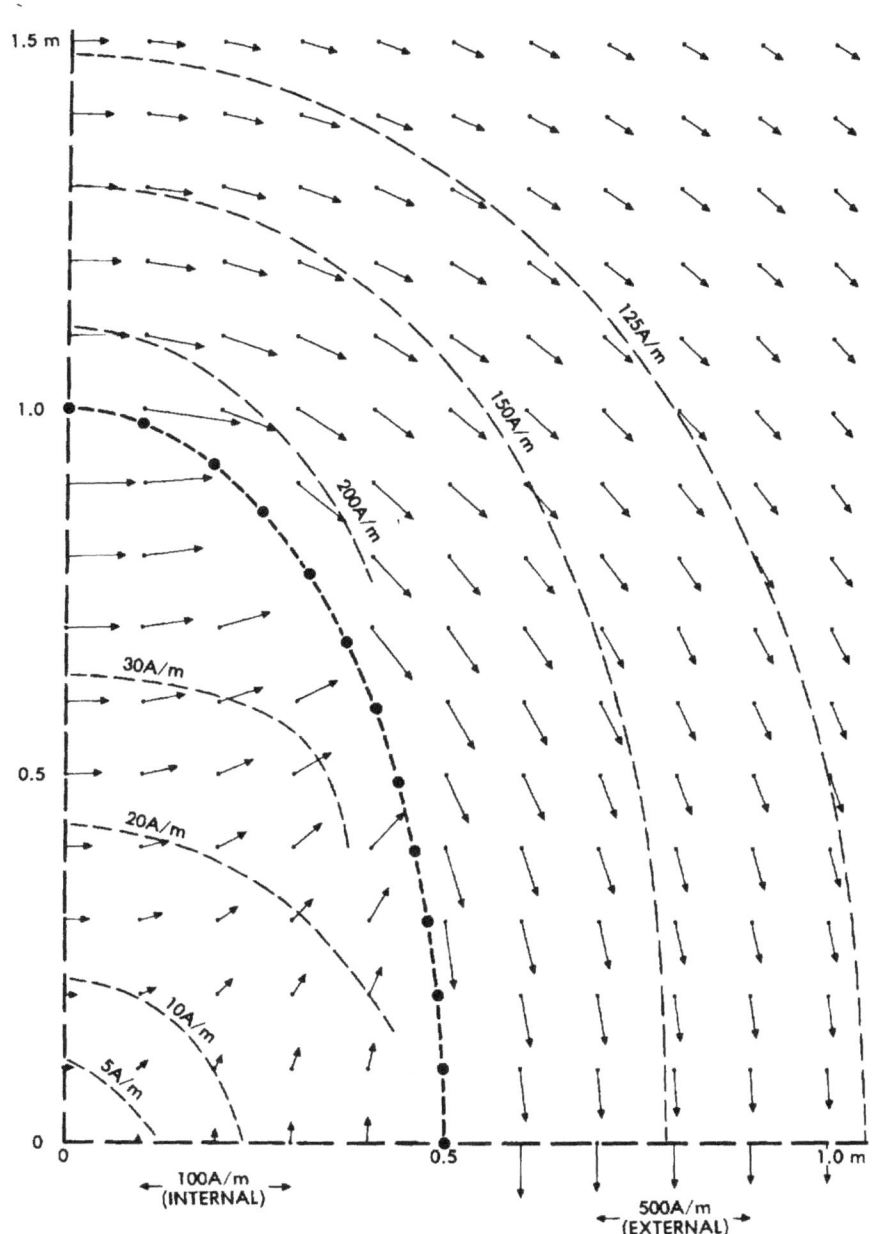

1.5 m

1.0

0.5

0

125A/m

150A/m

200A/m

30A/m

20A/m

10A/m

5A/m

0   0.5   1.0 m

100A/m
(INTERNAL)

500A/m
(EXTERNAL)

Figure 9.15 Field with equal current distribution.

289

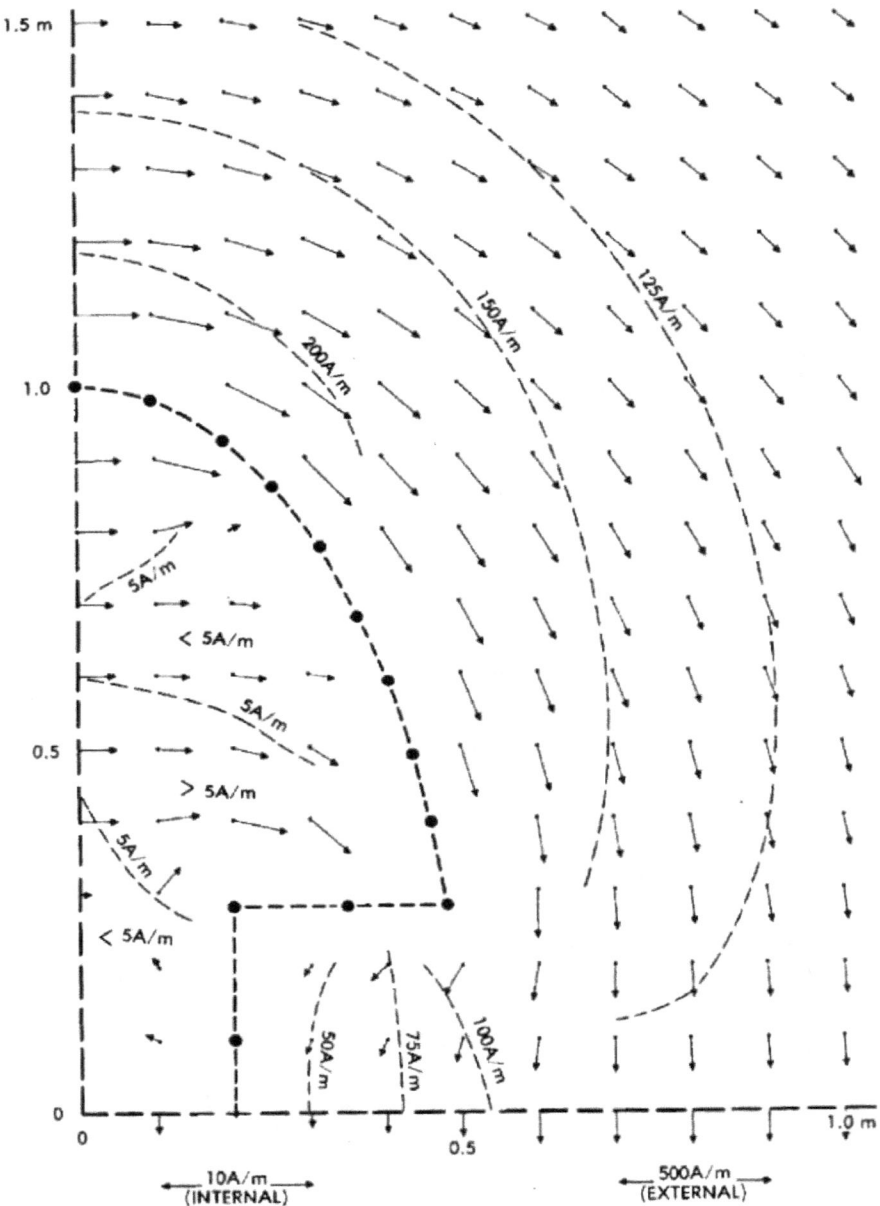

Figure 9.16 Recessed bays.

290

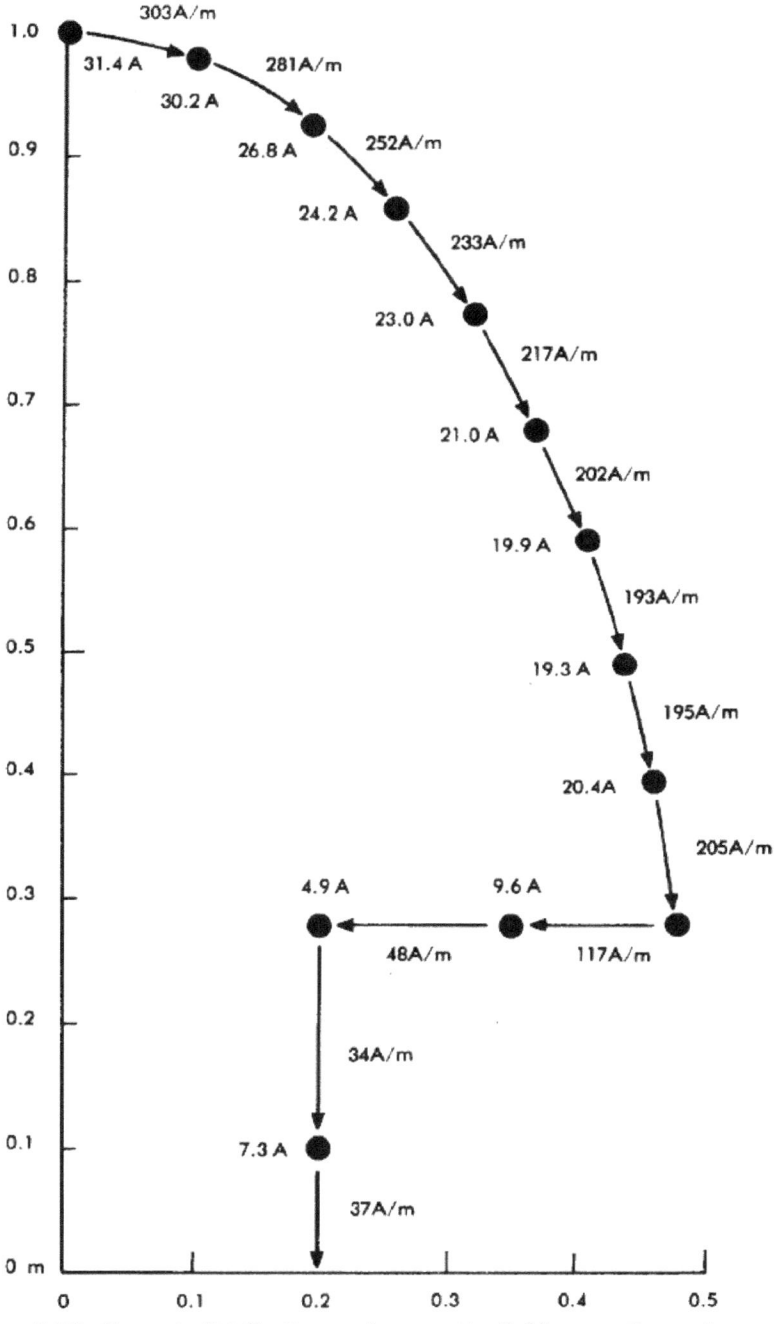

Figure 9.17 Current distribution and magnetic field strength at the surface (one quadrant only shown).

291

REFERENCES

9.1   V. Belevitch, "The lateral skin effect in a flat conductor," *Philips Technical Review,* 32, 6/7/8 (1971): 221-231.

9.2   H. Kaden, "Über den Verlustwiderstand von Hochfrequenzleitern," (On the Resistance of High-Frequency Conductors), *Archiv für Elektrotechnik,* 28 (1934): 818-25.

9.3   J. B. Scarborough, *Numerical Mathematical Analysis,* 6th ed. (Baltimore: Johns Hopkins University Press, 1966).

9.4   G. E. Forsythe and W. R. Wasow, *Finite-Difference Methods for Partial Differential Equations* (New York: John Wiley and Sons, 1960).

9.5   F. A. Fisher, J. G. Anderson, and J. H. Hagenguth, "Determination of Lightning Response of Transmission Lines by Means of Geometrical Models," *Power Apparatus and Systems,* 78, American Institute of Electrical Engineers, New York, New York (February 1960): 1725-36.

9.6   J. G. Anderson and J. H. Hagenguth, "Magnetic Fields Around a Transmission Line Tower," *Transactions of the AIEE, Part 3: Power Apparatus and Systems,* 77, American Institute of Electrical Engineers, New York, New York (February 1959): 1644-50.

CHAPTER 10
THE INTERNAL MAGNETIC FIELDS CREATED BY DIFFUSION

10.1 Circular Cylinders

Consider Figure 10.1 in which a current, I, is entering a circular cylinder. The cylinder is considered long compared to other dimensions, so that there are no end effects, but short compared to the electrical wavelength of any of the frequency components of the current I. The return path for the current is not shown, but it is assumed to be sufficiently far away from the cylinder that there are no proximity effects. Also shown are two conductors, one external (1) and one internal (2) to the cylinder. These are connected to an end cap considered sufficiently massive that no electromagnetic fields penetrate the cap. At the other end of the cylinder are shown two voltages, V_1 measured from conductor 1 to the external surface of the cylinder, and V_2 measured from conductor 2 to the inner surface of the cylinder.

10.1.1 Resistance

The cylinder will have a dc resistance

$$R = \frac{\rho \ell}{A} = \frac{\rho \ell}{2\pi ra} \qquad (10.1)$$

where

ρ = resistivity
ℓ = length
A = cross-sectional area
r = radius
a = thickness ($a \ll r$)

If the cylinder has the following dimensions and is made of aluminum of the indicated resistivity

ℓ = 2m
r = 15.7 cm
a = 0.281 mm (0.015")
ρ = 2.69×10^{-8} $\Omega \cdot$ m

then the dc resistance, R, will be 1.43×10^{-4} Ω. If we consider that the input current is 116 A, there will then be developed a voltage

$$e = IR = 116 \times 1.43 \times 10^{-4} = 0.0166 \text{ V} \qquad (10.2)$$

If the current of 116 A is established and allowed to flow for a time sufficiently long that steady state conditions are reached, the voltage developed along the cylinder will then be 16.6 mV. This same voltage drop would be measured by a conductor external to the cylinder or by one internal to the cylinder.

293

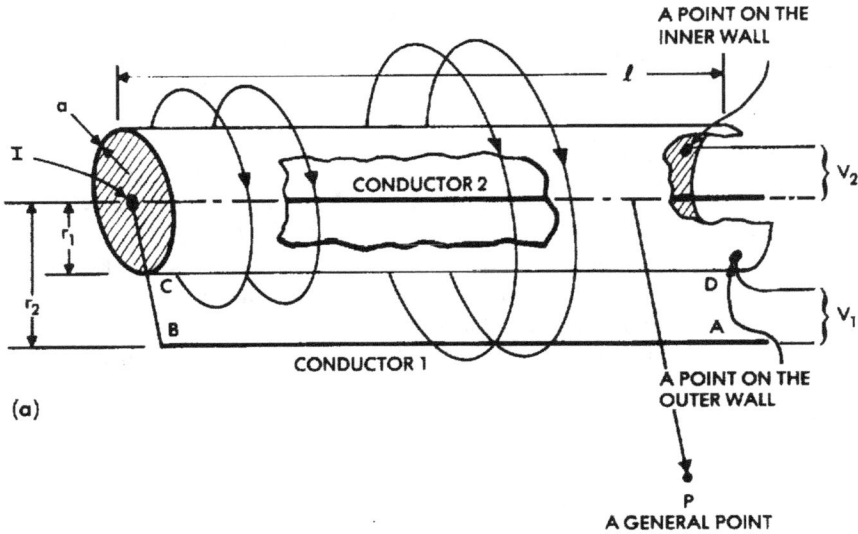

(a)

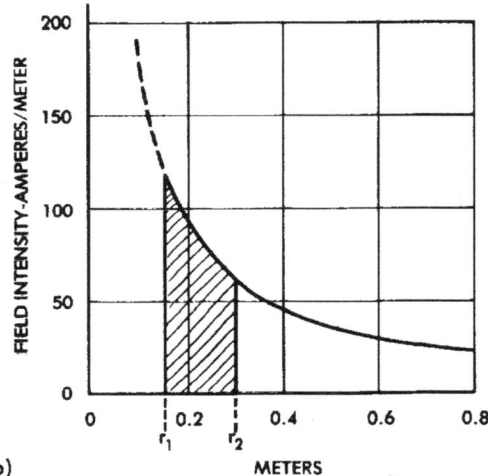

(b)

Figure 10.1 Magnetic fields around a circular cylinder.
(a) Geometry
(b) Field intensity vs radius (for I = 116 A)

10.1.2 The External Loop Voltage

Until such steady state conditions have been established, V_1 will not be equal to V_2, and neither of them will be equal to the steady state resistance voltage drop. Consider first voltage V_1. External to the cylinder the flow of current sets up a magnetic field of intensity.

294

$$H = \frac{I}{2\pi r} \qquad (10.3)$$

where
$$I = \text{current}$$
$$r = \text{radius}$$
$$H = \text{field strength}$$

having a pattern as shown in Figure 10.1(b). V_1 will then be

$$V_1 = \frac{d\phi}{dt} = \left[\mu_0 H\ell \log_\epsilon \left(\frac{r_2}{r_1}\right) \right] \frac{dI}{dt} \qquad (10.4)$$

and ϕ, the flux passing through the loop ABCD, is represented by the shaded area of Figure 10.1(b). The flux, ϕ, would be measured in webers. Remembering that

$$\mu_0 = 4\pi \times 10^{-7} \text{ A/m}$$

V_1 becomes

$$V_1 = \left[2 \times 10^{-7}\, \ell \log_\epsilon \left(\frac{r_2}{r_1}\right) \right] \frac{dI}{dt} \qquad (10.5)$$

If r_2 is 31.4 cm and the indicated current of 116 A rises to crest in an equivalent time of 0.25 μs (Figure 10.2), V_1 will then rise to an initial voltage of 129 V. As steady state conditions are reached and the external magnetic field ceases to change with time, V_1 will decay to its steady state value of 0.0166 V.

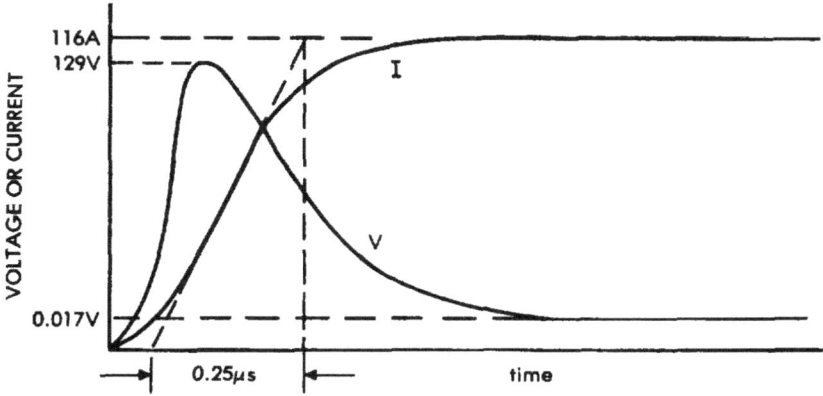

Figure 10.2 External voltage (not to scale).

This analysis ignores the skin effect, a phenomenon that causes the effective resistance to be higher under transient conditions than under steady state conditions. For conductors external to the cylinder, the increased resistance resulting from skin effect will be of little consequence compared to the much larger voltage induced by the changing magnetic field.

While conductors external to a current-carrying cylinder, such as an aircraft fuselage, are not common, they are not unknown. An example might be the cables on a missile or rocket that run between a control assembly in the nose and the engine controls at the tail. Of necessity, such cables must run, as shown in Figure 10.3, external to the fuel and oxidizer tanks. If the cables are not in a shielded cable tunnel, they will be exposed to the external magnetic field.

If the spacing between the wires and the surfaces of the vehicle is not large, Equation 10.5 may be somewhat simplified, since the magnetic flux density does not vary greatly with distance away from the surface of the vehicle and may be considered uniform.

$$V_1 = \left[\frac{2 \times 10^{-7} \ell d}{r} \right] \frac{dI}{dt} \tag{10.6}$$

10.1.3 The Internal Loop Voltages

Let us now consider the conditions internal to the cylinder, where conditions are very different. First of all, the magnetic field internal to the inner surface of the cylinder will be zero. This may be shown as in Figure 10.4. Let there be passed two planes parallel to the axis through any internal point, P, the planes separated by a small angle, a. The magnitudes of the currents in the two filaments are to each other as the ratio of the distances from point P.

$$\frac{i''}{i'} = \frac{p''}{p'} \text{ or } \frac{i''}{p''} = \frac{i'}{p'} \tag{10.7}$$

Furthermore, the magnetic intensity effects of the filaments at P have the ratio unity

$$\frac{H''}{H'} = \frac{\left(\frac{i''}{2\pi p''} \right)}{\left(\frac{i'}{2\pi p'} \right)} = 1 \tag{10.8}$$

These two filaments then create no field at any internal point; nor does the tube as a whole, as it is composed of filament pairs.

The second important factor is that the phenomenon of skin effect delays the buildup of current on the inner surface of the cylinder. The origin of the skin effect phenomenon is shown in Figure 10.5. If a magnetic field line is assumed to be suddenly established internal to the wall of the conducting cylinder, there will be induced eddy currents circulating around that field line. These eddy currents will induce a magnetic field of their own of polarity opposite to that set up by the external magnetic field. Only as the eddy currents decay will the magnetic field penetrate the wall of the cylinder.

The current density across the wall thickness at several times is shown in Figure 10.5(b). Initially, the current density for a step-function input current is entirely confined to the outer surface. At a slightly later time, t_1, the current

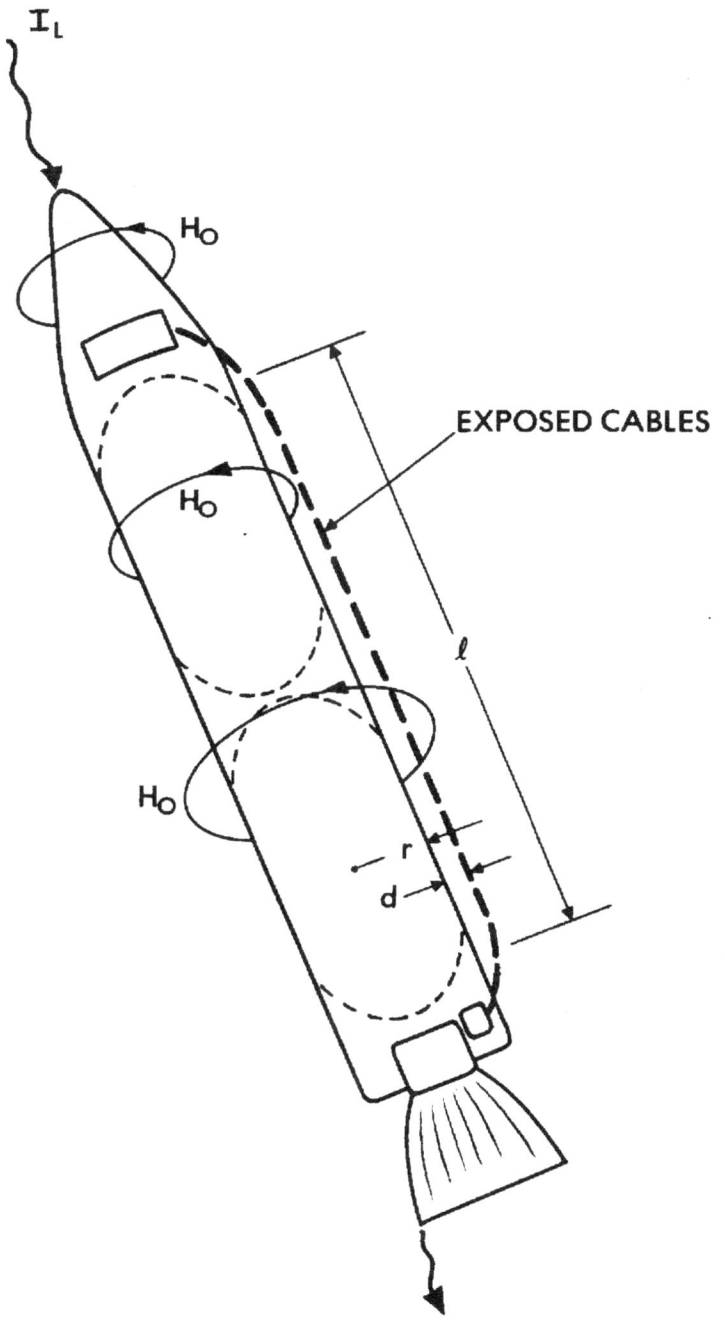

Figure 10.3 Exposed cable antennas.

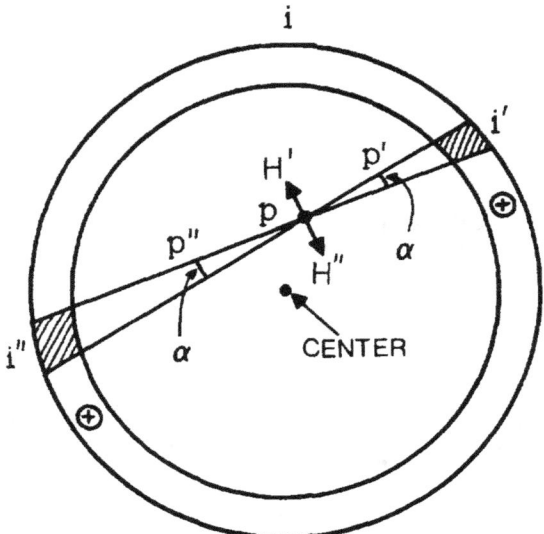

Figure 10.4 Cylindrical tube of current. Internal magnetic field intensity is zero.

density will still be high near the outer surface, but a small amount of current will have penetrated to the inner surface. As time increases, the current density on the outer surface will fall, and the current density on the inner surface will increase, until at t_{final} the current density will be uniform across the surface of the wall. At any time, the area under the current density curve multiplied by the peripheral distance around the cylinder will equal the input current. The resistive voltage drop along any surface (including any line internal to the wall of the cylinder) will be equal to the product of the current density at that point times the resistivity of the material. Accordingly, the voltage drop along the inner surface must follow the curve defining the buildup of current density on the inner surface.

10.1.4 The Characteristic Diffusion-Type Response

The nature of this response is shown in Figure 10.6 and is governed by the following equation:

$$V = IR \left\{ 1 - 2 \left[\epsilon^{-t/\tau} - \epsilon^{-4t/\tau} + \epsilon^{-9t/\tau} - \epsilon^{-16t/\tau} - - - \right] \right\} \qquad (10.9)$$

Figure 10.6 is plotted in terms of a characteristic time constant, τ, sometimes called the *penetration time constant*.

$$\tau = \frac{L}{\pi^2 R} \qquad (10.10)$$

298

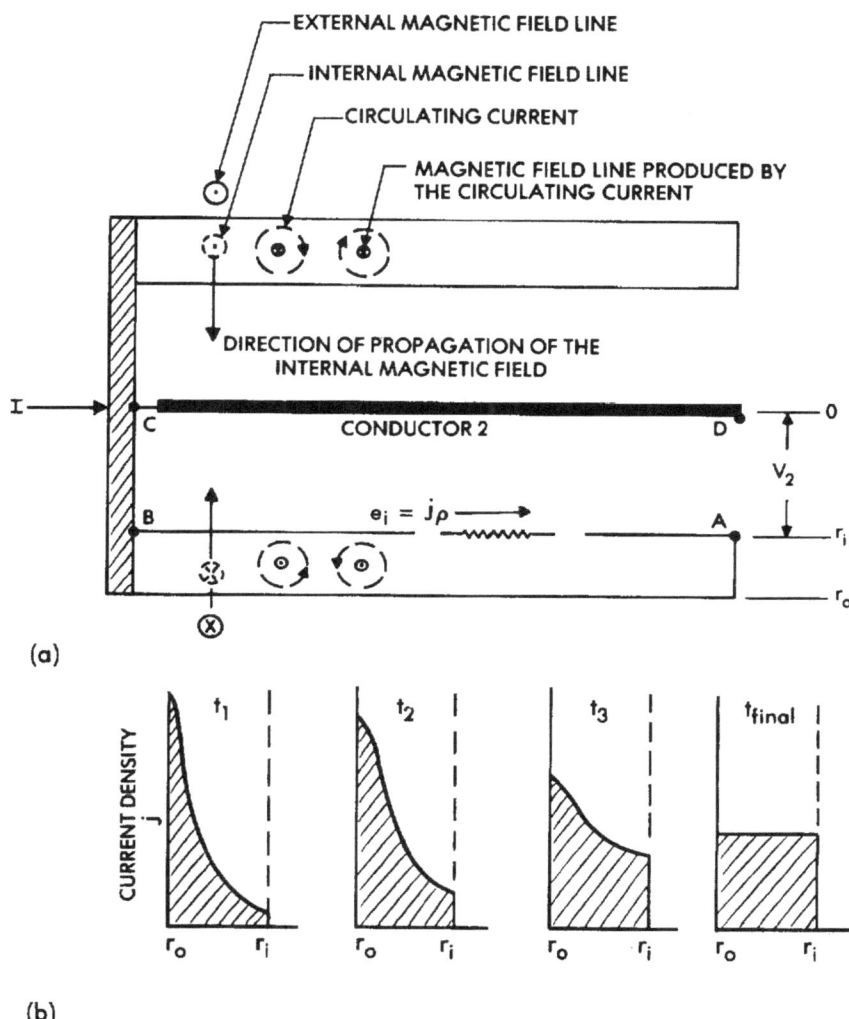

(a)

(b)

Figure 10.5 Factors governing the internal voltage.

    (a) Geometry and decaying eddy currents

    (b) Current density at different times

R, the dc resistance of the tube, is

$$R = \frac{\rho \ell}{A} = \frac{\rho \ell}{2\pi r a} \tag{10.11}$$

where ρ = resistivity of the material

                  a = thickness of the wall

                  r = radius of the tube

299

L, the internal inductance of the tube, is given by

$$L = \frac{\mu \ell a}{2\pi r} \qquad (10.12)$$

where μ = permeability of the material

Hence the penetration time constant is seen to be

$$\tau = \frac{\mu a^2}{\pi^2 \rho} \qquad (10.13)$$

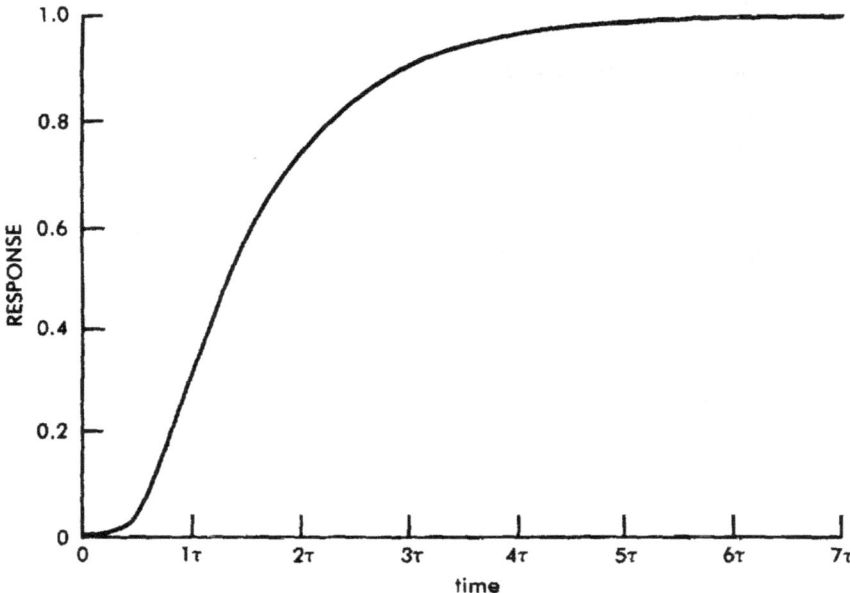

Figure 10.6 Diffusion-type response to a step function.

The response curve, shown in Figure 10.6, is called a *diffusion-type response* and is characteristic of many types of situations involving the transmission of energy through a distributed medium. An example would be the transfer of heat into a block if a heat flux were suddenly applied to one face of the block. Another example would be the transfer of electrical energy through an R/C or L/R ladder network. Figure 10.7 shows examples, with Figure 10.7(b) often being used as the best illustrative example of the phenomenon of skin effect. Several important observations might be made about the shape of the response curve shown in Figure 10.6. The first is that the response initially changes only slowly and thus has a zero first derivative, unlike a simple exponential response, which has a finite first derivative. The second is that the response approaches its final value much more slowly than does a simple exponential response. The third is that in three time constants the response has reached a large fraction of its final value. In three time constants the diffusion-type response reaches 90%, while the exponential response reaches 95% of its final value.

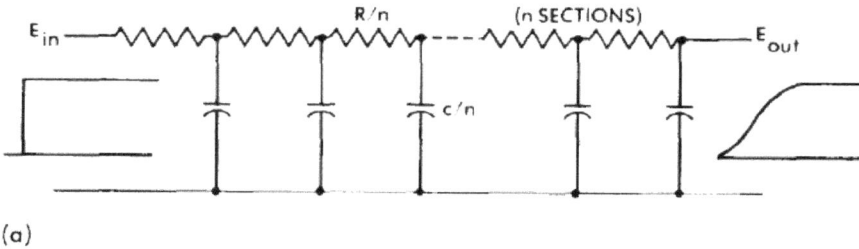

(a)

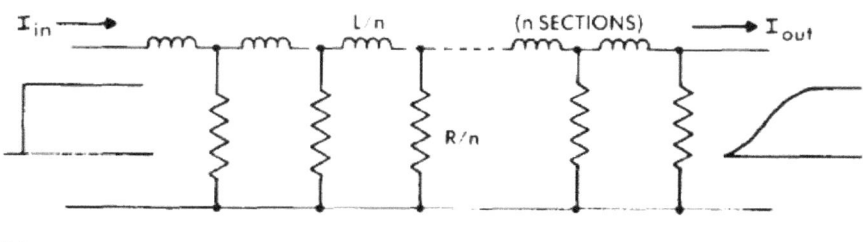

(b)

Figure 10.7 Ladder networks displaying a diffusion-type response.
(a) R/C network
(b) L/R network

With respect to Equation 10.13 it should be noted that the penetration time constant is directly proportional to the permeability of the material, inversely proportional to the resistivity of the material, and directly proportional to the square of thickness of the material. The relative permeability of structural material used in aircraft is almost always very nearly unity. Thickness and resistivity can vary over wide ranges.

For reference, Equation 10.13 is shown plotted in Figure 10.8 as a function of material thickness and the resistivity of the material. The resistivities of some typical metals are shown in Table 10.1. As an example, if we assume an aluminum alloy of resistivity twice that of copper and a skin thickness of 0.040 inches, the penetration time constant would be 3.9 μs. If a step-function current were established on the outside of a sheet of such metal, it would take 11.7 μs for the current density on the other side to buildup to 90% of its final value.

Returning to Figure 10.5, we are now in a position to evaluate V_2. The fundamental definition of the voltage between any two points is

$$V = \int P \, d\ell \qquad (10.14)$$

301

where P is the potential at any point. There is an *infinite variety of paths that* one might take going from point A to point D, but the simplest to consider is the path A-B-C-D. The path A-B is along the inner surface of the cylinder. Integrating along this path, the potential difference obtained is the internal voltage drop on the wall of the cylinder. Along path B-C there will be no potential difference, since we have assumed that the end cap is sufficiently massive that there will be no voltage drops along its inner surface. Along path C-D again there will be no potential difference, since we will assume that the conductor is not carrying any current. To the resistive drops will then be added the voltage produced around the loop by the passage of changing magnetic fields through the loop ABCD. As explained earlier, in a circular geometry there is no magnetic field internal to the cylinder. Consequently, there will be no magnetically induced component of voltage, and V_2 will have the shape shown on Figure 10.6, where the final value is given by the dc resistance drop along the cylinder.

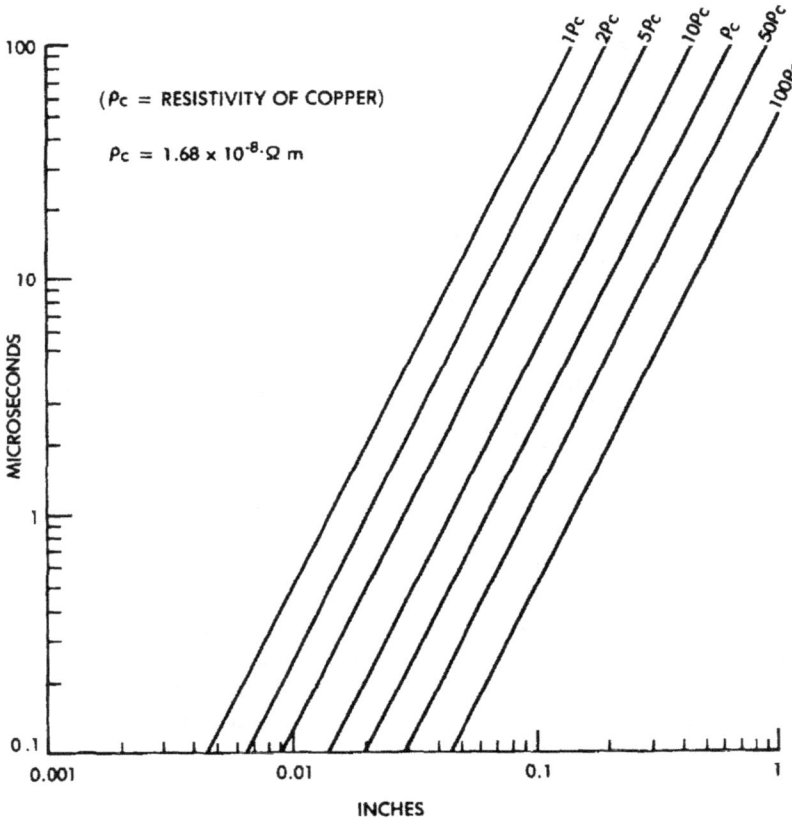

Figure 10.8 Skin thickness vs penetration time constant.

302

Table 10.1 RESISTIVITIES OF TYPICAL METALS

Material	Resistivity ohm-meters	Conductivity as a fraction of that of copper	Resistivity relative to copper
copper	1.68×10^{-8}	1.0	1.0
aluminum*	2.69×10^{-8}	0.62	1.6
magnesium**	4.46×10^{-8}	0.38	2.65
nickel	10×10^{-8}	0.17	6
Monel	42×10^{-8}	0.04	25
stainless steel	70×10^{-8}	0.024	42
Inconel	100×10^{-8}	0.017	60
titanium	180×10^{-8}	0.009	107

*Aluminum alloys range from 2.8 to 5.6×10^{-8} $\Omega \cdot$m; harder alloys generally have higher resistivities.

**Magnesium alloys containing aluminum and zinc range from 10 to 17×10^{-8} $\Omega \cdot$m.

10.2 Elliptical Cylinders

If we consider a cylinder of other than circular shape, conditions may be considerably different. Figure 10.9 shows an elliptical cylinder into which a step-function current is injected. As previously, we will assume that the cylinder is long enough that all end effects may be neglected, that it is short compared to the wavelengths of any frequency components of the injected current, and that the return path for the current is far enough removed that no proximity effects need be considered.

We will also treat the instantaneous current in the cylinder as being composed of the sum of a steady state and a transient component of current. The transient component takes the form of a circulating eddy current. For the following section it should be kept in mind that the eddy currents described represent only the transient component of current and that the total current at any point or time is the sum of the two components.

10.2.1 Eddy Currents and the External Magnetic Field

Under steady state conditions the current density along the wall of the cylinder will be governed by the dc resistance and, if we assume uniform wall thickness, will have a uniform current density. The current in the cylinder will produce a magnetic field. Most of the field lines will completely encircle the cylinder, as shown in Figure 10.9(a), but some, because of the uniform current density, will pass through the cylinder. The greater the eccentricity of the

303

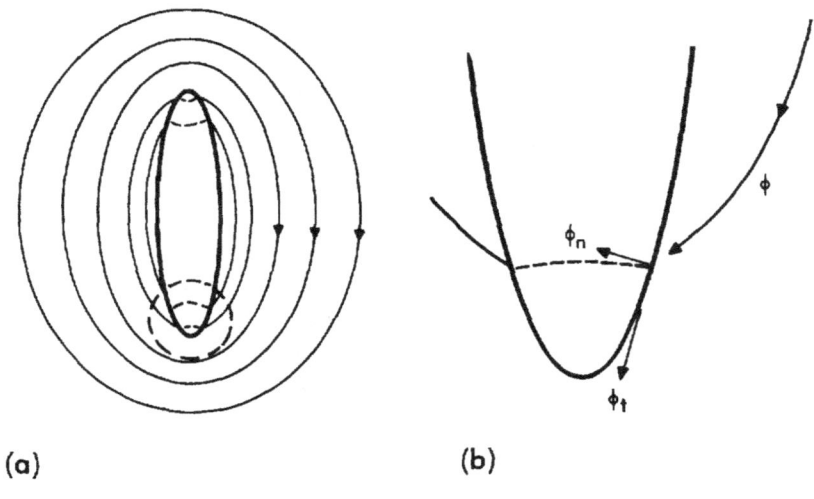

(a)   (b)

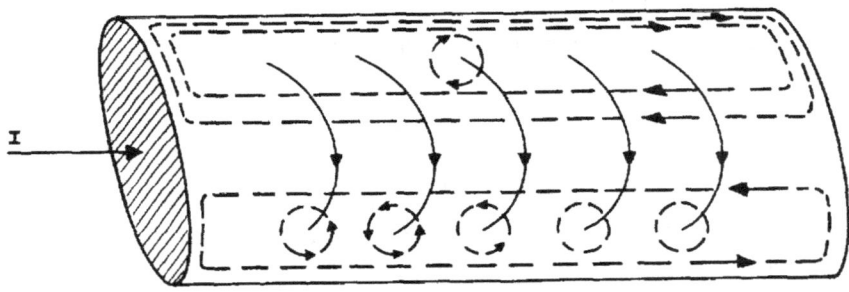

(c)

Figure 10.9 Magnetic fields around an elliptical cylinder.

    (a) Penetrating lines of flux
    (b) Detail showing resolution into components
    (c) Circulating currents induced by
        penetrating lines of flux

cylinder, the greater will be the number of lines of flux passing through the cylinder. Those lines that do pass through the walls of the cylinder will have important effects on the internal response of the cylinder prior to the time that steady state conditions have been established. One such penetrating line is shown in Figure 10.9(b). The vector defining that line may, at the point of entry, be resolved into two vector components, one normal to the surface, ϕ_n, and another tangential to the surface, ϕ_t. If the field line ϕ is suddenly established, it will induce a circulating, or eddy, current in the conducting sheet which it attempts to penetrate. The eddy current will produce a magnetic field of its own, and the intensity of the eddy current will be of a nature such that the magnetic field produced is exactly that required to cancel the normal component of the exterior field. If, as shown in Figure 10.9(c), a number of lines of magnetic flux attempt to penetrate the surface of the elliptical cylinder, the eddy currents produced by each line of flux combine to produce a circulating current. In an elliptical cylinder of this nature there will be four regions of circulating current, two on each of the two sides. These circulating currents will be of a nature such as to increase the current density at the edges of the cylinder and to reduce it along the center. They also cancel any penetrating magnetic field, forcing the field around the cylinder to be entirely tangential to the surface, at least initially.

The eddy currents cannot exist forever, since energy will be lost as the currents circulate through the resistance of the metal sheet. Accordingly, the current density at all points will vary with time, eventually becoming uniformly distributed in structures having uniform thickness and made of uniform resistivity materials. Figure 10.10 shows the manner in which the current density will vary. As the circulating current shown in Figure 10.10(c) decays to zero, the current at the edge will decay from its initial high value to the final resistively determined value, and the current at the center will increase from its initially low value. The current densities will change according to an essentially exponential pattern, though the transient increase in surface resistance produced by skin effects will prevent the circulating current from following a true exponential decay.

One expression giving the magnitude of this redistribution time constant that has appeared in the literature (Reference 10.1) is:

$$\tau = \frac{\mu A a}{\rho P} \tag{10.15}$$

where
A = enclosed area of structure
P = peripheral distance around the structure

The thickness of the wall, a, is assumed to be very small compared to other dimensions. For a rectangular box of sides height h and width d, Equation 10.15 becomes

$$\tau = \frac{\mu h d a}{2\rho (h+d)} \tag{10.16}$$

A physical treatment of the redistribution time constant can also be given. If the decay is treated as exponential, the factors governing that decay are the characteristic inductance and resistance of the path through which the circulating currents flow. These factors are illustrated in Figure 10.11. The circulating currents may be viewed as flowing in one direction along a strip along the edge of the cylinder and back along another strip towards the center of the cylinder. The rate at which the circulating currents decrease is governed by the inductance and resistance of the pair of hypothetical strips. Such a pair is shown

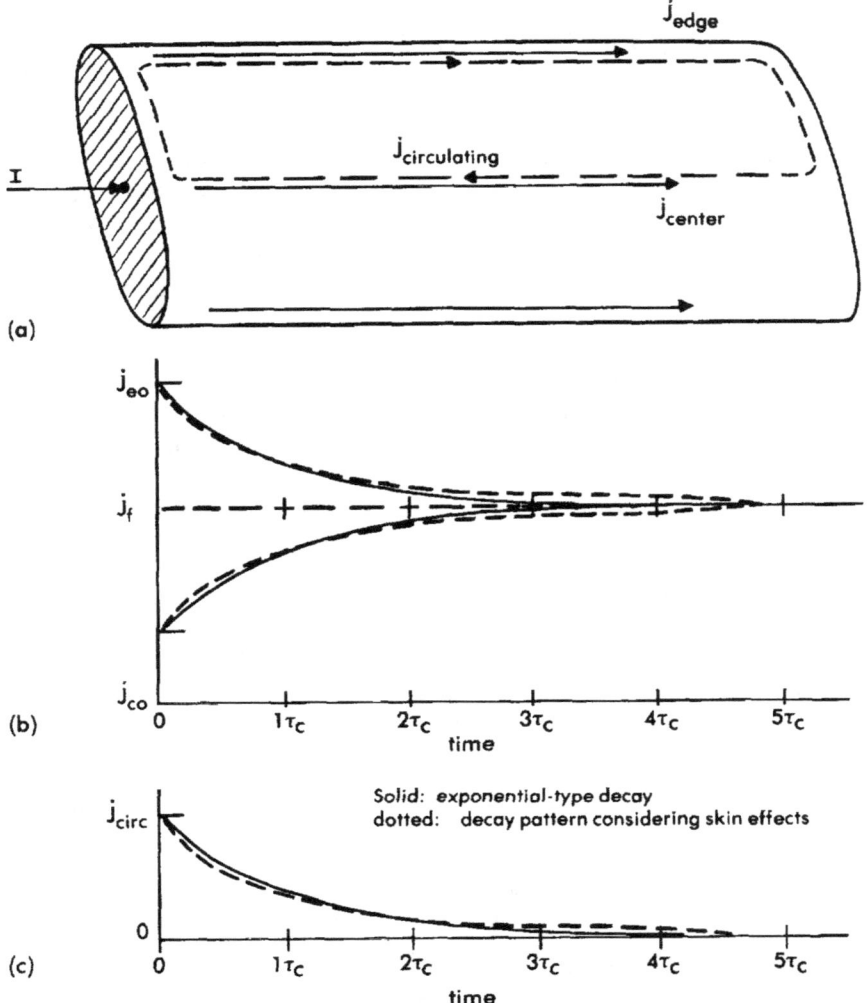

j_{edge}

$j_{circulating}$

I

j_{center}

(a)

j_{eo}

j_f

j_{co}

(b)

0        $1\tau_c$ $2\tau_c$ $3\tau_c$ $4\tau_c$ $5\tau_c$
                          time

j_{circ}

Solid: exponential-type decay
dotted: decay pattern considering skin effects

0

(c)

0        $1\tau_c$ $2\tau_c$ $3\tau_c$ $4\tau_c$ $5\tau_c$
                          time

Figure 10.10 Variation of current-density with time.
(a) Current components defined
(b) Edge and center currents
(c) Circulating currents

at the top of Figure 10.11(a). The resistance of the pair of strips will be

$$R = \frac{4\rho\ell}{aS} \qquad (10.17)$$

where

S = width
ℓ = length
a = thickness

306

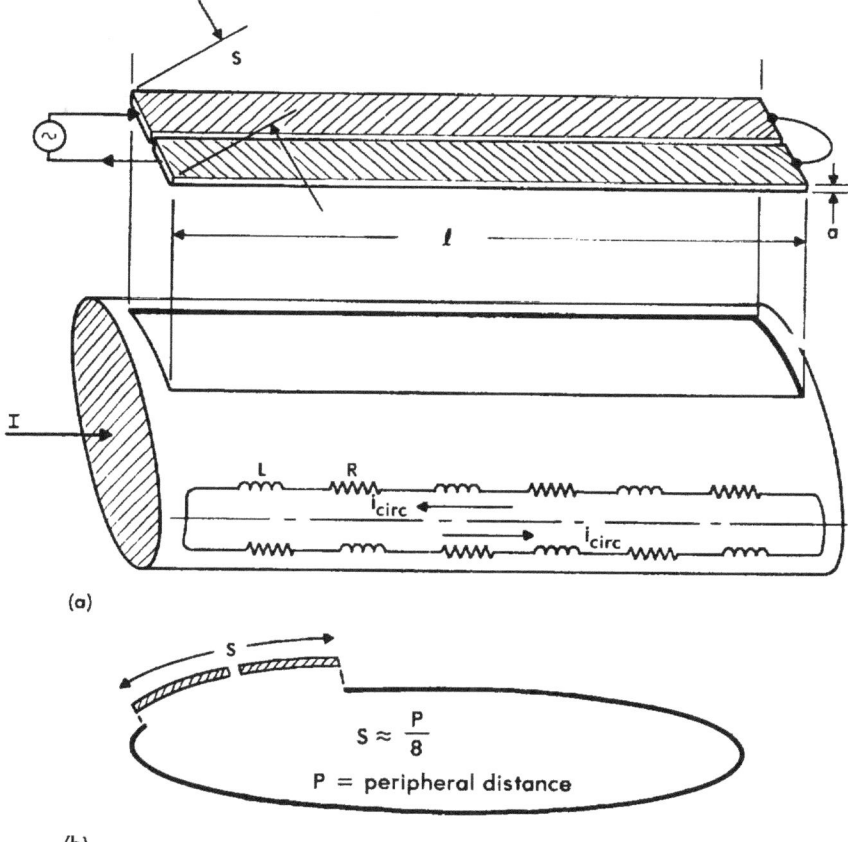

(a)

$$S \approx \frac{P}{8}$$

P = peripheral distance

(b)

Figure 10.11 Geometrical factors governing the decay of the circulating current.
      (a) The character of the L-R circuit
      (b) The width of the strip involved

The inductance of a similar geometry (Figure 10.12) is given by Grover (Reference 10.2) as

$$L = 0.4\ell \left[\log_\epsilon \frac{d}{B+C} + 1.5 + \log_\epsilon K - \log_\epsilon e \right] \qquad (10.18)$$

where K and e are factors relating to the geometrical mean radius of the strips. For the limiting case of Figure 10.12(b)

$$\log_\epsilon K = -.1137 \qquad\qquad (10.19)$$

$$\log_\epsilon e = 0 \qquad\qquad (10.20)$$

$$L = 0.555 \times 10^{-6}\ell \qquad\qquad (10.21)$$

307

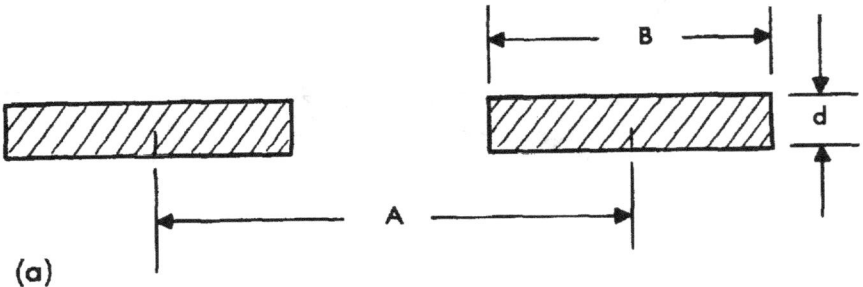

(a)

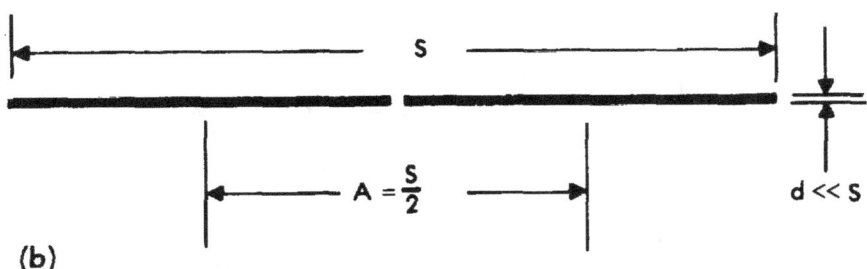

(b)

Figure 10.12 Inductance of parallel strips.
(a) General case
(b) Limiting case

The inductive time constant would then be

$$\tau = \frac{L}{R} = \frac{0.277 \times 10^{-6}\,aS}{\rho} \qquad (10.22)$$

and for aluminum of resistivity $2.69 \times 10^{-8}\ \Omega \cdot m$

$$\tau = \frac{L}{R} = 10.3\,aS \qquad (10.23)$$

The time constant governing the decay of the circulating currents is thus seen to depend upon the width of the region through which the lines of flux would tend to penetrate. Since the penetrating lines of flux would tend to be confined to the outer portion of the ellipse, it appears that the effective width would be of the order of one-eighth of the peripheral distance around the cylinder.

A numerical example is as follows: if the effective-width strip through the circulating currents flow is 0.123 m and the thickness is 0.038 cm (0.015 inches), the L/R time constant is 480 μs.

308

10.2.2 Eddy Currents and the Internal Magnetic Field

As the circulating currents die out and the external lines of flux penetrate the walls of the cylinders, there will be set up an internal magnetic field oriented as shown in Figure 10.13. In its latter stages the rate at which the internal field builds up will be dependent upon the rate at which the externally induced circulating currents die away. For an elliptical cylinder made of aluminum of thickness 0.038 cm, having a major axis of 47 cm, a minor axis of 9.4 cm, and a peripheral distance of 98.7 cm, the internal field will build up with a time constant of the aforementioned 480 μs.

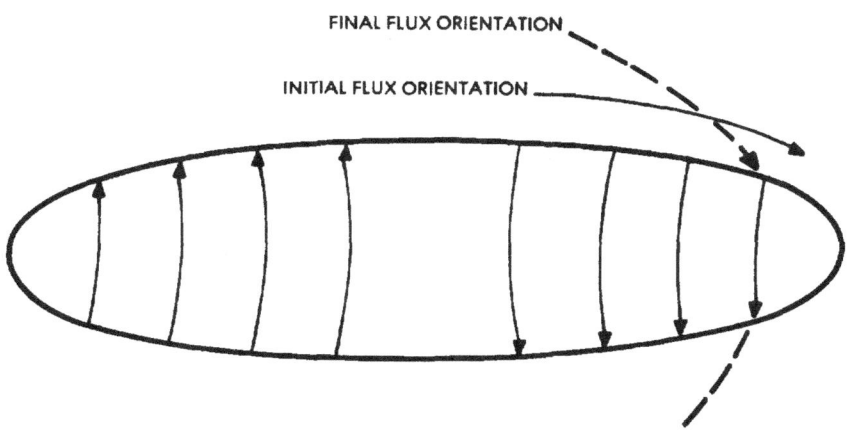

Figure 10.13 The internal magnetic field that arises as a result of flux pentration.

The early time buildup of the magnetic field will be strongly influenced by the phenomenon of skin effect, and the initial rate of buildup of magnetic field will now be seen to have an important effect on the voltages induced on circuits contained within such an elliptical cylinder. Figures 10.5 and 10.6 showed how an electric field will be built up along the inner surface of a circular cylinder. In a similar manner, an electric field will be built up along the inner surface of the elliptical cylinder, but, unlike the circular cylinder, the electric field intensity will be different at different points on the internal surface of the elliptical cylinder, since the initial current density on the outer surface is not uniform. The internal electrical field will be essentially proportional to the current density on the exterior surface, but delayed and distorted as a result of the pulse penetration time produced by eddy currents (shown in Figure 10.6). The internal electric field, E_i, being different at different points on the inner surface, will give rise to circulating currents on the inner surface. The currents may again be visualized as flowing on paired strips. There will be four pairs of strips, two on each of the sides of the cylinder. Each of these strips will have associated with it an inductance and resistance which may again be calculated from the

309

expressions given earlier. The equivalent circuit governing the rate of buildup of internal current then becomes that of Figure 10.14, which may be simplified to that of Figure 10.15. If e_{ie} were a step function, the buildup of current would follow the expression

$$i_{circ} = \frac{e_{ie}}{R} \left[1 - \epsilon^{-t/(L/R)} \right] \tag{10.24}$$

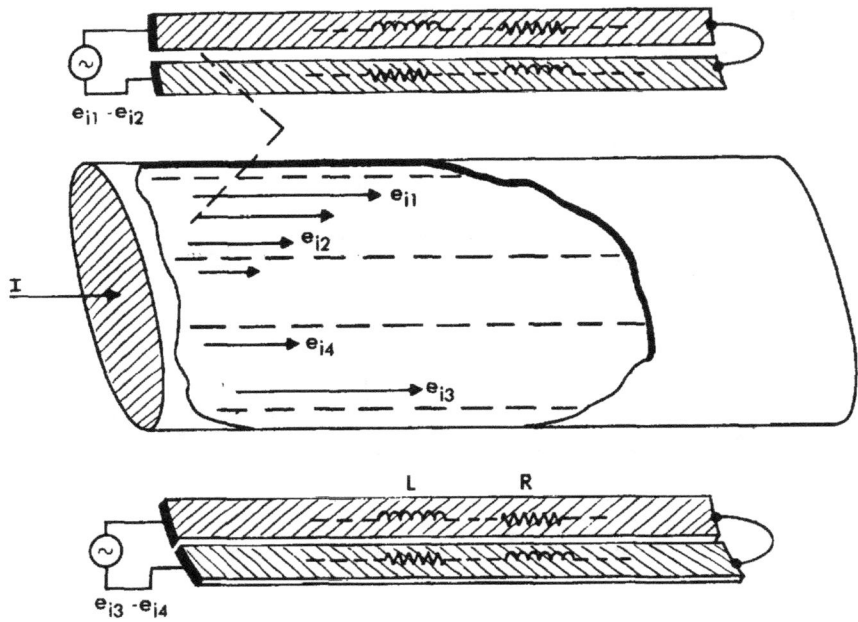

(a)

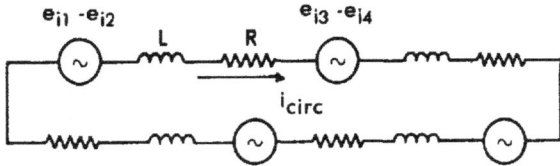

(b)

Figure 10.14 Factors governing the early time buildup of the internal magnetic field.
(a) Physical factors
(b) Equivalent circuit

310

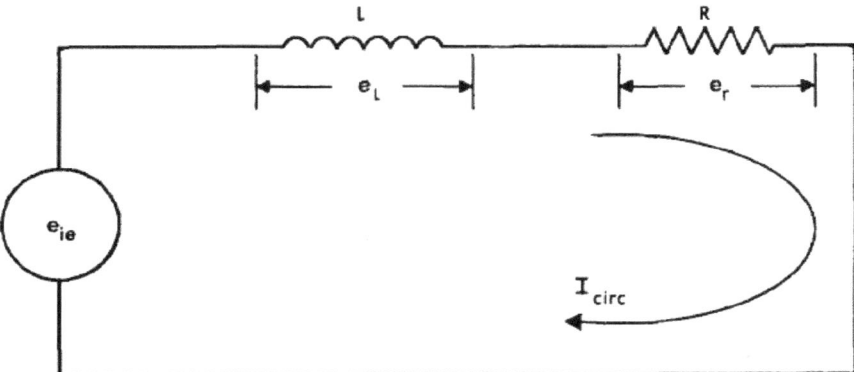

Figure 10.15

Figure 10.15 Final equivalent circuit governing increase of internal circulating circuit.

If the internal voltage were not a step function but were governed by some other function of time, such as the pulse penetration predicted by Equation 10.9, the internal current would be

$$i_{circ} = \int^{t} \frac{e_{ie}f(t)}{R} \frac{d}{dt}\left[1 - \epsilon^{-(t-\tau)/(L/R)} \right] d\tau \qquad (10.25)$$

The convolution integral of Equation 10.25 need not be evaluated analytically, since only the early time rate of change of current is important. So long as the current is small, the voltage across the resistive part of the circuit will be small and the voltage across the inductive part will be equal to the internal resistance drop. Under these conditions the internal current will be

$$I_{i_{circ}} = \frac{1}{L} \int e_{ie} \, dt \qquad (10.26)$$

The internal magnetic field will follow the same time pattern as that of the buildup of the internal circulating currents.

10.2.3 Internal Loop Voltages

We are now in a position to evaluate the voltages on conductors contained in a cylinder of noncircular geometry. Figure 10.16 shows an elliptical cylinder with two internal conductors, one adjacent to the surface and one in the center. Both are connected at one end to an end cap sufficiently massive that no voltage drops will appear along its inner surface. The other ends are open circuited. The usual assumptions about the length of the cylinder and the return path for the injected current are assumed to apply. Voltages V_1 and V_2 are shown, both being measured between their respective conductors and a point on the inner wall of the cylinder. Figure 10.16(b) shows that all of the internal flux will pass between conductor 2 and the inner wall of the cylinder, while only a small amount will pass between conductor 1 and the inner wall. Correspondingly, a large fraction of the internal flux passes through the plane defined by conductors 1 and 2.

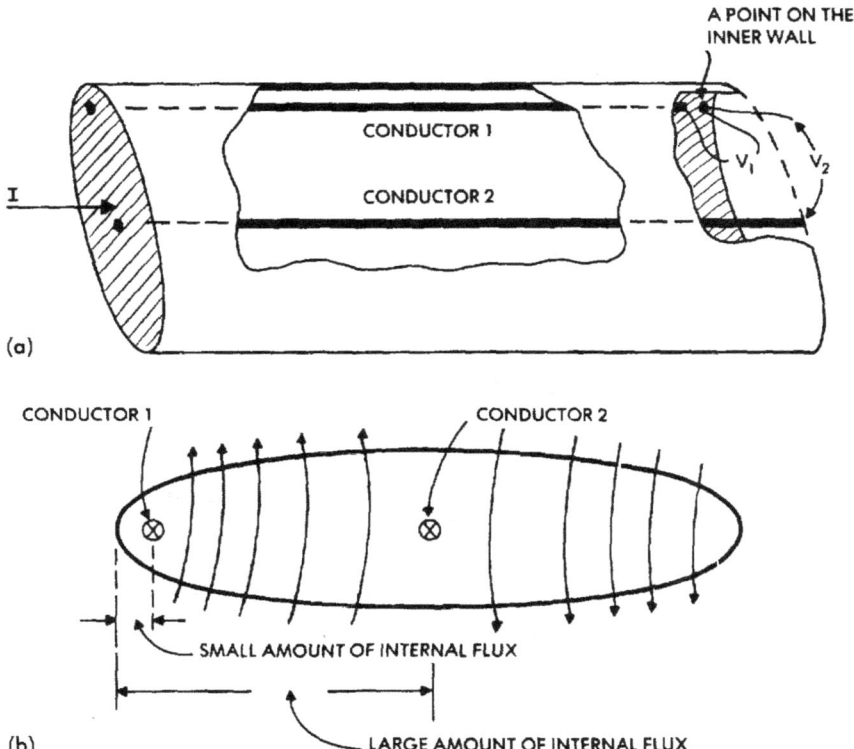

Figure 10.16 Factors governing the internal voltage.
(a) The geometry
(b) Internal flux linkages

The voltage between any two points is defined again as the line integral of the potentials around a closed path. Figure 10.17(a) and (b) shows the simplest paths to consider. V_1 would be the sum of the potentials developed around the loop ABCD. If there is no current along conductor 1, the potential along path A-B will be zero. The potential drop along the path B-C will be zero because of the assumptions regarding the end cap. The potential along the path C-D will then be the voltage drop produced by the inner current density times the

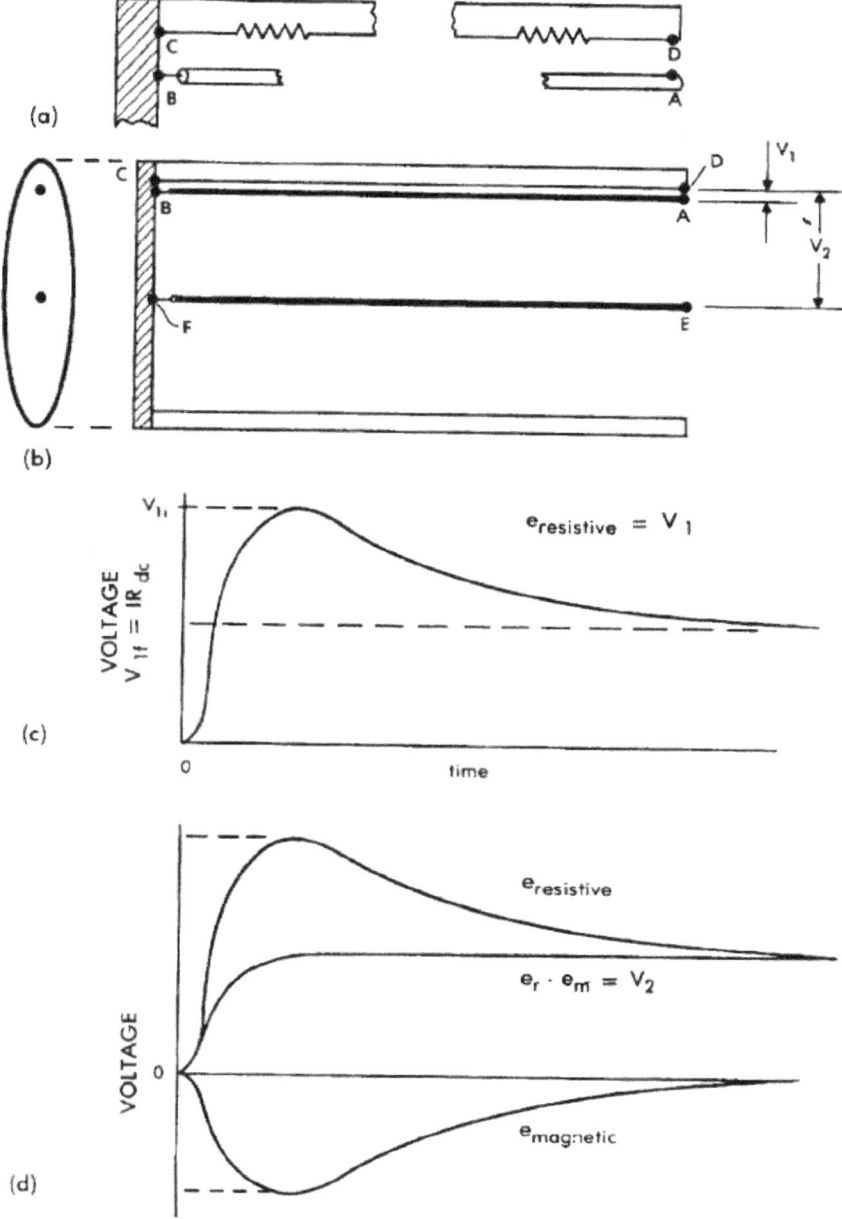

Figure 10.17 The internal voltages.

(a) Detail of the edge surface

(b) Paths of integration

(c) Components of V_1

(d) Components of V_2

resistivity of the material along the path C-D. To these potentials must be added the voltage induced magnetically by the changing magnetic flux passing through the loop defined by the points A, B, C, and D. If the spacing of the conductor to the wall is made vanishingly small, so that C-B and D-A become zero, there will be no magnetic flux; hence the voltage V_1 between points A and D will be only the resistive voltage drop along the path C-D. As in the cylindrical geometry case, the voltage for a step-function current injected into the exterior of the tube will build up according to the pattern shown in Figure 10.6. Its magnitude will be greater than the dc resistance drop by the ratio to which the initial current density on the exterior along the end of the ellipse exceeds the steady state current density.

V_2 will again be the sum of a resistive voltage drop and a magnetically induced drop, this time along the path E-F-C-D. The resistive component will be identical to the resistive component of V_1, the resistance drop along the path C-D. For V_2, however, there will be a nonzero magnetic component of voltage produced by the passage of a finite amount of magnetic flux through the finite loop EFCD. The magnetically induced component of voltage will be given by

$$e_m = \frac{d\phi}{dt} = K \frac{d}{dt} (I_{circ}) \qquad (10.27)$$

where K is a proportionality constant relating the flux produced in the loop EFCD to the internal current. In Equation 10.26, however, it was shown that the internal current was proportional to the integral of the internal resistance drop. This leads to the rather unusual observation that the magnetically induced component of voltage has, initially at least, the same waveshape as the component of voltage produced by the flow of internal current density through the resistance of the material. The long term response of the magnetically induced voltage will be different from the resistively generated component, since, as steady state conditions are reached and the internal magnetic field reaches its final value, its rate of change of the internal magnetic field will decrease to zero.

The amount of the magnetically induced voltage will depend upon the location of the conductor and upon the degree to which the initial distribution of magnetic flux around the outside of the cylinder differs from the final distribution of magnetic flux. Since the difference between the initial and the final flux patterns is greater for cylinders of high eccentricity than it is for cylinders of low eccentricity, it follows that the flatter the cylinder, the greater will be the influence of the magnetic component.

10.2.4 Experimental Verification

In order to demonstrate the above principles, a test was made in which current was injected into an elliptical cylinder and the voltage on several internal conductors measured. The physical arrangement of the test setup is shown in Figure 10.18. The cylinder had a major axis of 47 cm, a minor axis of 9.4 cm

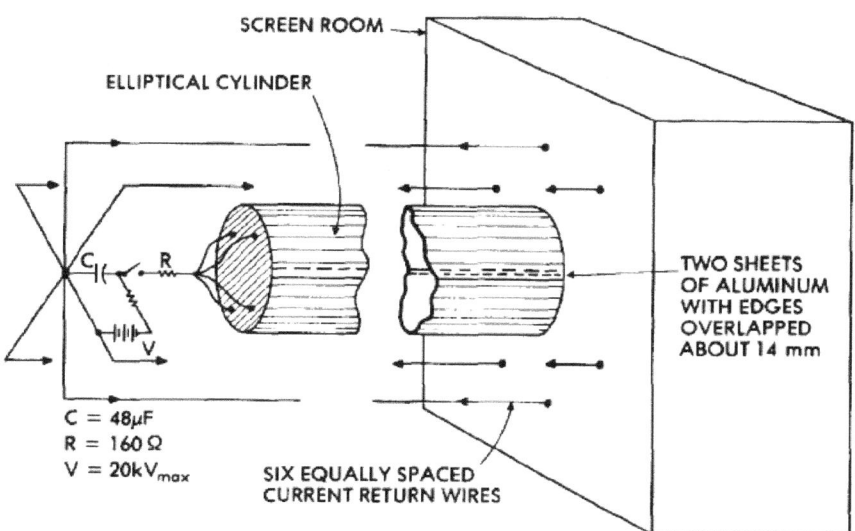

SCREEN ROOM

ELLIPTICAL CYLINDER

TWO SHEETS
OF ALUMINUM
WITH EDGES
OVERLAPPED
ABOUT 14 mm

C = 48μF
R = 160 Ω
V = 20kV$_{max}$

SIX EQUALLY SPACED
CURRENT RETURN WIRES

Figure 10.18 Setup of test to demonstrate internal voltages in an elliptical cylinder.

(major/minor = 5) and was made from two sheets of aluminum bent over a wooden framework with the overlapping edges stapled at close intervals to the internal wooden framework. The edges overlapped by about 1.4 cm. After stapling, the seam was covered with a 5 cm wide sheet of adhesive-backed aluminum foil. The thickness of the metal sheet was 0.38 mm (0.015 inches). One end of the cylinder was bolted to a metal plate facing the outer wall of a screen room. The other end of the cylinder was sealed with metal strips and connected to a pulse generator. The pulse generator consisted of a capacitor bank capable of being charged to a maximum of 20 kV. For most of the tests the capacitor was 48 μF and was discharged through 160 Ω into the end of the elliptical cylinder. Current was brought into the end of the cylinder through four conductors so arranged as to minimize any local end effects. Current was returned to the pulse generator through six equally spaced return wires spaced about 18 inches from the center of the cylinder. Multiple return wires were used both to minimize the external inductance through which the generator had to discharge and to minimize any possible proximity effects. Internal to the cylinder were placed eight conductors at different locations, the locations being as shown in Figure 10.19. Two of the conductors were taped against the inner surface of the metal sheet. One of these, conductor number 1, was located at the narrow end of the cylinder, while the other, conductor number 8, was located about halfway to the center. The remaining conductors were all located at different points along the major axis of the ellipse. The conductors were number 22AWG and insulated with the thickness of insulation normal to such conductors. Each conductor was connected at the current input end to the inner surface of the end cap through a resistor of 68 Ω, the purpose of the resistor

315

being to damp out any internal oscillations that might be developed. The conductors were terminated on switches so that measurements could be made to ground on any conductor or be made between any two conductors.

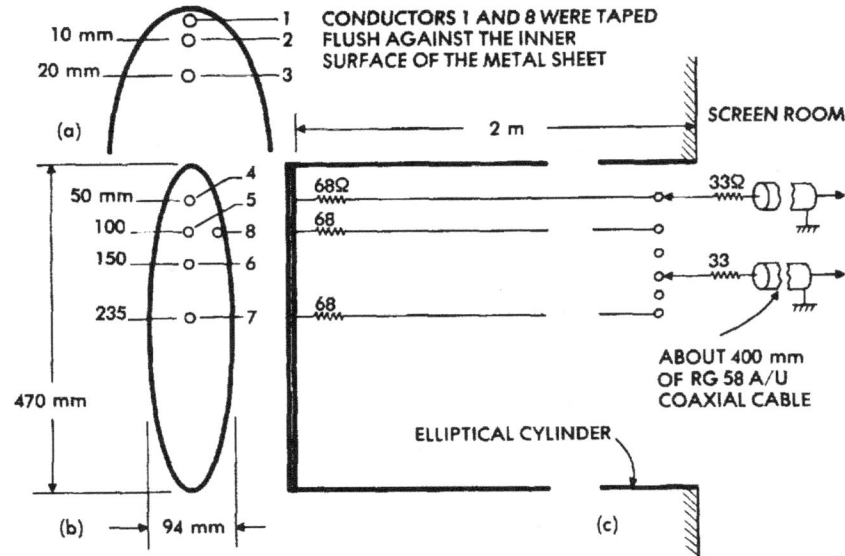

Figure 10.19 Arrangement of conductors inside elliptical cylinder.
    (a) Detail
    (b) End view showing locations of conductors
    (c) Electrical termination of internal conductors (typical)

The measurement principles are shown in Figure 10.20. Line-ground voltage measurements were made with the normal input preamplifiers of an oscilloscope. The oscilloscope was located in the screen room directly adjacent to the end of the elliptical cylinder and connected to the output jacks on the cylinder by only a short length of coaxial cable. Accordingly, the measurements would not be influenced by any loading effects of the measurement cables. Voltages between pairs of conductors were made with the oscilloscope preamplifier in the A minus B mode. Since the voltage between any pair of conductors would depend only upon the magnetically induced component of voltage, the resistive component being rejected by the differential amplifier, it was then possible to integrate the difference voltage and derive a signal proportional to the internal magnetic field. This was done using an operational amplifier connected as an integrator, shown in Figure 10.20(c).

Typical tests results are shown in Figures 10.21 and 10.22. Figure 10.21 shows, on four different time scales, the injected current and the line-to-ground voltage on two different conductors, conductor 1 located adjacent to the small radius end of the ellipse and conductor 7 located near the center. With the

316

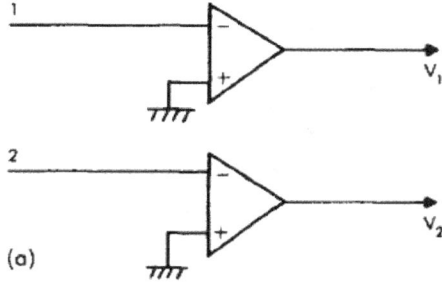

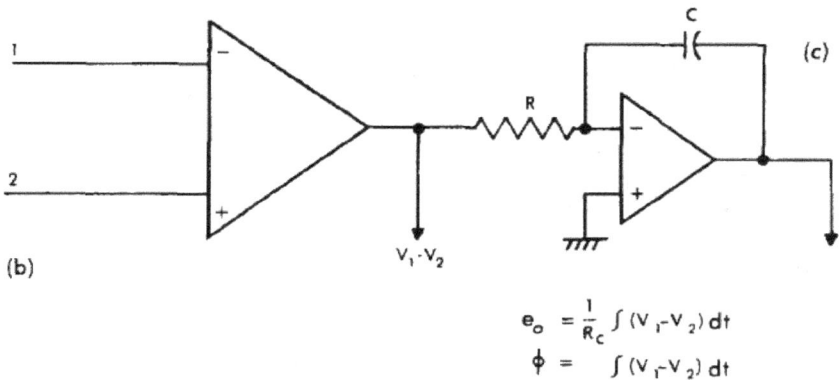

$$e_o = \frac{1}{R_c} \int (V_1 - V_2)\, dt$$

$$\phi = \int (V_1 - V_2)\, dt$$

Figure 10.20 Measurement principles.
    (a) Line-ground measurements
    (b) Line-line measurement
    (c) Magnetic field measurement

exception of the oscillograms displayed at 500 μs per division, the input current can be considered as a step-function current. In both cases the line-ground voltage is seen to display the pulse penetration-type response shown in Figure 10.6. The small bump at the leading end of the voltage on conductor 1 is primarily the result of leakage from the surge generator into the screen room housing the measuring instruments and to coupling from the trigger pulse used to fire the surge generator. In both cases the initial rise of voltage follows the same pattern and reaches essentially its final value in about 3 μs, a time in accordance with the calculated pulse penetration time constant of 1.2 μs discussed previously. At later times the voltage on conductor 1 is seen to decay back toward a lower final value in the manner shown in Figure 10.17. Likewise, the voltage on conductor 7 is seen to be nearly flat, rising only slightly, as shown in Figure 10.21. It is probably only a fortuitous combination of conductor location and the characteristics of the elliptical cylinder that result in the voltage

317

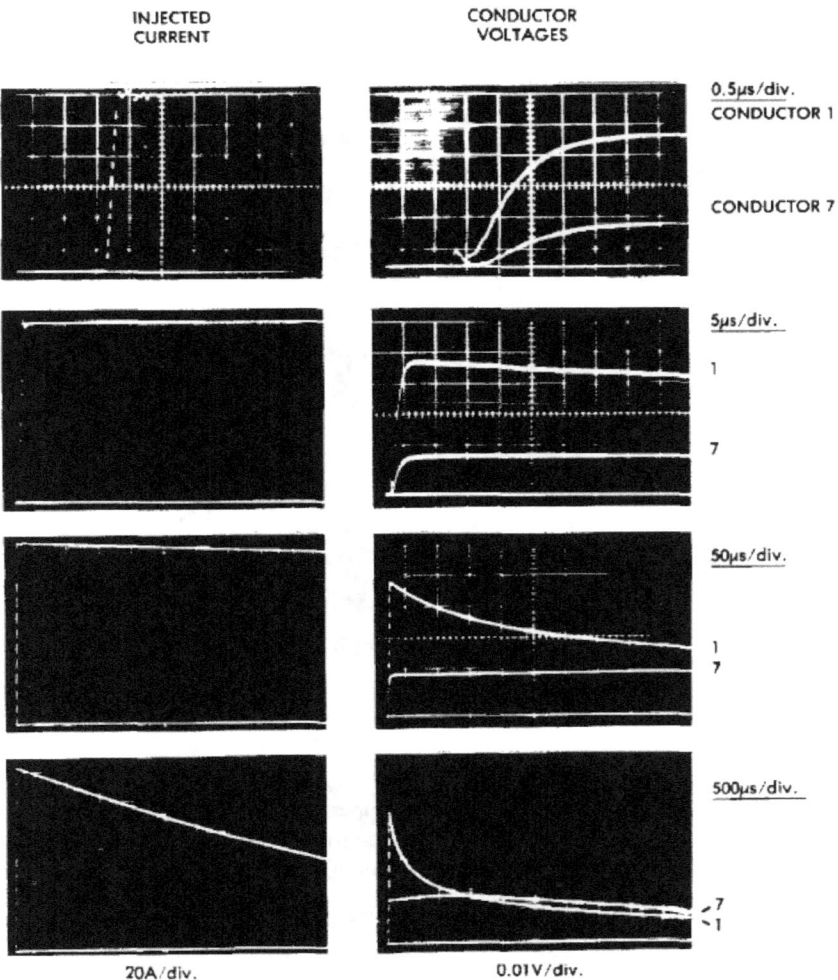

INJECTED
CURRENT

CONDUCTOR
VOLTAGES

0.5µs/div.
CONDUCTOR 1

CONDUCTOR 7

5µs/div.

1

7

50µs/div.

1
7

500µs/div.

7
1

20A/div.

0.01V/div.

Figure 10.21   Variation of conductor voltage with time.
(Leading edges of waveforms retouched for photographic clarity)

on conductor 7 being nearly flat. If the injected current were truly a step function, the two voltages would eventually become equal to the dc resistance drop. The bottom set of oscillograms indicates that the current significantly departs from a step-function current pattern at later times, and this is reflected in the long-time response of the two line-ground voltages. Just as the voltage on conductor 7 lagged behind the voltage on conductor 1 during the initial phase of the response, it lags behind during the final decay of the current. The redistribution time constant with which V_1 decays toward its final value is seen to be of the order of 300 µs, in accordance with the calculations predicated on the theoretical factors described in Figure 10.10.

318

VOLTAGE -
CONDUCTOR 1 - CONDUCTOR 7

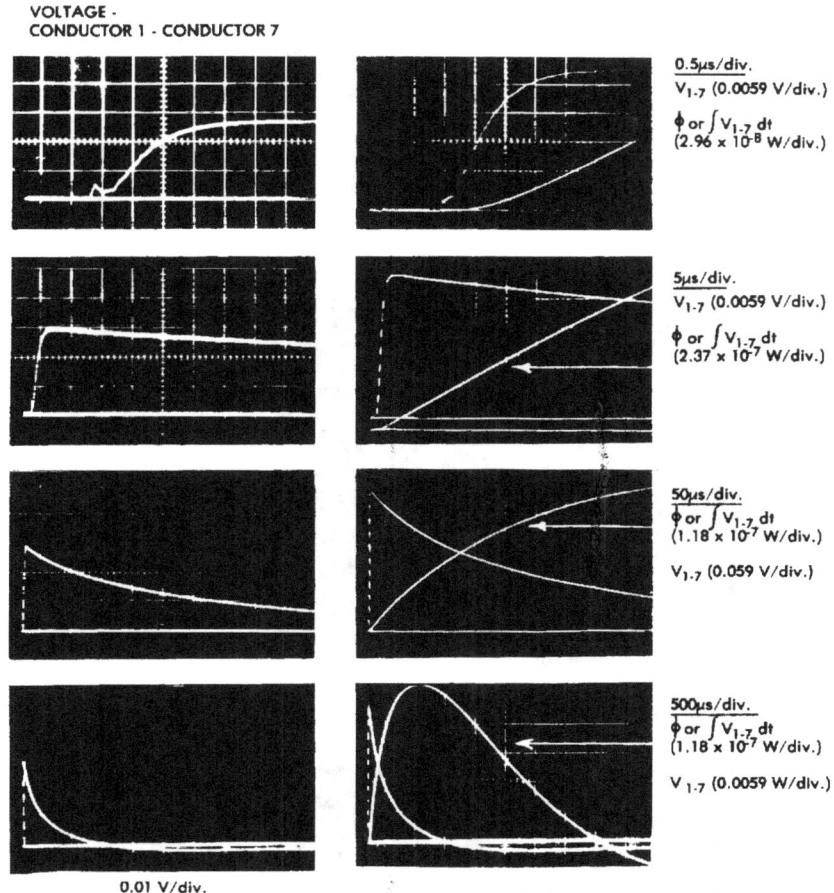

0.5μs/div.
V₁₋₇ (0.0059 V/div.)

ϕ or $\int V_{1-7}\, dt$
(2.96 × 10⁻⁸ W/div.)

5μs/div.
V₁₋₇ (0.0059 V/div.)

ϕ or $\int V_{1-7}\, dt$
(2.37 × 10⁻⁷ W/div.)

50μs/div.
ϕ or $\int V_{1-7}\, dt$
(1.18 × 10⁻⁷ W/div.)

V₁₋₇ (0.059 V/div.)

500μs/div.
ϕ or $\int V_{1-7}\, dt$
(1.18 × 10⁻⁷ W/div.)

V₁₋₇ (0.0059 W/div.)

0.01 V/div.

Figure 10.22 Difference voltages and total flux.
(Leading edges of waveforms retouched for photographic clarity)

Figure 10.22 shows the voltages between conductors (displayed on two separate oscilloscopes and hence displayed on two separate oscillograms) and the integral of the conductor-to-conductor voltage. This latter is of course proportional to the magnetic flux that builds up internally to the cylinder. Since, at early times, the line-to-ground voltages have the same waveshape but different amplitudes, it follows that the line-line voltage will also have that waveshape. Unlike the line-ground voltages, the line-line voltage decays toward zero. Since the input current was not a step-function current, the oscillograms displayed at 500 μs per division show the line-line voltage decaying to zero and then reversing as the input current decays and the internal magnetic fields attempt to follow the decaying external magnetic fields. The input current did not last long enough for the magnetic field to reach a steady state value, but the oscillograms do indicate that it seems to crest at about 850 μs. Why the

319

indicated magnetic field does not crest at the time the line-line voltage goes through zero, 1300 μs, is not known. Perhaps instrumentation errors in the integrator are to blame. At any rate, the decay time constant for the line-line voltage agrees with the value predicted by Equation 10.23.

With a somewhat different waveshape of injected current, Figure 10.23 shows the voltages as a function of the position of the conductor. The voltage on conductor 8 is not shown; however, it was virtually identical with that shown for conductor 5.

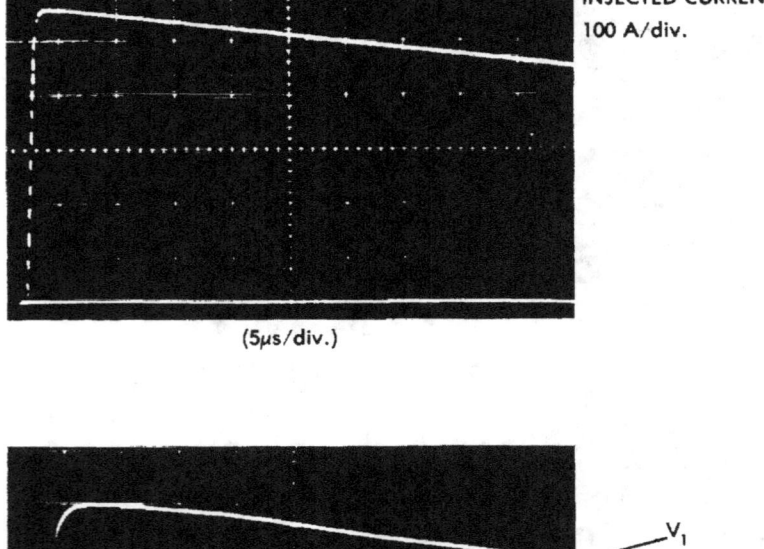

INJECTED CURRENT
100 A/div.

(5μs/div.)

(5μs/div.)

Figure 10.23 Voltages as a function of position.

10.3 Fields Within Cavities

It frequently happens that in the fuselage of an aircraft there will be a cavity that is effectively exposed to the external field on only one face, either because the inner walls of the cavity are thick enough to provide more shielding from the other parts of the external field or because the cavity is much closer to

320

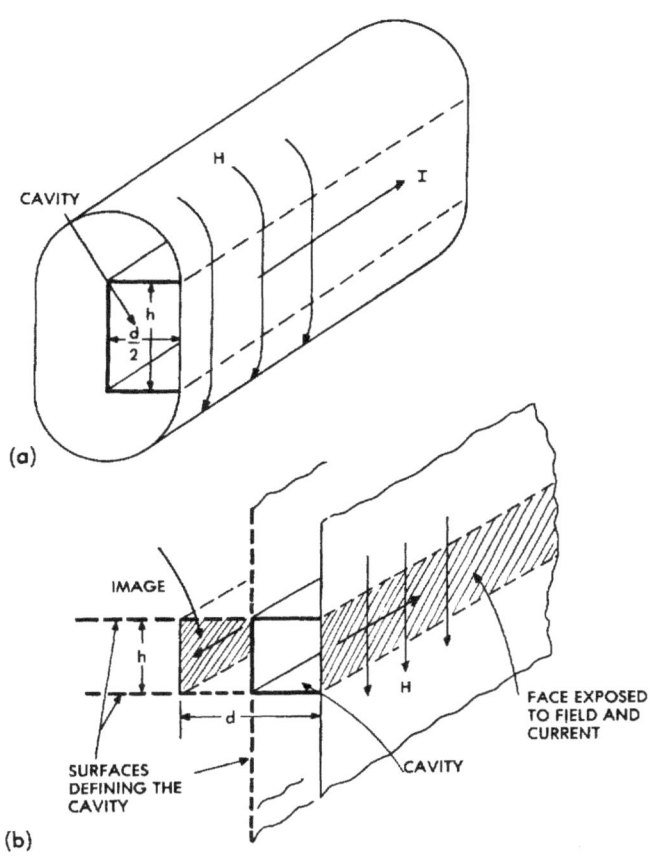

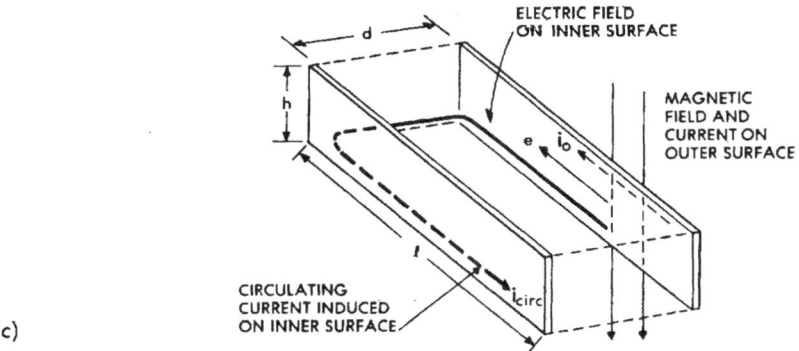

Figure 10.24 A cavity exposed to a field on only one side.

(a) Cavity and field orientations

(b) An image in the reflecting surfaces defining the cavity

(c) Current path defining the loop inductance

one of the external surfaces than it is to any of the other external surfaces. A typical cavity would be a gun bay or an electronic equipment bay located along the fuselage of the aircraft and accessible through access panels. For the moment the effects of the access panels will be ignored. The geometry is shown in Figure 10.24(a).

The cavity may be viewed as being formed from one face exposed to the external magnetic field and the current producing that field, and with its other faces formed by metal sheets, which, like the exposed surface, may be assumed to extend toward infinity. Mirrored in the reflecting surface defining the back of the cavity will be an image of the face of the cavity exposed to the magnetic field. The image will be spaced a distance, D, behind the face carrying the current, and in the image the reflection of the external current will be of opposite polarity. The electric field produced on the inner surface of the face exposed to the external current, shown in Figure 10.24(c), will act to force a current, I_{circ}, around the interior of the circuit defining the cavity. The length of the loop involved will be the same as the length of the cavity, and the width will be twice the depth of the cavity. The internal voltage gradient will be proportional to the product of the inner current density and the resistivity of the face carrying the external current and, as described earlier, will exhibit the characteristic pulse penetration buildup described in Figure 10.6. This voltage may be viewed as impressed across a loop or cavity inductance, Z

$$Z = R + j\omega L \qquad (10.28)$$

The resistance and inductance will both be governed by the effective characteristics of the loop defining the cavity.

Typical current paths and their characteristic impedances are shown in Figure 10.25. Some cavities may be viewed as being sufficiently long and narrow that they may be defined by parallel strips. Others are basically of rectangular or circular shape, or of some simple shape that may be approximated by an equivalent circular cylinder. The inductance and resistance of each configuration are shown. Each of the inductance equations (Reference 10.3) has a correction factor--$\log_e$ k, F', or K-- that relates to the shape of the enclosure. These correction factors are shown in Figure 10.26.

If the cavity is provided with a removable cover, and if this cover is in the external current flow or is exposed to the external magnetic field, the effects are as illustrated in Figure 10.27. If the cover is assumed to be of the same material and thickness as the rest of the face upon which it is mounted, the principal effects relate to the resistance of the fasteners used to hold the cover in place. The covers will seldom make good electrical contact to the rest of the surface except at the fasteners themselves. Accordingly, the external currents flowing in the face will be constricted in the vicinity of the fastener and pass from that face onto the cover through the fastener. The major effect of this constriction of current flow is to introduce a lumped resistance into the electrical circuit, although there is a certain amount of influence on the inductance of the circuit whenever the current is constricted. It will be seen, then, that the greater the

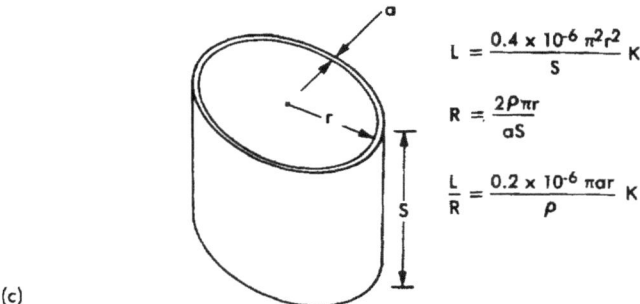

(a)

$$L = 0.4 \times 10^{-6} l \left[\log_e \frac{d}{S} + 1.5 + \log_e K \right]$$

$$R = \frac{2\rho l}{aS}$$

$$\frac{L}{R} = \frac{0.2 \, as \times 10^{-6}}{\rho} \left[\log_e \frac{d}{S} + 1.5 + \log_e K \right]$$

(b)

$$L = \frac{0.4 \times 10^{-6} l d}{S} F'$$

$$R = \frac{2\rho(l + d)}{aS}$$

$$\frac{L}{R} = \frac{0.2 \times 10^{-6} \, a}{\rho} \left(\frac{l d}{l + d} \right) F'$$

(c)

$$L = \frac{0.4 \times 10^{-6} \, \pi^2 r^2}{S} K$$

$$R = \frac{2\rho \pi r}{aS}$$

$$\frac{L}{R} = \frac{0.2 \times 10^{-6} \, \pi a r}{\rho} K$$

Figure 10.25 Typical current paths and characteristic impedances.
  (a) Long sheets
  (b) Rectangular box
  (c) Circular cylinder

number of fasteners, the less the restriction of current flow and the less resistance inserted into the current path. The resistance introduced by the fasteners is important because it is frequently much higher than the intrinsic resistance of the metal surface and because the resistance is not subjected to the

323

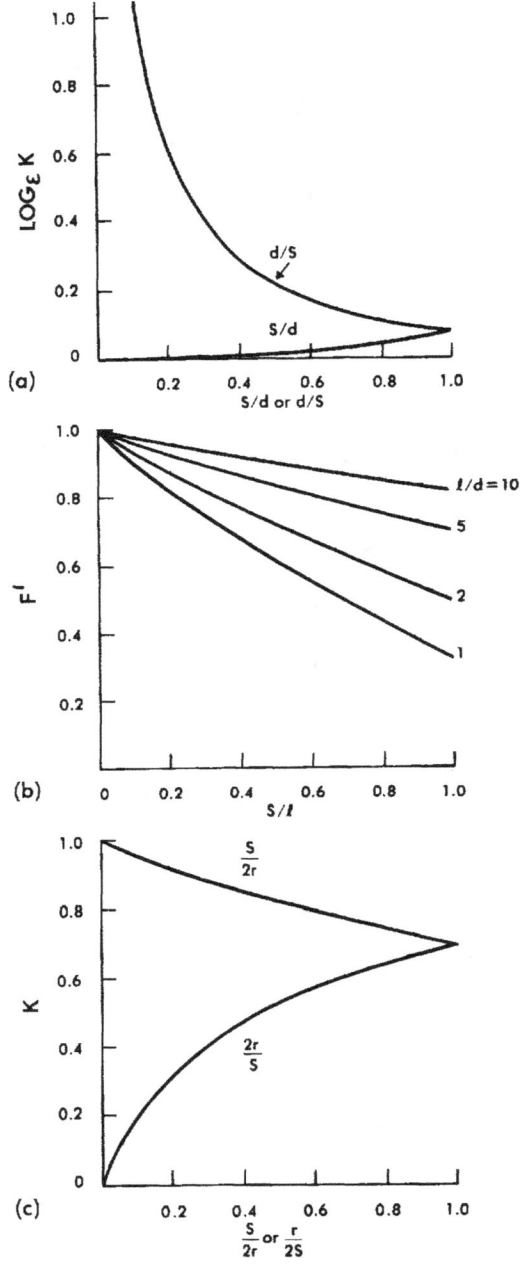

Figure 10.26 Correction factors for inductance.

    (a) Parallel strips

    (b) Rectangular boxes

    (c) Circular cylinders

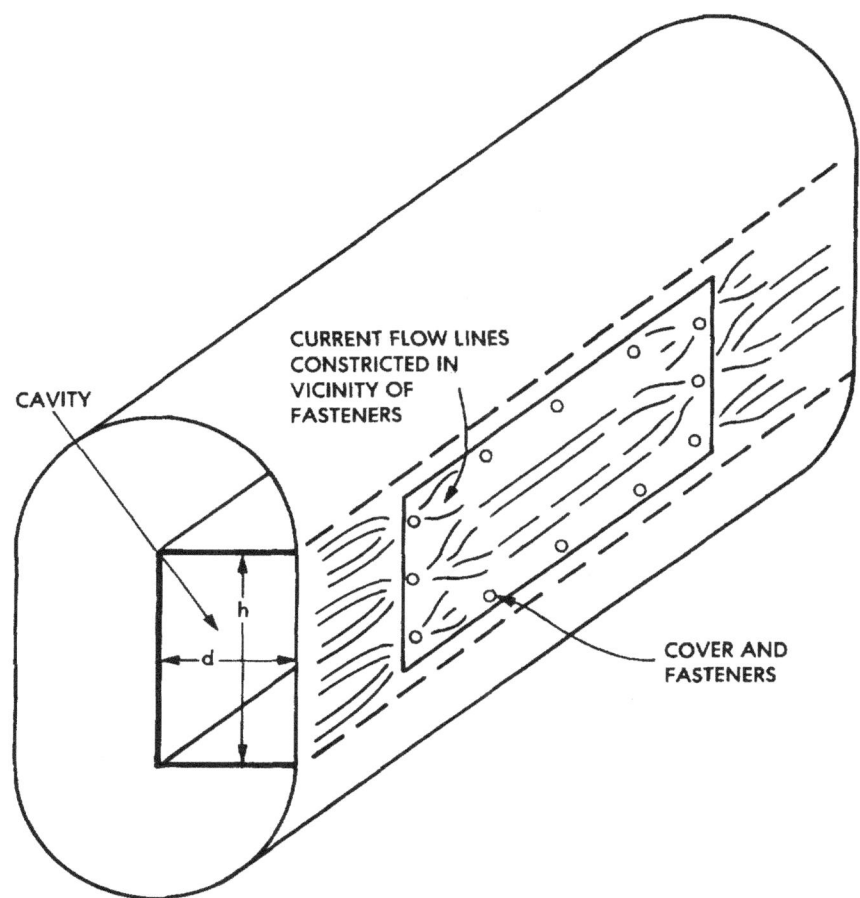

CAVITY

CURRENT FLOW LINES
CONSTRICTED IN
VICINITY OF
FASTENERS

COVER AND
FASTENERS

Figure 10.27 Effects of covers and fasteners.

skin effects that retard the buildup of current density on the inner surface.

An equivalent circuit of the cavity including the effects of fasteners is shown in Figure 10.28. R_f is the equivalent resistance of the fasteners. Circulated through this resistance is the undistorted current flowing in the exterior face of the fuselage. The sum of R_1 and R_2 is equal to the intrinsic resistance of the loop defining the cavity, the resistances given by the equations in Figure 10.25. The external current will develop a voltage across only a portion of this resistance, since it flows in only a portion of the loop. Letting R_1 be the resistance through which the external current flows, that resistance may be assumed to be subjected to the current as retarded by the pulse penetration time constant. The two voltages developed across these resistances then circulate current around the entire loop. The rate at which the current builds up will then be the rate at which the magnetic field inside the cavity builds up.

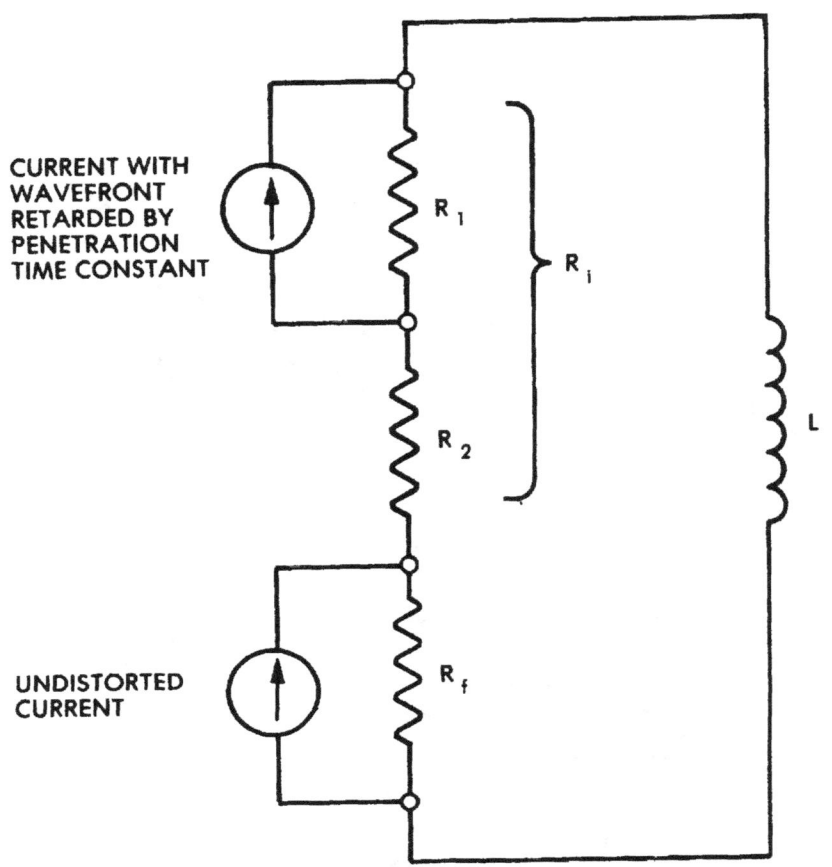

Figure 10.28 Equivalent circuit governing buildup of magnetic field inside a cavity.

REFERENCES

10.1 K. Khalaf-Allah, *Time Constant for Magnetic Field Diffusion into a Hollow Cylindrical Conductor,* UKAEA (United Kingdom Atomic Energy Authority) Research Group Report CLM-R 141, Culham Laboratory, Abingdon, Oxfordshire, England (1974).

10.2 F. W. Grover, *Inductance Calculations: Working Formulas and Tables* (New York: Dover, 1962).

10.3 Grover, *Inductance Calculations,* pp. 15-16 *et passim.*

CHAPTER 11
THE INTERNAL FIELDS COUPLED THROUGH APERTURES

11.1 Magnetic Fields

The most important mode by which magnetic fields are coupled from the exterior of the aircraft to the interior is coupling through magnetic apertures. There are two major reasons for the importance of aperture-coupled fields. The first is that some apertures may be quite large, windows being a prime example. The second reason is that, unlike fields coupled by diffusion through metal surfaces, the waveshape of the interior field is not retarded, but tends to be the same for small apertures as that of the external magnetic field. The most important consequence of the coupled magnetic fields relates to the voltages induced by such fields. A field of given peak intensity is more apt to cause trouble if it rises to its peak quickly than if it is delayed and distorted.

Physically, the basic problem is as shown on Figure 11.1. An external magnetic field tangential to a metal surface passes across an aperture. At the aperture some of the lines of magnetic force leak into the interior region. Analytically, the problem is to be able to determine the field strength at any point in the interior volume, preferably to be able to determine both the magnitude and the direction of the magnetic field.

Often the analytical problem is best solved by placing in the plane of the aperture an equivalent magnetic dipole, the fields around which may then be determined from dipole theory. The strength of the equivalent dipole is determined by the strength of the external magnetic field and by the size, shape, and orientation of the aperture. A geometry amenable to mathematical analysis (Figure 11.2) is that of an elliptical aperture in a conductive sheet of a size large enough to separate completely the interior volume in which the fields are to be calculated from the exterior volume in which the basic magnetic field exists. In the following analysis we will treat only fields tangential to the conductive surface, since, as explained in the previous sections, a magnetic field with components perpendicular to the conductive surface induces circulating currents that initially cancel the penetrating component.

Mathematically the problem is most directly solved if the external magnetic field is resolved into two components, H_x and H_y, lying along an axis directed through the major and minor axes of the elliptical aperture. These field components give rise to two equivalent magnetic dipoles, M_x and M_y, again oriented along the major and the minor axes of the aperture. The strength of these dipoles is a function of the strength of the external magnetic field that would exist in the plane of the aperture if the aperture were absent. The governing expressions (Reference 11.1) are

$$M_x = a_{11} H_x \qquad\qquad (11.1)$$

$$M_y = a_{22} H_y \qquad\qquad (11.2)$$

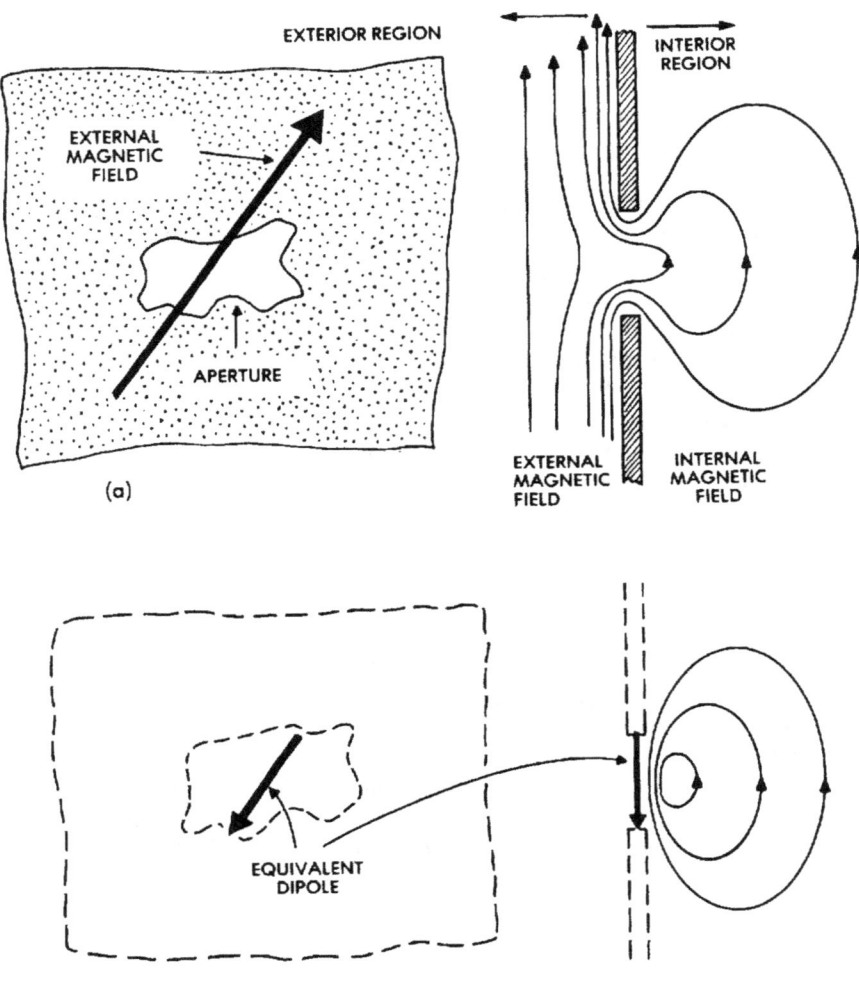

EXTERIOR REGION

EXTERNAL
MAGNETIC
FIELD

APERTURE

(a)

INTERIOR
REGION

EXTERNAL
MAGNETIC
FIELD

INTERNAL
MAGNETIC
FIELD

EQUIVALENT
DIPOLE

(b)

Figure 11.1 The aperture-coupling problem.
  (a) A field across an aperture
  (b) Equivalent dipole producing
      the same internal field

$$a_{11} = -\frac{2\pi}{3} \left[\frac{\ell_1}{2}\right]^3 \frac{e^2}{K(e^2) - E(e^2)} = \left[\frac{\ell_1}{2}\right]^3 a'_{11} \qquad (11.3)$$

$$a_{22} = -\frac{2\pi}{3} \left[\frac{\ell_1}{2}\right]^3 \frac{e^2(1-e^2)}{E(e^2) - (1-e^2)K(e^2)} = \left[\frac{\ell_1}{2}\right]^3 a'_{22} \qquad (11.4)$$

330

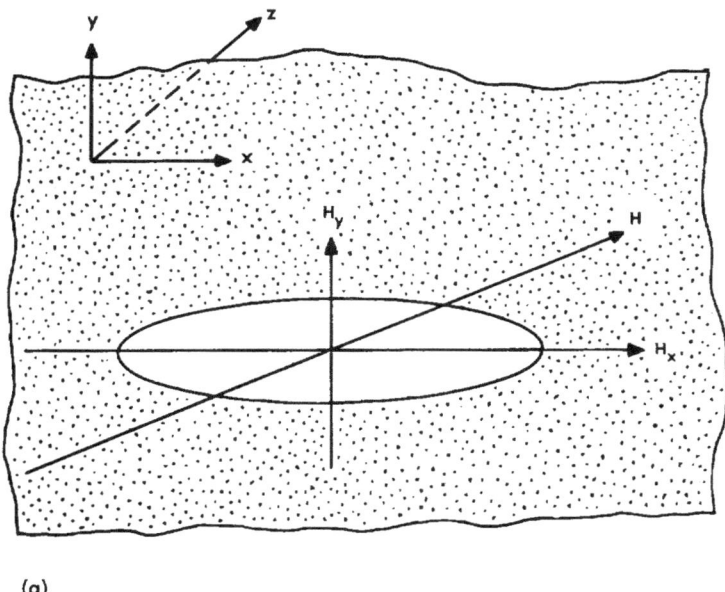

(a)

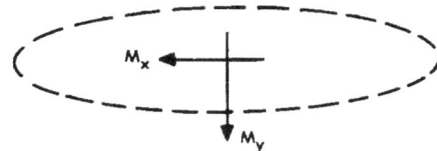

(b)

Figure 11.2   An elliptical aperture.

(a) Geometry and the components of the external magnetic field vector

(b) The equivalent magnetic dipoles

$$e^2 = 1 - \left[\frac{\ell_2}{\ell_1}\right]^2 \quad (\ell_2 < \ell_1) \qquad (11.5)$$

The expressions K_e and E_e are elliptical integrals of the first and second kinds, respectively. The quantities a'_{11} and a'_{22} are shown in Figure 11.3. Since the dipole strength is proportional to the cube of the major axis length of the aperture, it follows that large apertures will couple much more field into an inner volume than will small apertures, the strength of the dipole increasing faster than the area of the aperture increases.

The pattern of magnetic fields in the interior volume produced by a dipole

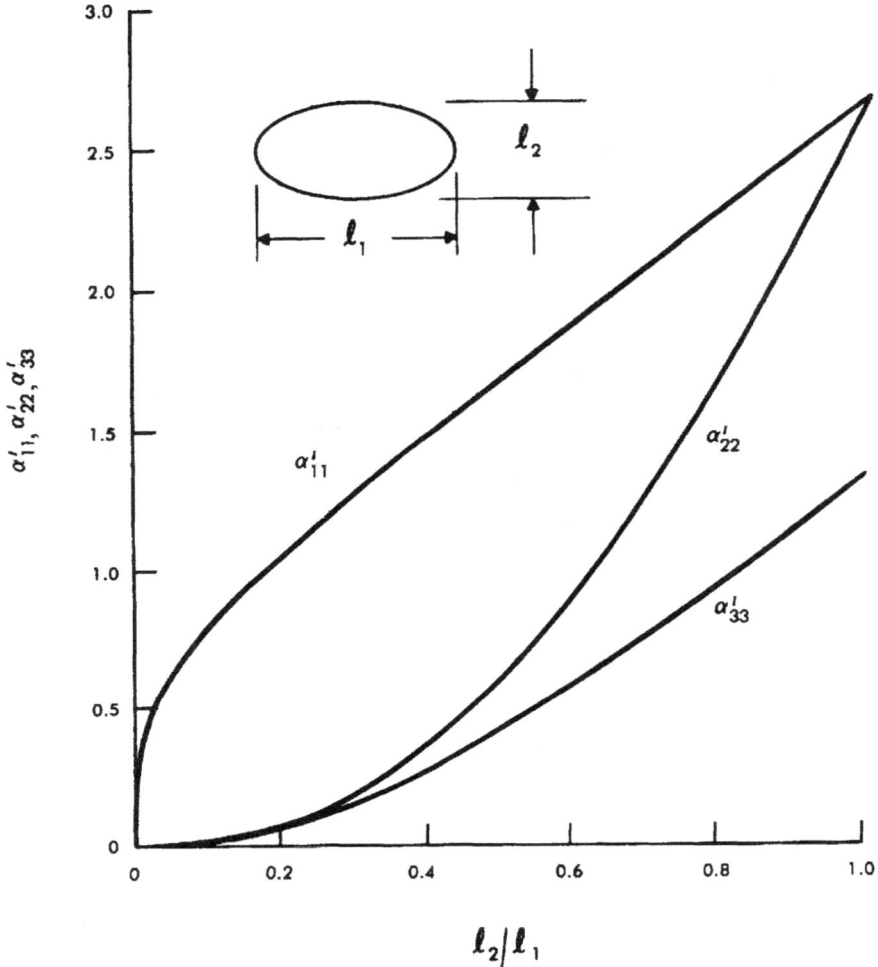

Figure 11.3 Shape factor for elliptical apertures.

lying along the X axis would appear as on Figure 11.4. Assuming the dipole to lie in the XY plane, the field patterns will be concentric closed loops lying in planes passed through the X axis. At any point, P, the total magnetic field will be tangential to the corresponding closed loop. The components of the magnetic field at point P––H_x, H_y, and H_z––(Reference 11.2) will be

$$H_x = \frac{M_x}{4\pi} \left[\frac{3X^2 - r^2}{r^5} \right] \qquad (11.6)$$

332

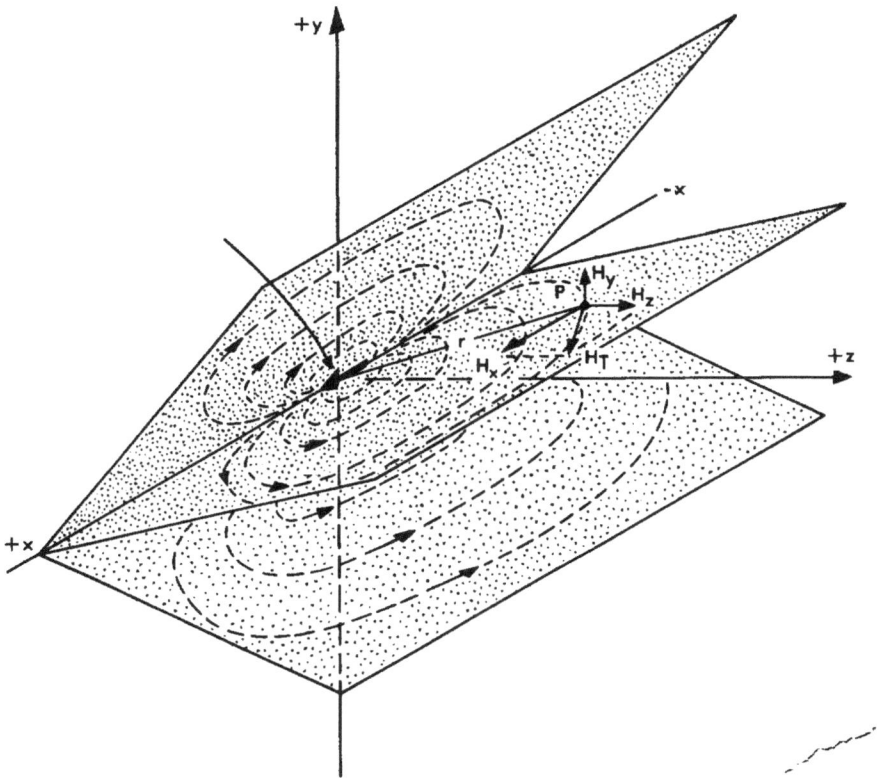

Figure 11.4 Field patterns produced by a dipole lying along the X axis.

$$H_y = \frac{M_x}{4\pi} \left[\frac{3XY}{r^5} \right] \tag{11.7}$$

$$H_z = \frac{M_x}{4\pi} \left[\frac{3XZ}{r^5} \right] \tag{11.8}$$

Typically, the external magnetic field would include an equivalent dipole oriented along the Y axis as well. This dipole would likewise produce a magnetic field at the point P, the components of which would be

$$H_x = \frac{M_y}{4\pi} \left[\frac{3XY}{r^5} \right] \tag{11.9}$$

$$H_y = \frac{M_y}{4\pi} \left[\frac{3Y^2 - r^2}{r^5} \right] \tag{11.10}$$

333

$$H_z = \frac{M_y}{4\pi} \left[\frac{3YZ}{r^5} \right] \qquad (11.11)$$

The total magnetic field strength at P, then, would be the sum of the components produced by the dipoles lying along the X and Y axes.

$$H_x = H_{xx} + H_{xy} \qquad (11.12)$$

$$H_y = H_{yx} + H_{yy} \qquad (11.13)$$

$$H_z = H_{zx} + H_{zy} \qquad (11.14)$$

The total field strength at point P would be the vector summation of the X, Y, Z components of the field.

$$H_T = \sqrt{H_x^2 + H_y^2 + H_z^2} \qquad (11.15)$$

The field strength is seen to be inversely proportional to the cube of the distance from the aperture to the point in question. Accordingly, the aperture-coupled fields will be localized in space near the aperture. Small apertures are thus less troublesome than large apertures, both because of the decreased magnetic field strength associated with the corresponding dipole and because of the lesser distance one must be removed from the aperture before the field strength becomes negligible compared to the field strength closer to the aperture.

The above formulation has one major restriction: it is valid only in the low-frequency region, low-frequency implying that the aperture is small compared to the wavelengths of the field under consideration. Coupling through apertures increases with increasing frequency, as witness the quarter-wave resonant slot antenna. The formulation also treats only the internal magnetic fields produced by external magnetic fields and does not treat the electric field produced by a changing magnetic field or the magnetic field produced by a changing electric field. Taylor (Reference 11.1) treats the interaction in more detail. This interaction between changing electric and magnetic fields is greater at high frequencies and, for the frequency spectrum associated with lightning, is possibly of importance only at those frequencies for which the aperture is not electrically small.

The above formulation is also valid only when the point at which the field is to be calculated is at a distance from the aperture that is large compared to the dimensions of the aperture. If an attempt is made to calculate the fields close to the aperture, the results are inaccurate. If the distance from the point to the aperture goes to zero, the calculated magnitude of fields becomes infinite, whereas the field itself can never in fact become larger (barring reflections) than the external field. It can be shown, in fact, that the actual field strength in the plane of the aperture, as distinct from the strength of the equivalent dipole, will be half the field strength that would exist if the aperture were not there. This intractable behavior of the calculated magnitude of the field near the aperture arises as a result of a simplifying assumption commonly used in dipole analysis. This assumption, illustrated in Figure 11.5(a), is that point P is sufficiently far from the center of the dipole that each of the radial distances r_1 and r_2 from the ends of the dipole to the point P may be represented by the average radial distance, r, and that each of the angles θ_1 and θ_2 may be represented by the angle θ.

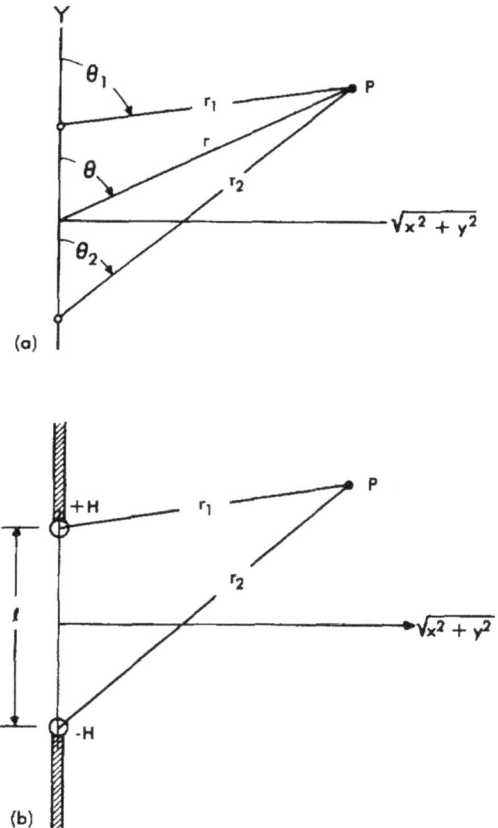

Figure 11.5 Dipole approximations.
(a) Elementary dipole
(b) Dipole of finite size

With a great increase in numerical complexity, it is possible to develop a formulation for the fields at point P that, as shown in Figure 11.5(b), allows the dipole to have a finite length. Such a formulation allows one to determine with

335

better, though not perfect, accuracy the field strength close to the aperture. The governing equations (Reference 11.3) presented here, without proof, are

$$H_x = G_1 \times F_1 + G_2 \times F_2 - G_3 \times F_3 \qquad (11.16)$$

$$H_y = G_4 \times F_1 + G_5 \times F_2 - G_6 \times F_4 \qquad (11.17)$$

$$H_z = G_7 \times F_1 + G_8 \times F_2 \qquad (11.18)$$

where

$$G_1 = \frac{3K_1 (X - \ell_1/2)}{C_1^{1.5}} \qquad (11.19)$$

$$G_2 = \frac{3K_2 X}{C_2^{1.5}} \qquad (11.20)$$

$$G_3 = \frac{K_1 \ell_1}{C_1^{1.5}} \qquad (11.21)$$

$$G_4 = \frac{3K_1 Y}{C_1^{1.5}} \qquad (11.22)$$

$$G_5 = \frac{3K_2 (Y - \ell_2/2)}{C_1^{1.5}} \qquad (11.23)$$

$$G_6 = - \frac{K_2 \ell_2}{C_2^{1.5}} \qquad (11.24)$$

$$G_7 = \frac{3K_1 Z}{C_1^{1.5}} \qquad (11.25)$$

$$G_8 = \frac{3K_2 Z}{C_2^{1.5}} \qquad (11.26)$$

$$F_1 = \left[\frac{X\ell_1}{C_1}\right] + 1.45833 \left[\frac{X\ell_1}{C_1}\right]^3 \cdots$$

$$+ 1.804688 \left[\frac{X\ell_1}{C_1}\right]^5 + 2.094727 \left[\frac{X\ell_1}{C_1}\right]^7 \qquad (11.27)$$

$$F_2 = \left[\frac{Y\ell_2}{C_2}\right] + 1.45833 \left[\frac{Y\ell_2}{C_2}\right]^3 \cdots$$

$$+ 1.804688 \left[\frac{Y\ell_2}{C_2}\right]^5 + 2.094727 \left[\frac{Y\ell_1}{C_2}\right]^7 \qquad (11.28)$$

$$F_3 = 1 + 2(X\ell_1)^2 + 5(X\ell_1)^4 + 7(X\ell_1)^6 \qquad (11.29)$$

$$F_4 = 1 + 2(Y\ell_2)^2 + 5(Y\ell_2)^4 + 7(Y\ell_2)^6 \qquad (11.30)$$

$$K_1 = \frac{a_{11} H_x \text{ (ext)}}{4\pi\ell_1} \qquad (11.31)$$

$$K_2 = \frac{a_{22} H_y \text{ (ext)}}{4\pi\ell_2} \qquad (11.32)$$

Frequently a calculation made to determine the effects of coupling through apertures needs to take into account the presence of reflecting surfaces. The simplest case, shown in Figure 11.6, treats one reflecting surface of infinite dimensions and parallel to the surface containing the aperture. The total field at any point between the two surfaces is the sum of the fields produced by an infinite array of images of the original magnetic dipole. The field strength from each of the images would again be determined from Equations 11.6 through 11.11 or 11.16 through 11.32. Fortunately, in most cases only a few of the images need be considered because of the dependence of field strength on the cube of the distance to the point under consideration. In principle, additional reflecting surfaces could be included in the formulation to define completely an interior volume. Such multiple reflecting surfaces are shown on Figure 11.7.

Typically, one is concerned, not with the magnitude of the fields themselves, but with the effects of the fields on electrical circuits. A typical problem, as shown in Figure 11.8, might be determining the voltage produced by an aperture-coupled magnetic field between a conductor and a metal surface containing the aperture. That voltage would be determined by the total magnetic flux passing through the plane defined by the conductor and its projection onto the metal surface. This total magnetic field would be the summation, both along the length and across the width of the loop so defined, of the magnetic field strength normal to the loop. If the field strength were to vary considerably with position along the loop, it would be necessary to evaluate the field strength at a number of points across the area of the loop and to perform some sort of graphical integration to determine the total magnetic flux passing through the loop. The numerical calculations, while straightforward in principle, would be of sufficient complexity to make the problem insoluble in any practical sense if all the calculations had to be made by hand. In order to overcome the numerical

337

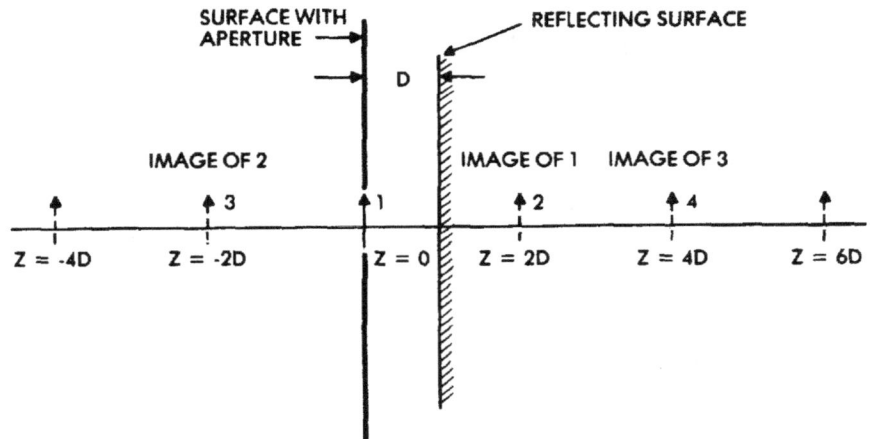

Figure 11.6 Reflecting surface.

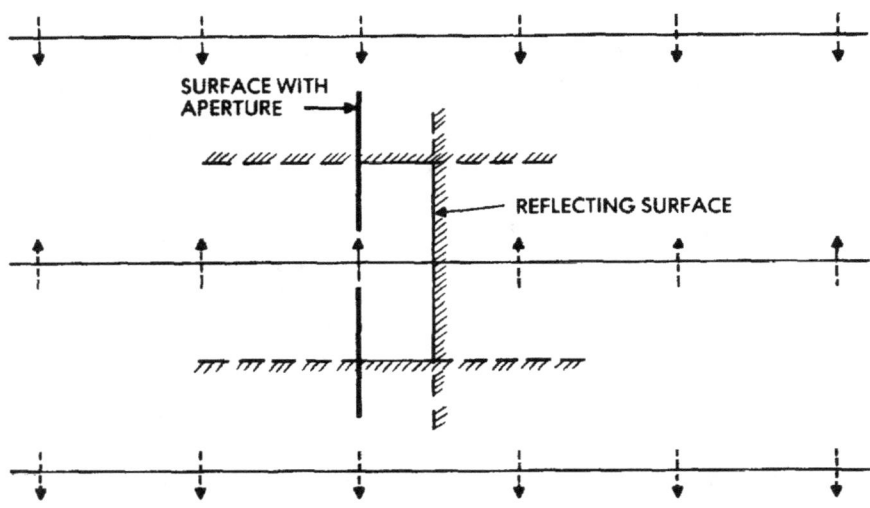

Figure 11.7 Multiple reflecting surfaces.

problems, a computer program, APERTURE, has been written and is described in the literature (Reference 11.4). This program allows one to define an arbitrary elliptical aperture in an X-Y plane. The orientation of the aperture and the orientation and strength of the external field are defined as input quantities and do not necessarily have to be oriented along the X and Y reference axes. The program allows one to tabulate the interior field strength at whatever points are desired, in either rectangular or polar coordinates. The program also allows one to determine the total magnetic flux passing any arbitrary four-sided plane

338

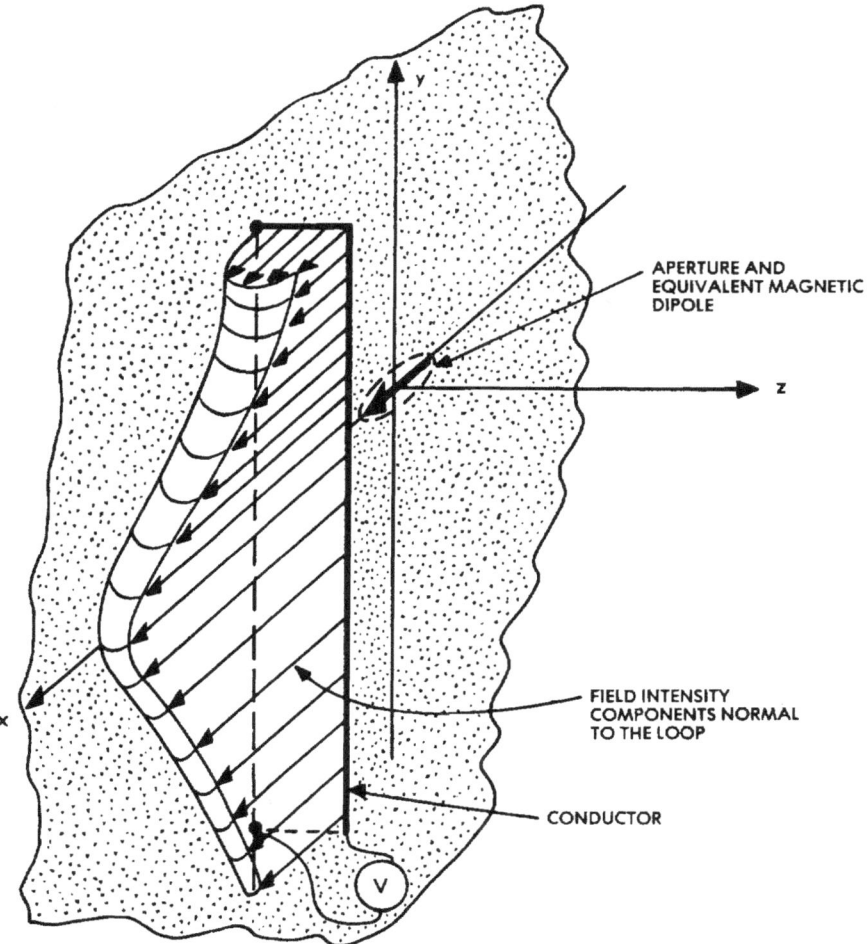

Figure 11.8 Determination of the voltage produced by aperture-coupling.

surface. The program calculates the field strength normal to that surface and performs the graphical integration suggested in Figure 11.8 and necessary to determine the total magnetic flux passing through the plane of the loop.

If the aperture under consideration is not elliptical, a corresponding equivalent elliptical aperture can generally be specified, as shown in Figure 11.9. The equivalent elliptical aperture would have the same area and the same eccentricity as those of the aperture under study. If the aperture were of an irregular shape, its area might have to be determined numerically.

An example of an aperture-coupled field is shown in Figure 11.10 in which the elliptical fuselage described earlier has been approximated by two sheets of infinite size, one containing the aperture and another serving as the reflecting surface. The figure shows only the top half of the field pattern, the

339

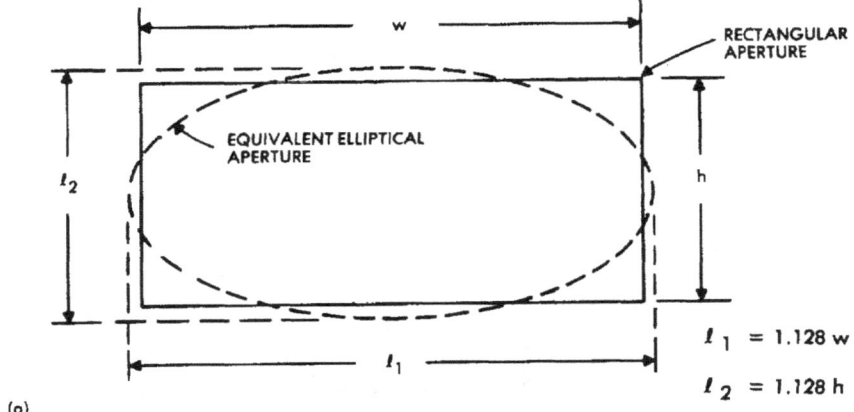

$l_1 = 1.128\,w$

$l_2 = 1.128\,h$

(a)

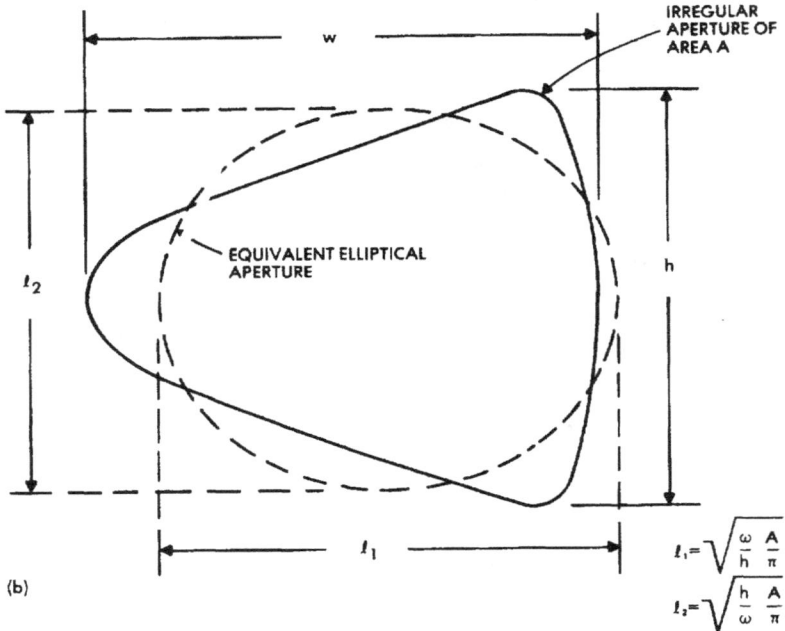

$l_1 = \sqrt{\dfrac{\omega}{h}\dfrac{A}{\pi}}$

$l_2 = \sqrt{\dfrac{h}{\omega}\dfrac{A}{\pi}}$

(b)

Figure 11.9 Equivalent apertures.
    (a) Rectangular aperture
    (b) Irregular aperture

field in the bottom half being symmetrical. The aperture considered was 0.2 m high by 0.1 m wide, the long axis of the ellipse being oriented at right angles to the plane of the figure. The field strength at the surface of the aperture, assuming the aperture not to be there, was 167 A/m, the field strength that would be produced by the aforementioned current of 1000 A flowing axially along the elliptical fuselage. The field strength is seen to decay very rapidly with distance away from the aperture. In this particular case, the reflecting surface was sufficiently far from the aperture under study that the field through most of the interior volume would have been virtually the same had the reflecting surface been disregarded. Doubling the size of the aperture leads to the field pattern shown in Figure 11.11. Roughly in accordance with simplest theory, doubling the dimensions of the aperture leads to an 8-fold increase in the field strength at points removed from the aperture. Looked at in another light, if one considers the volume in which the magnetic field is of a given strength or greater, doubling the lineal dimensions of the aperture will double the volume through which the given field exists.

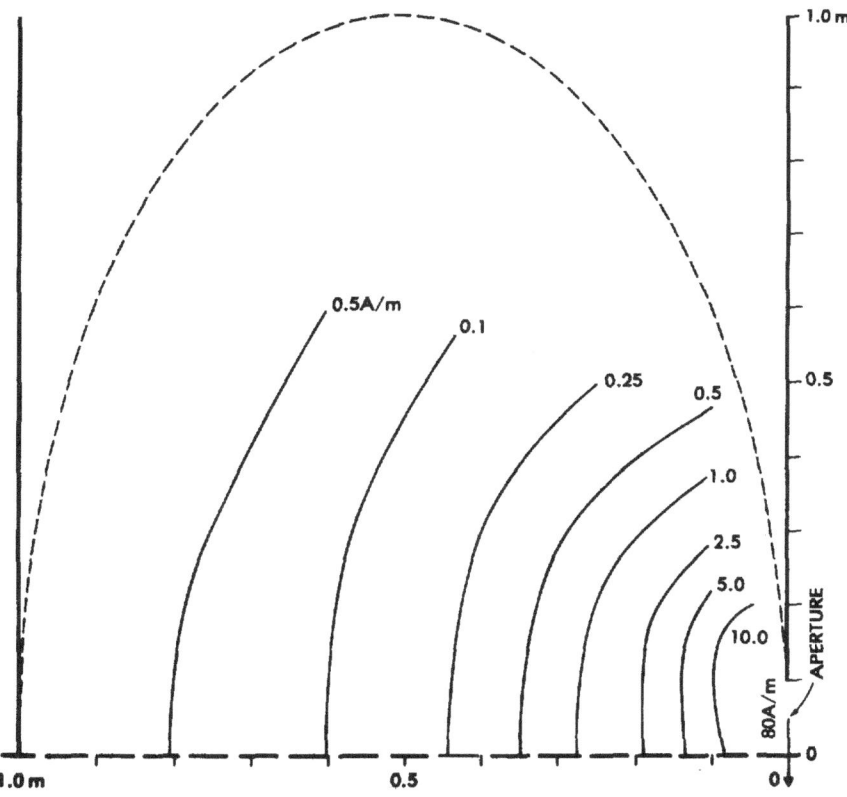

Figure 11.10 Fields coupled through an aperture.
Major axis oriented vertically
(0.2 x 0.1 m aperture)

341

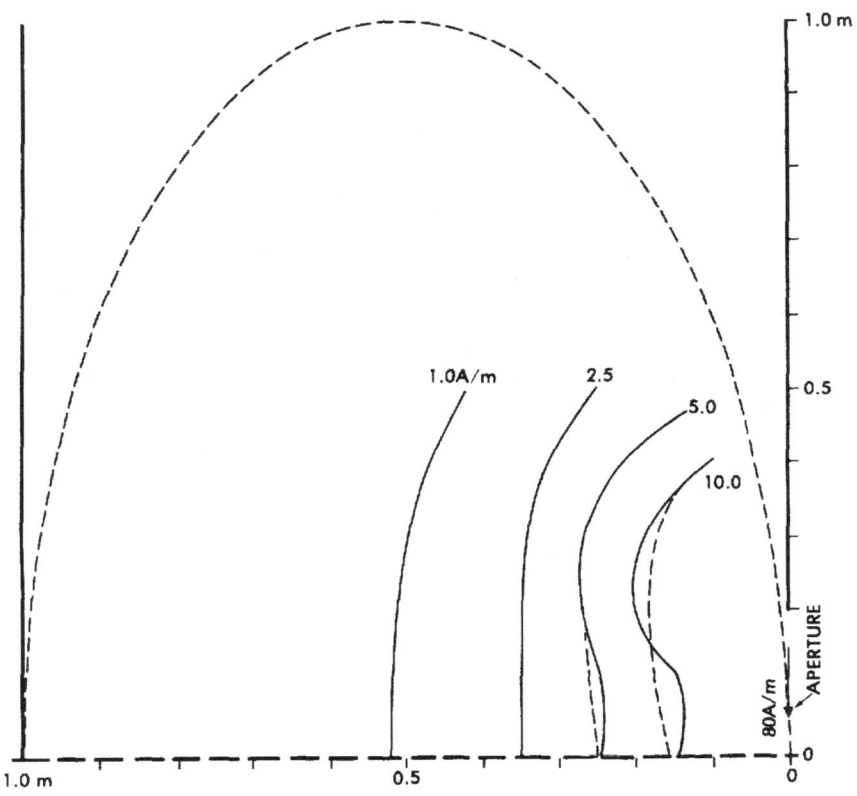

Figure 11.11 Fields coupled through an aperture.
Major axis oriented vertically
(0.4 x 0.2 m aperture)

11.2 Electric Fields

Unlike magnetic fields, which may penetrate into an interior region either through apertures or by diffusion through continuous walls, electrical fields, as a practical matter, can couple only through apertures. Any metallic surface provides sufficient shielding against electric fields that considerations of direct penetrations of the field through a metal surface can be neglected. Electric fields of practical consequence can exist only perpendicular to a metallic surface, since in a manner analogous to penetrating magnetic fields, electric fields of oblique incidence have all but the perpendicular component of the field shorted out by the low impedance of the metallic surface. The appropriate geometry for electric field coupling is then as shown on Figure 11.12. The electric fields of the interior region may be calculated as though they were produced by an electric dipole of strength (Reference 11.5):

342

$$D = a_{33} E_z \tag{11.33}$$

$$a_{33} = -\frac{2\pi}{3} \left[\frac{\ell_1}{2}\right]^3 \frac{(1-e^2)}{E(e^2)} = \left[\frac{\ell_1}{2}\right]^3 a'_{33} \tag{11.34}$$

The quantity a'_{33} has previously been presented in Figure 11.3. Subject to the provisions that the dimensions of the aperture be small compared to the wavelengths of the highest frequency component in the external electric field and that electric fields produced by changing magnetic fields can be neglected, the electric field in the interior region becomes

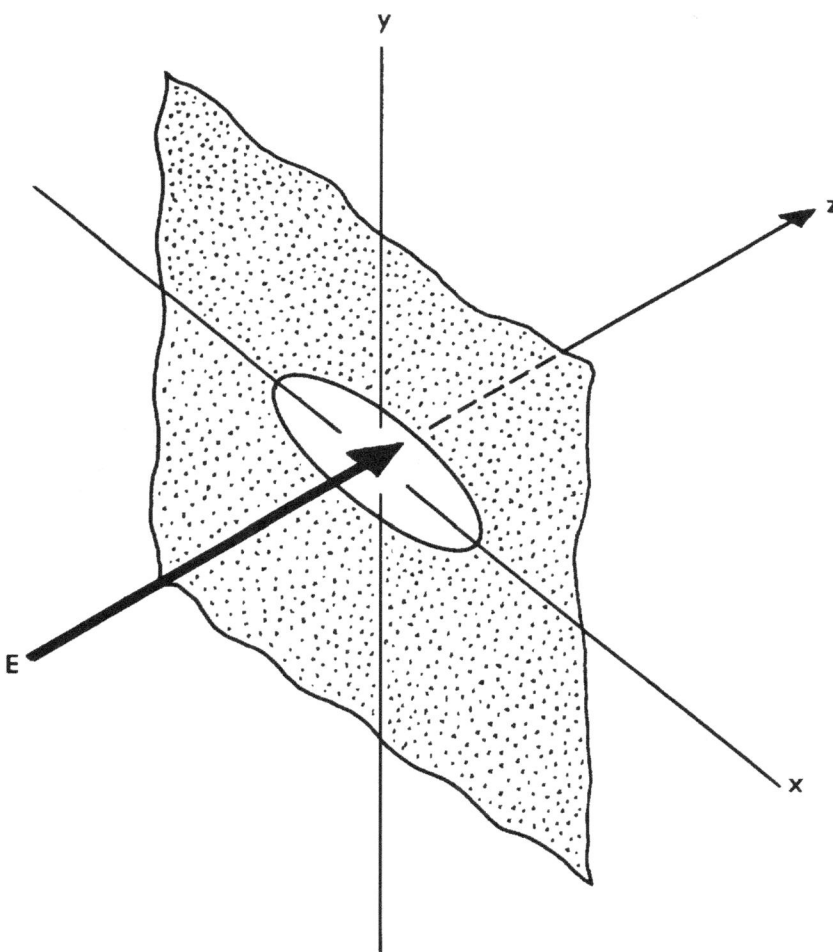

Figure 11.12   Geometry appropriate to electric field coupling.

$$E_x = \frac{D_o}{4\pi} \; \frac{3XZ}{r^5} \qquad\qquad (11.35)$$

$$E_y = \frac{D_o}{4\pi} \; \frac{3YZ}{r^5} \qquad\qquad (11.36)$$

$$E_z = \frac{D_o}{4\pi} \; \frac{3Z^2 - r^2}{r^5} \qquad\qquad (11.37)$$

These expressions are based upon the further assumption of a pure dipole and, as with Equations 11.6 through 11.11 for the magnetic field formulation, would indicate infinite field strengths as one approached the plane of the aperture, whereas in fact the field in the plane of the aperture would be one-half the field strength that would exist at that point if the aperture were absent. A power series expansion of the field from a dipole of finite size could be prepared but, as of this writing, has not been. The bookkeeping routines necessary to handle electric fields are virtually identical with those used for magnetic fields and could be incorporated into the APERTURE program, although, as of this writing, they have not been so included.

11.3 Magnetic Field Zones

From the foregoing it can be seen that the task of mapping the magnetic field inside a vehicle and actually calculating the voltage on any particular circuit could be a formidable task. A possible solution to the problem lies in recognizing that aircraft, or at least aircraft within a particular category, tend to possess characteristic zones. On a fighter aircraft, for example, the cockpit can be regarded as a magnetically open zone exposed primarily to aperture-coupled magnetic fields. Within reasonable limits, all fighter aircraft probably have approximately the same magnetic field in the cockpit. Another type of equipment zone characteristic of fighters would be equipment bays located in the forward section and behind the radome. All such equipment bays tend to be alike, the differences, perhaps, relating mostly to the type of fasteners used to hold the covers in place. The structure within a wing is a type of magnetic field structure fundamentally different from either the cockpit or the forward equipment bays. Accordingly, it would seem possible to divide an aircraft into a relatively small number of typical zones, to assign a ruling or characteristic magnetic field intensity to those zones, and to provide rather simplified tables of nomograms listing the characteristic transient likely to be induced in wiring of a given length.

This concept of dividing an aircraft into shielding zones was first used on the *Space Shuttle* (Reference 11.6). The zones so defined are shown in Figure 11.13 (Reference 11.7). The electromagnetic fields assigned to each of these zones were initially determined by "seat of the pants" engineering by a group of engineers knowledgeable in the field. These magnetic fields were then refined

344

during the course of an extensive analytical investigation. The magnetic field amplitudes assigned to each of these zones, based on the analytical study, are given in Table 11.1 (Reference 11.8). The fields were divided into two components, an A-component referring to fields coupled through apertures and a B-component referring to fields coupled by diffusion through metal surfaces. The A-component of the field would tend to have the same rapidly changing waveshape as the external magnetic field, while the B-component would have a much slower waveshape. In the *Space Shuttle* study the waveforms of the different components were taken to be as shown on Figure 11.14 (Reference 11.9). In each case the field intensity was based on a worst case 200 kA lightning

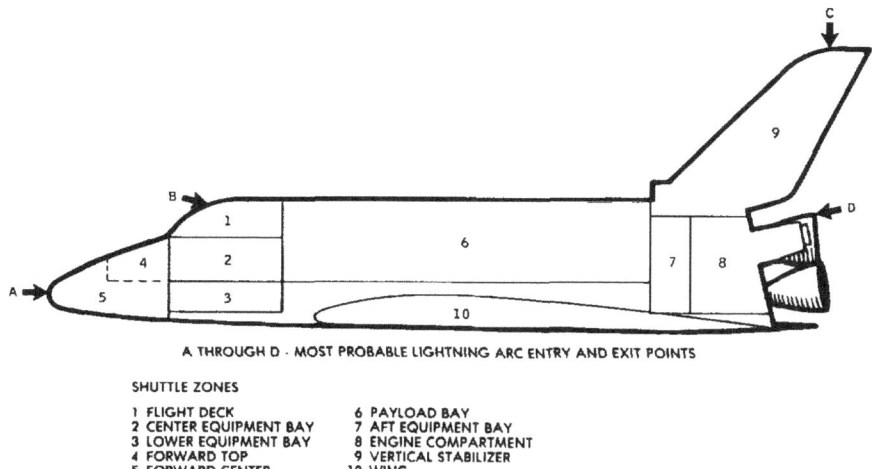

A THROUGH D - MOST PROBABLE LIGHTNING ARC ENTRY AND EXIT POINTS

SHUTTLE ZONES

1 FLIGHT DECK	6 PAYLOAD BAY
2 CENTER EQUIPMENT BAY	7 AFT EQUIPMENT BAY
3 LOWER EQUIPMENT BAY	8 ENGINE COMPARTMENT
4 FORWARD TOP	9 VERTICAL STABILIZER
5 FORWARD CENTER	10 WING

Figure 11.13 Shielded zones within the *Orbiter* structure.

current passing through the *Orbiter* vehicle. The field amplitudes of Table 11.1 were the maximum amplitudes calculated for any of the possible lightning current entry or exit points. While no particular claim for accuracy can be made about any individual point within the *Orbiter*, the field amplitudes at least seem reasonable.

The concept of dividing an aircraft structure into different magnetic field zones and assigning a ruling magnetic field strength to each zone, while imperfectly formulated at present, is fundamentally no different from the civil engineering practice of designing a structure to withstand a standard (generally worst case) wind loading. While the wind loading may differ widely at different points on the structure, the task of calculating the wind loading on each and every structural member would probably be sufficiently expensive that it would offset the savings that one might realize by tailoring each structural member to its own specific wind loading.

**Table 11.1 MAGNETIC FIELDS IN DIFFERENT ZONES
OF THE *SPACE SHUTTLE***

Zone	Aperture Fields A-component (A/m)	Diffusion Fields B-component (A/m)
1	1200	800
2	60	200
3	0	200
4	50	150
5	50	100
6	280	300*
		150**
7	50	570
8	200	680
9	200	3700
10	65	300

*Payload
**No payload

The task of dividing aircraft into typical magnetic field zones and of calculating and assigning the appropriate field strength to each of the zones is probably of sufficient complexity that it should be done by specialists in the field.

Analytical studies of field intensity could well be supplemented by experimental studies in which currents were circulated through an aircraft and the magnetic fields inside measured. Aircraft used for such studies should be in realistic condition in regard to the mechanical soundness of the structures, particularly relating to the access panels, but otherwise they need not contain complete electronic systems. They should not, however, have been cannibalized unduly, leaving an excessive number of empty bays where equipment might normally have been expected to help provide some electromagnetic shielding capability. The use of older aircraft for such studies might well be realistic in that the electromagnetic shielding capability of a structure is in large measure determined by the electrical characteristics of the various access panels and fasteners. Because these tend to degrade with age, an older aircraft might be a more realistic test vehicle than a new and carefully maintained aircraft.

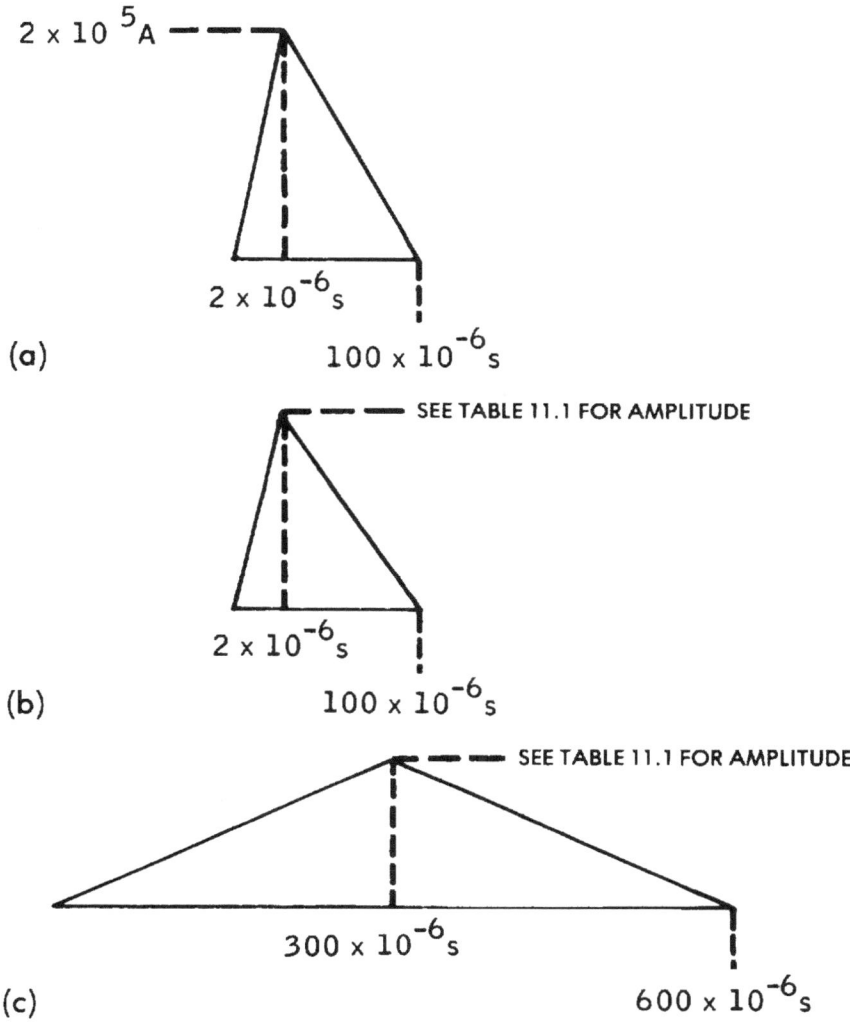

2×10^5 A

2×10^{-6} s

100×10^{-6} s

(a)

SEE TABLE 11.1 FOR AMPLITUDE

2×10^{-6} s

100×10^{-6} s

(b)

SEE TABLE 11.1 FOR AMPLITUDE

300×10^{-6} s

600×10^{-6} s

(c)

Figure 11.14 Waveforms of aperture- and diffusion-coupled magnetic fields.
    (a) Lightning current
    (b) Aperture-coupled field, A-component
    (c) Diffusion-coupled field, B-component

REFERENCES

11.1 C. D. Taylor, "Electromagnetic Pulse Penetration Through Small Apertures," *IEEE Transactions on Electromagnetic Capability*, EMC-15, 1, Institute of Electronic and Electrical Engineers, New York, New York (February 1973): 17-26.

11.2 C. D. Taylor, "Electromagnetic Pulse Penetration Through Small Apertures," Interaction Note 74, *Electromagnetic Pulse Interaction Notes*, 5, Air Force Weapons Laboratory, Kirtland Air Force Base, Albuquerque, New Mexico (March 1973).

11.3 K. J. Maxwell, F. A. Fisher, J. A. Plumer, and P. R. Rogers, *Computer Programs for Prediction of Lightning Induced Voltages in Aircraft Electrical Circuits*, Air Force Flight Dynamics Laboratory, Air Force Systems Command, Wright-Patterson Air Force Base, Ohio (February 1975).

11.4 K. J. Maxwell *et al, Computer Programs*, Section 3.

11.5 C. D. Taylor, "Electromagnetic Pulse Penetration."

11.6 *Space Shuttle Lightning Protection Criteria Document*, JSC-07636, National Aeronautics and Space Administration, Lyndon B. Johnson Space Center, Houston, Texas (September 11, 1973).

11.7 *Space Shuttle Program Lightning Protection Criteria Document* JSC-07636, Revision A, National Aeronautics and Space Administration, Lyndon B. Johnson Space Center, Houston, Texas (November 4, 1975), p. F-7.

11.8 *Space Shuttle Program Lightning Protection Criteria Document*, Revision A, p. F-6.

11.9 *Space Shuttle Program Lightning Protection Criteria Document*, Revision A, p. F-8.

CHAPTER 12
VOLTAGES AND CURRENTS INDUCED ON WIRING

12.1 Elementary Considerations

The fields by themselves are of little concern; of primary concern are the voltages and currents produced by the changing magnetic fields. In principle, these may be calculated from the geometry of the aircraft wiring and knowledge of the strength and orientation of the internal magnetic and electric fields. While calculation of the voltages and currents on actual wiring in aircraft may never be practical, because of the mechanical complexity of the wiring harnesses, calculations may be performed for simplified geometries. Such calculations will illustrate the scope of the problem and indicate the direction of practices that minimize the voltages and currents, and hence minimize the risk of circuit damage or upset.

The simplest geometry to consider is that of a conductor, or group of conductors, placed adjacent to a metal surface and exposed to a uniform magnetic field oriented to produce maximum voltage in the wiring. The geometry is shown in Figure 12.1(a).

If only common-mode voltages are considered and only cable systems short enough that transmission line effects need not be considered, the induced voltage will be

$$e = \frac{d\phi}{dt} = \mu_0 A \frac{dH}{dt} \tag{12.1}$$

where A = area of the loop involved in meters squared
μ_0 = $4\pi \times 10^{-7}$ in henries per meter (permeability of free space)
ϕ = total flux linked in webers
H = magnetic field intensity in amperes per meter
t = in seconds
e = in volts

Expressed in inch units:

$$e = 8.11 \times 10^{-10} \, \ell h \, \frac{dH}{dt} \tag{12.2}$$

where ℓ = length of cable bundle in inches
h = height above ground plane in inches
H = magnetic field intensity in amperes per meter
t = in seconds

349

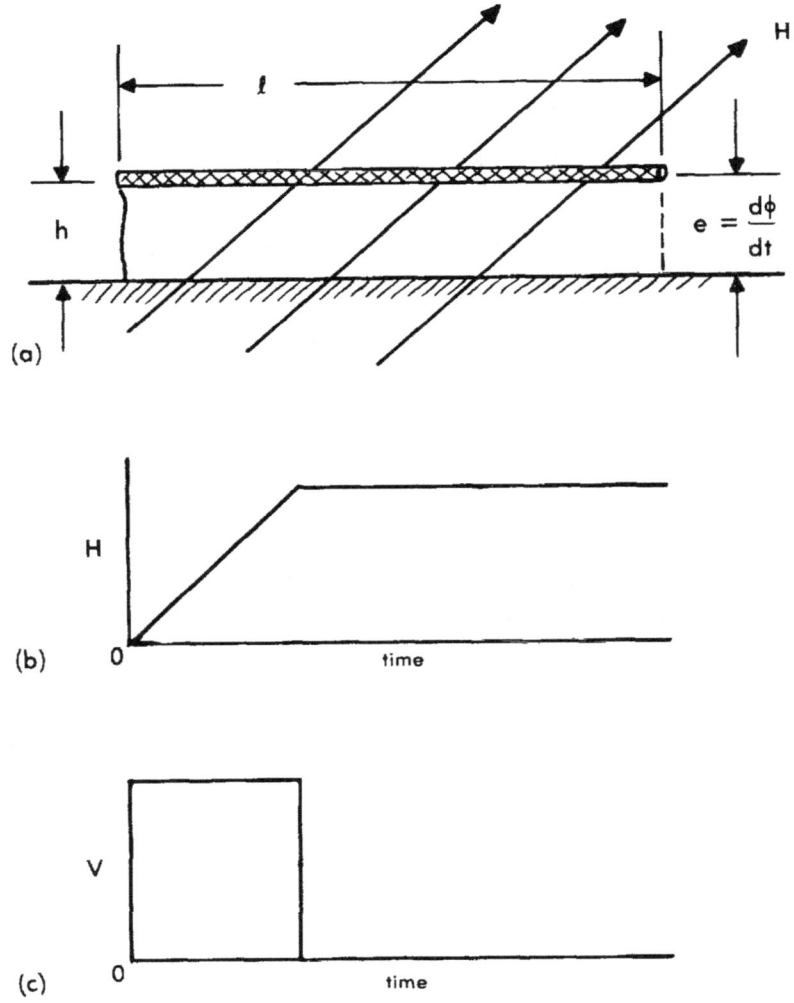

Figure 12.1 Response to a changing magnetic field: open circuit voltage.

(a) Geometry
(b) Magnetic field waveshape
(c) Voltage waveshape

It must be emphasized that the voltage so calculated is that existing between the entire group of conductors (comprising the cable) and the vehicle structure. The voltage will divide between the loads at the ends of the cable inversely as the impedance of the loads. For worst case analysis, consider one end of the cable grounded with the other end open-circuited. All the voltage so developed will appear at the open circuit end of the cable.

Line-to-line or circuit voltages will be less, generally by a factor of 10 to 200, or 20 to 46 dB, down from the common-mode voltages because individual conductors are usually close together and are often twisted, thus reducing the total loop area.

The maximum cable current is that which flows when both ends of the cable are connected to the vehicle structure through a low or zero impedance. Such an impedance may be an overall shield grounded at each end, or it may be a group of semiconductor circuits, each having low input impedance. In the first case, the current will flow on the overall shield with the current on the input circuits determined by the shielding properties of the shield. In the second case, the current will flow directly through the input semiconductors and their bias sources.

The short circuit current that flows (Figure 12.2) may be determined from the familiar expression

$$e = L \frac{di}{dt} \qquad (12.3)$$

whence

$$I = 1/L \int e\, dt \qquad (12.4)$$

where
- I = amperes
- L = self-inductance of cable in henries
- e = open circuit induced voltage in volts
- t = time in seconds

Cable inductance may be estimated from the expression

$$L = 2 \times 10^{-7} \log_e \frac{4h}{d} \; H/m \qquad (12.5)$$

or

$$5.08 \times 10^{-9} \log_e \frac{4h}{d} \; H/inch \qquad (12.6)$$

where
- h = height above a ground plane
- d = conductor diameter

The induced voltage, e, which drives the current, is proportional to the cable height, but the cable inductance, which impedes the flow of current, is proportional to the logarithm of the cable height.

12.2 Transmission Line Effects

12.2.1 Response to Magnetic Fields

Conductors always have associated with them some distributed capaci-

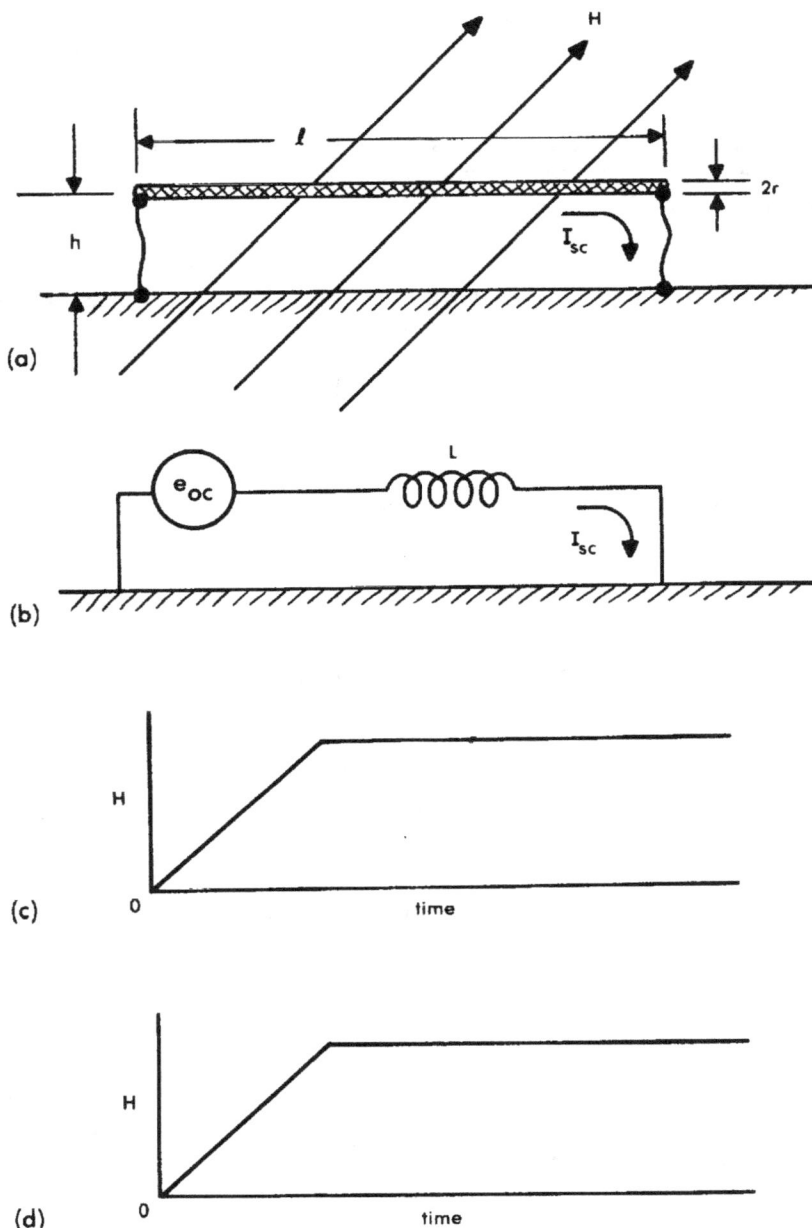

Figure 12.2 Response to a changing magnetic field: short circuit current.

(a) Geometry
(b) Equivalent circuit
(c) Magnetic field waveshape
(d) Current waveshape

tance and inductance, the values of which are determined by the size of the conductors and the distance of the conductors from adjacent ground planes and other conductors. When these are considered, the effect of a changing magnetic field is to produce an oscillatory open circuit voltage. As shown in Figure 12.3, these oscillations will be superimposed on an underlying voltage proportional to the rate of change of the magnetic field. When the internal magnetic field is of complex waveshape (as is the usual case) and not the idealized ramp function shown in Figures 12.1 and 12.3, the resulting open circuit voltage may be of a very complex nature. A few examples will be shown in Chapter 14. While the maximum voltage may be difficult to predict, given the complex nature of the superimposed oscillations, the amplitude of the envelope can at least be approximated from Equations 12.1 and 12.2. The frequency of the superimposed oscillations tends to be inversely proportional to the conductor length. Conductors, such as shields, grounded at one end tend to ring as quarter-wave dipoles: for example, a conductor 10 m long tends to ring at 7.5 MHz. Even this simple relationship is difficult to apply, since one conductor is seldom free of the influence of adjacent conductors.

Aircraft wiring is grouped into bundles, the bundles usually containing both short and long conductors. The assembly, even if exposed to a magnetic field of simple waveshape will oscillate in a complex manner. Generally there will be one dominant frequency with several other frequencies, usually higher, superimposed. Each will have its own characteristic decrement. Almost the only reliable generalization is that the cables associated with large aircraft will be longer than the cables associated with small aircraft and will characteristically oscillate at lower frequencies. On fighter aircraft, measurements of induced voltages have shown the characteristic frequencies to be in the range 1 to 10 MHz.

Currents measured on bundles of conductors tend to be oscillatory, like the voltages, as long as the conductors are part of a wiring group employing a single-point ground concept. If the conductors are part of a wiring group employing a multiple-point ground concept, the conductor currents tend not to be oscillatory but to follow the underlying shape of the internal magnetic field.

12.2.2 Response to Electric Fields

When the aircraft is struck by lightning, it is forced to assume the electric field potential of the lightning flash at that point, perhaps 100 MV with respect to ground. The magnitude of voltage with respect to ground is of no real importance, but the electric field intensity at the surface of the aircraft is of importance. This electric field intensity depends upon the total amount of charge stored on the aircraft surface and the size of the aircraft surface, and could, in principle, be extremely high. In practice, the electric field intensity at the surface cannot exceed more than about 500 kV/m before intense electrical discharges take place from the aircraft surface into the surrounding air. The electric field strength will be greatest about points of small radius of curvature. The discharges from those points will tend to reduce the electric field strength at

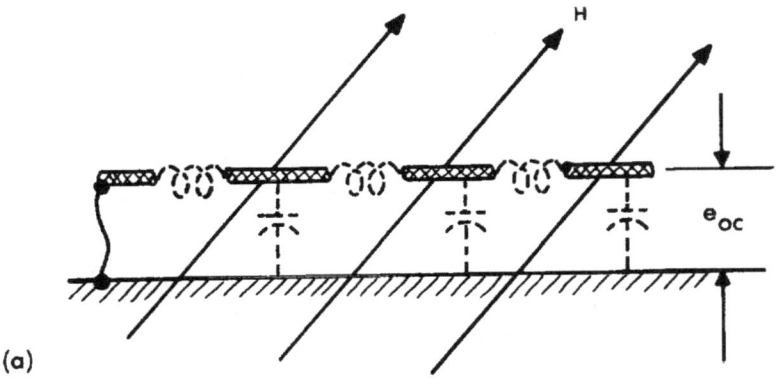

(a)

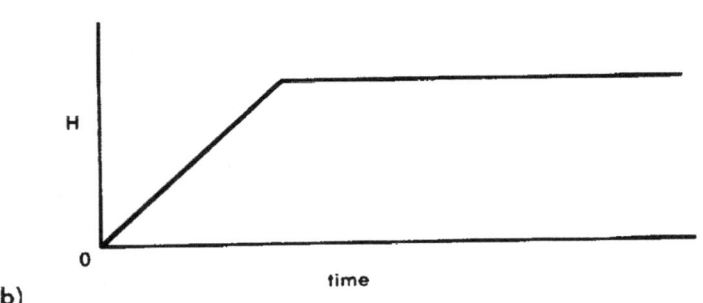

(b)

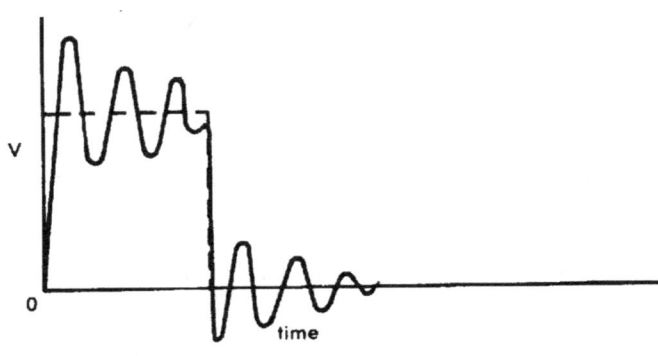

(c)

Figure 12.3 Oscillatory voltage response excited by a changing magnetic field.
(a) Geometry
(b) Magnetic field waveshape
(c) Voltage waveshape

354

other points on the aircraft. As a result, the maximum field stress to which most points of the aircraft can be raised is probably on the order of 100 kV/m. This electric field will be radial with respect to the metal surface of the aircraft. In places where there are apertures, the electric field lines will pass through the apertures and will produce electric fields on the inner surface of the aircraft.

The response of a conductor to a suddenly applied electric field is shown in Figure 12.4. If a conductor is a height, h, above a metal surface upon which an electric field is incident, the conductor will assume the potential of the field at that point:

$$e = hE_i \qquad (12.7)$$

This holds true only if the conductor is isolated from ground. If the conductor is completely grounded, as shown in Figure 12.4(b), the voltage on the conductor will be zero. If we were dealing only with static electric fields, any ground on the conductor would reduce the conductor voltage to zero. However, here we are dealing with conductors having distributed inductance and capacitance, and exposed to a suddenly applied electric field. Accordingly, the electric field applied at point 1 will raise the conductor to the voltage $e = hE_i$, and the voltage will maintain itself at that level until such time as a voltage wave has had time to propagate from point 1 to the ground conductor at the left and back to point 1. Thus, at point 1 the conductor experiences a square-wave voltage lasting for twice the travel time required for the wave to travel from point 1 to the grounded end and back. The electric field applied at point 2 likewise raises the conductor at point 2 to $e = hE_i$, and it again remains there until a wave has had time to propagate to the grounded end and back. Since less of the conductor is involved, this voltage lasts for a shorter period of time. Nearing the limit, the field applied at point 4 can raise the conductor to a voltage of $e = hE_i$ for only a very short period of time.

As the voltage waves that are produced return to the open end of the conductor, they again experience a reflection. The voltage waves reflect back and forth between the open and grounded ends, reversing their polarity at the grounded end and doubling in amplitude at the open end. Thus, the voltage at the open end is oscillatory (Figure 12.5), the amplitude of the oscillation decreasing with time as energy is lost in the resistance of the conductor or radiated into surrounding objects.

Associated with these traveling waves of voltage are traveling current waves. The degree to which the waves are excited depends not only upon the magnitude of the incident electrical field but upon its rate of change. Fields with longer rise times are proportionately less effective in exciting oscillations. If the conductor of Figure 12.4 is viewed as the shield on a group of wires, the current and voltage on the shield will be oscillatory. The commonly accepted practice of grounding shields at only one end is based on the premise that the electric fields to which the shielded conductors will be exposed are of such low frequency that all points of the shield may be held at ground potential with only the one ground point. When the rapidly changing electric field associated with lightning

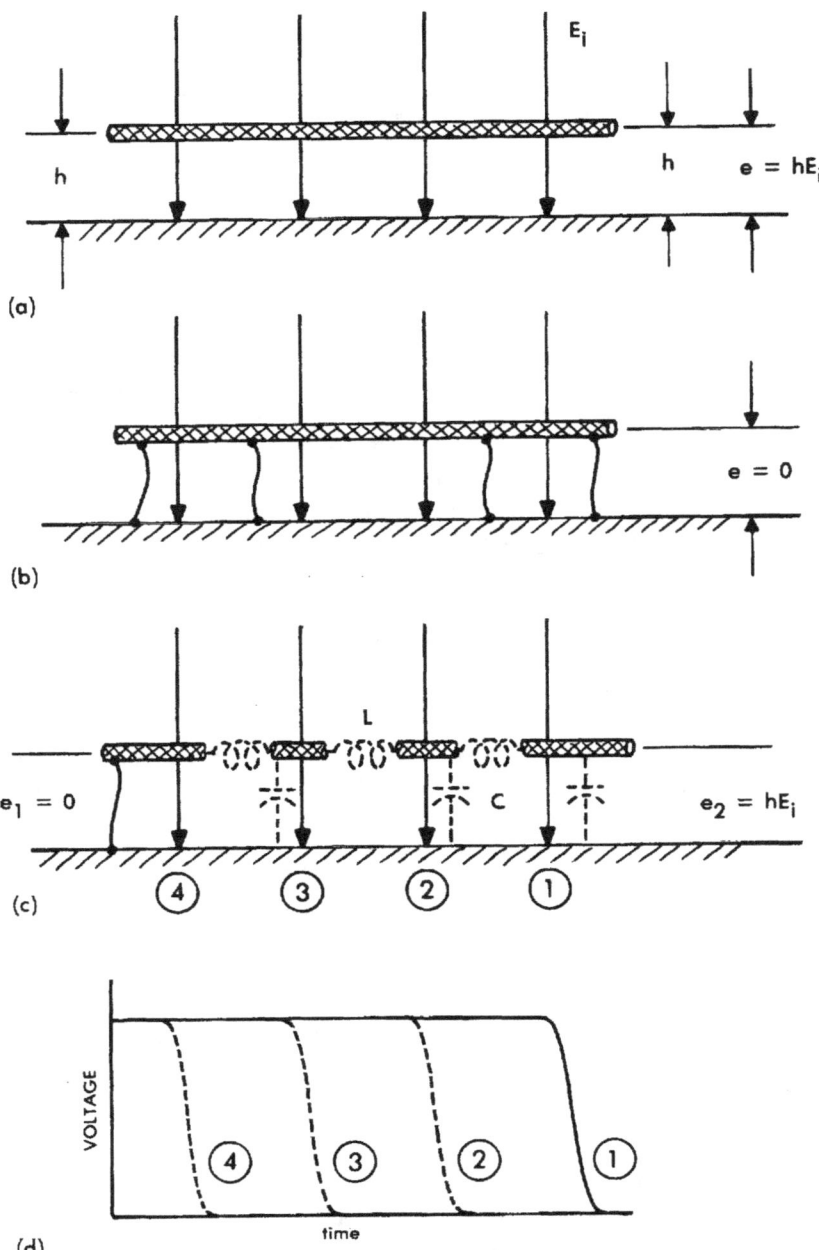

Figure 12.4 Response to a suddenly applied electric field.

(a) Ungrounded conductor
(b) Well-grounded conductor
(c) Conductor grounded at one end
(d) Waveshape at indicated points

356

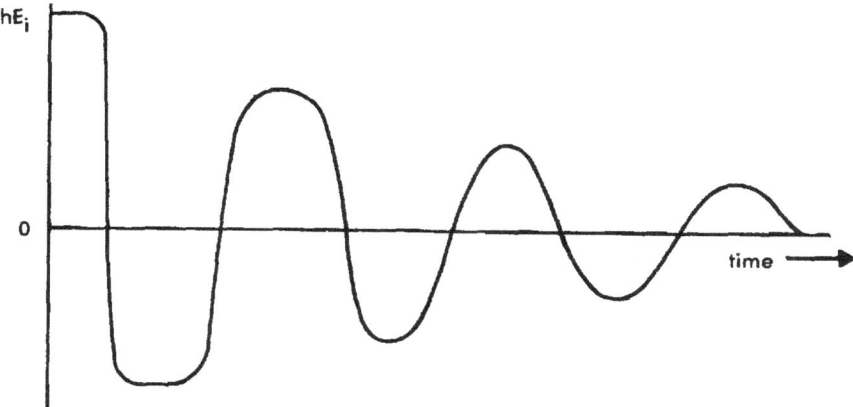

Figure 12.5 Oscillatory response excited by a suddenly applied electric field.

is present, frequently this condition is not met.

12.3 Aids for Calculation of Voltage and Current

In Section 11.3 there was introduced the concept of magnetic field zones, according to which in any particular area of the aircraft the total magnetic field may be viewed as the direct superposition of two components, one the result of coupling through apertures and the other the result of diffusion-type coupling. If the assumption of a ruling field is made, there may be derived some simplified relationships between the magnetic field and the resulting induced voltages and currents. Additional assumptions are the following:

- The conductor is of length ℓ, diameter d, and spaced a height, h, above a ground plane.
- The magnetic field is oriented to produce maximum voltages in the conductor.
- One end of the conductor is grounded.
- The magnetic fields are of the shape shown in Figure 11.14.

These are the same assumptions as those illustrated in Figures 12.1 and 12.2, and upon which Equations 12.1 and 12.2 were based. Under these assumptions

$$e_{oc} = K_1 \ell h H \qquad (12.8)$$

where

$$K_1 = 0.63 \text{ if } \ell \text{ and h are in meters}$$
$$K_1 = 0.63 \times 10^{-4} \text{ if } \ell \text{ and h are in centimeters}$$
$$K_1 = 0.41 \times 10^{-3} \text{ if } \ell \text{ and h are in inches}$$

357

In all cases H is expressed in amperes per meter. The waveshape of the open circuit voltages would typically be proportional to the derivative of the H field, and hence oscillatory.

If we assume the conductor to be grounded at each end

$$I_{sc} = K_2 hH \qquad (12.9)$$

where K_2 is given by either Figure 12.6 or 12.7, according to the units used for measurement of the conductor. Conductor length does not influence short circuit current. The waveshape of the short circuit current tends to be the same as that of the incident magnetic field.

The height, h, of the cable bundle above a ground plane is difficult to specify because the ground plane is seldom purely a plane surface and because cable bundles are frequently strapped directly to a supporting structure. For purposes of this analysis, assume

- That the height, h, is measured to the nearest substantial metallic structural member
- That if the cable bundle is laid directly on that member, h equals one-half the cable diameter
- That if the cable bundle is elevated above the metallic structural member, h equals the clear height above the member plus one-half the cable diameter. (If the cable height differs along its length, use an average height.)

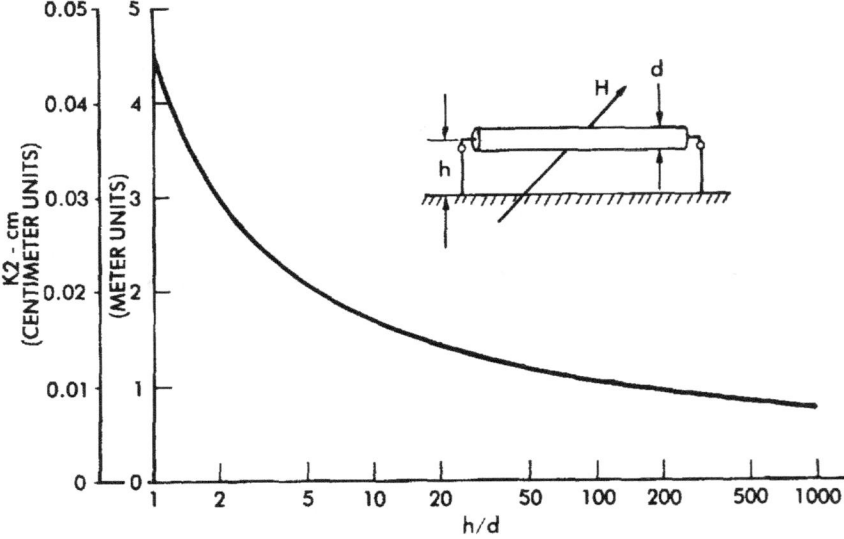

Figure 12.6 K_2 – metric units.

358

12.4 Voltage and Currents in the *Space Shuttle*

Based upon Equations 12.1 to 12.8, Figure 11.14, and Table 11.1, one may calculate the voltages and currents that would be developed on typical wiring. The results are shown on Table 12.1 (Reference 12.1).

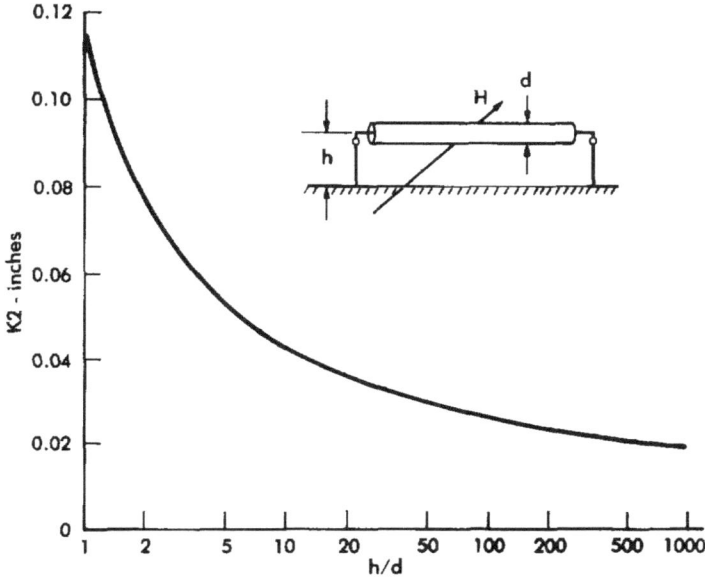

Figure 12.7 K_2 – inch units.

Table 12.1 OPEN CIRCUIT VOLTAGE AND SHORT CIRCUIT CABLE CURRENT IN THE VARIOUS ZONES OF THE *SPACE SHUTTLE*

Zone	h = 1 in. Voltage (volts)	(0.0259 m) Current (amperes)	h = 2 in. Voltage (volts)	(0.0508 m) Current (amperes)	h = 5 in. Voltage (volts)	(0.1270 m) Current (amperes)	h = 10 in. Voltage (volts)	(0.254 m) Current (amperes)
1	76.90	133.3	153.8	184.9	384.6	321.8	769.1	521.1
2	3.92	7.10	7.84	9.42	19.59	16.39	39.18	26.54
3	0.085	0.154	0.170	0.204	0.425	0.356	0.850	0.576
4	3.25	5.89	6.51	7.82	16.27	13.62	32.54	22.05
5	3.23	5.85	6.46	7.76	16.14	13.51	32.29	21.87
6	17.99	32.59	35.99	43.26	89.96	75.28	179.9	121.9
7	3.43	6.21	6.87	8.26	17.16	14.36	34.33	23.26
8	13.05	23.64	26.11	31.38	65.27	54.62	130.5	88.44
9	14.33	25.96	28.66	34.45	71.65	59.96	143.3	97.09
10	4.27	7.74	8.55	10.28	21.37	17.88	42.75	28.96

*All values based on cable length of 157.48 inches (4 m) and diameter of 1 inch (0.0254 m). For other lengths, scale the voltage proportionately.

359

REFERENCE

12.1 *Space Shuttle Program Lightning Protection Criteria Document*, JSC-07636, Revision A, National Aeronautics and Space Administration, Lyndon B. Johnson Space Center, Houston, Texas (November 4, 1975) p. F-6.

CHAPTER 13
EFFECTS OF SHIELDS ON CABLES

13.1 Introduction

If electronic equipment is to be operated in a region where there are changing electromagnetic fields and if experience or analysis indicates that the currents and voltages induced by those fields may be harmful, the most straightforward approach toward achieving transient compatibility is to shield interconnecting wiring from those fields. In aircraft, shielded conductors are most commonly used to control low-level and low-frequency interference signals, For many good and legitimate reasons shields on cables are often grounded at only one end. If the wiring is subjected to the rapidly changing electromagnetic fields associated with lightning, however, shields grounded at only one end may not be effective, and in some cases can make the surge voltages even larger than they would be if the conductors were not shielded in the first place. In this section of the report we shall discuss the ways in which shielded cables reduce surge voltages, some of the ways in which common practice regarding the use of shields may be at variance with the use required for the control of lightning effects, and some of the ways in which this apparent conflict of use may be resolved. We shall also discuss how the noise currents flowing on shields of cables may be related to the noise signals coupled onto the signal conductors.

13.2 Some Elementary Illustrations

To illustrate some of these effects let us consider a series of tests (Reference 13.1) that were made on a 5m long length of RG-58/U coaxial cable. The cable was placed adjacent to a metal ground plane and a magnetic field passed between the cable and the ground plane. The magnetic field was not a distributed field but was confined to the core of a pulse injection transformer in the manner shown in Chapter 17. Measurements were made of the voltage between the center conductor and ground at each end of the cable. These voltages thus represent the common-mode voltages that would exist. The first set of results is shown in Figure 13.1. In an unshielded conductor or one in which the shield on a conductor is not used, equal and opposite voltages appear at the two ends if all of the loading impedances are balanced. The voltages at the two ends of the conductor are of equal amplitude but opposite polarity, as would be produced if the conductor were considered to have an equivalent voltage generator at its center, as shown in Figure 13.1(c). If the load impedances at the ends are unbalanced by the addition of a 50 Ω resistor at one end (Figure 13.2), the total voltage induced around the loop remains unchanged, but most of it appears at the end with the highest impedance. With reference to the equivalent circuit of Figure 13.2(c), the fact that there is any voltage at V_2 implies the existence of some capacitive loading as well as the desired resistance load; otherwise there would be no voltage across V_2.

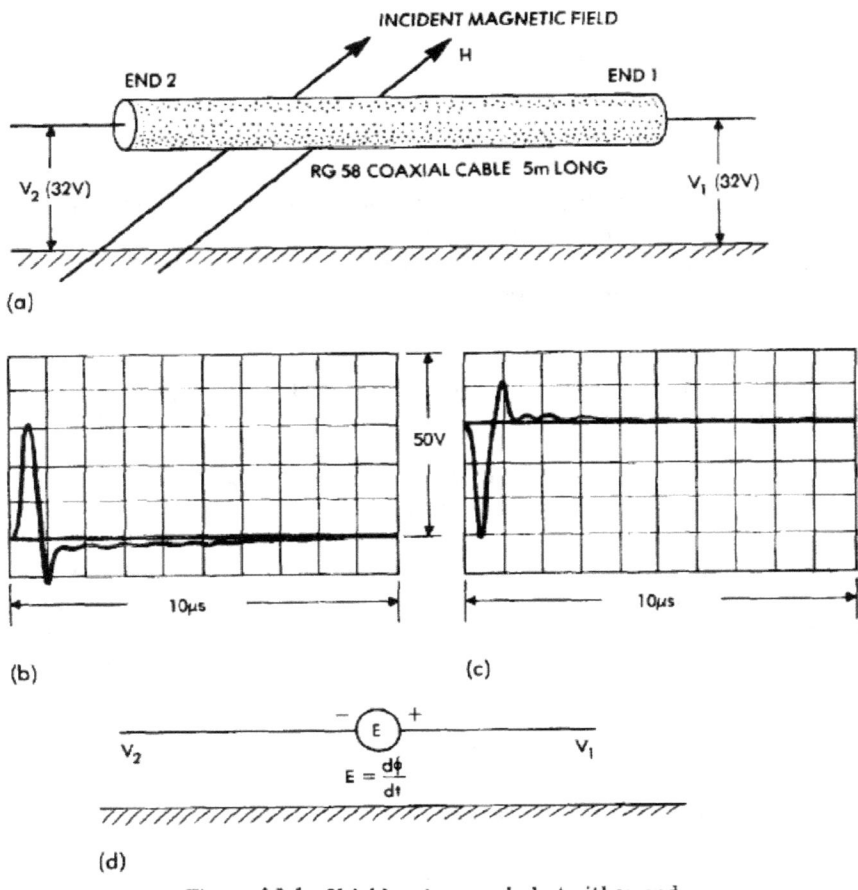

Figure 13.1 Shield not grounded at either end.
    (a) Test conditions
    (b) V_1
    (c) V_2
    (d) Equivalent circuit

Grounding the shield at one end, shown in Figure 13.3, does not significantly affect the common-mode voltage at the other end. The changing field induces a voltage between the open end of the shield and ground. The conductor is exposed to the same field, and there is thus the same voltage developed between the ends of the conductors as that between the ends of the shield. Because of the unbalanced load impedances, all this voltage must appear between the conductor and ground at end 1. Leaving aside considerations of unequal load impedance, the shield can reduce voltage at one end only by increasing it at the other end. This effect is shown in Figure 13.4. The reason the shield reduces the line-to-ground voltage at V_2 is that the capacitance between

362

shield and conductor at the grounded end loads the signal conductor, just as did the 50 Ω resistor on Figure 13.3.

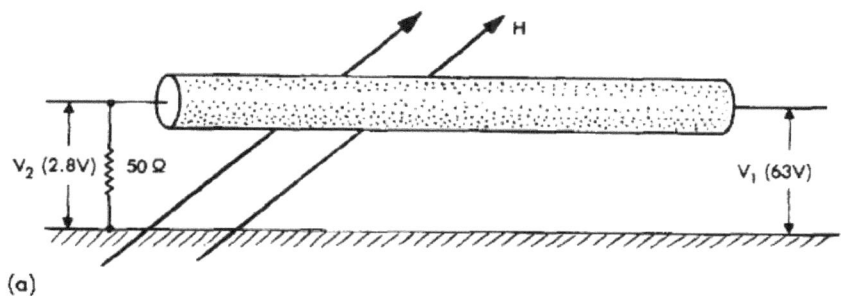

(a)

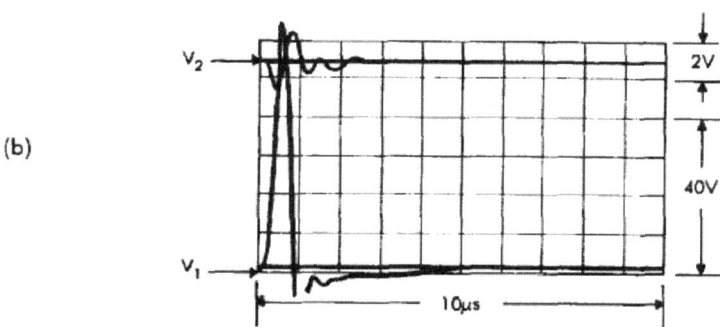

(b)

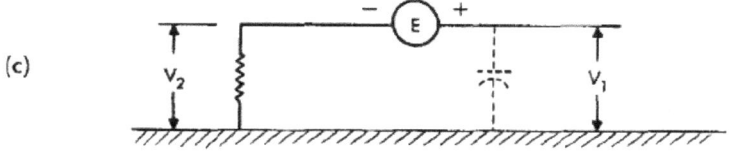

(c)

Figure 13.2 Unequal load impedances.
(a) Test conditions
(b) Voltages
(c) Equivalent circuits

Grounding the shield at both ends, as shown in Figure 13.5, produces an entirely different response. If the shield is grounded at both ends, the voltage induced by the changing magnetic field in the shield produces a flow of current through the shield.

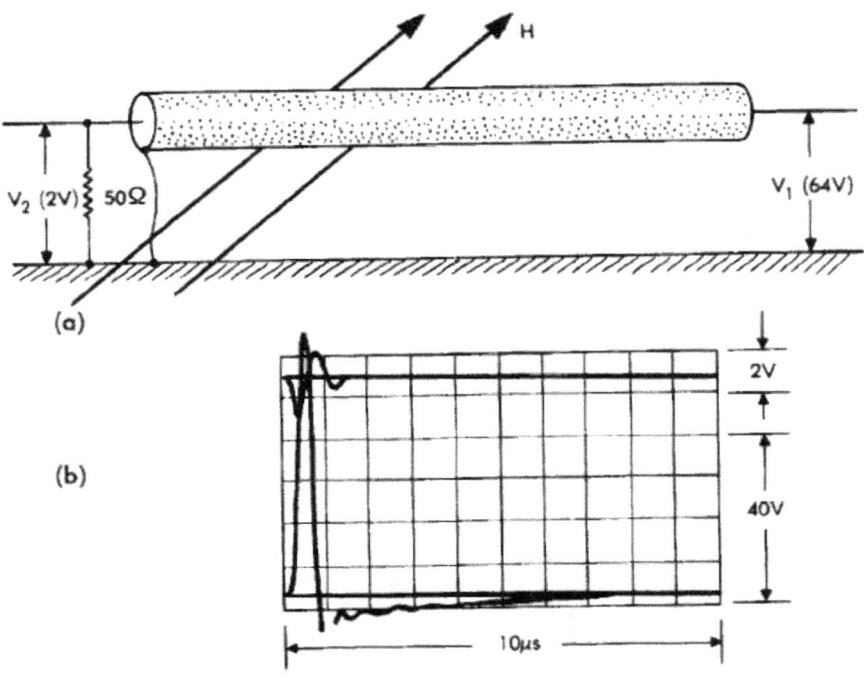

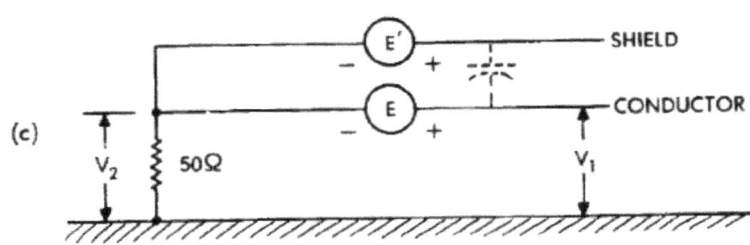

Figure 13.3 Shield grounded at one end.
(a) Test conditions
(b) Voltages
(c) Equivalent circuit

$$E' = \frac{d\phi}{dt} = \mu A \frac{dH}{dt} \qquad (13.1)$$

$$I = \frac{1}{L_s} \int E'dt = \frac{\mu A}{L} H \qquad (13.2)$$

where E' = voltage induced between ends of the shield

364

I = current on the shield
L = self-inductance of the shield

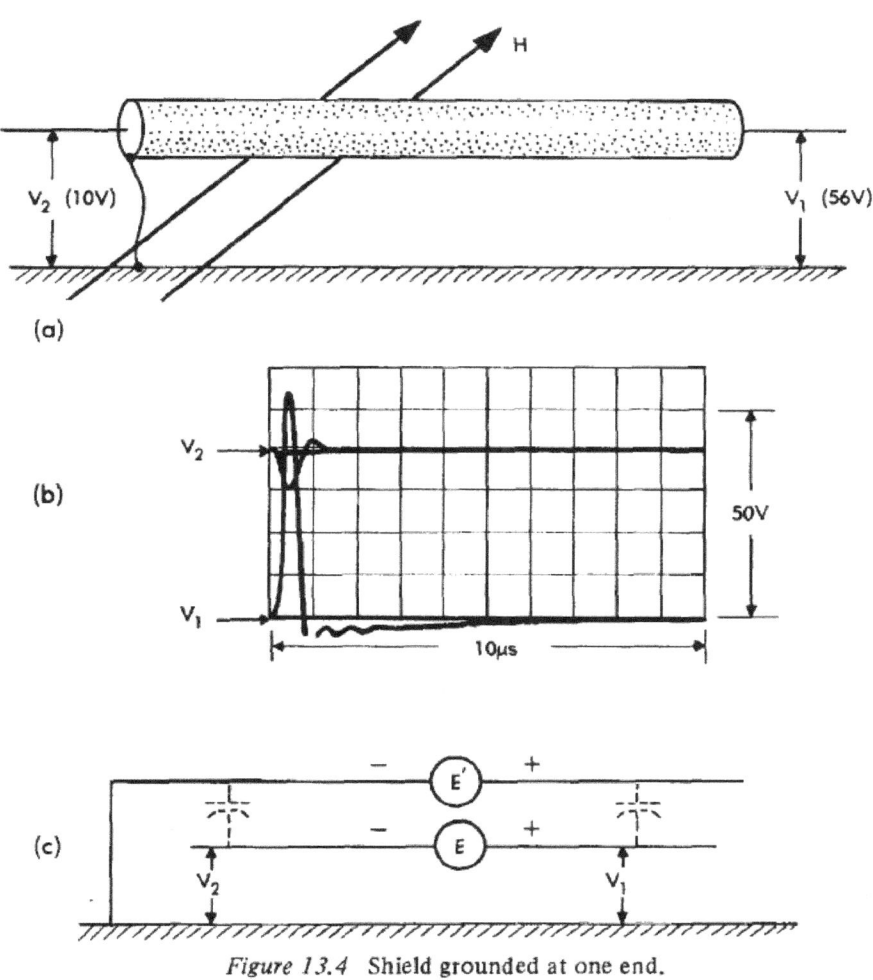

(a)

(b)

(c)

Figure 13.4 Shield grounded at one end.
(a) Test conditions
(b) Voltages
(c) Equivalent circuits

This shield current reduces the voltage induced between the signal conductor and ground. The reduction in voltage can be viewed equally well from two different viewpoints.

The first is that the shield current produces a magnetic field that tends to cancel the incident field. From this viewpoint the voltages on the signal conductors to ground respond only to the difference between the incident and

365

the canceling fields. Alternatively, the reduction in voltage can be viewed as the effect of the mutual inductance between the shield and the signal conductor. This latter approach is illustrated by the equivalent circuit of Figure 13.6(c). The voltage between the ends of the conductor is the sum of that resulting from the voltage induced in the signal conductor, E, and that coupled through the mutual inductance between the shield and the signal conductor. Since the shield is grounded at both ends and able to carry a current, the voltage appearing across

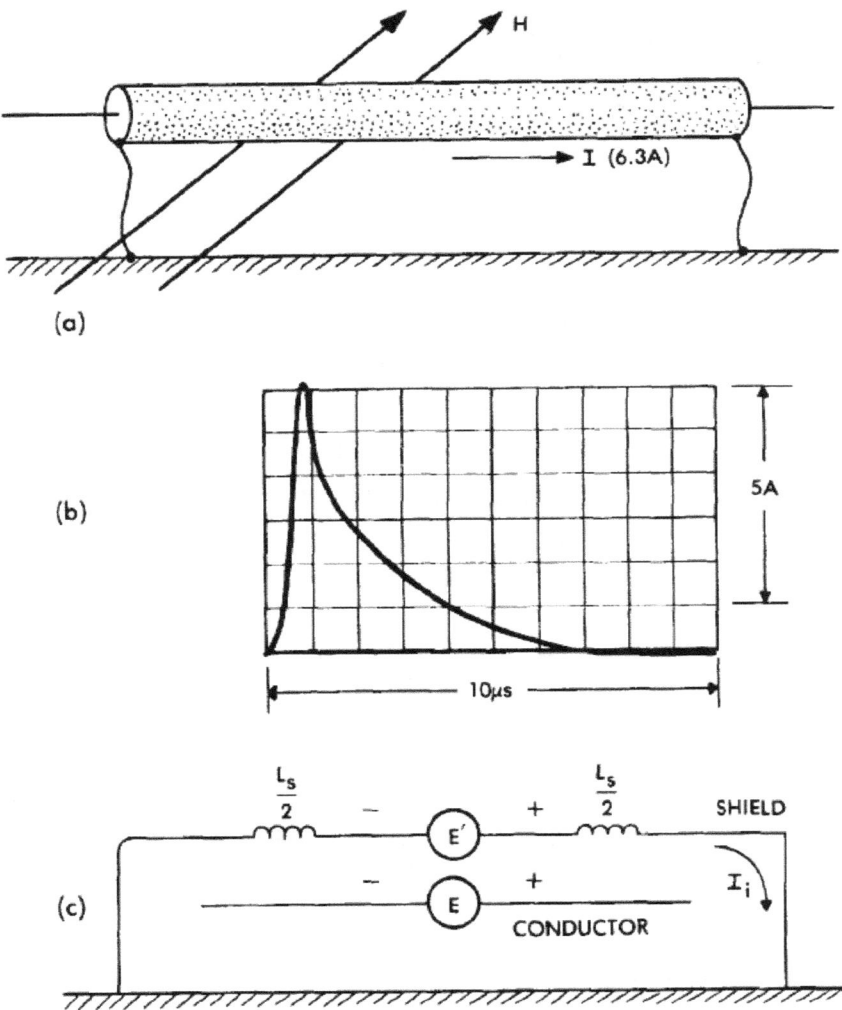

Figure 13.5 Shield grounded at both ends.
    (a) Test conditions
    (b) Shield current
    (c) Equivalent circuit

366

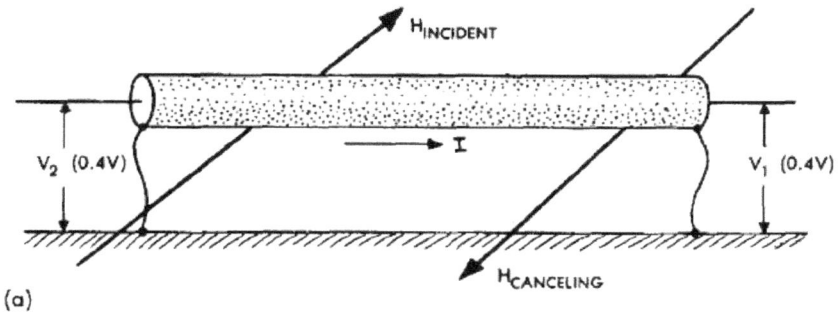

(a)

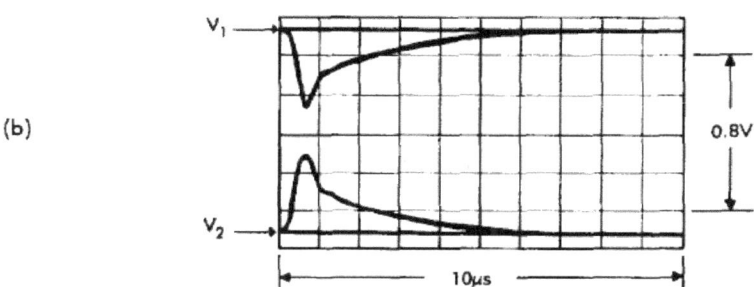

(b)

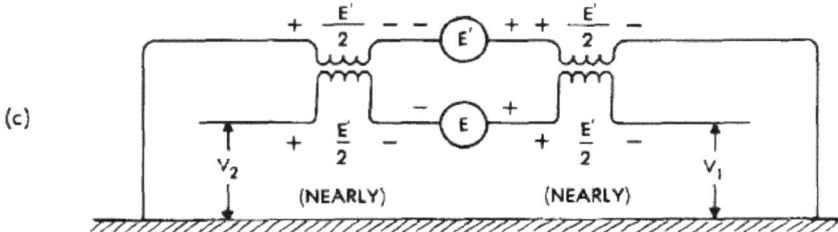

(c)

Figure 13.6 Shield grounded at both ends.

(a) Test conditions
(b) Conductor voltages
(c) Equivalent circuit

the primary inductance of the equivalent transformer is equal to the voltage E′ induced between the ends of the shield by the incident magnetic field. Since the mutual coupling between the shield and the signal conductor is very nearly unity, the voltage induced in the secondary, or signal conductor, side of the equivalent transformer is about equal to the voltage originally induced in the shield. Accordingly, the voltages appearing at the ends of the conductor are much lower than they would be if the shield could not carry current.

An alternative form of the equivalent circuit of Figure 13.6 is shown in Figure 13.7. There are two parts of the circuit, one relating the magnetic field to the current flowing on the shield and one relating the current on the shield to the voltage developed on the conductor. Both the measurements and the circuit indicate that voltage on the center conductor is affected by the relative impedance of the loads on the ends. A low impedance at one end pulls the voltage at that end down but raises the voltage at the other end. M_{12} and R represent transfer quantities, the values of which depend upon the type of shield.

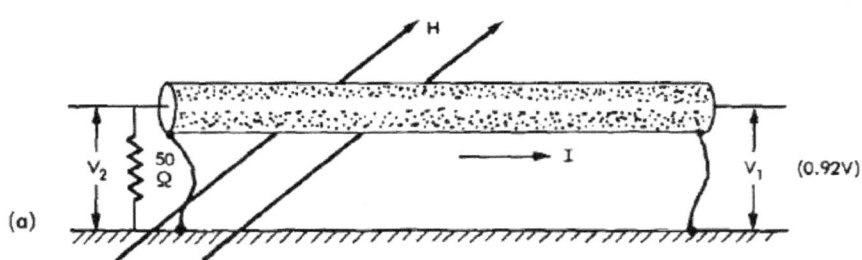

(a)

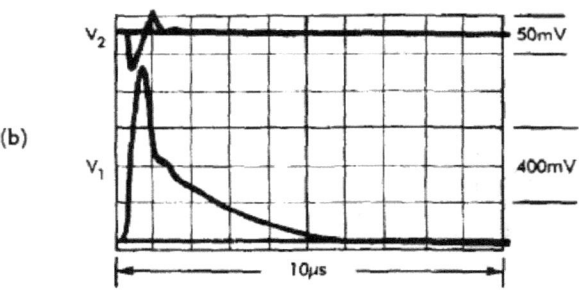

(b)

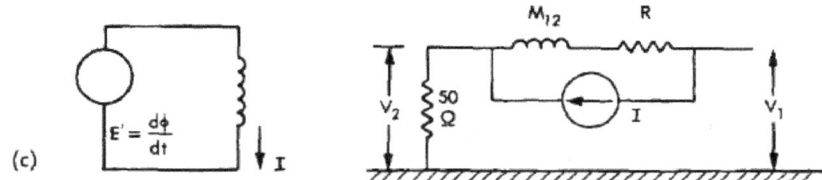

(c)

Figure 13.7 Shield grounded at both ends.
(a) Test conditions
(b) Conductor voltages
(c) Equivalent circuit

368

13.3 Transfer Quantities

If the mutual inductance between the shield and the signal conductor were equal to the self-inductance—unity coupling—the flow of current on the shield of the cable would not cause any voltage to be developed between the shield and the signal conductor. This section of the report will now discuss the factors that prevent the coupling between shield and signal conductor from being perfect. Consider first a shield in which only resistance effects are treated, shown in Figure 13.8. In this condition an external noise current flowing through the resistance of the cable shield produces an electric field on the internal surface of the shield. The nature of the coupling between this internal field and the signal conductor depends upon the connections at the ends of the cable and upon the distributed inductance and capacitance of the circuit internal to the cable. As a simple case, imagine the left-hand end of the signal conductor to be shorted to the cable shield. The internal electric field would then be completely coupled to the signal conductor, and the voltage between the signal conductor and the shield at the right-hand end of the cable would be equal in

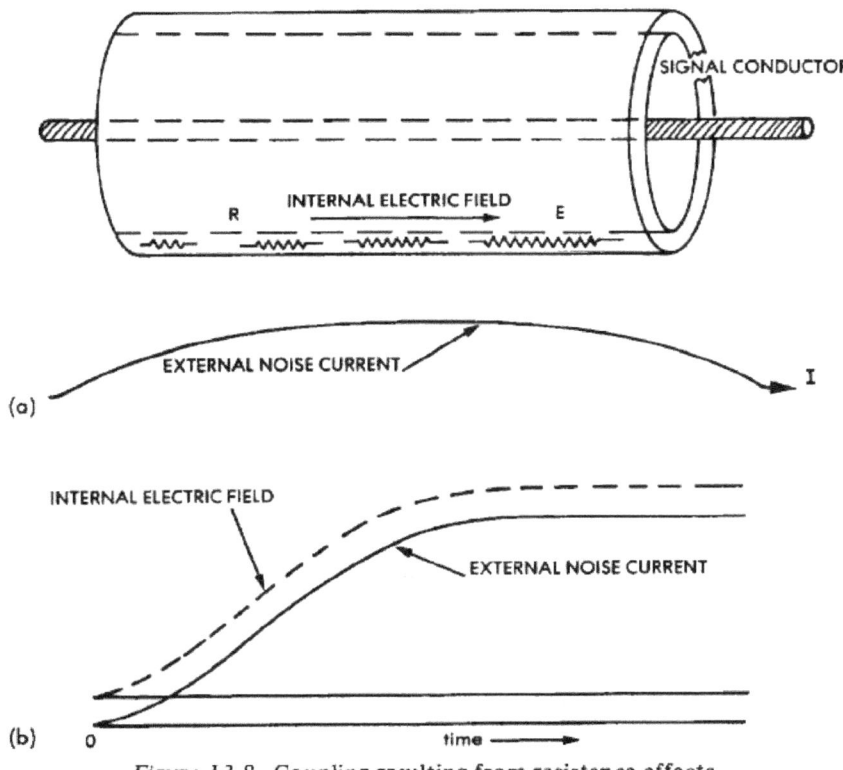

Figure 13.8 Coupling resulting from resistance effects.
    (a) Current and field polarities
    (b) Current and field waveshapes

magnitude to the total internal electric field along the length of the shield. This simple type of coupling is valid for low-frequency or slowly changing external noise currents. For noise currents sufficiently slow that this model applies, the waveshape of the internal electric field is the same as the external noise current, being related by Ohm's Law:

$$E = IR \tag{13.3}$$

13.3.1 Physical Phenomena

Generally the coupling is not as simple as this. If the overall shield is a solid-walled cylinder, as shown in Figure 13.9, the internal electric field depends upon the product of the resistivity of the shield material and the density of the current on the internal surface of the shield. As discussed in Section 10.1, this internal current density, J_i, will not, in general, be the same as the density of the current on the external surface of the cylinder. Because of the phenomenon of skin effect, the current density on the inner surface of the shield will rise more slowly than does the external noise current. The rate at which the current density on the inner surface increases will be directly proportional to the permeability of the field material and to the square of the wall thickness and inversely proportional to the resistivity of the wall material. Cables with solid-wall shields and cable trays of solid metal with tightly fitting covers will typically exhibit this type of behavior.

A different type of behavior is commonly observed on cables with braided shields, shown in Figure 13.10. The shields do not provide a perfect conducting cylinder, for they have a number of small holes which permit leakage. The holes between the individual bundles of wire forming the shield can be approximated as a group of diamonds, the size and orientation of which depend upon the weave angle of the braided shield. The long axis of the triangles may be oriented along the shield, as at end A of the cable shown in Figure 13.10(a), or, if the weave angle is great, the long axis of the hole may be oriented circumferentially to the cable, as at end B of the cable. The external noise current produces a circumferential magnetic field around the outside of the shield. As a first approximation, one may visualize the external field lines looping in and out of the holes in the braided shield. There will be more leakage in holes with long axes oriented circumferentially to the end of the shield (end B of the cable, large weave angle) than there will be if the long axes of the holes are oriented along the shield, (end A of the cable, small weave angle). Viewed from the end of the cable, as in Figure 13.10(b), the field that leaks into the cables produces a net magnetic field circumferentially around the inside surface of the shield. This magnetic field produces an internal electric field of an amplitude dependent upon the degree of leakage into the shield and upon the rate of change of the magnetic field. The rate of change of this internal magnetic field is to a first approximation proportional to the rate of change of the external noise current. The total internal electric field, then, is proportional to the resistance of the cable shield, to the rate of change of external noise current, and to the number of holes in the cable.

370

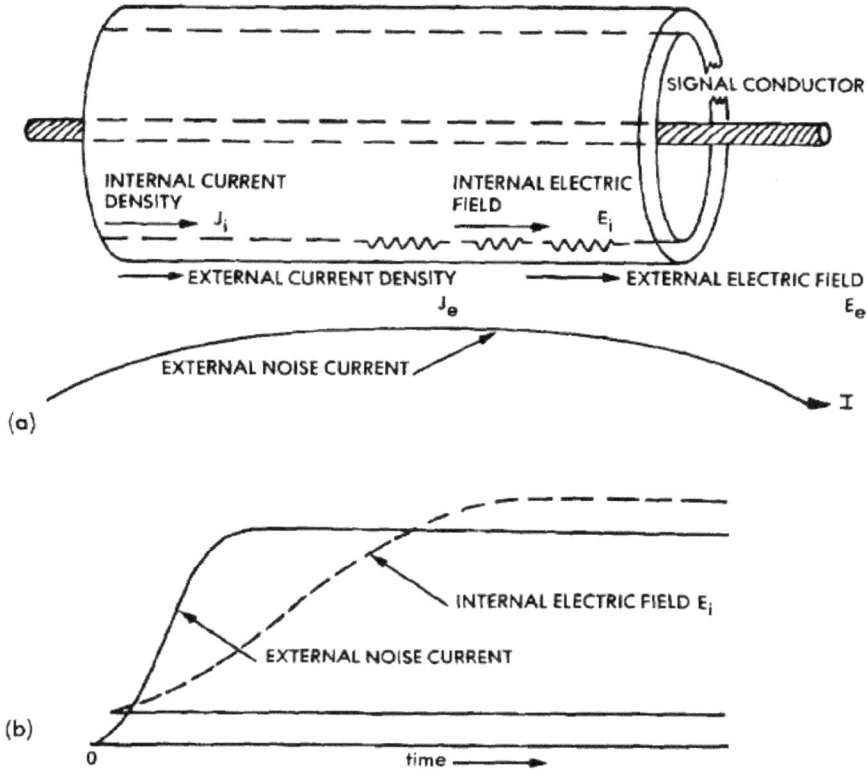

INTERNAL CURRENT DENSITY J_i

SIGNAL CONDUCTOR

INTERNAL ELECTRIC FIELD E_i

EXTERNAL CURRENT DENSITY J_e

EXTERNAL ELECTRIC FIELD E_e

EXTERNAL NOISE CURRENT

I

(a)

INTERNAL ELECTRIC FIELD E_i

EXTERNAL NOISE CURRENT

(b)

0

time →

Figure 13.9 Coupling via diffusion through a solid wall.
(a) Current and field polarities
(b) Current and field waveshapes

Frequently braided cables exhibit a phenomenon, shown in Figure 13.11, in which the net polarity of the internal magnetic field is opposite to the polarity of the external magnetic field. The result of this phenomenon is that a change in external noise current induces an internal electric field of the opposite polarity. The reasons for this behavior are not clear to this author, but the phenomenon seems to depend upon the thickness of the braided covering of the shield and the total diameter of the shield, as well as upon the degree of tightness of weave of the shield.

Some illustrative examples of the behavior of braided shields are shown in Figures 13.12 and 13.13. The figures show the test geometry and the conductor-to-ground voltage produced by currents having waveforms of several different times to crest. The 47 Ω resistors shown provide the termination for the signal conductor and prevent the voltage at the other end from being obscured by the internal oscillations of the cable. Figure 13.12 shows a double-shielded cable with the voltage measured from the inner shield to ground. This particular cable exhibits the phenomenon shown in Figure 13.8. In Figure

371

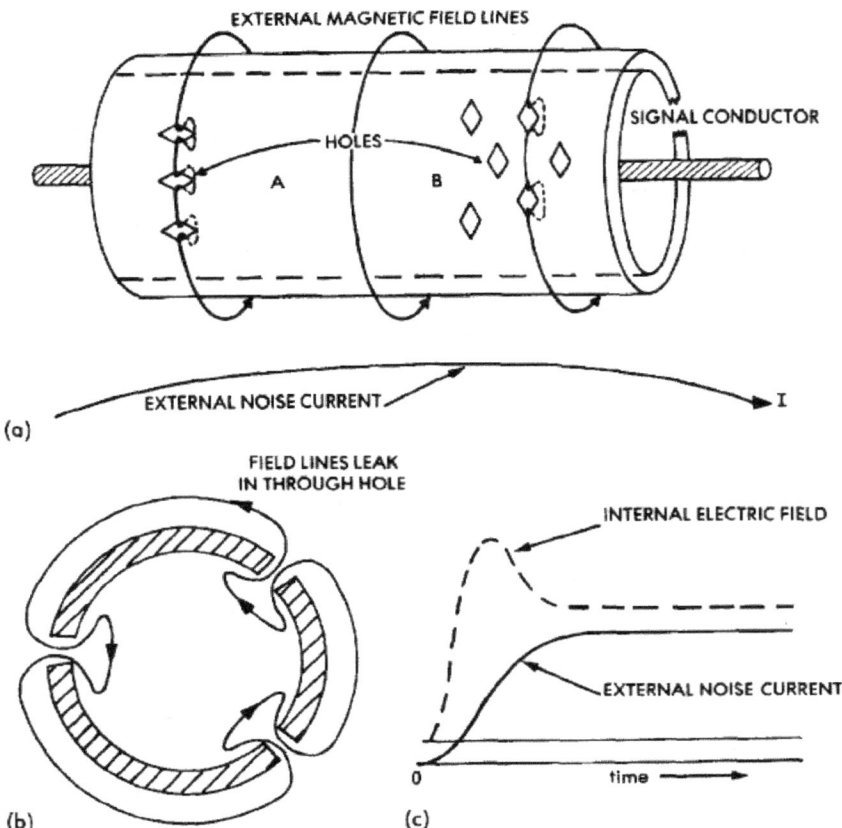

Figure 13.10 Coupling via magnetic leakage through holes.
    (a) Current and field polarities
    (b) Field leakage through holes: end view
    (c) Current and field waveshapes

13.13 a cable similar to RG-8/U cable is shown. This cable exhibits the negative coupling inductance shown in Figure 13.11.

Another source of coupling is via capacitive leakage through the holes in the shield, as shown in Figure 13.14. If the shield is held at ground potential by deliberate grounds, if the signal conductor is held near ground potential by external loads, and if the shielded cable is subjected to an external and changing electric field, dielectric flux will pass through the holes in the cable from the external source and onto the signal conductor. The flow of these dielectric, or displacement, currents through the external load impedances produces a voltage between the signal conductor and the shield. An alternate source of coupling, shown in Figure 13.14(b), involves the voltage on the shield itself. Noise currents flowing through the external impedance of the shield may, if the shield is not perfectly grounded, produce a voltage between the shield and any external

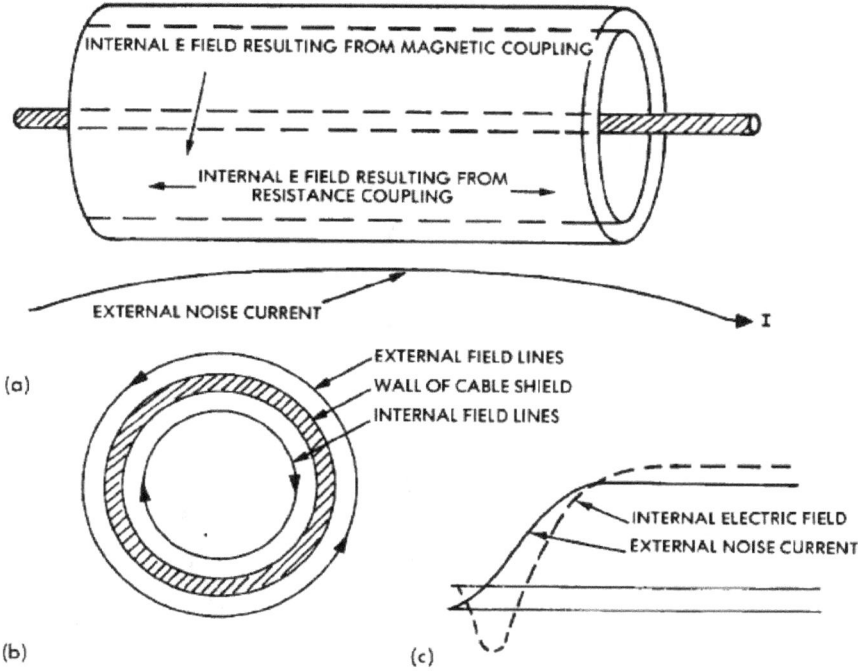

Figure 13.11 Magnetic coupling giving effect of a negative coupling inductance.

    (a) Current and field polarities

    (b) Direction of field

    (c) Current and field waveshapes

ground structure. External impedances between the signal conductor and the shield, as well as the inherent capacitance between the shield and the signal conductor, force the signal conductor to assume nearly the same potential as the shield. Because the signal conductor is then at a potential different from the surrounding ground, dielectric flux can pass from the signal conductor through the holes in the shield and to ground, the dielectric currents again giving rise to a voltage between the signal conductor and the shield.

In most cases the effects of magnetic leakage are probably more important than the effects of capacitive leakage.

13.3.2 Calculation of Transfer Quantities

The factors that affect the voltage which may appear between a signal conductor and the shield of a cable, as distinct from the voltages that may appear between the signal conductor and ground, are, then, the resistance of the shield, the degree to which the shield allows magnetic fields to leak to the inside, and the degree to which the shield allows electric fields to leak to the inside. These latter two characteristics may be expressed in terms of an equivalent

373

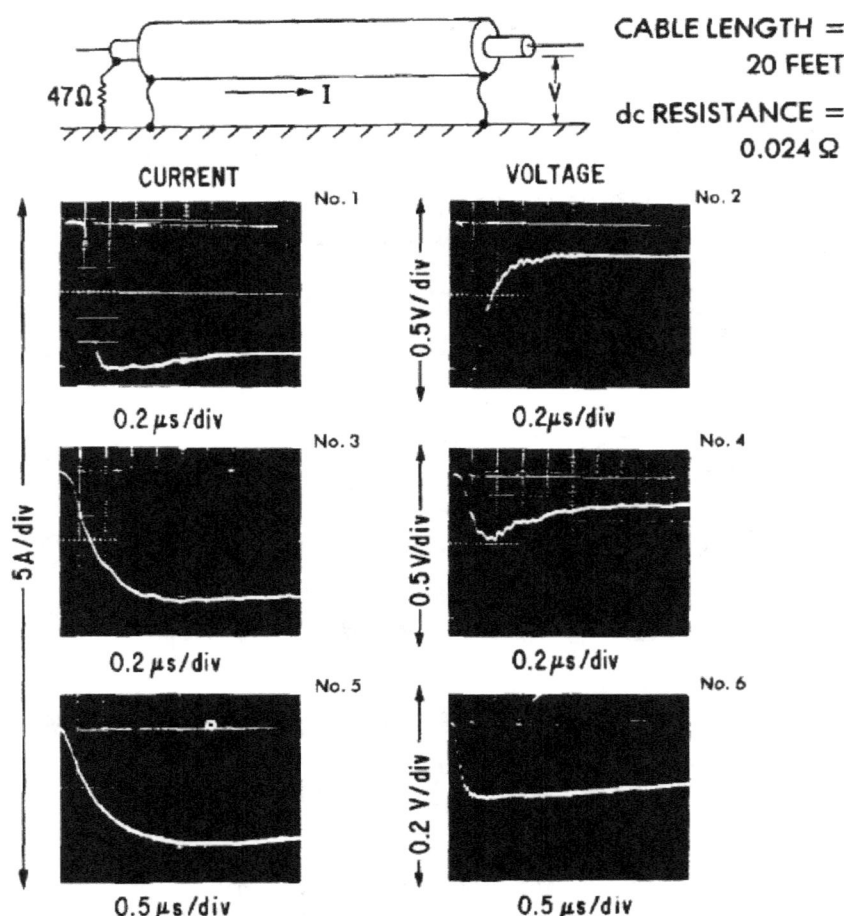

Figure 13.12 Coupling from overall shield to center shield of a double-shielded cable.

transfer inductance of the shield and an equivalent transfer capacitance. These effects may be expressed in terms of the equivalent circuits shown in Figure 13.15. Figure 13.15(a) shows the equivalent circuit by which noise current on the shield of the cable may be related to the voltage induced between the signal conductor and the shield. In the frequency domain, this may be expressed as

$$V_s = I_n(Z_d + j\omega M_{12}) \qquad (13.4)$$

As will be seen momentarily, the term Z_d may frequently be treated simply as the dc resistance of the cable shield, shown in Figure 13.15(b), in which case the relation between current and voltage becomes simply

$$V_s = I_n(R_{dc} + j\omega M_{12}) \qquad (13.5)$$

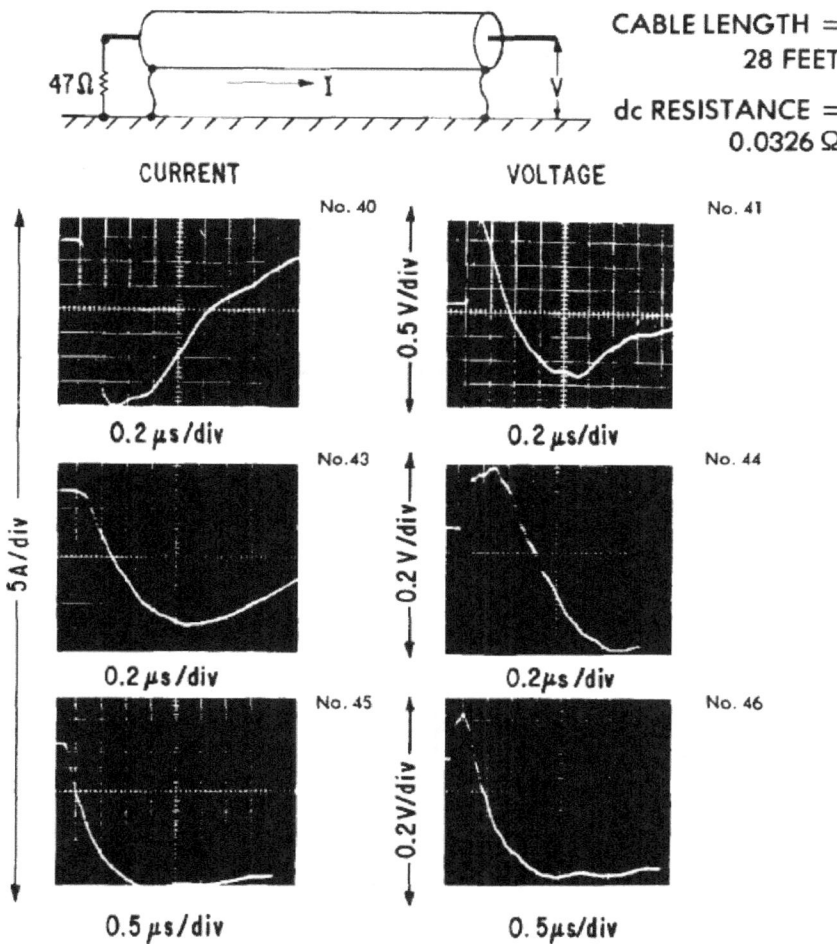

CABLE LENGTH =
28 FEET

dc RESISTANCE =
0.0326 Ω

CURRENT   VOLTAGE

Figure 13.13 Coupling from overall shield to center conductor of a cable similar to RG-8/U.

Treatment of capacitive effects yields the equivalent circuit shown in Figure 13.15(c), in which the relation between loop current induced on the signal conductor and external noise voltage becomes

$$I_s = E_n j\omega C_{12} \tag{13.6}$$

Vance (Reference 13.2) gives an excellent dissertation on how the transfer inductance M_{12} and transfer capacitance C_{12} may be related to the characteristics of the braided shield. Many of the equations and figures that follow are reproduced from his work.

A braided shield would, if rolled out on a plane, typically have the

375

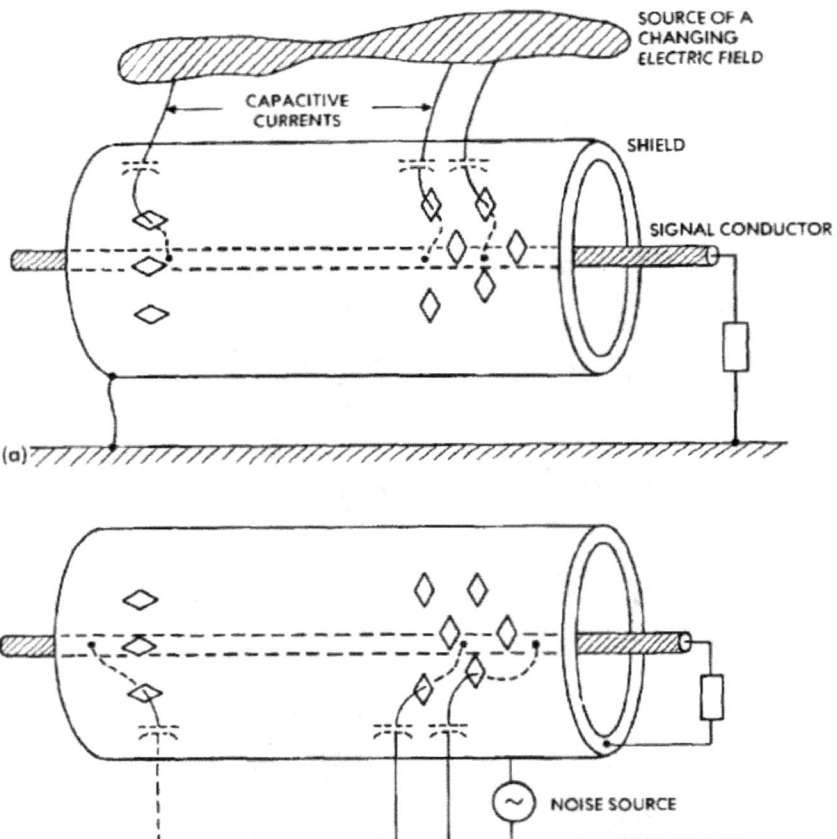

Figure 13.14 Coupling via capacitive leakage through holes.
(a) Coupling from an external source
(b) Coupling from voltage on the shield

appearance shown in Figure 13.16 (Reference 13.3). The characteristics of the shield may be described in terms of the following symbols:

a — the radius of the shield
C — the number of carriers, or bands of wire, that make up the shield
N — the number of ends, or the number of wire strands, in each carrier
P — the number of picks, or carrier crossings, per unit length
d — the diameter of the individual wires

Some important quantities that may be defined in terms of the above characteristics are

$$\text{Weave angle: } \alpha = \tan^{-1} \frac{4\pi a P}{C} \qquad (13.7)$$

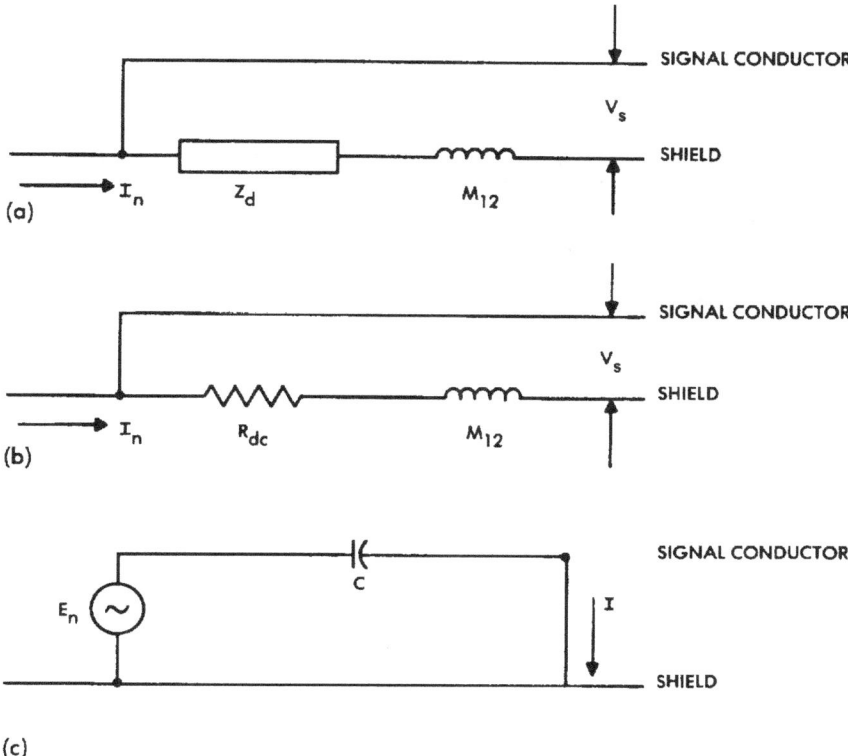

Figure 13.15 Equivalent circuits.
        (a) Transfer impedance
        (b) Transfer impedance simplified
        (c) Transfer admittance

$$\text{Fill: } F = \frac{PNd}{\sin \alpha} = \frac{Nd}{W} = \frac{NdC}{4\pi a \cos \alpha} \tag{13.8}$$

$$\text{Optical coverage: } K = 2F - F^2 \tag{13.9}$$

$$\text{Volume of metal: } v = \pi^2 adF \tag{13.10}$$

The optical coverage, K, relates to the size of the small holes not covered by the carriers. The greater the optical coverage, the smaller the total area of the holes with respect to the total area covered by the shield and, other things being equal, the better the electrical performance of the shield. The weave angle affects the orientation of the diamond-shaped holes. Since holes oriented along the axis of the cable are less effective in coupling energy into the interior than are holes oriented circumferentially around the cable, it follows that, other

377

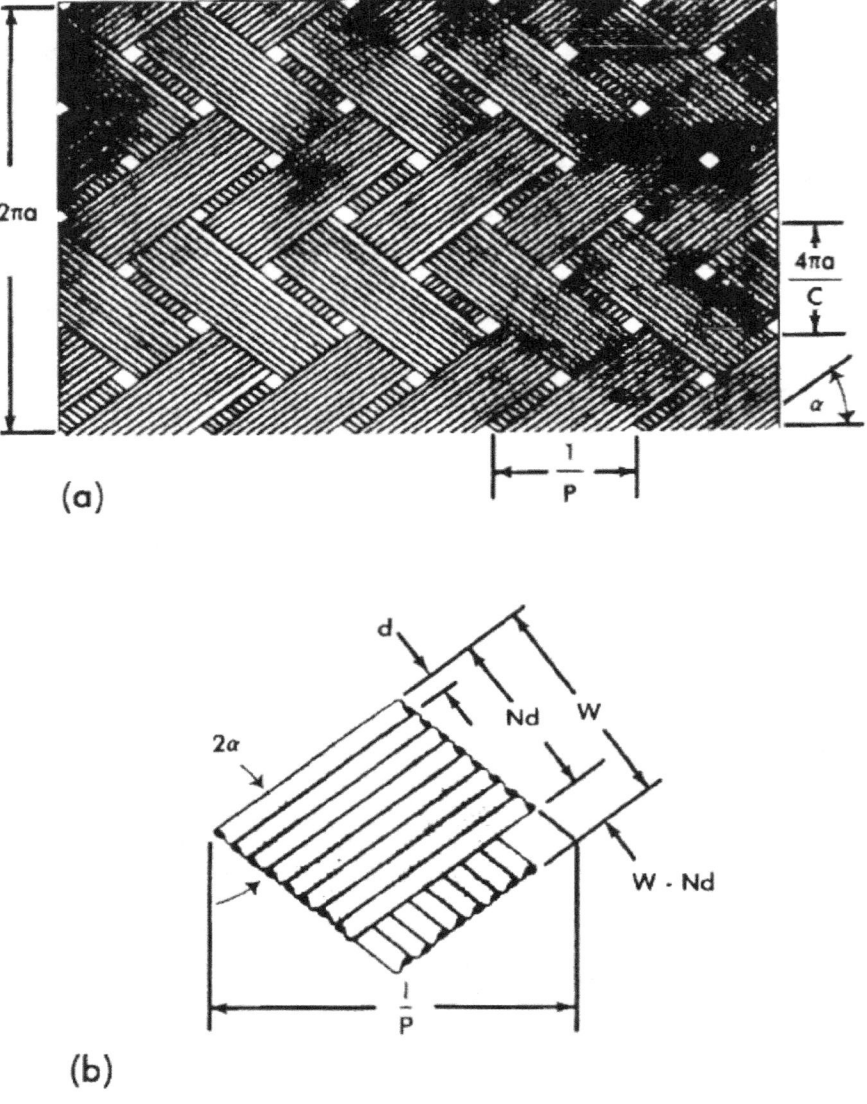

Figure 13.16 Properties of a typical braided shield.
(a) Braid pattern developed on a plane
(b) Properties of a typical braided shield

things being equal, a shield with a small weave angle will give better electrical performance than a shield with a large weave angle.

Presented here without proof, the elements of the equivalent circuit of Figure 13.15(a) are

$$Z_d \approx \frac{4}{\pi d^2\, NC\sigma \cos a} \frac{(1+j)\,d/\delta}{\sinh(1+j)\,d/\delta} \qquad (13.11)$$

$$M_{12} \approx \frac{\pi\mu_0}{6C}\,(1-K)^{3/2}\,\frac{e^2}{E(e)-(1-e^2)\,K(e)} \qquad (a<45°) \qquad (13.12)$$

$$\approx \frac{\pi\mu_0}{6C}(1-K)^{3/2}\,\frac{e^2/}{K(e)-E(e)}\frac{1-e^2}{} \qquad (a<45°) \qquad (13.13)$$

In the above expressions

σ = conductivity of the wires

$\mu_0 = 4\pi \times 10^{-7}$

K(e) and E(e) = complete elliptic integrals of the
first and second kinds

$$\delta = 1/\sqrt{\pi f \mu \sigma}, \qquad (13.14)$$

$$e = \sqrt{1-\tan^2 a} \qquad (a<45°) \qquad (13.15)$$

$$= \sqrt{1-\cot^2 a} \qquad (a>45°) \qquad (13.16)$$

For many analyses Z_d may be taken, with sufficient accuracy, as the dc resistance of the shield. This dc resistance is also slightly dependent on the coverage.

$$R_0 \approx \frac{4}{\pi d^2\, NC\sigma \cos a} = \frac{1}{\pi^2\, ad\, \sigma\, F \cos^2 a} \qquad (13.17)$$

For the specific cable with construction as indicated, Figure 13.17 (Reference 13.4) shows how the transfer impedance varies with frequency for different degrees of optical coverage. Several observations might be made about this figure. The first is that at low frequencies, below about 500 kHz, the transfer impedance is basically just the dc resistance of the cable shield. The second is that for frequencies above about 2 MHz the transfer impedance is primarily reactive, being represented by the transfer inductance of the cable shield. The third is that for the most common types of shields (those with optical coverage of less than 90%) the dip in transfer impedance below the dc resistance (the dip being a result of skin effects) is not great. It is for this reason that the equivalent circuit of Figure 13.15(b), in which the frequency-dependent properties of the shield resistance are ignored, is adequate for most engineering analyses.

The transfer admittance of the braided-wire shield, governing the degree to which the shielded cable may be affected by external electric field, is

$$Y_t = j\, C_{12} \qquad (13.18)$$

$$C_{12} \approx \frac{\pi C_1 C_2}{6\epsilon C} (1-K)^{3/2} \frac{1}{E(e)} \qquad (a < 45°)$$

$$\approx \frac{\pi C_1 C_2}{6\epsilon C} (1-K)^{3/2} \frac{\sqrt{1-e^2}}{E(e)} \qquad (a > 45°) \qquad (13.19)$$

C_1 is the capacitance per unit length between the internal conductors and the shield, C_2 is the capacitance per unit length between the shield and its external current return path, and ϵ is the permittivity of the insulation inside the cable.

The preceding equations indicate that both the mutual inductance and the mutual capacitance depend on the weave angle of the shield. For a constant coverage and for a specific size cable, the variations of M and C with weave angle are as shown in Figure 13.18 (Reference 13.5). The mutual inductance per unit length increases rapidly with increasing weave angle, since, as explained earlier, apertures with their major axes oriented parallel to the lines of magnetic flux are

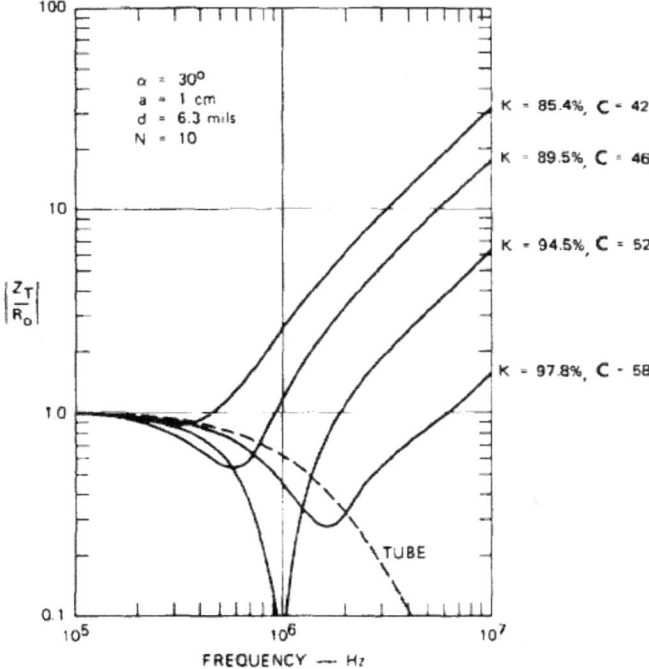

Figure 13.17 Transfer impedance of a braided-wire shield.

380

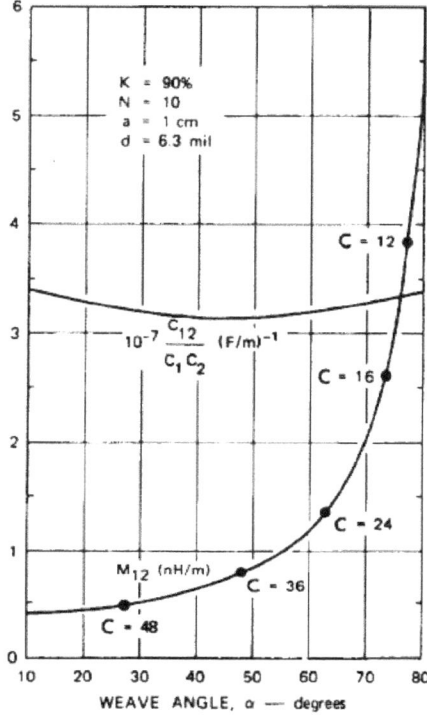

Figure 13.18 Variation of mutual inductance and mutual capacitance with weave angle for constant coverage.

more effective in allowing flux to leak inside than are apertures of the same area but with their major axes oriented at right angles to the lines of flux. The mutual capacitance varies only slightly with weave angle, since the major factor governing the mutual capacitance is simply the projected area of the apertures.

To simplify the task of estimating M and C from the preceding equations, the functions of the eccentricity of the apertures are plotted in Figure 13.19 (Reference 13.6) and the variation in $(1\text{-}K)^{3/2}$ plotted in Figure 13.20 (Reference 13.7).

Tape-wound shields, as shown in Figure 13.21, are often used where flexibility of the shielded cable is required. The shield may be formed either from a narrow metal sheet spiraled around the core or from a carrier of fine wires, again spiraled around a core. Flexible armor and flexible conduit, which is normally used primarily for mechanical protection, may also be analyzed as tape-wound shields. The tape-wound, or spiral-wound, shield is a rather poor shield for preventing coupling of current from the shield onto the internal conductor because the shield tends to behave as a solenoid wound about the internal conductors. There is thus a rather large mutual coupling term relating the internal voltage to the shield current. This is particularly true if the tape is

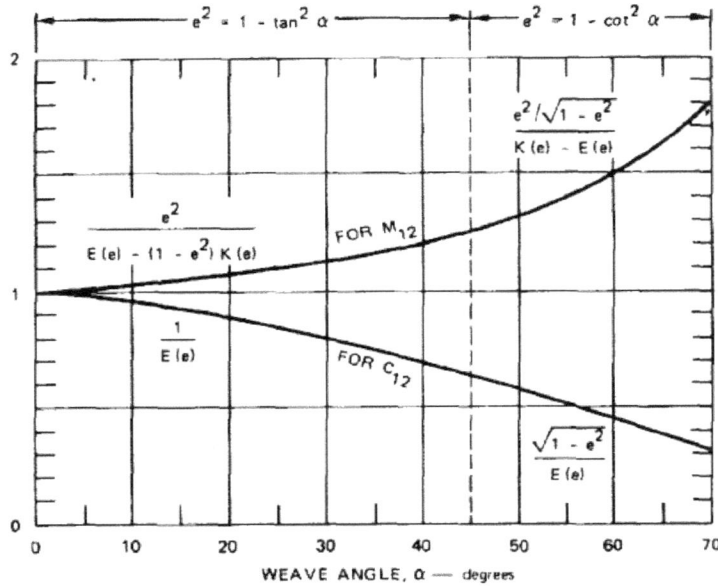

Figure 13.19 Variation of the eccentricity functions for braided-wire shields.

wound without any overlap because in this case the entire cable current is forced to spiral around the core. Vance analyzes a 1 cm radius cable wound with .25 mm thick, 1 cm wide copper tape, the tape applied without overlap. The dimensions of the cable are thus similar to those for the braided-wire shield shown in Figure 13.18. The mutual inductance of the tape-wound shield was calculated as 3.9 μH/m of the order 10^4 greater than the mutual inductance obtained for a braided-wire shield.

The thin-walled tubular shield consists of a metal tube of uniform cross section and uniform wall thickness. Coupling through the shield can occur only by diffusion of the electromagnetic fields through the walls of the tube. The transfer impedance of thin-walled tubes such as these, as derived by Schelkunoff, (Reference 13.8) is

$$Z_a = \frac{1}{2\pi r \sigma t} \frac{(1 + j)t/\delta}{\sinh{(1 + j)}t/\delta} \qquad (13.20)$$

where r is the radius of the shield, a is its wall thickness, σ is the conductivity of the shield, and δ is the skin depth in the shield given by

$$\delta = \frac{1}{\sqrt{\pi f \mu \sigma}} \qquad (13.21)$$

It is assumed that the wall thickness, a, is small compared to the radius, r, of the tube and that the radius, r, is small compared to the smallest wavelength of

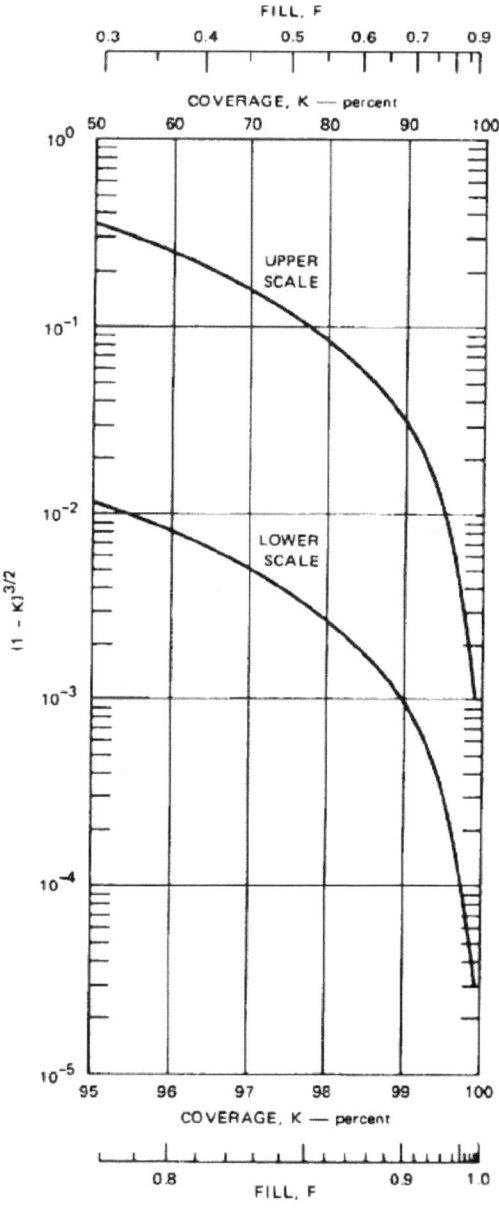

Figure 13.20 Variation of $(1-K)^{3/2}$ with shield coverage K.

interest. It is also assumed that the shield is made from a good conductor (metal) so that the displacement current in the shield material is negligible compared to the conduction current.

Figure 13.21   Tape-wound shield.

The maximum transfer impedance is found at dc or low frequencies, or frequencies low enough that the thickness of the shield is small compared to the skin depth. That impedance is the dc resistance

$$R_0 = \frac{1}{2\pi r \sigma a} \qquad (13.22)$$

The transfer impedance decreases with increasing frequency. Accordingly, the internal electric field will, in response to a step-function current, increase according to the pulse penetration time constant predicted by Equation 10.9 and shown in Figure 10.6 of Section 10.1.

In the absence of any end effects or openings through which magnetic fields can leak, the mutual inductance term M_{12} of Equation 13.4 will be zero.

Cable trays are most often used for mechanical protection of wires, but if viewed as shields, they may also provide electrical protection. The characteristics of a cable tray that make for good electrical protection are the same as those of any other shield:

- It must be able to carry current along its axis.
- It should be of low-resistance material.
- It should completely surround the conductors, implying that the tray should be fitted with a conducting cover.
- It should have a minimum number of openings through which magnetic fields may leak.
- It should have as few joints as possible, and such joints should be made in such a manner as to provide minimum resistance and leakage of magnetic fields.

The transfer characteristics of the tray by itself would be about the same as those of the solid tubular shields discussed previously. This comparison assumes the tray to be of solid metal and fitted with a well-sealed cover. Since covers never form perfect seals, the pulse penetration time constant would probably not be as long as predicted by Equation 10.9. Even with an imperfect cover, the transfer impedance Z_T would not be greater than the dc resistance.

Trays are most commonly built in short sections and joined by splices or transition sections. Such sections frequently provide for thermal expansion and contraction and are, at any rate, seldom designed either to provide good electrical continuity or to protect against magnetic leakage. When joints are considered, the transfer characteristic of the tray is frequently found to depend almost entirely on the treatment of the joints. The transfer characteristics can seldom be determined by calculation; they must be determined experimentally.

13.4 Multiple Conductors and Shields

If there are multiple conductors in the core, each conductor will develop nearly the same voltage between that conductor and the shield. This holds true whether the conductor is located adjacent to the shield or in the center of the bundle of conductors comprising the core. Accordingly, the voltage between any pair of conductors in the core will be small. Present analytical tools do not seem of sufficient accuracy to predict with any assurance the magnitude of line-line voltages; they are best determined by actual measurement. Line-line voltages are much more strongly influenced by load impedances to which the conductors are connected than by the position of the conductors within an overall shield.

13.4.1 Physical Phenomena

One very common situation involving multiple conductors within an overall shield is a configuration in which a shielded conductor is itself contained in an overall shield. Figure 13.22 shows such a condition and shows some of the basic considerations with which one must deal in order to predict the response of this two-cable network to a noise current flowing on the outer shield. The total circuit response is a combination of the internal field of the outer shield and the internal field of the inner shield plus the coupling from the internal field of the inner shield to the signal conductor contained within the inner shield. These internal fields are created by noise current flowing on the outer shield plus noise current flowing on the inner shield. The noise current on the inner shield depends upon the internal field of the outer shield and on the impedance presented to signals traveling on the outer surface of the inner shield and returning along the inner surface of the outer shield.

The most important factor concerning this latter impedance is whether or not the internal shield is grounded at both ends. Figure 13.22(b) shows the internal shield connected to the outer shield at the left-hand end and a switch that would allow connection to be made between the shields at the right-hand end. It is furthermore assumed that the impedance of any connection between shields is zero, a situation that definitely may not be true in practice.

If the inner shield is not grounded at the right-hand end, the internal field of the outer shield (E_{i-os}) produces a voltage difference, V_{is-os}, at the right-hand end of the cable. Neglecting the distributed capacitances between the inner shield and the outer shield, the noise current on the inner shield, I_{is}, is then zero. If the inner shield is connected to the outer shield at both ends and the switch, S, is closed, the voltage difference between the two ends of the cable is reduced to zero. The internal field of the outer shield, however, will induce a circulating current, I_{is}, between the inner shield and the outer shield. The magnitude of this current is governed by the strength of the field E_{i-os} and the impedance around the loop between the inner shield with return on the outer shield. This impedance, at least for many cables, is predominantly inductive. The waveshape of the current on the inner shield, I_{is}, is then proportional to the time integral of the voltage developed along the inner surface of the outer shield. The current on

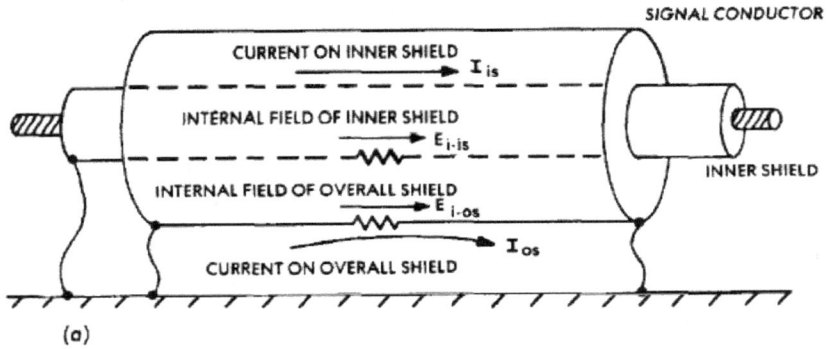

(a)

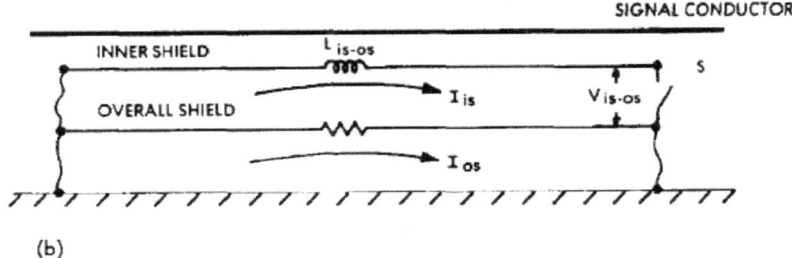

(b)

Figure 13.22 Double-shielded cable.
(a) Current and field polarities
(b) Current on inner shield and forced
  through $L_{is\text{-}os}$

the inner shield thus typically has a slower rise time and a longer duration than does the field along the inner surface of the outer shield. This current, I_{is}, on the internal shield then produces an electric field along the surface of the inner shield, $E_{i\text{-}is}$.

13.4.2 Examples of Response of Multiple-Shielded Cables

It will be apparent that there are a number of combinations of shield connections and termination impedances which are possible, all of them affecting the overall response of the cable system. Some of these effects have been demonstrated during an extension of the tests described in Figure 13.1 through 13.6. The piece of RG-58/U coaxial cable used for those tests was modified by pulling over the outer jacket another length of flexible copper braid. The test connections and results are shown on Figure 13.23. The equivalent circuit has three separate components—separate in the sense that each, for all practical purposes, may be treated independently. The first part of the circuit determines the voltage induced along the inner surface of the outer

386

shield. This is governed by the outer current flowing through R_1 and M_1, the transfer factors of the outer shield. In the second stage of the problem, this inner shield voltage is impressed across the inductance existing between the inner and outer shields in order to determine the current flowing on the inner shield. In the third state, the current on the inner shield is passed through R_2 and M_2, the transfer parameters of the inner shield.

The resultant voltages are then the voltages that appear between line and ground on the central signal conductor. The amplitudes and shapes of these

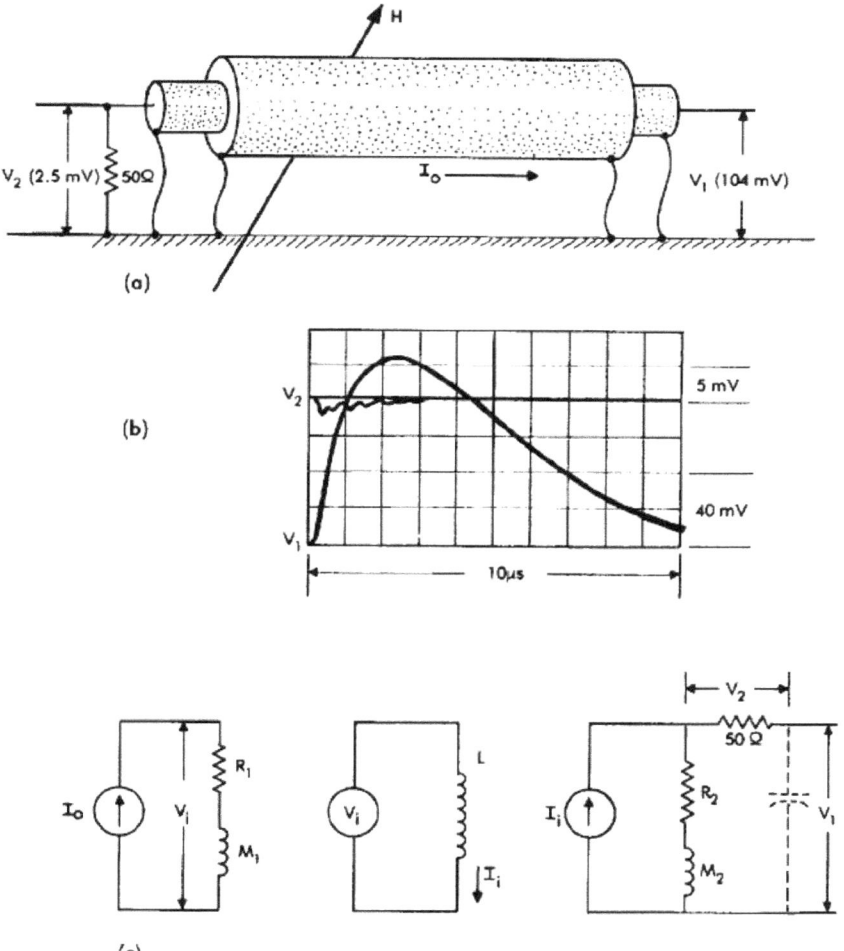

Figure 13.23 Multiple-shielded cable.
(a) Test conditions
(b) Conductor voltages
(c) Equivalent circuits

voltages are also shown in Figure 13.23. More effects have been illustrated by a series of measurements on a rather complex multiconductor cable. The overall makeup of the cable itself is shown in Figure 13.24. It consists of eight shielded quads surrounding a shielded center complex of nine conductors. The entire group of conductors is contained within an overall shield. The overall shield and the shields on the individual complexes are insulated from each other.

Figure 13.25 shows the waveshape of the current injected onto the overall shield of the cable. This current was transformer-coupled to the shield, allowing the overall shield to be grounded at both ends.

Figure 13.26 shows voltages measured on three of the different shielded

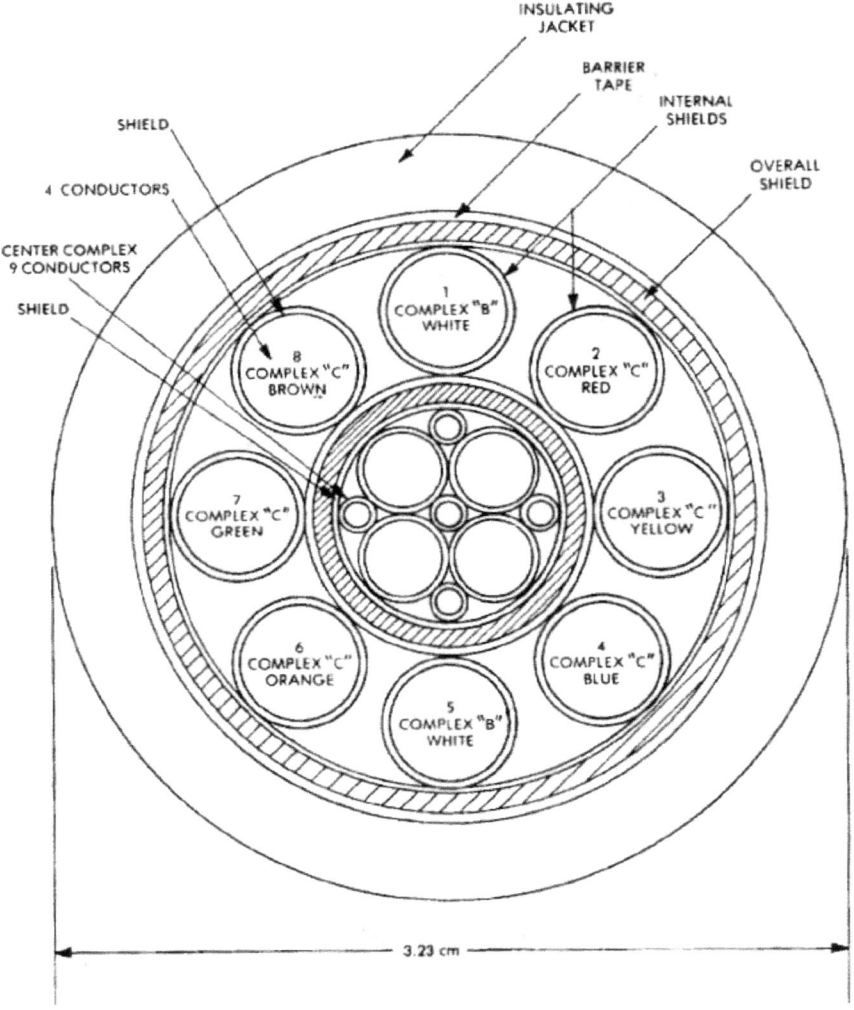

Figure 13.24 Construction of a multiconductor cable.

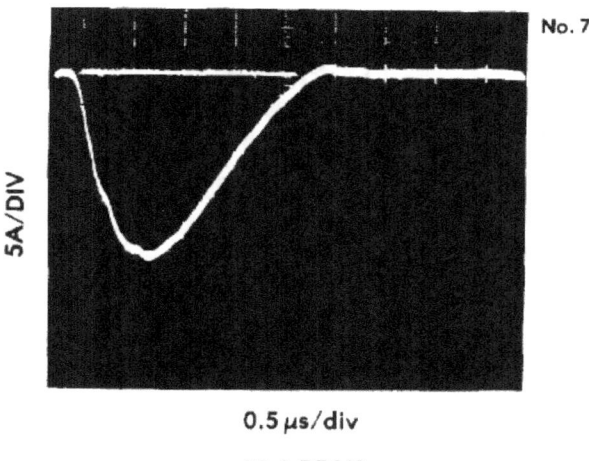

5A/DIV

No. 7

0.5 µs/div

18 A PEAK

Figure 13.25 Current on overall shield.

complexes within the overall shield. Measurements were made on the shields of two of the eight shielded quads and on the shield of the center complex. With high-impedance terminations at each end of the cable, the voltage from each of the individual shields to ground was the same. Voltages on individual conductors within the shielded groups would be the same as the voltage on the shield of that group. The position of a conductor within an overall shield thus has only a second-order effect on the coupling from the internal electric field of the shield to the signal conductor.

Figure 13.27 shows shield voltages measured with a low-impedance load connecting the center shield to ground. A load impedance of 47 Ω approximately matches the surge impedance of the transmission line existing between any of the shields on the internal groups and overall shield. It prevents the development of oscillations which would otherwise be superimposed on an underlying waveform, like those shown in Figure 13.26. Note that the voltage on the center shield is now increased to nearly twice what it was with a high-impedance load on the ends of the shields. Note, furthermore, that the voltage on the other shields remains nearly the same as it was with all shields at the left open circuited. The slight increase of voltage noted on the unconnected shields results from capacitive coupling at the left end between the unconnected shields and the one center shield that is held nearly at ground potential by the 47 Ω load. Oscillogram No. 11 on Figure 13.28 shows the nature of these superimposed oscillations if the shield on the center complex is grounded directly. The voltage, V, shown measured at the right-hand end of the cable was in fact measured on one of the conductors within the center complex. The same voltage would be measured on any of the other conductors or on the shield itself.

If the ground on the center shield is moved to the right-hand side, the

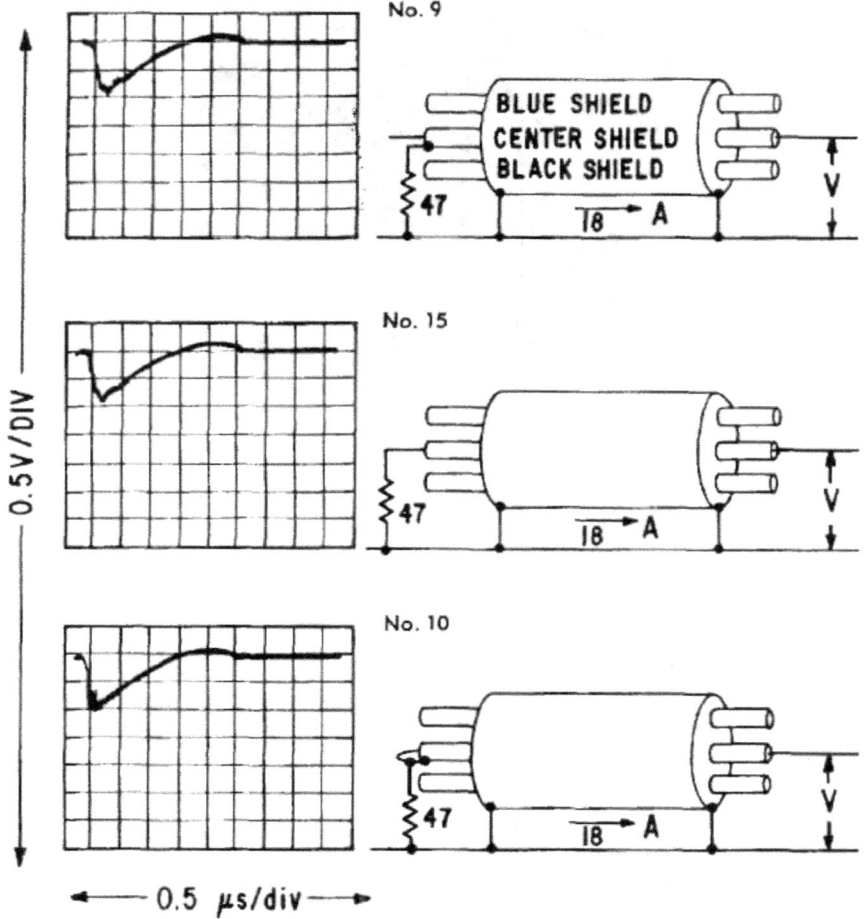

Figure 13.26 Signal voltages vs source connections.

voltage between a signal conductor and ground at the right-hand side is then reduced to zero because the capacitive coupling between the center conductor and center shield tries to hold the center conductor at the potential of the center shield, which, of course, is zero. A voltage of magnitude and waveshape as shown on oscillogram No. 11, but of opposite polarity, would then appear at the left-hand end of the cable.

If both ends of the center shield are connected to ground, the voltage on the signal conductor is then held at zero at each end of the cable. Note that in this case a ground loop is formed between the two shields without the harmful effects commonly attributed to ground loops.

Some additional considerations regarding the effect of connections to the signal conductors and the shield around the signal conductors are shown in

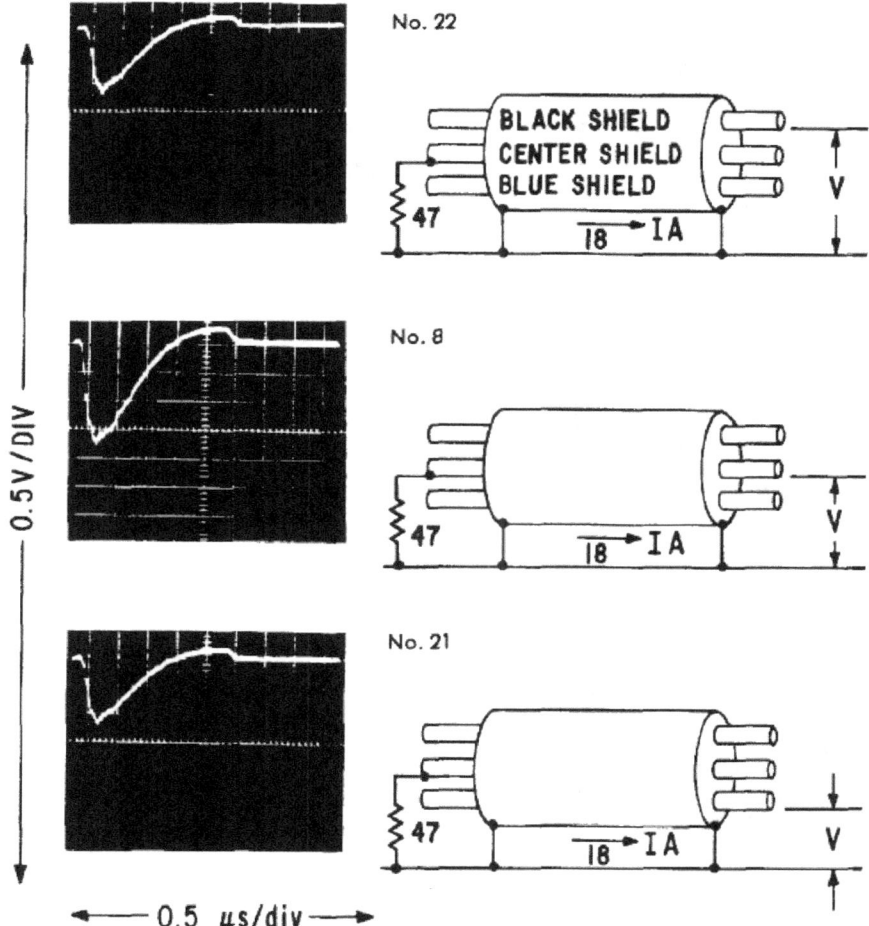

Figure 13.27 Signal shield voltages: low source impedance on one shield.

Figure 13.29. Compare oscillogram No. 17 of this figure to oscillogram No. 11 in Figure 13.28. Connecting the signal conductor to ground via a low impedance at the same end as that at which the shield around that signal conductor is connected to ground does not change the voltage appearing at the other end of the cable on that signal conductor. If the ground on that shield is moved to the right-hand of the cable, however, the voltage on the signal conductor at the right-hand end of the cable remains virtually unchanged. This was not the case when there was no low-impedance termination on the signal conductor at the left-hand end of the cable. Compare oscillogram No. 16 of Figure 13.29 with oscillogram No. 14 of Figure 13.28. Finally, note that a ground on the internal shield at each end keeps the voltage on the signal conductor low even with a low-impedance connection to ground on the other end of the signal conductor.

391

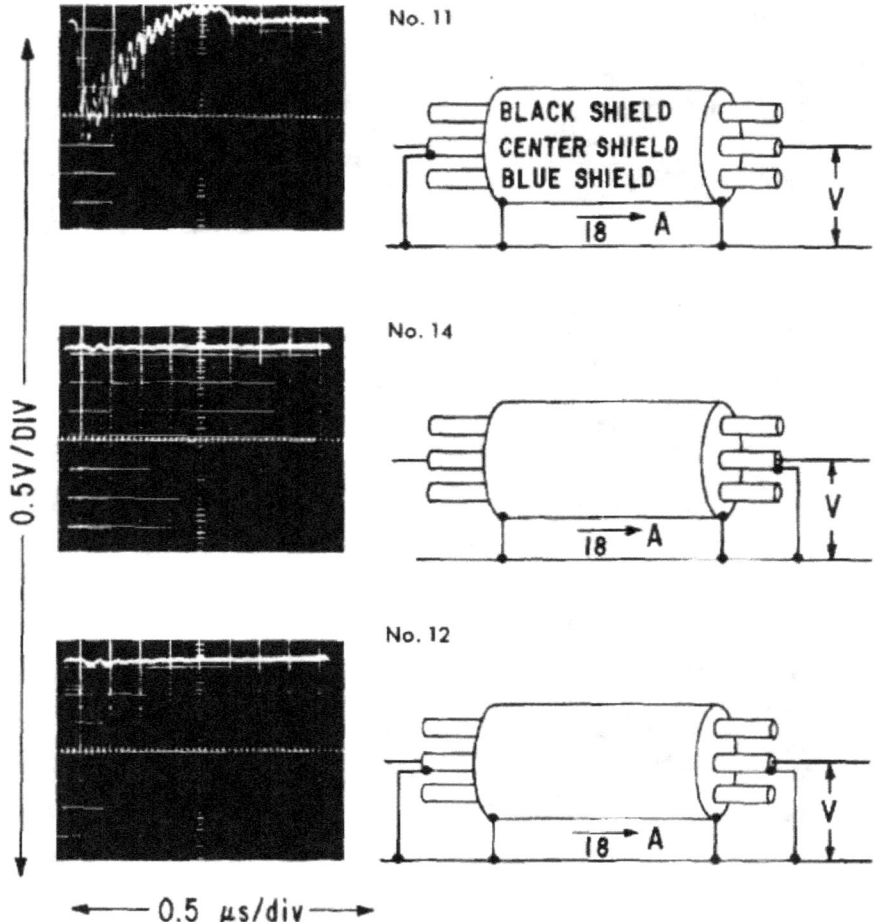

No. 11

BLACK SHIELD
CENTER SHIELD
BLUE SHIELD

No. 14

No. 12

0.5 V/DIV

0.5 μs/div

Figure 13.28 Signal voltage vs grounding of signal shield: high source impedance – high receiving impedance.

Compare oscillogram No. 18 in Figure 13.29 to oscillogram No. 12 in Figure 13.28.

The significant feature of these measurements is that rather complex phenomena can occur on multiple-shielded cables when the grounding of the inner shield is different from the grounding of the overall shield. Furthermore, the voltages on signal conductors can often be significantly lowered if the internal shield is grounded at each end, and the corresponding voltages are scarcely reduced at all if the inner shield is not grounded at each end. Most commonly, of course, an inner shield is grounded at only one end for control of low-frequency or low-level interference. Multiple grounding of the internal shields and the production of ground loops, then, is not automatically harmful; it may, in fact, be helpful.

392

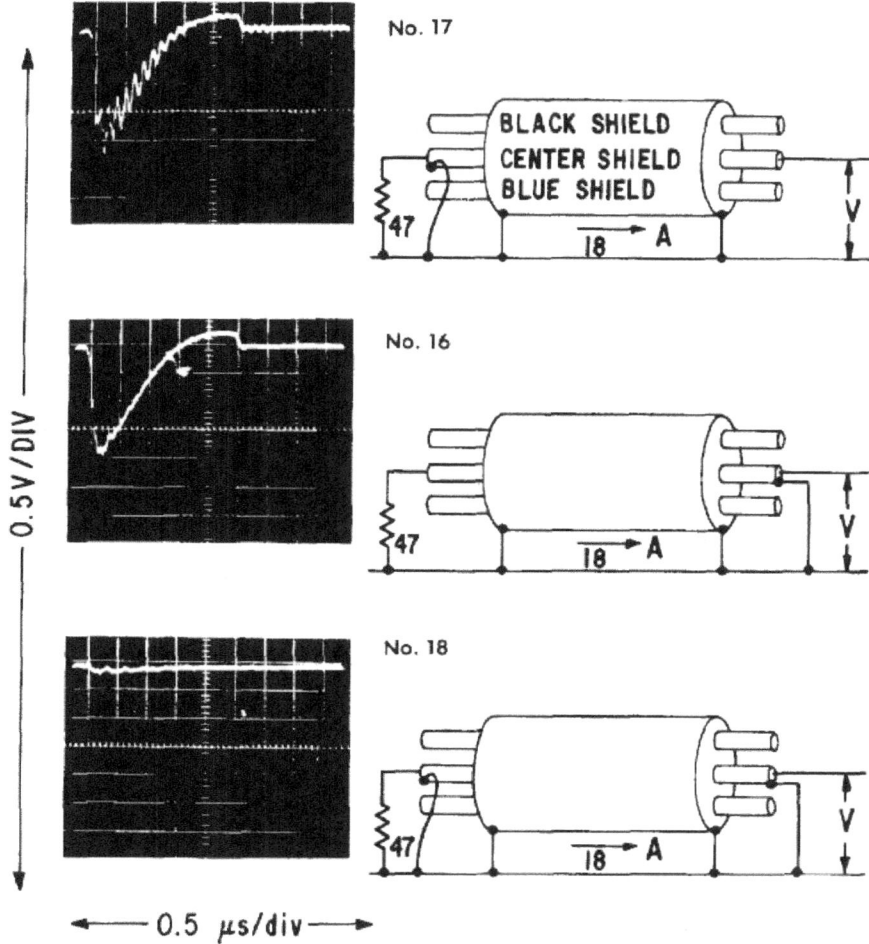

Figure 13.29 Signal voltage vs grounding of signal shield: low source impedance — high receiving impedance.

13.5 Transfer Characteristics of Actual Cables

The transfer impedance of a shielded cable is not a factor commonly specified either by the manufacturer or by procurement specifications. Even among cables of the same nominal type, this transfer impedance may vary considerably between the cables supplied by different manufacturers. The most straightforward method of determining the transfer characteristic of a shielded cable, particularly if the cable involves multiple shields and multiple load impedances on conductors within the shield, is to make actual measurements of the conductor voltages that are produced by currents which are circulated through the shield. The techniques by which transient currents may be injected

393

into the shields of cables through coupling transformers are described in Chapter 17.

If actual measurements of coupling effects are not available, the transfer characteristics of many cables can be estimated from other published literature. One such summary giving the parameters of coaxial cable shields is shown in Table 13.1 (Reference 13.9).

13.6 Transfer Characteristics of Connectors

The transfer impedance of a cable connector or splice can be represented by

$$Z_T = R_0 + j\omega M_{12} \qquad (13.23)$$

where R_0 is the resistance measured across the connector and M_{12} is a mutual inductance between the external shield circuit and the internal conductors of the cable. The value of Z_T is usually not calculable, but it can be measured by passing current through a cable sample containing the connector and measuring the open circuit voltage induced on conductors inside the shield. Some typical values of R_0 and M_{12} have been measured on cable connectors and are listed in Table 13.2.

The transfer impedance of a connector is a lumped element in the cable circuit, in contrast to the distributed nature of the transfer impedance of the cable shield. The effect of leakage through the connector can thus be represented by a discrete voltage source

$$V = I_0 Z_t \qquad (13.24)$$

where I_0 is the shield current.

Because most connectors that would be used for shielded cables have essentially 100% optical coverage, their transfer admittance is usually negligible. In addition, most bulkhead or panel-mounting connectors are located at points where the shield voltage is a minimum, so that excitation of the internal conductors by the transfer admittance is small even if the transfer admittance itself is not small.

13.7 Grounding of Shields

It should be emphasized that the transfer parameters of Table 13.2 refer only to the properties of the connector itself. Additional transfer impedances may be produced by the manner in which the cable shield is connected to the connector or by the manner in which the cable connector is mounted to a bulkhead. Even slight inattention to detail may introduce into the circuit transfer impedances far greater than the impedance of the connector or possibly far greater than that of the rest of the cable shield. A common treatment of a shield at a connector is to insulate the shield with tape and connect it to the back shell through a pigtail. Such treatment is shown in Figure 13.30(a). Equally

Table 13.1 COAXIAL CABLE SHIELD PARAMETERS.*

Cable Type (RG-)		Strand Diameter d (inches)	Outside Diameter (inch)	Carriers, C	Ends, N	Picks, P (inch⁻¹)	Weave Angle α (degrees)	Fill, F	Coverage, K (%)	Weight per Foot (lbs)	Stranding Factor	DC Resistance, R_o (mΩ/m)	Leakage Inductance, M_{12} (nH/m)	Leakage Capacitance, $C_{12}/C, C_2$ (m/F)
6	I	0.0063	0.189	16	9	5.90	25.0	0.790	95.61	0.019	1.54	6.6	0.42	2.9×10^7
6	AI	0.0063	0.189	24	6	8.90	24.9	0.790	95.58	0.019	1.54	6.6	0.28	1.9
6	O	0.0063	0.214	16	9	8.79	37.7	0.886	96.23	0.022	1.96	7.5	0.36	1.9
6	AO	0.0063	0.214	24	6	13.00	37.6	0.885	96.18	0.022	1.98	7.5	0.25	1.3
11		0.0071	0.292	24	8	6.50	27.5	0.799	95.96	0.033	1.59	4.0	0.25	1.6
22	I	0.0063	0.291	24	8	9.10	35.9	0.763	95.27	0.028	1.95	5.5	0.34	1.9
22	O	0.0063	0.316	24	8	12.00	45.9	0.842	97.50	0.033	2.45	6.4	0.14	0.6
23	I	0.0063	0.304	24	9	10.50	48.2	0.799	95.95	0.039	2.82	5.9	0.29	1.2
25	I	0.0063	0.298	16	13	5.00	31.4	0.786	95.44	0.029	1.74	4.8	0.46	2.8
25	AI	0.0063	0.298	24	9	6.00	26.0	0.776	94.98	0.029	1.69	4.4	0.34	2.3
25	O	0.0063	0.355	16	15	5.00	35.8	0.887	96.29	0.036	1.88	4.4	0.35	1.9
25	AO	0.0063	0.355	24	11	5.00	25.7	0.799	95.96	0.035	1.54	3.7	0.25	1.6
35		0.0071	0.470	24	10	9.00	48.8	0.850	97.74	0.056	2.71	4.3	0.12	0.5
58		0.0050	0.120	12	9	7.70	27.7	0.745	93.57	0.089	1.71	14.2	1.0	6.6
58	A	0.0050	0.120	16	7	10.30	27.7	0.775	94.92	0.018	1.65	13.7	0.53	3.5
59		0.0063	0.150	16	7	8.20	27.6	0.790	95.14	0.015	1.63	8.6	0.49	3.2
59	A	0.0063	0.150	24	5	12.30	27.6	0.835	97.29	0.016	1.53	8.1	0.14	0.9
62		0.0063	0.151	16	7	8.20	27.8	0.776	94.98	0.015	1.65	8.7	0.52	3.4
62	A	0.0063	0.151	24	5	12.30	27.8	0.831	97.15	0.016	1.54	8.1	0.15	1.0
63		0.0071	0.295	16	12	4.30	27.6	0.792	95.66	0.033	1.61	4.0	0.42	2.7
63	A	0.0071	0.295	24	8	6.50	27.8	0.793	95.71	0.033	1.61	4.0	0.27	1.8

A – Alternate
A1 – First alternate
A2 – Second alternate

I – Inner
O – Outer
S – Shield (insulated from braid)

* Braid actually consists of flat tape conductors. Strand diameter and ends are estimated for the equivalent cross-sectional area of the tape conductor; therefore, F, K, and stranding factor are approximate.

Table 13.1 COAXIAL CABLE SHIELD PARAMETERS. (CONTINUED)

Cable Type (RG-)		Strand Diameter d (inches)	Outside Diameter (inch)	Carriers, C	Ends, N	Picks, P (inch⁻¹)	Weave Angle, α (degrees)	F_il, F	Coverage, K (%)	Weight per Foot (lbs)	Stranding Factor	DC Resistance, R₀ (mΩ/m)	Leakage Inductance, M₁₂ (nH/m)	Leakage Capacitance, C₁₂/C₁C₂ (m/F)
65		0.0071	0.295	24	8	6.50	27.8	0.793	95.71	0.033	1.61	4.0	0.27	1.8
108		0.0050	0.164	16	6	10.00	36.4	0.546	79.36	0.009	2.83	17.6	4.6	25.0
114		0.0063	0.295	24	8	7.00	29.4	0.718	92.07	0.020	1.83	5.1	0.70	4.4
119	–	0.0071	0.337	24	10	5.40	26.4	0.862	98.10	0.041	1.45	3.1	0.08	0.5
119	O	0.0063	0.367	24	8	10.60	46.5	0.737	93.06	0.033	2.86	6.5	0.65	2.8
122		0.0050	0.099	16	6	12.90	28.9	0.891	96.82	0.008	1.63	16.2	0.37	2.4
122	A	0.0050	0.099	24	5	12.20	19.2	0.978	99.48	0.010	1.21	12.0	0.01	0.08
130		0.0100	0.487	24	8	6.30	39.9	0.786	95.41	0.076	2.16	2.3	0.33	1.7
142	–	0.0050	0.121	16	7	11.50	30.6	0.791	95.61	0.010	1.71	14.1	0.43	2.7
142	O	0.0050	0.141	16	7	14.50	40.7	0.778	95.09	0.011	2.23	16.0	0.55	2.8
144		0.0063	0.290	24	8	9.20	36.1	0.787	95.47	0.029	1.95	5.5	0.32	1.7
156	–	0.0063	0.290	24	8	11.20	41.6	0.851	97.77	0.031	2.10	6.0	0.11	0.55
156	O	0.0070	0.333	24	8	9.20	39.9	0.803	96.13	0.034	2.11	4.7	0.26	1.3
156	S	0.0063	0.413	24	8	14.00	57.3	0.838	97.38	0.043	4.10	8.3	0.17	0.51
157	–	0.0063	0.465	24	9	12.90	58.0	0.856	97.92	0.049	4.16	7.5	0.12	0.36
157	O	0.0070	0.500	24	10	7.30	44.5	0.779	92.67	0.046	2.69	4.1	0.70	3.1
157	S	0.0063	0.580	24	9	13.50	64.5	0.848	97.70	0.050	6.35	9.2	0.16	0.34
174		0.0040	0.063	16	4	16.30	24.4	0.630	86.33	0.003	1.91	36.5	2.3	15.8
179		0.0040	0.066	16	5	12.00	19.2	0.729	92.65	0.004	1.54	28.1	0.88	6.6
181	–	0.0063	0.215	24	7	8.90	27.7	0.836	97.30	0.023	1.53	5.7	0.14	0.9
181	O	0.0100	0.490	24	8	7.00	43.1	0.820	96.76	0.090	2.28	2.4	0.20	0.95
189	–	0.0100	0.635	48	7	6.00	27.2	0.918	99.33	0.114	1.38	1.1	0.008	0.06
189	O	0.0100	0.600	48	6	7.00	32.7	0.778	95.07	0.104	1.81	1.4	0.17	1.0
192	–	0.0100	1.725	48	11	5.87	53.3	0.806	96.22	0.265	3.47	1.1	0.14	0.49
192	O	0.0095	1.780	48	11	5.95	54.5	0.764	94.43	0.182	3.88	1.1	0.25	0.85
192	S	0.0100	1.890	48	11	6.03	56.4	0.796	95.84	0.289	4.11	1.2	0.17	0.53
193	–	0.0100	1.725	48	11	4.15	43.5	0.664	88.68	0.220	2.66	0.89	0.66	3.1
193	O	0.0095	1.780	48	11	5.50	52.3	0.726	92.58	0.275	3.69	1.2	0.39	1.4
193	S	0.0100	1.890	48	9	7.70	67.6	0.781	95.20	0.264	6.83	1.7	0.23	0.54

Table 13.1 COAXIAL CABLE SHIELD PARAMETERS. (CONTINUED)

Cable Type (RG-)		Strand Diameter d (inches)	Outside Diameter (inch)	Carriers C	Ends N	Picks P (inch^{-1})	Weave Angle α (degrees)	Fill F	Coverage K (%)	Weight per Foot (lbs)	Stranding Factor	DC Resistance R (mΩ/m)	Leakage Inductance M$_{12}$ (nH/m)	Leakage Capacitance C$_{12}$/C$_1$C$_2$ (pF/F)
194	I	0.0100	1.725	48	11	4.15	43.5	0.664	88.66	0.220	2.86	0.89	0.66	3.1
194	O	0.0095	1.700	48	11	5.50	52.3	0.728	92.50	0.225	3.69	1.2	0.11	0.38
210		0.0063	0.151	16	7	8.20	27.8	0.776	94.98	0.015	1.65	8.7	0.52	3.4
210	A	0.0063	0.151	24	5	12.30	27.8	0.831	97.15	0.016	1.54	8.1	0.15	0.96
211		0.0080	0.625	36	10	5.64	32.1	0.844	97.56	0.082	1.65	1.7	0.09	0.48
211	A	0.0080	0.625	48	8	5.60	25.2	0.843	97.53	0.082	1.45	1.5	0.06	0.40
212		0.0063	0.189	16	9	5.90	25.0	0.790	95.61	0.019	1.54	6.6	0.42	2.9
212	AI	0.0063	0.189	24	6	8.80	24.9	0.790	95.58	0.019	1.54	6.6	0.28	1.9
212	O	0.0063	0.214	16	9	8.70	37.7	0.806	96.23	0.022	1.98	7.5	0.36	1.9
212	AO	0.0063	0.214	24	6	13.00	37.6	0.805	96.18	0.022	1.98	7.5	0.25	1.3
213		0.0071	0.292	24	8	6.50	27.5	0.799	95.96	0.033	1.59	4.0	0.25	1.6
214		0.0063	0.292	24	6	16.60	52.9	0.786	95.44	0.029	3.50	9.9	0.37	1.3
214	O	0.0063	0.317	24	7	15.40	53.0	0.850	97.75	0.034	3.25	8.5	0.13	0.45
217		0.0071	0.380	24	10	5.40	29.1	0.788	95.49	0.042	1.66	3.2	0.30	1.9
217	O	0.0071	0.405	24	8	10.60	49.3	0.794	95.75	0.045	2.96	5.4	0.32	1.3
218		0.0100	0.690	24	14	3.10	30.0	0.869	98.29	0.118	1.53	1.2	0.07	0.44
218	AI	0.0100	0.690	36	9	4.00	26.4	0.811	96.41	0.118	1.54	1.2	0.14	0.92
218	A2	0.0100	0.690	48	7	5.60	27.5	0.849	97.72	0.115	1.59	1.1	0.05	0.35
220		0.0100	0.925	36	12	3.50	30.0	0.840	97.44	0.151	1.59	0.91	0.09	0.53
220	A	0.0100	0.925	48	9	4.20	27.5	0.820	96.76	0.148	1.55	0.89	0.09	0.59
227		0.0063	0.189	16	9	5.90	25.0	0.790	95.61	0.019	1.54	6.6	0.42	2.9
227	AI	0.0063	0.189	24	6	8.80	24.9	0.790	95.58	0.019	1.54	6.6	0.28	1.9
227	O	0.0063	0.214	16	9	8.70	37.7	0.806	96.23	0.022	1.98	7.5	0.36	1.9
222	AO	0.0063	0.214	24	6	13.00	37.6	0.805	96.18	0.022	1.98	7.5	0.25	1.3
223		0.0050	0.120	12	9	9.00	31.5	0.775	94.95	0.010	1.77	14.8	0.72	4.4
223	AI	0.0050	0.120	16	7	11.50	30.4	0.795	95.90	0.010	1.69	15.5	0.43	2.3
223	O	0.0050	0.140	12	9	10.00	38.1	0.729	92.63	0.010	2.22	16.0	1.3	7.0
223	AO	0.0050	0.140	16	7	15.00	41.5	0.793	95.71	0.011	2.25	16.2	0.45	2.7
225		0.0063	0.298	24	6	16.60	52.7	0.788	95.52	0.029	3.46	9.8	0.36	1.3

Table 13.1 COAXIAL CABLE SHIELD PARAMETERS. (CONTINUED)

Cable Type (RG-)		Strand Diameter d (inches)	Outside Diameter (inch)	Carriers, C	Ends, N	Picks, P (inch⁻¹)	Weave Angle, ρ (degrees)	Fill, F	Coverage, K (%)	Weight per Foot (lbs)	Stranding Factor	DC Resistance, R_e (mΩ/m)	Leakage Inductance M_{12} (nH/m)	Leakage Capacitance $C_{12}/C_1 C_2$ (m/F)
225	O	0.0063	0.315	24	7	15.40	52.9	0.852	97.80	0.033	3.22	8.5	0.12	0.44
226	1	0.0063	0.375	24	10	10.50	46.8	0.907	99.14	0.042	2.35	5.2	0.03	0.12
226	O	0.0063	0.400	24	8	10.50	48.6	0.706	91.33	0.035	3.24	6.6	0.92	3.7
301		0.0050	0.190	16	10	8.00	32.1	0.752	93.84	0.014	1.86	10.0	0.73	4.4
302		0.0050	0.151	16	7	11.50	36.0	0.684	99.04	0.010	2.23	15.0	1.5	8.4
303		0.0050	0.121	16	7	11.50	30.6	0.791	95.61	0.010	1.71	14.1	0.43	2.7
304	1	0.0063	0.190	24	5	14.50	37.6	0.749	93.71	0.213	2.12	9.0	0.52	2.8
304	O	0.0063	0.215	24	6	11.50	34.4	0.769	94.57	0.021	1.91	7.2	0.40	2.3
316		0.0040	0.063	16	5	4.50	7.2	0.273	92.32	0.004	1.41	26.8	0.88	7.7
326	1*	0.0035	0.550	24	27	6.46	43.3	0.890	98.80	0.047	2.12	5.9	0.05	0.22
326	O*	0.0035	0.566	24	27	6.46	44.1	0.877	98.49	0.043	2.21	6.0	0.06	0.30
328	1	0.0100	1.085	48	9	5.50	38.5	0.795	95.80	0.167	2.05	1.0	0.14	0.75
328	O	0.0070	1.125	48	12	6.70	45.0	0.796	95.85	0.111	2.51	1.7	0.15	0.66
328	S	0.0100	1.225	48	9	5.60	42.4	0.748	93.63	0.177	2.45	1.1	0.28	1.3
329	1	0.0100	0.390	24	7	5.90	32.3	0.772	94.80	0.050	1.82	2.4	0.38	2.3
329	O	0.0070	0.430	24	9	9.20	46.9	0.794	95.74	0.043	2.70	4.7	0.31	1.3
391		0.0063	0.307	24	7	16.30	53.8	0.891	98.82	0.034	3.21	8.1	0.05	0.17

Table 13.2 RESISTANCE AND MUTUAL INDUCTANCE OF CABLE CONNECTORS

Connector	Identification	R_o (ohms)	M_{12} (H)
Multipin Aerospace connectors (Threaded)	Burndy NA5-15863	0.0033	5.7×10^{-11}
	Deutch 38068-10-5PN	0.15	2.5×10^{-11}
	Deutch 38068-18-31SN	0.005	1.6×10^{-10}
	Deutch 38060-22-55SN	0.023	1.1×10^{-10}
	Deutch 38068-14-7SN	0.046	5.0×10^{-11}
	Deutch 38060-14-7SN	0.10	8.2×10^{-11}
	Deutch 38060-14-7SN	0.023	6.7×10^{-11}
	Deutch 38068-12-12SN	0.0033	3.0×10^{-11}
	Deutch 38068-12-12SN	0.012	1.3×10^{-11}
	Deutch 38068-12-12SN	0.012	1.3×10^{-11}
	Deutch 38060-12-12SN	$<$0.001	2.5×10^{-12}
	Deutch 38068-12-12SN	0.014	3.5×10^{-11}
	AMP	0.0067	1.6×10^{-11}
	AMP	0.0067	1.5×10^{-11}
	AMP	0.0033	1.9×10^{-11}
Type N	UG 21B/U-UG58A/U	*	*
Type BNC (Bayonet)	UG 88C/U-UG1094/U	0.002	$4-8 \times 10^{-11}$
Anodized	MS 24266R-22B-55	5×10^4	$\omega M < R_o$ @ 20 MHz
Open shell	MS 3126-22-55	0.5-1	$\omega M < R_o$ @ 20 MHz
Split shell	MS 3100-165-1P MS 3106A-	0.001	$\approx 20 \times 10^{-11}$

*Too small to measure in presence of 4 inches of copper tube used to mount connector.

common practice is to insulate a panel connector from a panel with an insulating block and to ground the panel connector either to the panel through a pigtail or, more commonly, to an internal ground bus. A very common treatment of a shield at a connector, shown in Figure 13.30(c), is for the shield to be connected to one of the connector pins and grounded internally through a pigtail, either to the panel or to an internal ground bus.

The coupling introduced by any of these configurations can be studied in terms of the self-impedance of the conductor used for the pigtail and the mutual inductance between a portion of the pigtail and the conductors in the connector.

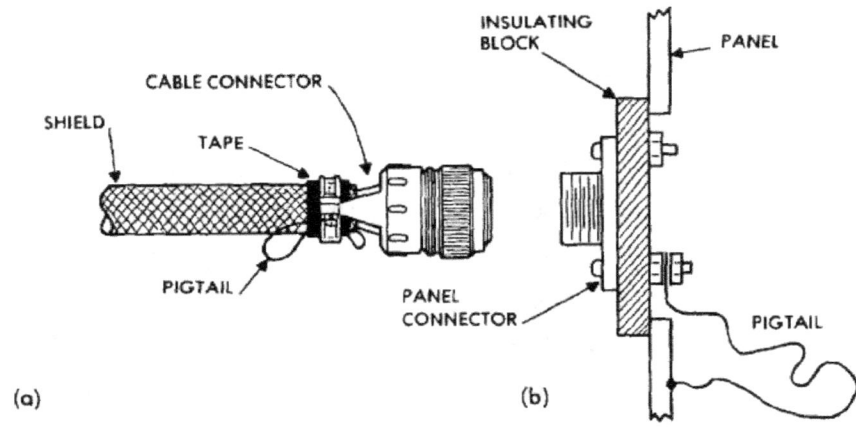

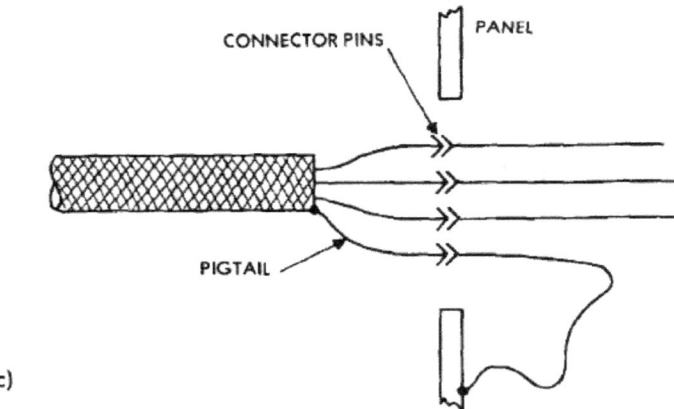

Figure 13.30 Common treatment of shields at connectors.
   (a) Pigtail connection to a backshell
   (b) Pigtail grounding to a panel connector
   (c) Shield carried on a connector pin

The flow of current through the self-inductance of the pigtail produces a voltage

$$V = j\omega L \qquad (13.25)$$

This voltage, less that induced in the mutual inductance between the pigtail and the conductors in the connector, will add directly to that produced by the flow of current along the shield.

The self-inductance of a straight conductor is

$$L = 2\ell \left[\log_\epsilon \frac{2\ell}{r} - \frac{3}{4}\right] \quad \text{nH/cm} \qquad (13.26)$$

where ℓ is the length of the conductor and r is its radius, both measured in centimeters. The inductance of a typical pigtail will not be significantly different even if it is not straight.

The mutual inductance between the two conductors is

$$M = 2\ell Q \; \text{nH/cm} \qquad (13.27)$$

where the factor Q, shown in Figure 13.31, is a function of the ratio of the length of conductors to their spacing. If $d \ll \ell$, then

$$M = 2\ell \left[\log_\epsilon \frac{2\ell}{d} - 1 + \frac{d}{\ell} - \frac{d^2}{2\ell} \cdots\right] \quad \text{nH/cm} \qquad (13.28)$$

Mutual inductance so defined is not the same as the mutual inductance referred to in Equations 13.4 and 13.12. This latter is best referred to as a *transfer inductance*, which one would normally want to be small. The corresponding quantity for the parallel conductors would be L minus M. One would normally want the mutual inductance (Equation 13.27) between two conductors to be large.

A numerical example will illustrate the problems associated with pigtail grounds. Assume a shield is carried, as in Figure 13.30(c), through a panel on a pigtail and a set of connector pins. Assume the total length of the pigtail-pin combination to be 7 cm, the spacing between the pigtail and a signal conductor to be, on the average, 0.05 cm, and the pigtail to be made from number 22AWG wire having a radius of 0.0323 cm. The self-inductance of the pigtail would then be 74.5 nH, the mutual inductance between the two would be 65.0 nH, and the difference, the factor equivalent to the transfer inductance for the shielded cable, would be 9.5 nH. A rather mediocre shielded braid would have an inductance on the order of 1 nH/m, and a large-diameter, tightly woven braid might have an inductance on the order of 0.25 nH/m. The amount of voltage injected into a signal circuit by the current flowing through this one relatively short pigtail, then, would be as much as that coupled onto the signal conductor by 10 to 40 m of cable shield braid. This point may be illustrated by Figure 13.32, an extension of the test series previously discussed in reference to Figures

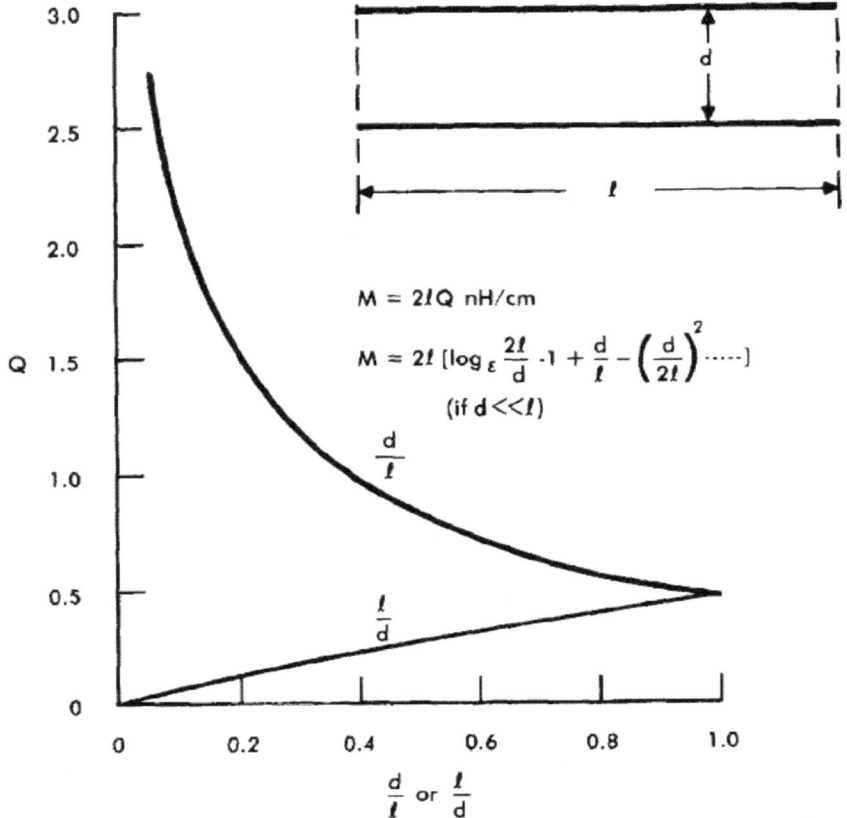

The figure shows a graph with Q on the vertical axis (ranging from 0 to 3.0) and d/ℓ or ℓ/d on the horizontal axis (ranging from 0 to 1.0). Two curves are plotted, labeled d/ℓ and ℓ/d. At top is a diagram of two parallel conductors separated by distance d and of length ℓ, with the equations:

$$M = 2\ell Q \text{ nH/cm}$$

$$M = 2\ell \left[\log_\varepsilon \frac{2\ell}{d} \cdot 1 + \frac{d}{\ell} - \left(\frac{d}{2\ell}\right)^2 \cdots \right]$$

$$(\text{if } d \ll \ell)$$

Figure 13.31 Mutual inductance.

13.1 through 13.6. One end of the cable was grounded through a 33-inch length of wire, while the other end was connected to the ground plane directly. Voltage V_1 was increased from 0.4 V (Figure 13.6) to 13.5 V solely because of the inductive drop in the ground lead.

There are two fundamental drawbacks to pigtail grounding. The first, shown in Figure 13.33(a), is that the shield current, being constricted to a path of small diameter, sets up a more intense magnetic field at the surface of the conductor than it would if the conductor were larger. Because of the higher flux density, the self-inductance of the conductor is larger. A more intense magnetic field of itself would not really be harmful if all the flux so set up would link the signal conductors as well, because then the mutual inductance would also be high, and the difference, L minus M, would be low. If the shield current is confined to a small-diameter path, the field intensity rapidly falls off with distance away from the conductor, and hence the mutual inductance will always be smaller than the self-inductance.

If the shield current is carried to ground on a shell concentric with the

402

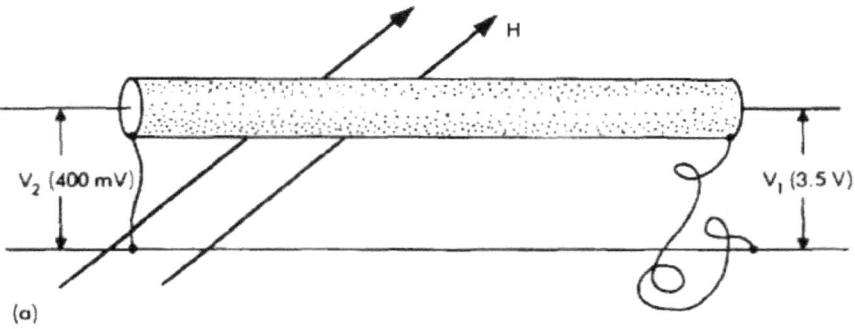

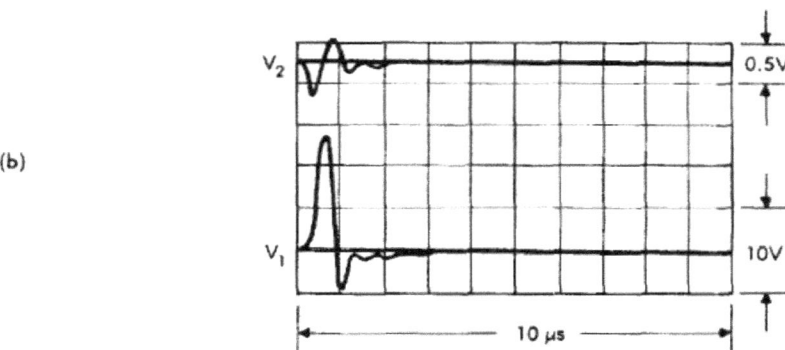

Figure 13.32 Effects of inductance in a ground lead.
(a) Circuit
(b) Voltage

conductors within the shield, shown in Figure 13.33(b), there is much less voltage introduced into the conductors for the following reasons:
- The length of the path through which the shield current must flow is shorter.
- The field intensity external to the shield is reduced by virtue of the inherently larger diameter of the path upon which the current flows.
- The field intensity inside the shield is low--nearly zero.

The shorter, and larger diameter, current path implies less self-inductance. The absence of magnetic flux within the shield implies that the signal conductors are exposed to as much flux as is the shield: that M = L, and L minus M = 0.

403

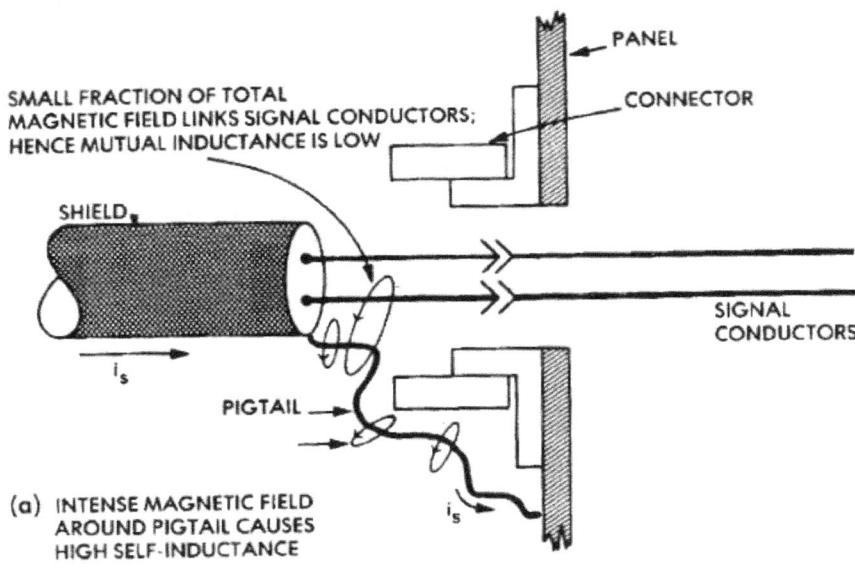

SMALL FRACTION OF TOTAL
MAGNETIC FIELD LINKS SIGNAL CONDUCTORS;
HENCE MUTUAL INDUCTANCE IS LOW

PANEL

CONNECTOR

SHIELD

i_s

PIGTAIL

SIGNAL
CONDUCTORS

(a) INTENSE MAGNETIC FIELD
AROUND PIGTAIL CAUSES
HIGH SELF-INDUCTANCE

i_s

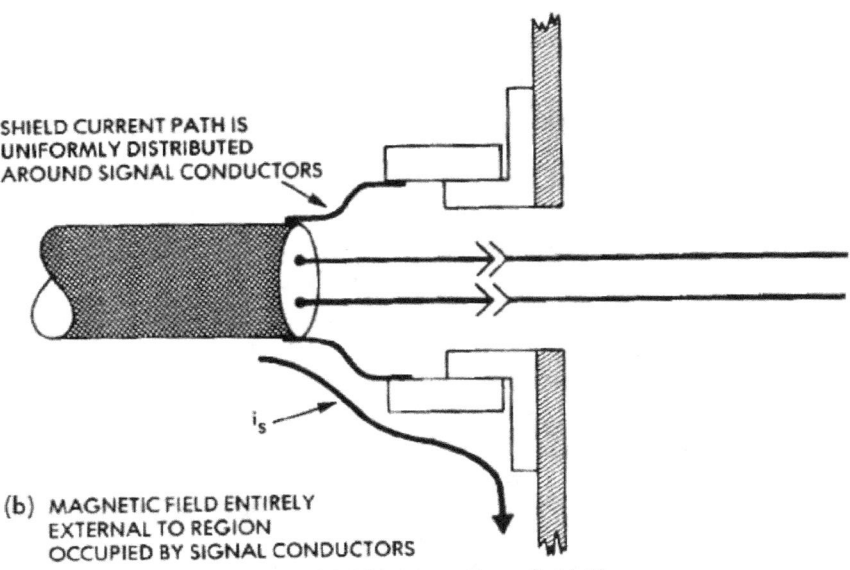

SHIELD CURRENT PATH IS
UNIFORMLY DISTRIBUTED
AROUND SIGNAL CONDUCTORS

i_s

(b) MAGNETIC FIELD ENTIRELY
EXTERNAL TO REGION
OCCUPIED BY SIGNAL CONDUCTORS

Figure 13.33   Grounding of shields.
(a) Pigtails
(b) 360° peripheral

404

REFERENCES

13.1   F. A. Fisher, *Effects of q Changing Magnetic Field on Shielded Conductors*, Lightning Protection Note 75-2, Internal General Electric Memorandum, High Voltage Laboratory, Corporate Research and Development, General Electric Company, Pittsfield, Massachusetts (2 June 1975).

13.2   Edward F. Vance, *Coupling to Cables*, DNA Handbook Revision, Chapter 11, Stanford Research Institute, Menlo Park, California (December 1974).

13.3   Vance, p. 11-125.

13.4   Vance, p. 11-128.

13.5   Vance, p. 11-129.

13.6   Vance, p. 11-130.

13.7   Vance, p. 11-131.

13.8   S. A. Shelkunoff, "Electromagnetic Theory of Coaxial Transmission Lines and Cylindrical Shields," *Bell System Technical Journal*, 13 (1934): 532-79.

13.9   Vance, pp. 1-133 to 11-136.

13.10 Vance, p. 11-166.

CHAPTER 14
EXAMPLES OF INDUCED VOLTAGES
MEASURED ON AIRCRAFT

14.1 Introduction

There have been several sets of tests made on aircraft in which simulated lightning currents were injected into the aircraft and the resultant voltages and currents on the aircraft wiring measured. Space precludes giving in this report a comprehensive summary of the test results, and the user is referred to the reports on the individual tests. A few examples of the measurements will be given in the following sections, and an attempt will be made to indicate the most important facts learned during the tests.

14.2 Wing from F-89J Aircraft

The first set of tests to be discussed (Reference 14.1) was one in which high lightning-like currents were injected into one wing of an F-89J aircraft. During the test, represented in Figure 14.1, the wing was fastened onto a screened instrument enclosure, which may be viewed as representing the fuselage of the aircraft. Lightning-like currents of up to 40 000 A were injected into the wing or into the external wing tip tank from a high-current surge generator, allowed to flow along the wing to the outer wall of the screened instrument enclosure, and then to ground. An example of one of the types of current injected into the wing is shown in Figure 14.1(b). In order to obtain maximum current, the surge generator was operated in a mode that essentially allowed the production of only one cycle of a damped oscillatory current, unlike a typical lightning current, which would rise to crest rather fast and decay at a much slower rate. The shape of the current wave must be considered when observing the waveshape of some of the voltages that will be discussed. In particular, note that at about 20 μs there appears a major discontinuity in waveshape. This discontinuity in current waveshape is reflected in the induced voltages.

Within the wing there were a number of electrical circuits, such as those to navigation lights, fuel gauges, pumps, relays, and switches indicating position of flaps. Some of these ran in the leading edge of the wing and were well shielded from many electromagnetic effects, while others ran along the trailing edge between the main body of the wing and the wing flaps. These latter were most exposed to the electromagnetic fields. All of the circuits were relatively independent of each other; they were not, as a general rule, bundled together in one large cable bundle, a practice that provides maximum coupling from one circuit to another and makes analysis difficult.

The first circuit that will be discussed, shown in Figure 14.2, was a circuit supplying power to a position light mounted on the external fuel tank. An electrical diagram of the circuit, shown in Figure 14.2(b), shows that the circuit consisted of one wire supplying power to the filament of the position light with the return circuit for the light being through the wing structure. Accordingly, if

407

the lightning current contacts the external tank, that circuit will be influenced by the resistance R_1 of the hangers fastening the tank to the wing, by R_2, the inherent resistance of the wing, and by magnetic flux arising from the flow of current.

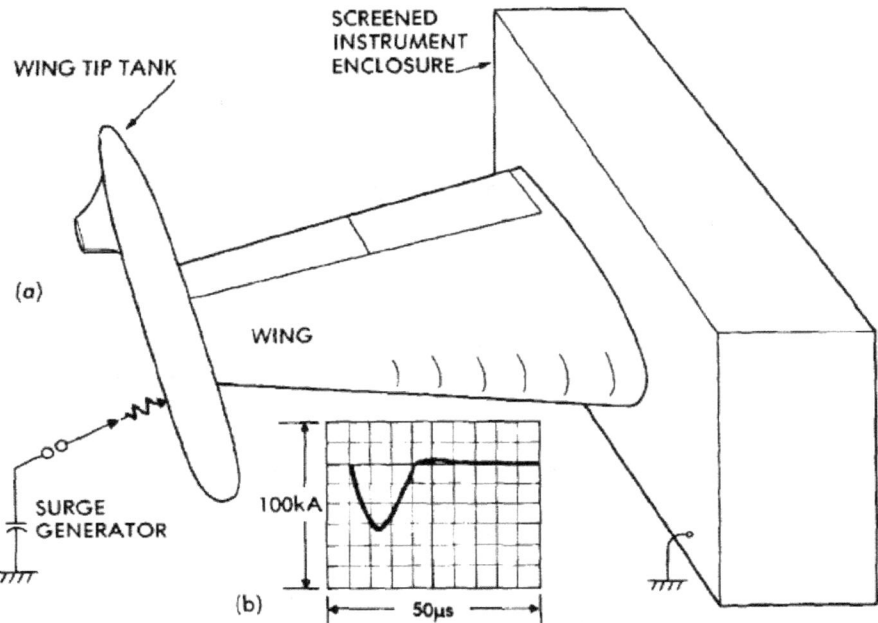

Figure 14.1 High current injection tests on the wing from an F-89J.

    (a) Test arrangement

    (b) Waveshape of injected current

       Typical results for the stroke position shown are given in Figure 14.2(c) and 14.2(d). The open circuit voltage is seen to rise rapidly to its crest and to decay more rapidly than does the injected current shown in Figure 14.1(b). As a result, the open circuit voltage was responding primarily to magnetic flux.

       When the conductor was shorted to ground at the instrument enclosure, the short circuit current rose to its crest in approximately the same length of time as did the injected current and displayed much the same waveshape as the injected current.

       Figure 14.2(e) shows an approximate equivalent circuit that might be derived. L_1 and R_1 represent a transfer impedance between the current flowing in the wing and the voltage developed on the circuit. L_2 and R_2 represent the inherent inductance and resistance of the wires between the fuselage and the light. The transfer inductance and resistance, which it should be emphasized do not necessarily represent any clearly definable resistance or inductance of the wing, are merely those values which, when operated upon by the external

408

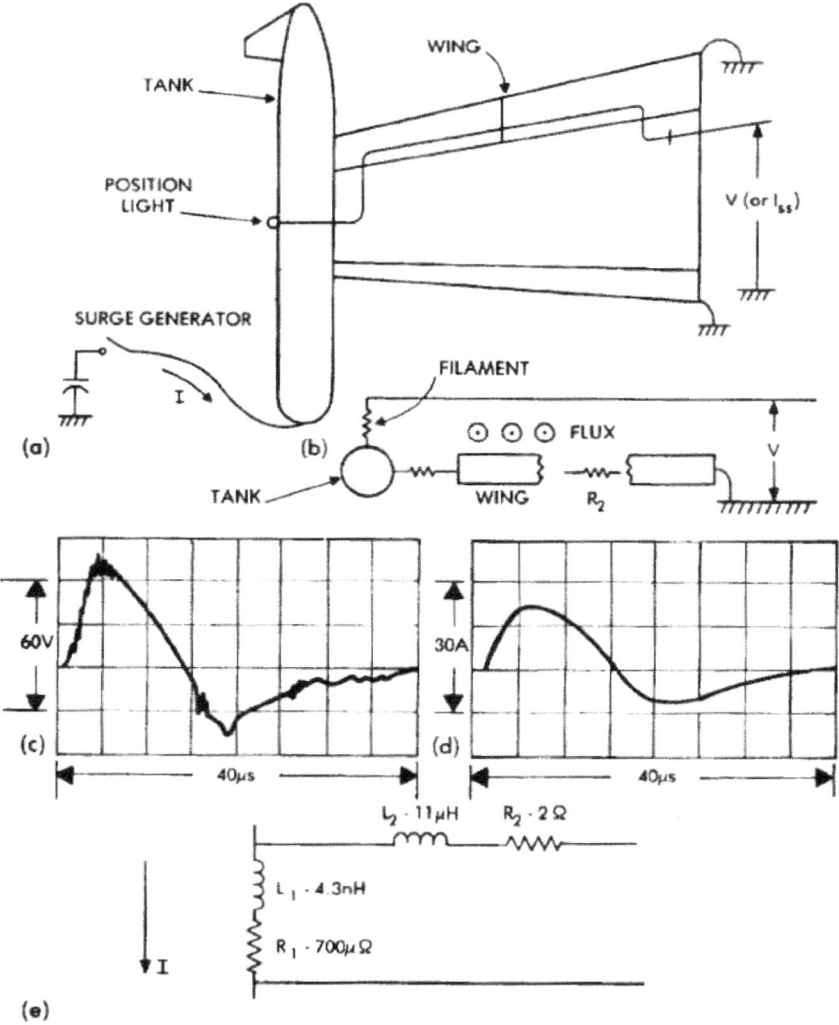

Figure 14.2 A wing tip circuit.
    (a) Circuit orientation
    (b) Electrical details of circuit
    (c) Open circuit voltage
    (d) Short circuit current
    (e) Equivalent circuit

lightning current, produced the observed open circuit voltage.

    A different type of circuit is that shown in Figure 14.3. In this circuit a conductor ran through the leading edge of the wing and terminated in an open circuit on a pylon mounted underneath the wing. In the electrical detail circuit shown in Figure 14.3(b), it can be seen that this circuit would not respond to

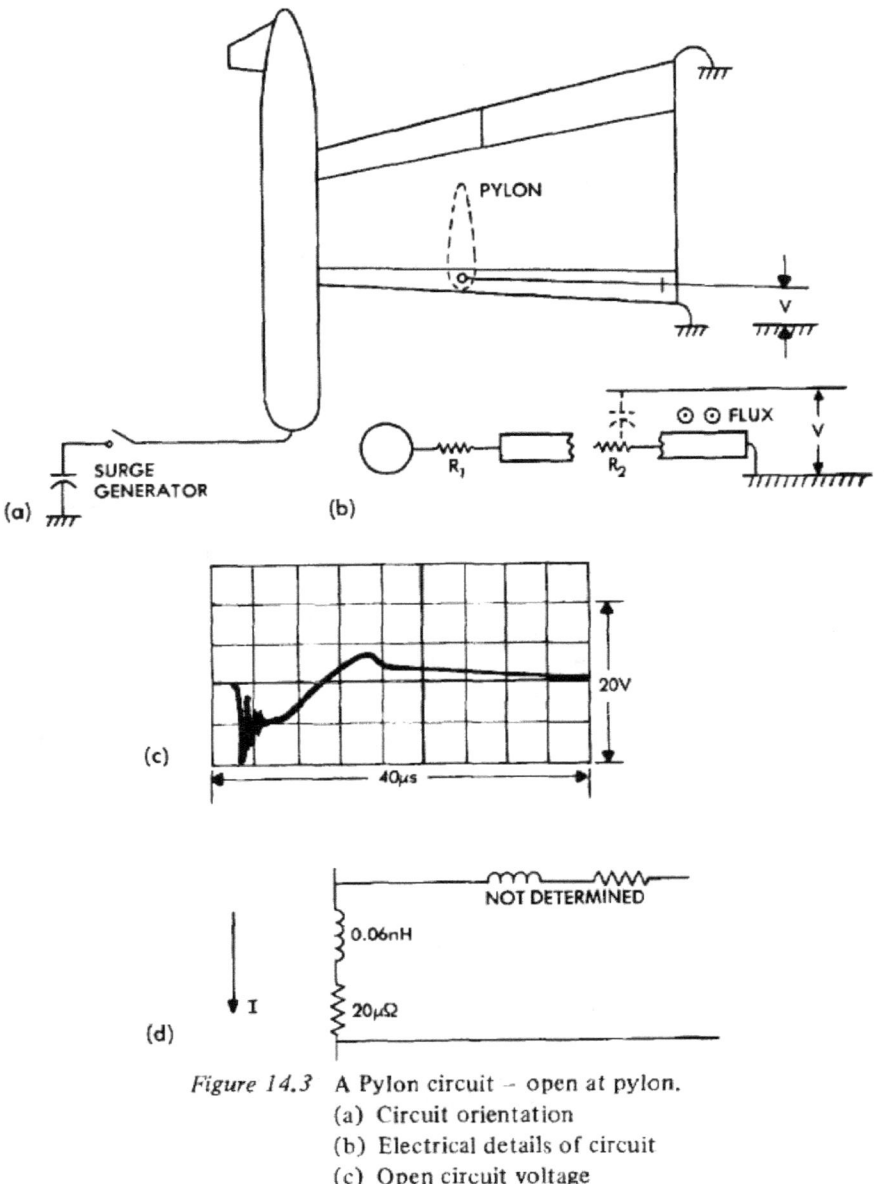

Figure 14.3 A Pylon circuit – open at pylon.

  (a) Circuit orientation
  (b) Electrical details of circuit
  (c) Open circuit voltage
  (d) Equivalent circuit

the voltage developed across the resistance between the tank and the wing. The circuit would respond in some measure to some fraction of the wing resistance and to some fraction of the magnetic field set up by the flow of current in the wing, but since the circuit was only capacitively coupled to the wing, the total coupling impedance should have been, and was, less than that of the circuit shown in Figure 14.2. The purpose of the conductor shown in Figure 14.3 was to supply power to a relay and to explosive bolts in the pylon used to hold a weapon. In Figure 14.3 the pylon was not installed, so there was no load on the conductor. Figure 14.4 shows the results when that pylon was installed, when the conductor was connected to a relay with a return through the aircraft structure, and when the lightning flash was allowed to contact the pylon. The combination of a structural return path for the circuit and a lightning flash terminating upon the pylon and thus including the resistive drop across R_3, the resistance between the pylon and the wing, served to make the voltage much greater than it was when the conductor was open circuited. No attempt was made to completely analyze from which area the total amount of magnetic flux was coming or whether the flux ϕ_1, representing that in the pylon, or ϕ_2,

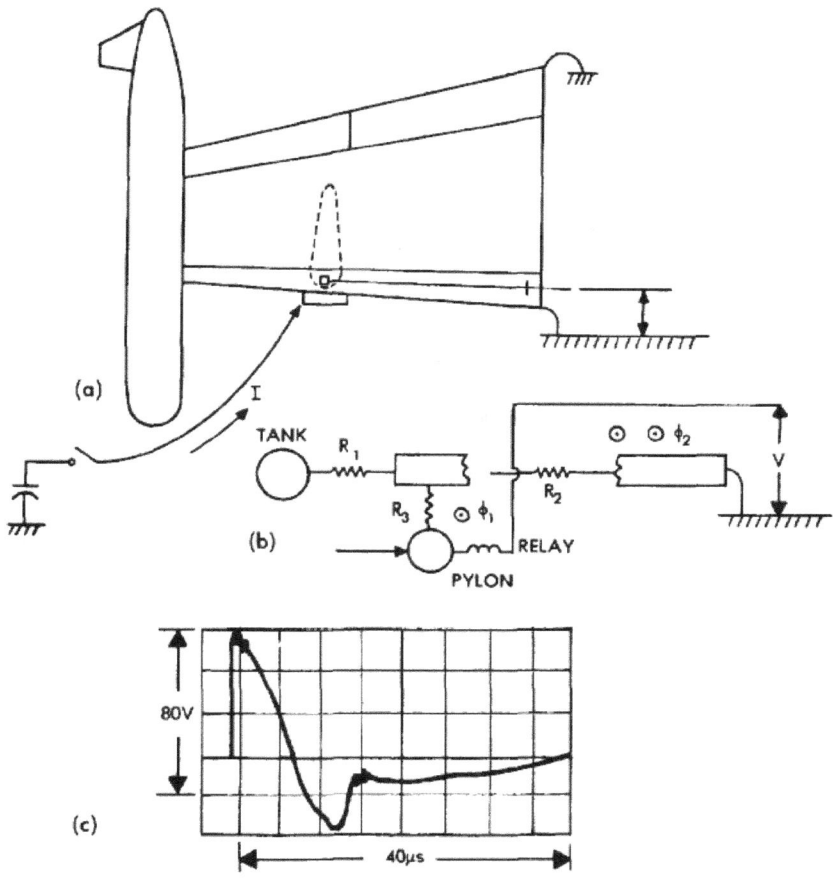

Figure 14.4   A Pylon circuit — loaded at pylon.
(a) Circuit orientation
(b) Electrical details of circuit
(c) Open circuit voltage

411

representing the flux within the wing, was the larger.

A fourth circuit, and one which illustrates the interaction between the lightning-developed voltage and the circuit response is shown in Figure 14.5. In this circuit a slot antenna excited by a grounded stub and fed from a length of 75 Ω coaxial cable was installed in the leading edge of the wing. The shorted stub that excited the slot antenna was the predominant area intercepting the magnetic flux produced by the lightning current in the wing. As in the three previous examples, the basic voltage developed in this antenna and shown in Figures 14.5(c) or 14.5(e) followed the same pattern as that shown for the other circuits. The rapid transition on the leading edge of the voltage, however, was capable of exciting an oscillation within the coaxial cable feeding the antenna. When the antenna circuit was terminated in a resistor matching the surge impedance of the cable, the higher frequency ringing oscillation disappeared, leaving only the underlying response of the antenna to the magnetic field surrounding the wing.

Several significant things were learned during this test series. The first was that the voltages induced in a typical circuit within the wing consisted of a magnetically induced component and a component proportional to the resistance of the current path. The location at which the lightning flash contacted the wing had an important effect on the magnitude of voltages developed on different circuits. None of the circuits displayed the easily understandable response of the conductors within the elliptical cylinder, and in none of the circuits could a clearly definable transfer impedance that was easily relatable to the structural geometry of the wing be derived. Nearly all the voltages and currents measured on the circuits could be explained in terms of the simple equivalent circuit shown in Figure 14.6, but there did not seem to be any easy way in which the magnitudes of the transfer impedance could be related to the physical geometry of the wing or to the location of the conductors within the wing. It was possible, however, to say that conductors located in the forward portion of the wing were better shielded, and consequently had lower transfer impedances, than those along the trailing edge. Likewise, it could be observed that circuits which did not have any electrical return through the wing structure had lower transfer impedances than those circuits which did have a return through the wing structure.

These equivalent circuits may be based upon the self-inductance of the wing, the self-inductance of the circuit within the wing, and the mutual impedance between the wing and the internal conductor. Alternatively, they may be based upon a transfer impedance, as shown in Figure 14.5(c). In the former the self-inductance of the wing would be nearly equal to the mutual inductance between the wing and the internal conductors. Their difference would be equal to the transfer inductance in the latter approach. Since the values of the transfer impedance will be much smaller than the values representing the self-impedance of the circuit, the latter leads to the more easily handled formulation. The latter approach is also compatible with the concept of mutual inductance or transfer impedance discussed for shielded cables in Chapter 13.

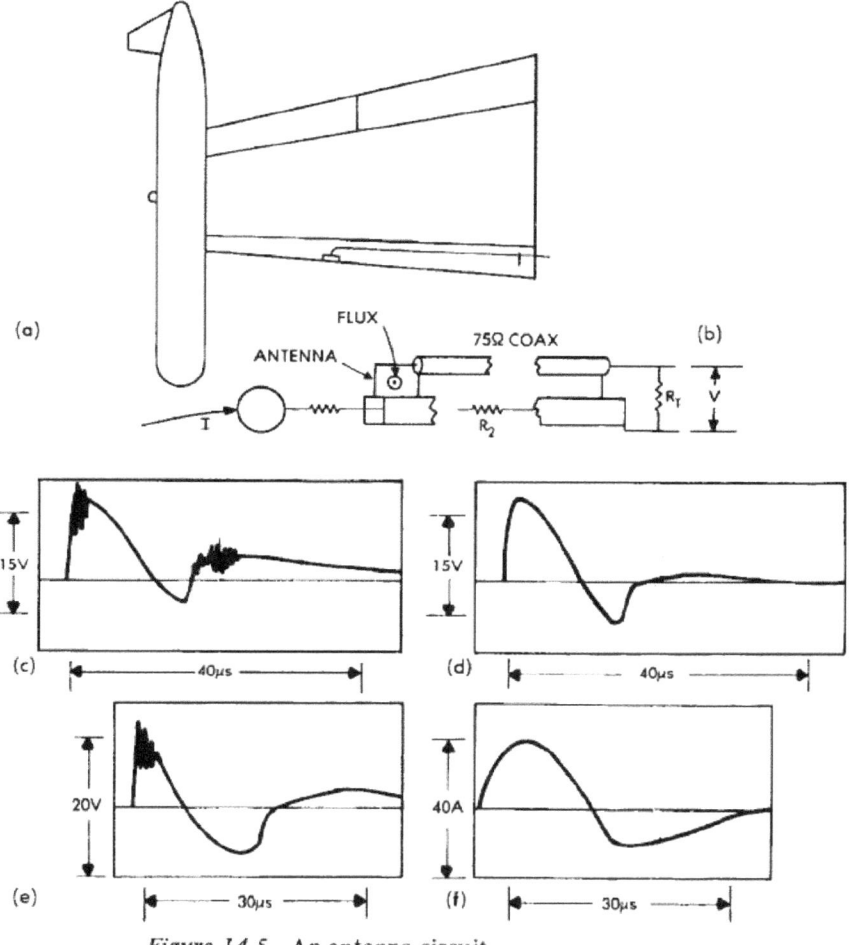

Figure 14.5 An antenna circuit.
    (a) Circuit orientation
    (b) Electrical details of circuit
    (c) Open circuit voltage – $R_T = \infty$
    (d) Open circuit voltage – $R_T = 75\Omega$
    (e) Open circuit voltage – $R_T = \infty$
    (f) Short circuit current – $R_T = 0$

14.3 F-8 Digital-Fly-By-Wire (DFBW)

The second set of measurements (Reference 14.2), about which some discussion will follow, was that made on an F-8 aircraft fitted with a fly-by-wire control system. The fly-by-wire controls, shown in Figure 14.7, consisted of a primary digital system, a backup analog system, and a common set of power actuators operating the control surfaces. The major components of the control

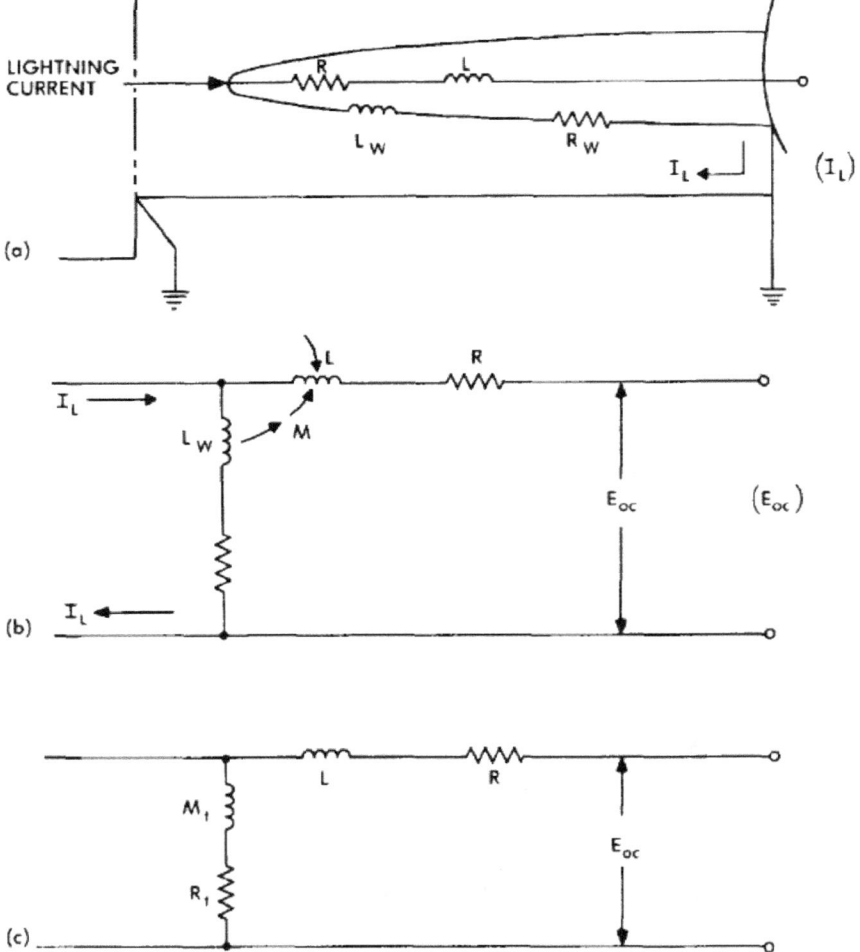

Figure 14.6 Equivalent circuits.

(a) and (b) Circuits based on self- and mutual inductances of the wing

(c) Circuit based on an equivalent transfer admittance:
$$Z = R_t + l_w M_t$$

system were located in three locations: the cockpit, where sensors coupled to the control stick provided signals for the control systems; an area behind the cockpit, where there was located the digital computer; and a compartment behind and below the cockpit on the left side of the aircraft. This latter compartment was one that would normally have been occupied by guns; accordingly, it will be referred to as the "gun bay." In this gun bay were located the interface and control assemblies.

Several hydraulic actuators were located at each of the major surfaces.

414

These were interconnected to the fly-by-wire control systems through wire bundles that ran under the wings. The control systems did not depend upon the aircraft structure as a return path; the system was considered to have a single-point ground, and that single-point ground was located at a panel in the gun bay. By and large, none of the control wiring in the aircraft was shielded.

In contrast to the tests on the wing of the F-89J aircraft, in which high-amplitude currents were injected into the wing from a high-power surge generator, the tests on the complete F-8 aircraft were made with what has been called the *transient analyzer,* a portable and relatively low-power surge generator capable of injecting currents of a waveform similar to that found in lightning but of a much lower, and nondestructive, level. During the tests on the F-8, the injected current was of the order of 300 A. Several different current waveshapes were employed; the one to which most frequent reference will be made in this abbreviated set of test results is that described as the *fast waveform,* a current rising to crest in about 3 μs and decaying to half value in about 16 μs. These waveforms are shown in Figure 14.8.

The first set of measurements to which reference will be made was that on a set of spare conductors running between an interface box in the gun bay and a disconnect panel located near the leading edge of the vertical stabilizer. The routings of the circuit and the waveforms are shown in Figures 14.9 and 14.10. The voltage measured between the conductor and ground consisted of a high-amplitude oscillatory component and a lower amplitude but longer duration component. The oscillatory component was excited by magnetic flux leaking inside the aircraft, while the longer duration component was produced by the flow of current through the structural resistance of the aircraft. Since the voltages induced by the leakage of magnetic flux were proportional to the rate of change of that flux, it followed that the oscillatory component would have been more pronounced for faster currents injected into the aircraft than it was

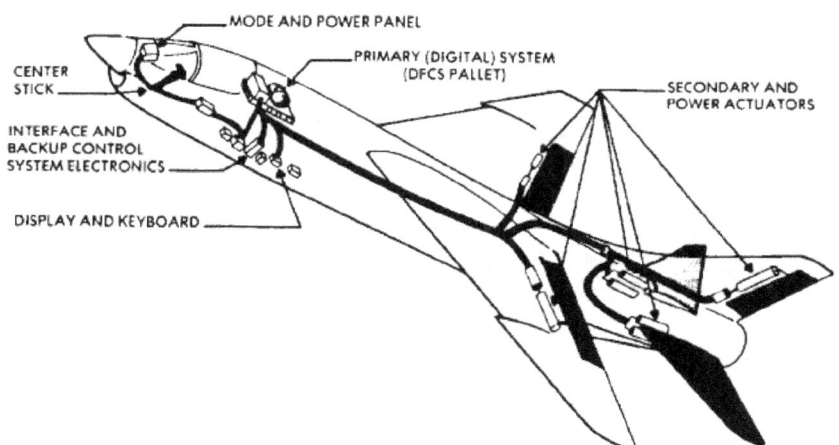

Figure 14.7 Location of fly-by-wire control system hardware and wiring bundles in F-8 aircraft.

415

for slower currents. This effect was noted as indicated by the oscillograms. Voltages measured between conductors, shown in Figure 14.10, were much smaller than the voltages measured between either of the conductors and the airframe. Such results would be expected on a well-balanced circuit.

SLOW WAVEFORM

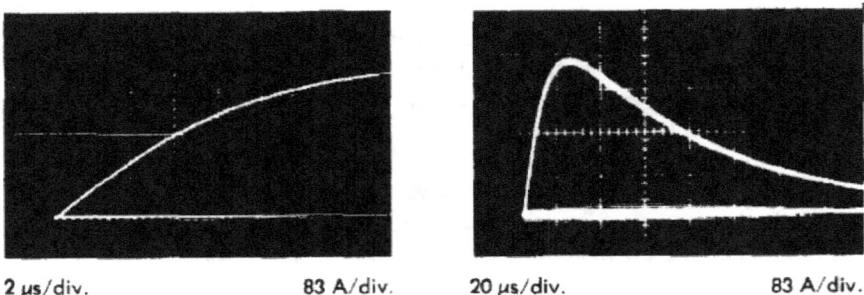

2 µs/div.   83 A/div.   20 µs/div.   83 A/div.

FAST WAVEFORM

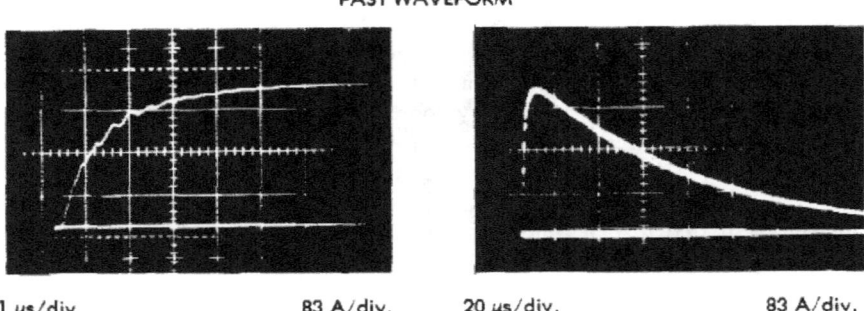

1 µs/div.   83 A/div.   20 µs/div.   83 A/div.

VERY FAST WAVEFORM

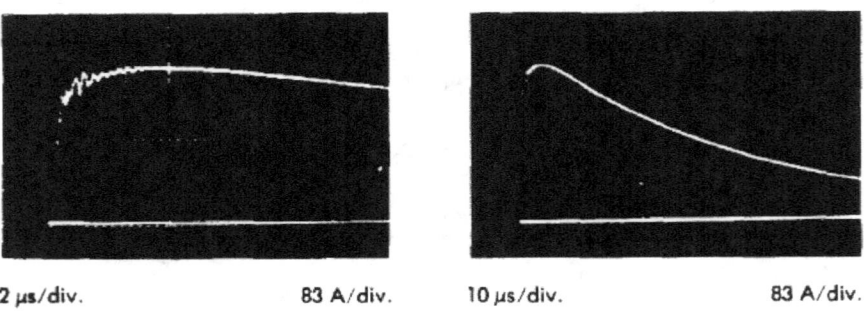

2 µs/div.   83 A/div.   10 µs/div.   83 A/div.

Figure 14.8 Simulated lightning test waveforms.

NOTE: "___ µs/div." is horizontal time scale and "___ A/div." is vertical amplitude scale.

416

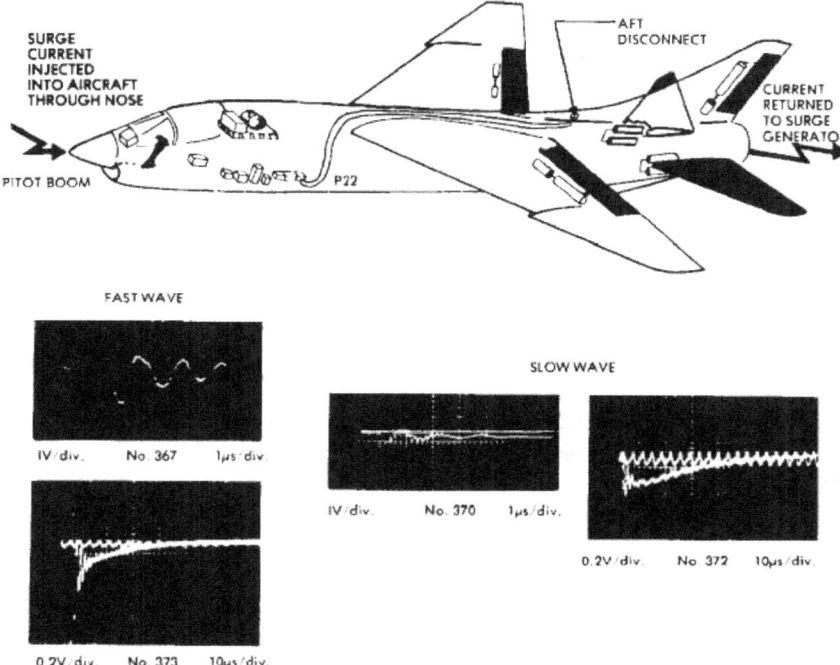

Figure 14.9 Spare conductor measurements on P22 (Pin 24 to airframe).

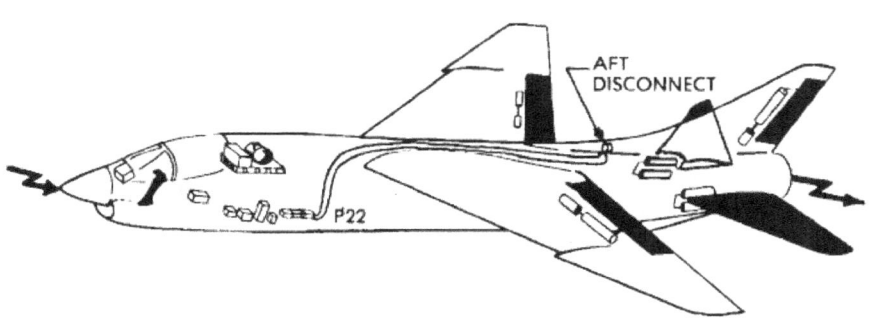

FAST I$_L$ WAVEFORM

PINS 24-25

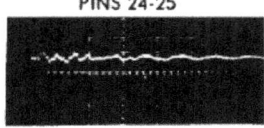

1V/div. No. 368 1 μs/div.

Figure 14.10 Spare conductor measurement on P22 (line-to-line, Pins 24 to 25).

The second set of measurements was made on a circuit running from the interface control unit and the gun bay to a wing position indicator switch located underneath the leading edge of the wing. The wing on the F-8 aircraft could be raised or lowered around a pivot point towards its rear in order to change the angle of attack during landing and takeoff. The purpose of the switch was to indicate the position of the wing. The voltages induced on that switch circuit are shown in Figure 14.11. The voltages measured from line to airframe were higher than those measured from line to line, but it is significant that the line-line voltages, while of a somewhat different waveshape, were not much lower than the line-to-ground voltages. The reason for this lay in the fact that the load impedances in the interface box in the gun bay were different from each other on the two sides of the circuit. One side connected to a power supply bus, while the other side probably connected to an emitter follower. Another significant feature about these voltages was that they were again of an oscillatory nature. They were apparently excited by the leakage of magnetic flux inside the aircraft and were not excited by the drop in potential along the structural resistance of the aircraft.

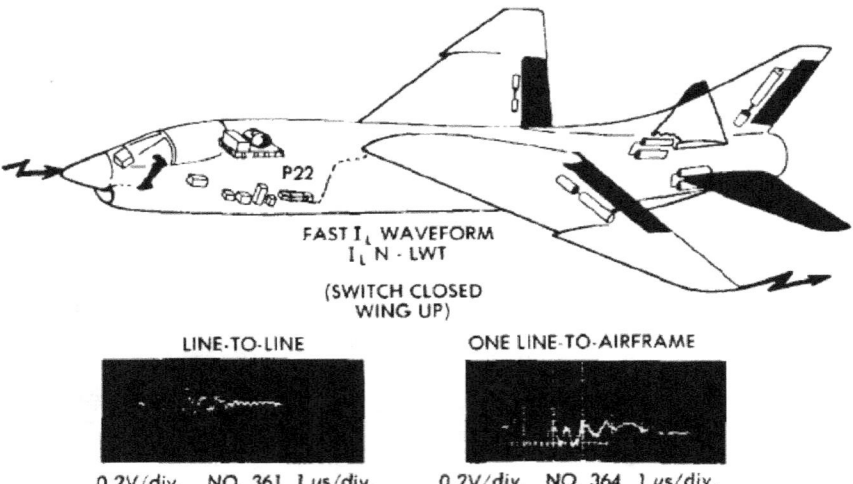

Figure 14.11 Voltages induced in wing position indicator switch circuit at open plug P22.

Figures 14.12 and 14.13 show voltages measured on two different circuits going to actuators, one (Figure 14.12) going to the left pitch actuator and the other (Figure 14.13) going to the left roll actuator. In both cases the voltages displayed were the output of the driver amplifier used to control the servo valve in the actuator. Both of these were differential measurements, line-line voltage measurements. The significant feature about these measurements was, again, that the characteristic response was oscillatory and apparently excited by the leakage of magnetic flux inside the aircraft. There was some dependence of the voltage

418

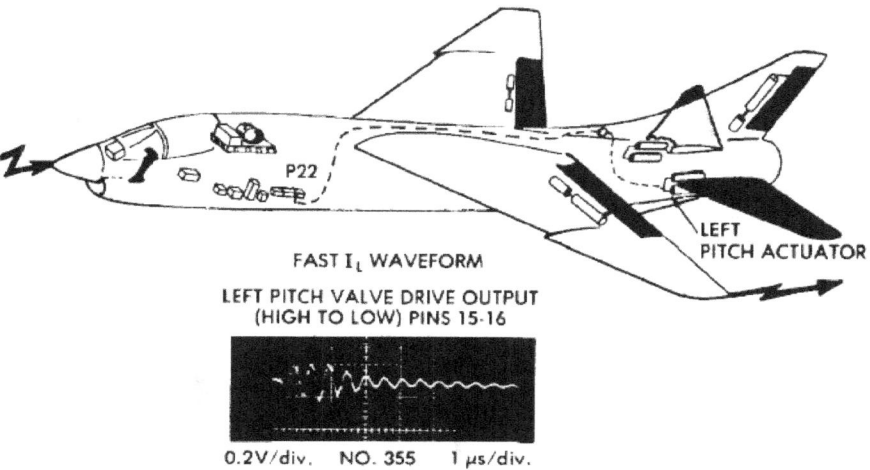

Figure 14.12 Left pitch valve drive output (high to low) at plug P22 (system battery powered).

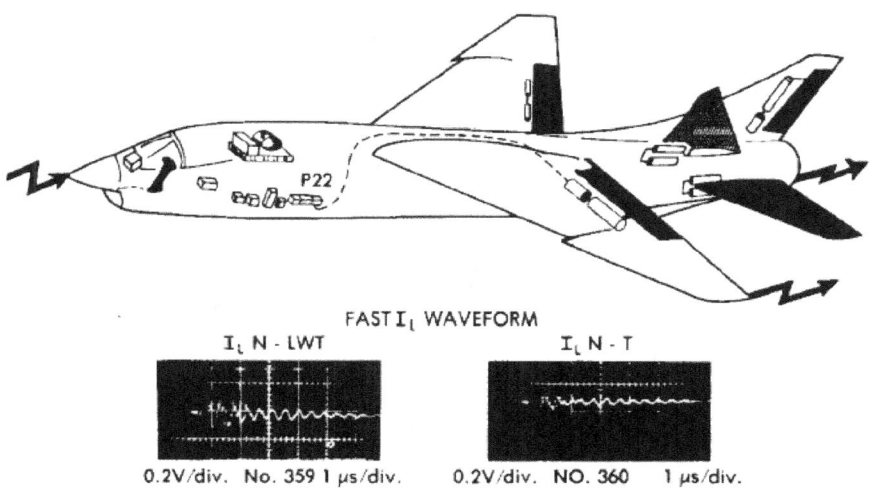

Figure 14.13 Voltages induced in left roll valve drive output circuit (Pins 44-45) at open plug P22.

on the path of current followed through the aircraft, but the dependence was not large. Both lightning current paths produced about the same peak voltage of transient.

On the F-8, as is typical of most aircraft, the control wires were laced together into fairly large bundles. The routing of some typical bundles in the gun bay housing the backup and interface electronic control boxes is shown in Figure 14.14. While it was not possible to measure the current on individual

419

wires within these cable bundles because of limitations of measurement technique and because of the large number of wires within the bundles, it was possible to measure the total current flowing on the various bundles. This was done by clamping around the bundle a current transformer having a split core. Typical results of these cable measurements are shown in Figures 14.15 and 14.16. The bulk cable currents were also found to be oscillatory, just as were the voltages on conductors described earlier. Since the flight control wiring did not make use of multiple ground points within the aircraft, it follows that none of the currents in these cable bundles would exhibit any of the long time response characteristic of multiple-grounded conductors.

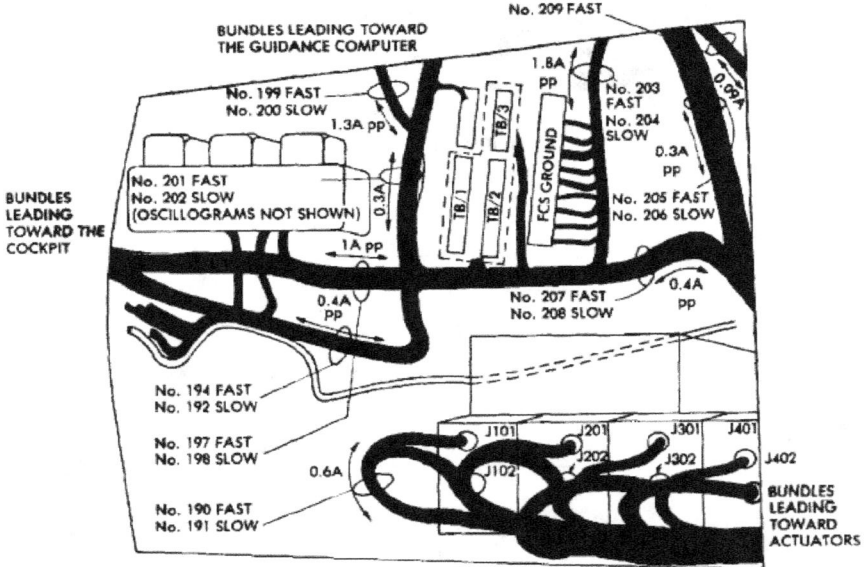

Figure 14.14 Cable bundles within the gun bay.

Figure 14.17 shows a statistical distribution of the peak amplitude of currents in all of the cable bundles upon which measurements were made. The distribution is shown both for the actual current amplitudes injected into the aircraft and in terms of what those currents would be if the results were scaled up to currents representative of actual lightning flashes. In terms of an average-amplitude lightning flash, 30 000 A, the total current on most cable bundles would have been on the order of 20 to 100 A.

Measurements were made of the amplitude and waveshape of the magnetic field at a number of points in and around the aircraft. One location upon which attention was concentrated was the cockpit, since the cockpit is an inherently unshielded region and one in which many control wires would be subjected to changing magnetic fields. The positions at which fields were measured, the peak amplitude of the fields, and the predominant orientation of the fields are shown

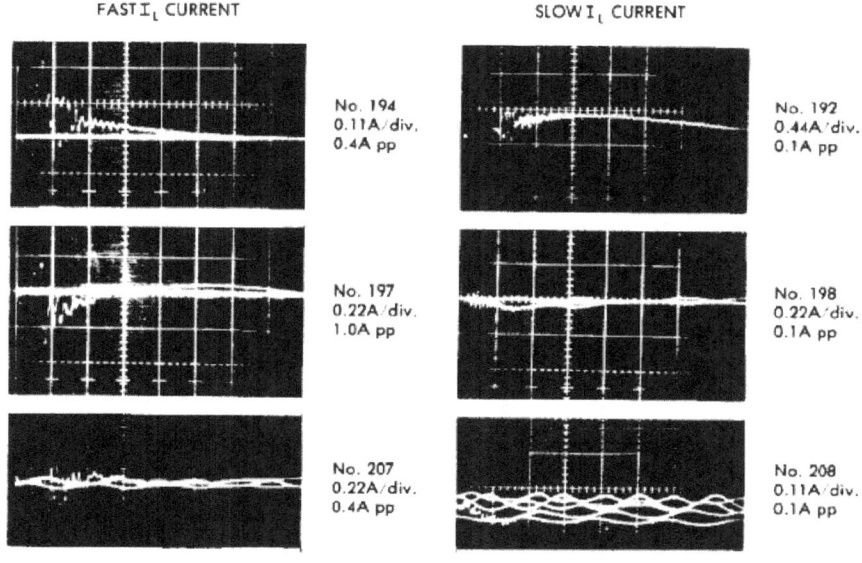

Figure 14.15 Currents on cable bundles leading toward cockpit and left-hand instrument panel.

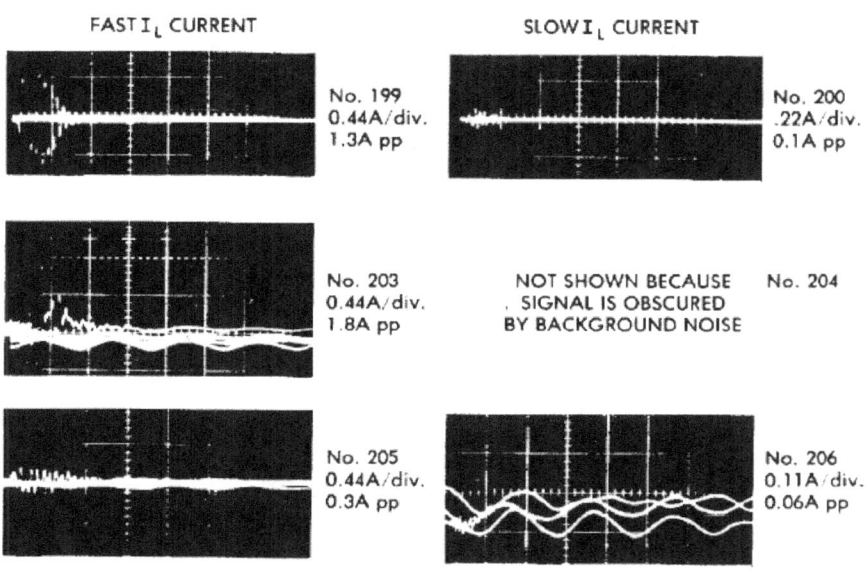

Figure 14.16 Currents on cable bundles leading toward the area behind the cockpit.

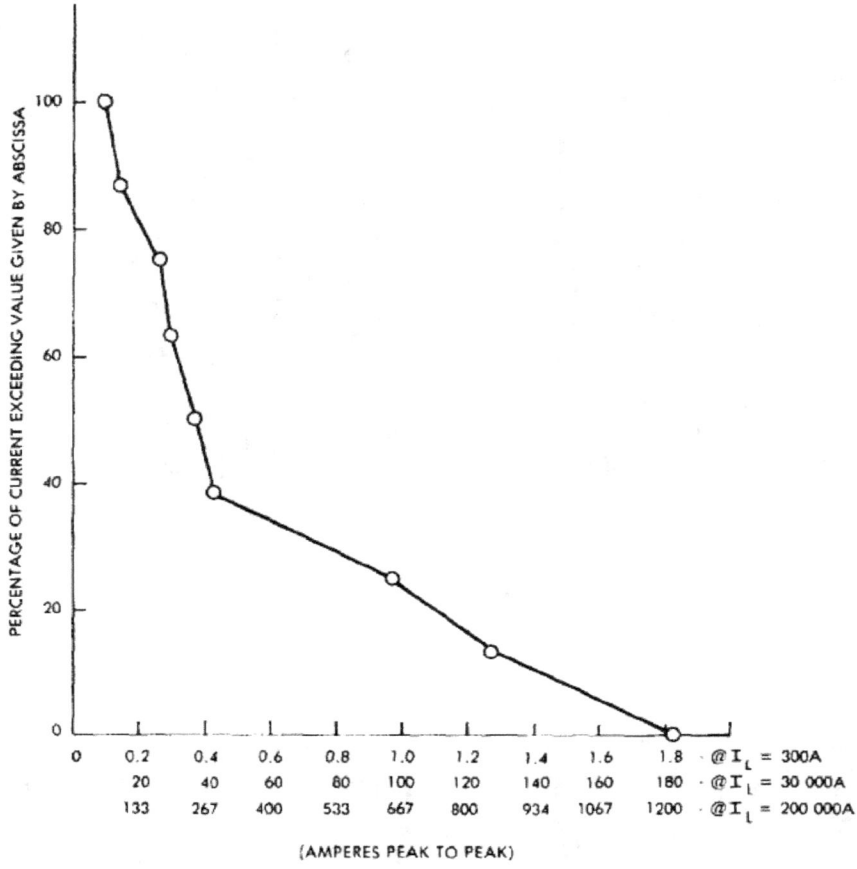

Figure 14.17 Distribution of amplitudes of cable bundle currents (measured in left gun bay).

in Figure 14.18. The magnetic fields were measured with a probe which had a characteristic time constant of about 4 μs. When exposed to fields changing in times less than 4 μs, it would respond to the absolute magnitude of the field intensity, and, when exposed to fields changing in times longer than 4 μs, it would respond to the rate of change of the magnetic field. A few typical measurements of field waveshape are shown in Figure 14.19. The most significant feature about these measurements is that there was no orientation of the magnetic field probe that resulted in a zero output, which indicated that the orientation of the magnetic field was not uniform with respect to time. The field produced at any one point was the sum of the field produced by the total flow of current through the aircraft and that produced by oscillatory current in the various structural members as the current in those structural members changed with time. Similar effects can be expected on other aircraft, and, accordingly,

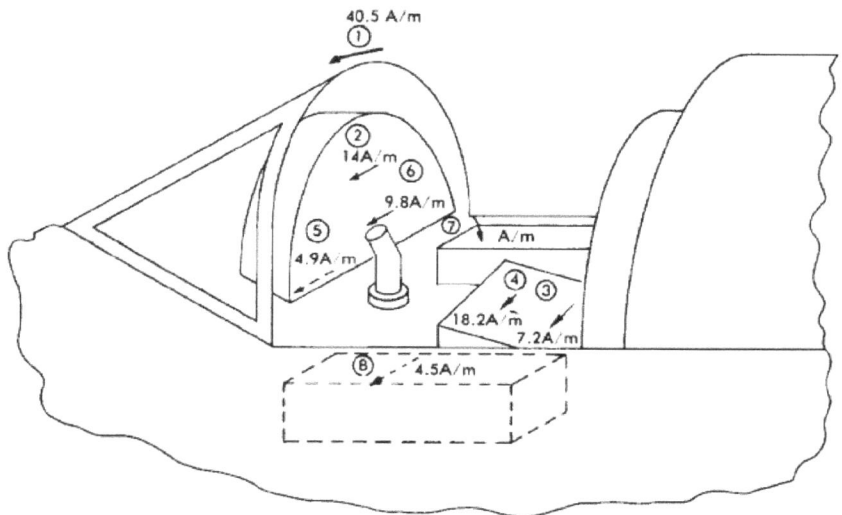

Figure 14.18 Magnetic field measurements in the cockpit.

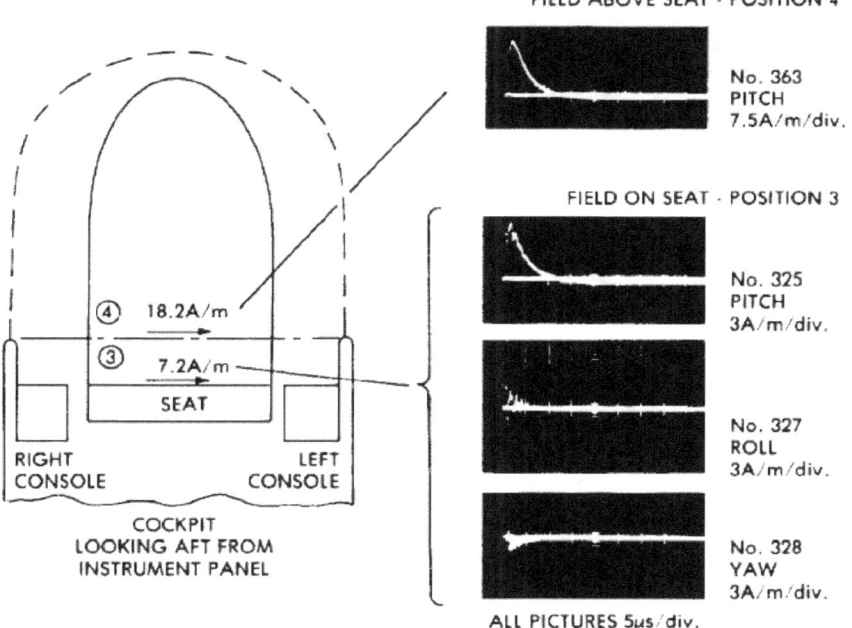

Figure 14.19 Magnetic fields near pilot's seat.

there will seldom be any clear-cut manner in which electrical wiring can be oriented so as to pick up the minimum value of magnetic field and hence have minimum induced voltages. One might best assume that the magnetic field at any point will always be oriented in the worst case.

Some measurements of the magnetic field within the gun bay housing the backup control system and the interface electronics package are shown in Figure 14.20. Clearly, the magnetic field inside the gun bay compartment was of lower amplitude than that of the field external to the gun bay. The waveshapes of the field are more difficult to understand. First of all, it must be kept in mind that the probe was responding to the rate of change in magnetic field after about 4 μs and was responding to the field itself for times shorter than about 4 μs. Accordingly, the oscillograms displaying the field inside the gun bay indicated that there was, first of all, a component of magnetic field that rose to its crest about as fast as did the outside field. From then on, the field continued to increase at a slower rate, the final value of the field not being indicated by the oscillograms. The fact that the field continued to rise, but at a slower rate, was in agreement with the behavior predicted in Section 10.2.4, where it was predicted that the field would increase with a time constant characteristic of the

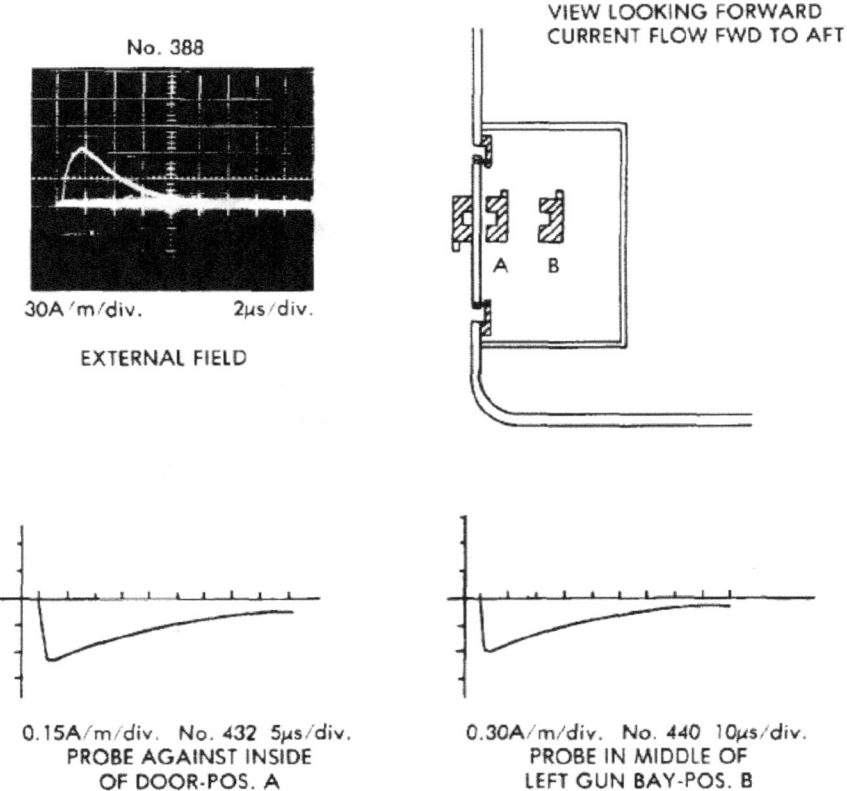

No. 388

30A'm/div.   2μs/div.

EXTERNAL FIELD

VIEW LOOKING FORWARD
CURRENT FLOW FWD TO AFT

A   B

0.15A/m/div. No. 432 5μs/div.
PROBE AGAINST INSIDE
OF DOOR-POS. A

0.30A/m/div. No. 440 10μs/div.
PROBE IN MIDDLE OF
LEFT GUN BAY-POS. B

INTERNAL FIELD

Figure 14.20   Magnetic fields inside gun bay.

424

internal inductance and resistance of the cavity in which the fields were to be measured.

Fields inside a battery compartment located aft of the gun bay are shown on Figure 14.21. The measurements showed first some oscillatory magnetic field, followed by a field which rose to crest at a time much longer than the crest time or even the duration of the lightning current that was injected into the

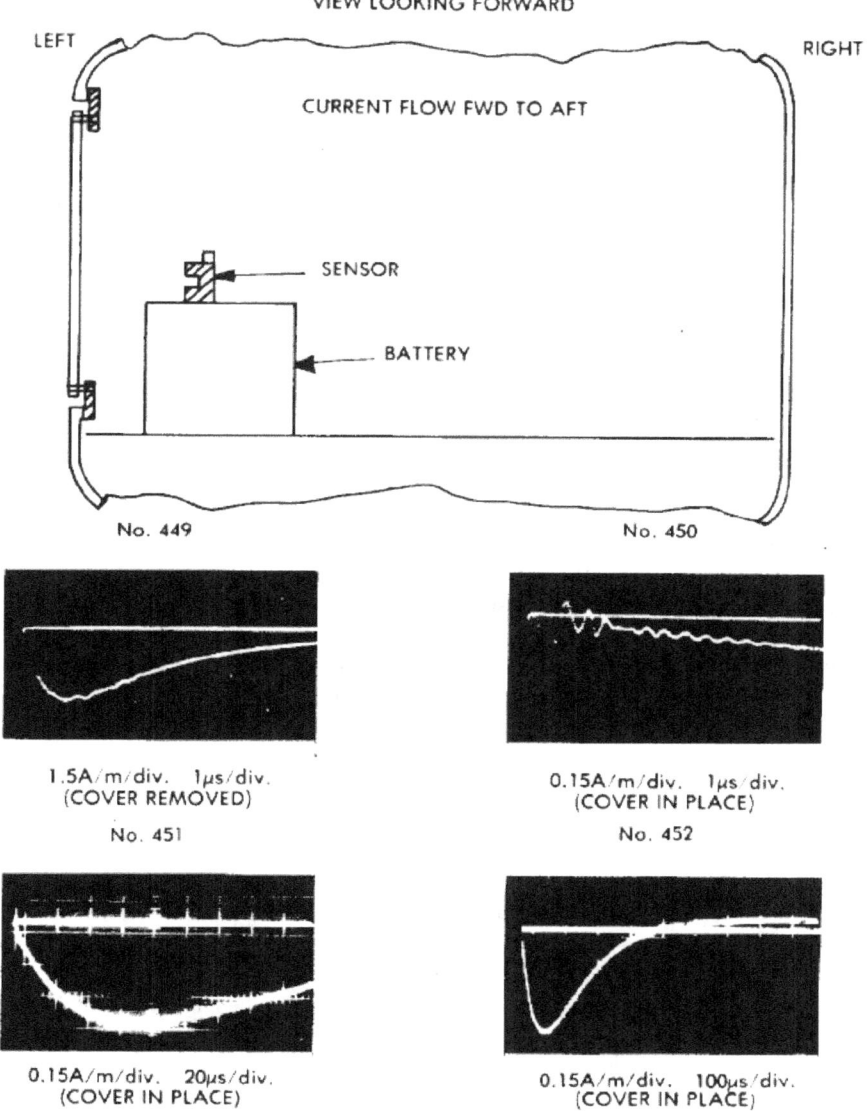

Figure 14.21 Magnetic fields inside the battery compartment (probe time constant = 4 μs).

aircraft. Oscillogram No. 452 in Figure 14.21 indicated the rate of change of field as falling to zero at about 400 μs. This would indicate that the field itself reached its crest value in about 400 μs. The different nature of the response of the fields inside the gun bay and the battery compartment can be explained in terms of the types of covers and fasteners used on the two compartments. The cover over the battery compartment was held in place by fasteners spaced about every 3.8 cm, and hence made good contact to the rest of the airframe. The use of multiple fasteners resulted in minimum constriction of the current flow and minimum resistance to the circulating current within the battery compartment. The gun bay covers, on the other hand, since they were originally designed for fast access time, had far fewer fasteners. The fasteners in the gun bay, in fact, were spaced about every 30 cm. Accordingly, the cover on the gun bay provided a much greater constriction of current flow and a much greater resistance than did the cover on the battery compartment. Both of these factors were of a nature to allow the field within the gun bay to reach its peak value faster than did the field within the battery compartment.

Some of the significant points about the results of tests on this aircraft may be summarized here. The first was that the use of a single-point ground system most emphatically did not eliminate all transient voltages produced by the flow of lightning current through the structure of the aircraft. The second was that the characteristic response of the wiring, both for voltage and for current, was a damped oscillation with a frequency in the range 1 to 5 MHz. The very abbreviated series of test results just presented does not indicate the fact clearly, but the frequency of oscillation depended considerably on the length of the circuit involved: the longer the physical length of the wires, the lower the oscillatory frequencies. This was not a clear-cut rule, since the response of any one circuit was greatly influenced by the high degree of coupling between all of the different circuits. The third significant point is that the total current on any cable bundle was of the order of 20 to 100 A for an average lightning flash. This bulk cable current was again oscillatory with a frequency tending to correspond to the length of the cable bundle. The equipment bays in this aircraft, not being designed for electromagnetic shielding qualities, allowed significant amounts of magnetic flux to develop within those bays. This is particularly true of those bays intended for ease of access. As a rule of thumb, it might be expected that on aircraft the equipment bays housing electronic equipment might well be fitted with covers designed more with ease of access in mind than with magnetic shielding qualities in mind. It must thus be expected that those enclosures for which the electromagnetic shielding should in principle be the greatest may well be those enclosures having the poorest shielding.

No attempt was made to determine in this aircraft the degree to which any circuit voltages could be reduced by the use of shielded conductors. It was noted, however, that on those circuits in which a shield was used and in which the shield was grounded at more than one point, the current on that shield tended more to have the slower double exponential waveshape of the external lightning current than to have the high-frequency oscillatory current excited on all of the unshielded cable bundles.

REFERENCES

14.1 K. J. Lloyd, J. A. Plumer, and L. C. Walko, *Measurements and Analysis of Lightning-Induced Voltages in Aircraft Electrical Circuits,* NASA CR-1744, National Aeronautics and Space Administration, Lyndon B. Johnson Space Center, Houston, Texas (February 1971).

14.2 J. A. Plumer, F. A. Fisher, and L. C. Walko, *Lightning Effects on the NASA F-8 Digital-Fly-By-Wire Airplane,* NASA CR-2524, prepared by the High Voltage Laboratory, Environmental Electromagnetics Unit, Corporate Research and Development, General Electric Company, Pittsfield, Massachusetts, for the National Aeronautics and Space Administration, Lewis Research Center, Cleveland, Ohio (March 1975).

CHAPTER 15
DESIGN TO MINIMIZE INDIRECT EFFECTS

15.1 Some Premises and Goals for Design

Any design program aimed at minimizing or eliminating the indirect effects of lightning needs to be founded on certain premises of which the following might be examples. The first premise is that it is desirable to achieve a design that prevents the indirect effects from causing irreversible physical damage. The second premise is that it is essential to eliminate that interference which provides an imminent hazard to the safety of the vehicle or its crew or one that presents a severe risk of preventing the completion of the aircraft's mission. (It is, of course, desirable to minimize all interference.) For example, indirect effects that cause warning lights to appear might be acceptable whereas indirect effects that lead to the tripping of circuit breakers would be unacceptable, even if it were possible to reset the circuit breakers. Immediately after a lightning flash the pilot of the aircraft might have enough things to do that he should not be called upon to reset circuit breakers. Further, interference that leads to the scrambling of one channel of a redundant digital control system is probably acceptable, but interference that causes computers to shut down is unacceptable, particularly if the computers are shut down in a disorderly manner which results in the internally stored programs becoming scrambled. A third premise is that it is more productive to design electronic equipment that can accept transient signals on input and output leads than it is to initiate a retrofit program to provide protection to an existing system. A fourth premise is that it is more practical to design an electronic system around the capabilities of existing and proven protective devices or techniques than it is to develop and retrofit new and improved protective techniques to an electronic system designed without consideration of the transients that might be produced by lightning. A fifth premise is that trade-offs must be made between the cost of providing electronic equipment capable of withstanding lightning-induced transients and the cost of shielding equipment and interconnecting wiring from the electromagnetic effects of lightning. A sixth premise is that designers should take as much advantage as possible of the inherent shielding that aircraft structures are capable of supplying and should avoid placing equipment and wiring in locations that are most exposed to the electromagnetic fields produced by lightning.

15.2 Improvement Through Location of Electronic Equipment

While it is recognized that the designer may not have much choice in the matter, it is often possible to make improvements in the resistance to indirect effects by locating electronic equipment in regions where the electromagnetic fields produced by lightning current are lowest and by avoiding the placement of equipment in the region where the electromagnetic fields are highest. For example, since the most important type of coupling from the outside electromagnetic environment to the inside of the aircraft is through apertures, it follows that equipment should be located as far from major apertures as possible.

Further, since access doors with their imperfectly conducting covers are a major source of electromagnetic leakage, it follows that equipment should be located as far from such access doors as possible. In practice this may be more easily said than done, however, because frequently the purpose of access doors is to provide ready access to electronic equipment.

One main goal should be to locate electronic equipment toward the center of the aircraft structure rather than at the extremities, since the electromagnetic fields tend to cancel toward the center of any structure.

Another main goal should be to locate electronic equipment away from the outer skin of the vehicle – particularly away from the nose of the aircraft, where the radius of curvature of a prime current-carrying path is smallest. Furthermore, if possible, electronic equipment should be located in shielded compartments.

15.3 Improvement Through Location of Wiring

The designer has somewhat more control over the routing of wiring used to interconnect equipment than he does over the location of equipment itself. Wiring should be located away from apertures and away from regions where the radius of curvature of structural members or the outer skin is smallest. Moreover, wiring should be located as close to a ground plane or structural member as possible. If the structural member is shaped or can be shaped to provide a trough, the member will provide more inherent shielding than it will if the wiring is placed on the edge of the member. Some examples of typical structural members and the best places for bundles of wiring are shown in Figure 15.1. In each case the structural member is assumed to be carrying current along its axis.

Some basic principles to follow are these:

1. The closer a conductor is placed to a metallic ground plane, the less is the flux that can pass between that conductor and the ground plane.
2. Magnetic fields are concentrated around protruding structural members and diverge in inside corners. Hence, conductors located atop protruding members will intercept more magnetic flux than conductors placed in corners, where the field intensity is weaker.
3. Fields will be weaker on the interior of a U-shaped member than they will be on the edges of that member.
4. Fields will be lowest inside a closed member.

Some examples of cable routing are shown on Figure 15.2. Along the interior of a structure a cable clamped to stiffeners, as at position 1, will effectively be spaced away from the metal skin by the height of the stiffeners. A conductor along the outside edge of the U-shaped member, as shown by conductor 2 in Figure 15.2(a), may or may not be better placed than conductor 1: effectiveness depends on how closely the conductor is attached to the side of

430

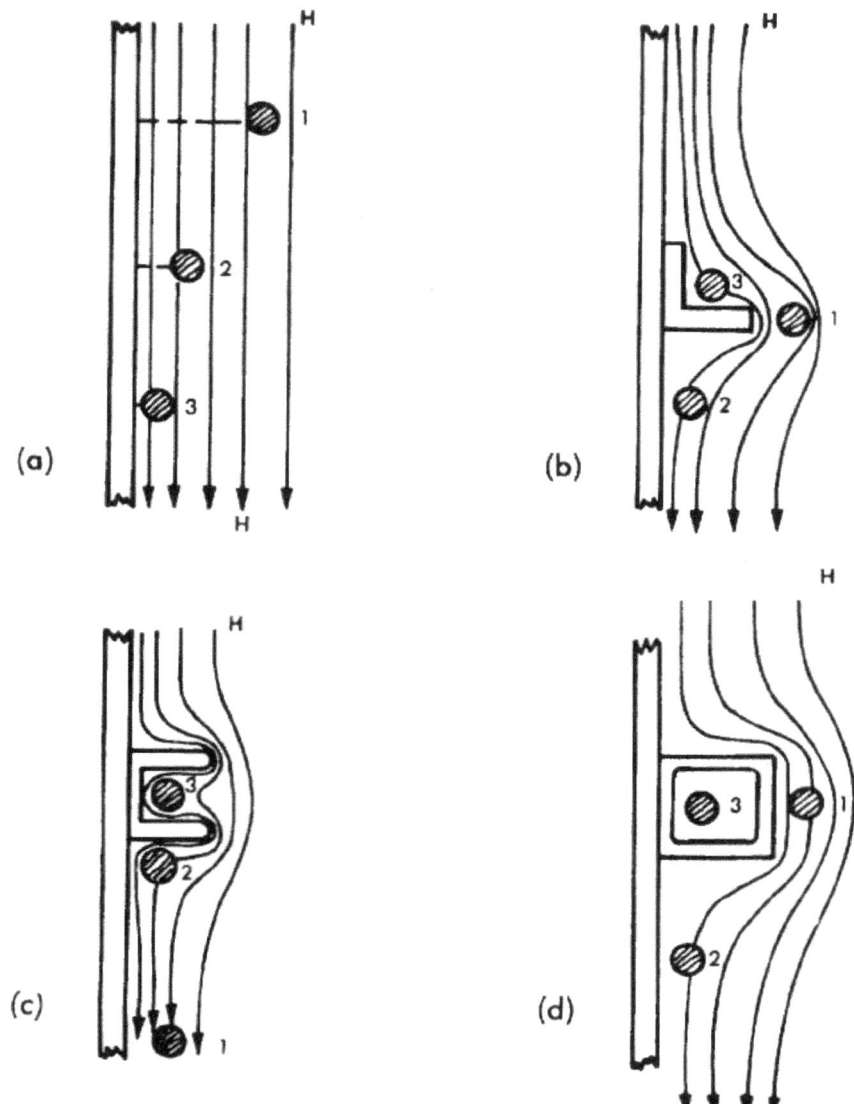

Figure 15.1 Flux linkages vs conductor position.

(a) Conductors over a plane

(b) Conductors near an angle

(c) Conductors near a channel

(d) Conductors near a box

In each case pictured

Conductor 1 – highest flux linkages: worst

Conductor 2 – intermediate linkages: better

Conductor 3 – lowest linkages: best

431

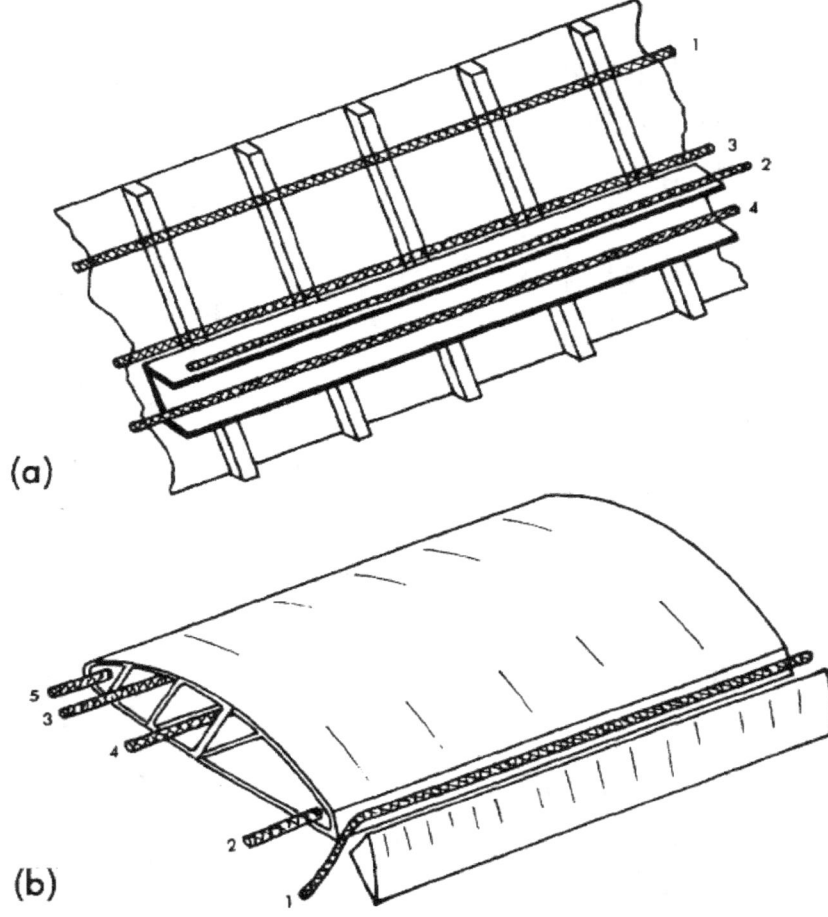

(a)

(b)

Figure 15.2 Conductor routing.
(a) A fuselage structure
(b) A wing structure
In each case pictured
Conductor 1 – highest flux linkages: worst
Conductor 4 – lowest flux linkages: best

the U-shaped member. Conductor 3, placed along the edge where the U-shaped member is attached to the stiffeners would probably be in a lower field environment than would either conductor 1 or conductor 2. Conductor 4, located on the interior of the U-channel, would be in the lowest field region and hence in the most effective position. Similar considerations apply to conductors located in structures like wings or stabilizers.

A conductor located along the outside trailing edge of the wing, as shown by conductor 1 in Figure 15.2(b), will pick up much more flux than will any conductor located on the inside, probably by several orders of magnitude. Hence conductor 2, located on the inside of the trailing edge, would be better placed. Conductors 3 and 4, in that order, would be in the regions of lowest magnetic flux. A conductor that could be run inside a major structural member, as shown by conductor 4, will be exposed to a minimum amount of magnetic flux. Conductor 5, located at the forward edge of the wing, would be in a well-shielded region if the forward edge of the wing were metal, but it would be in a high field region and therefore vulnerable if the forward edge were a nonmetallic covering. However, even if the covering were metallic, conductor 5 would not be in as protected a region as that of either conductor 3 or conductor 4.

Windshield posts, shown in Figure 15.3, tend to concentrate the current flowing on the exterior surface of the vehicle, particularly if a flash is swept back, contacting the windshield post directly or the eyebrow region above the windshield. Since the current is concentrated, the magnetic field intensity inside the crew compartment tends to be very high. The situation is aggravated by the fact that the windshields, unlike other regions where the field might have to diffuse through the metal surfaces, act as large apertures and so allow the internal magnetic flux to build to its peak values very rapidly. Instruments and wiring on the control panels are thus in a region of inherently high magnetic field strength. Conductors that run from overhead control panels (position A) to other instruments (position B) are often run along the windshield center posts. They are thus in a region of the most concentrated magnetic fields likely to be found on an aircraft, and accordingly they may have induced in them the highest voltages.

15.4 Improvement Through Shielding

In order to make an electronic system immune to the effects of lightning, it is almost always necessary to make judicious use of shielding on interconnecting wiring. Figure 15.4 shows some of the basic considerations. In Figure 15.4(a) an unshielded conductor, being exposed to the full magnetic field inside the structure, will have high voltages developed across the high-impedance terminations. The presence of a shield grounded at only one end will not significantly affect the magnitude of the voltage induced by changing magnetic fields, although a shield may protect against changing electric fields. While a shield may keep the voltage at the grounded end low, it will allow the voltage on the signal conductors to be high at the unshielded end.

Shielding against magnetic fields requires the shield to be grounded at both ends, as shown in Figure 15.4(b), in order that it may carry a circulating current. It is the circulating current that cancels the magnetic fields that produce common-mode voltages. There is some merit in grounding such a shield at multiple points, since frequently the cable will be exposed to a significant amount of magnetic field over only a small portion of its total length. If the shield is multiple-grounded, the circulating currents will tend to flow along only one portion of the cable whereas, if it is grounded at only the two ends, current is constrained to flow the entire length of the cable. There is likewise some virtue in staggering the spacing between multiple ground points on a cable shield,

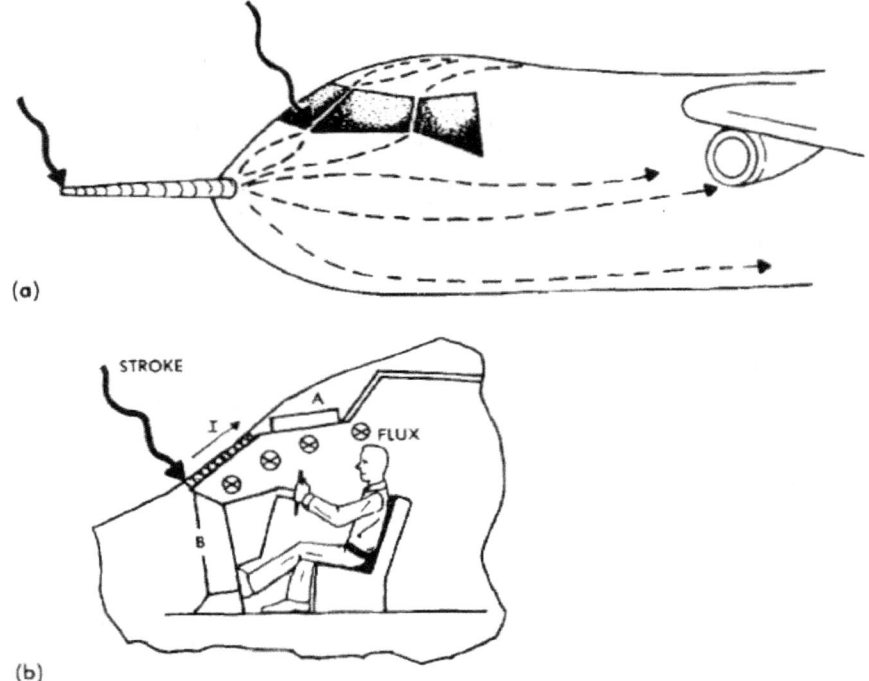

(a)

(b)

Figure 15.3 Current flow along windshield posts.
  (a)   External current flow
  (b)   Internal magnetic fields

since it is theoretically possible that uniform grounding can lead to troublesome standing waves if the shield is illuminated by a sustained frequency interference source.

Of the different types of shields, shown in Figure 15.4(c), the solid shield inherently provides better shielding than does a braided shield, and a spiral-wrapped shield can be far inferior to a braided shield in performance. In severe environments braided shields using two overlapping courses of braid may give shielding performance approaching that of a solid-walled shield.

Conduits, shown in Figure 15.4(d), may or may not provide electromagnetic shielding. Conduits in aircraft tend to be used more for mechanical protection than for electrical protection of conductors. Only if the conduit is electrically connected to the aircraft structure will it be able to carry current and thus provide shielding for the conductors within. Conduits for mechanical protection frequently are physically mounted in clamps that use rubber gaskets to prevent mechanical vibration and wear. Such clamps, of course, insulate the conduit and prevent it from having any magnetic shielding capability. Clearly, nonmetallic conduits will not provide electromagnetic shielding.

The requirement that a shield intended for protection against lightning effects must be grounded at both ends raises the perennial controversy about

434

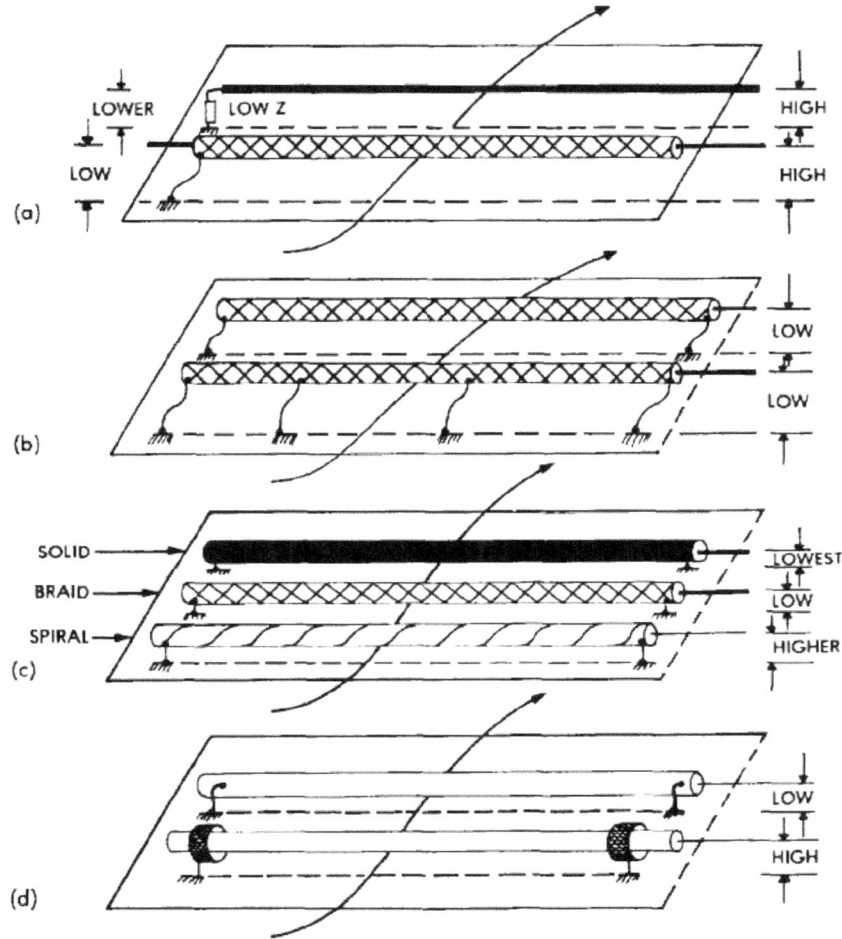

Figure 15.4 Types of shields.

    (a) Unshielded vs shielded

    (b) Multiple-grounded

    (c) Solid- vs braid- vs spiral-wrapped

    (d) Conduits

single- versus multiple-point grounding of circuits. For many, usually legitimate, reasons low-level circuits need to be shielded against low-frequency interference. Most commonly, and usually legitimately, the shields intended for such low-frequency interference protection are grounded at only one end. But in this practice a fundamental concept often overlooked is that the physical length of such shields must be short compared to the wavelength of the interfering signals. Lightning-produced interference, however, is usually broad band and includes significant amounts of energy at quite high frequencies, frequencies higher than those the typical low-frequency shields are intended to handle. The dichotomy between the requirements for shielding against high-frequency, lightning-produced interference and those for shielding against everyday, low-frequency interference is usually too great for both sets of requirements to be met by the use of only one shield system.

Most commonly both sets of requirements can be met only by having one shield system to protect against low-frequency interference and a second shield system to protect against lightning-generated interference. The lightning shield can usually consist of an overall braided shield on a group of conductors with this overall shield being grounded to the aircraft structure at least at the ends. Within the overall shield may be placed whatever types of circuits are needed. Frequently these circuits will have a shielded conductor of their own. In a coordinated shielded system the designers of individual circuits should have the option of grounding such inner shields as their own requirements dictate, but they should not have the power of dictating the treatment of the overall shield. Such an overall shield (OAS) is shown in Figure 15.5.

The method of grounding this overall shield can have a great impact on its effectiveness in protecting against lightning-generated interference. Figures 15.5(b) through 15.5(e) show several methods of grounding such a shield when that shield is placed over a group of conductors being brought into an equipment case. For best performance the OAS should be terminated on the back shell of a connector specifically designed for such termination. The shield should make a 360° circumferential connection to the back shell of the connector. The connector shell itself should be designed to have low dc resistance to its mating panel connector. Most commonly, such low-resistance mating requires the use of grounding fingers within the connector shell. Connectors without such grounding fingers frequently have high resistance between the mating shells, since the shells are frequently coated with an insulating coating to reduce problems of corrosion. The shell of the panel connector should also provide a 360° peripheral connection to the metal equipment case. Providing such a connection frequently requires that paint or other coatings on the case of the equipment be removed and the bare metal exposed.

In the absence of a 360° connector, an external pigtail is often used for grounding the OAS, as shown in Figure 15.5(c). Such pigtails are definitely inferior to the 360° connector because they force an interfering current on the shield to be concentrated through the pigtail, and hence to provide a much greater degree of magnetic coupling to the core conductors than does the distributed current flow on a properly designed connector back shell. If such a pigtail is used, it should be as short as possible and should terminate on the outside of the equipment case. A pigtail of only a very few inches may introduce more leakage from the shield onto the inner conductors than does a several-foot section of the shield itself.

The practice of grounding an overall shield to the inside surface of an equipment case through a pigtail and a set of contacts in the connector is less effective than the use of an external pigtail, partly because the length of the

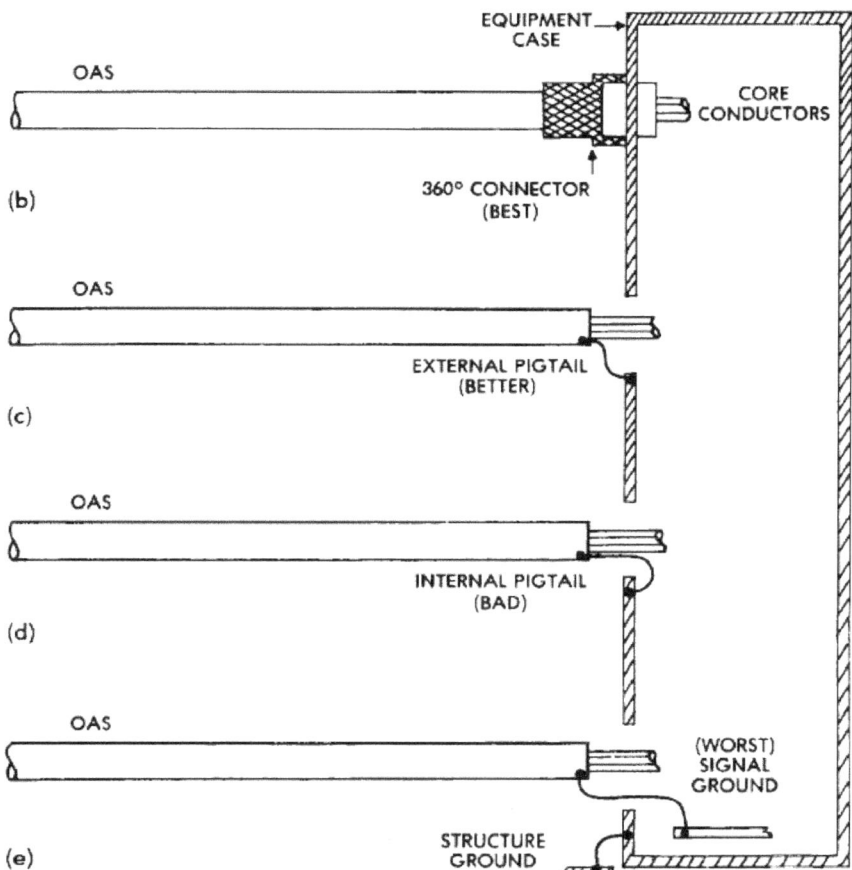

Figure 15.5 Types of grounding for shields.

pigtail is inherently longer and partly because it brings currents directly to the inside of the case. Such grounding of an overall shield should be avoided wherever possible and, in particular, must be avoided whenever the overall shield runs through a region where it will intercept a significant amount of energy from the external electromagnetic field.

In no case should an overall shield be connected to a signal ground bus.

437

15.5 Improvement Through Circuit Design

One of the most important considerations in the control of lightning-related interference through proper circuit design lies in the fundamental observation that a device with a broad band width can intercept more noise energy than can a narrow band-width device. Some of the considerations that derive from this observation are shown in Figure 15.6. The noise produced by lightning has a broad frequency spectrum. Considering for the moment only the spectrum of the lightning current, the observation is frequently made that most of the energy associated with the lightning current is contained in the low-frequency region, below 10 or 20 kHz. Before any sense of security is derived from that observation, it should be remembered that equipment is damaged or caused to malfunction in accordance with the total amount of energy intercepted. In a lightning flash there may be plenty of energy in the megahertz and multimegahertz region to cause interference. The energy that is available for damage or interference may well be concentrated in certain frequency bands by the characteristic response of the aircraft or the wiring within the aircraft.

Without reference to any specific frequency regions, however, the energy spectrum of the lightning-generated interference on electrical wiring within an aircraft will still be a broad spectrum. A receptor with a broad pass band, shown in Figure 15.6(a), will inherently collect more energy than will a receptor with narrow pass band, shown in Figure 15.6(b). The narrower the pass band, the better. In this respect analog circuits have an inherent advantage over digital circuits, since a narrow-pass band digital circuit is almost a contradiction in terms. If possible, circuits should not have a pass band that includes dc, shown in Figure 15.6(c), because, when dc is excluded, the circuits will inherently be able to reject more of the energy associated with the flow of current through resistance of the structure.

The studies of types of interference produced in aircraft by the flow of lightning current have shown that the lightning energy excites oscillatory frequencies on aircraft wiring, particularly if the wiring is based on a single-point ground concept. Those characteristic frequencies have tended to be in the range of several hundred kilohertz to a few megahertz. If at all possible, the pass bands of electronic equipment should not include these frequencies, as does the hypothetical pass band shown on Figure 15.6(d). Higher or lower pass bands would inherently be better than the one shown. As an extreme example, shown in Figure 15.6(e), fiber optic signal transmission operating in the infrared region avoids the frequency spectrum associated with lightning-generated interference almost completely.

Basic considerations about circuit design and signal transmission are shown in Figure 15.7. First, as shown on Figure 15.7(a), signal circuits should avoid the use of the aircraft structure as a return path. If the structure is used as a return path, the resistively generated voltage drops will be included in the path between transmitting and receiving devices. On the other hand, signal transmission over a twisted-pair circuit with signal grounds isolated from the aircraft structure tend

438

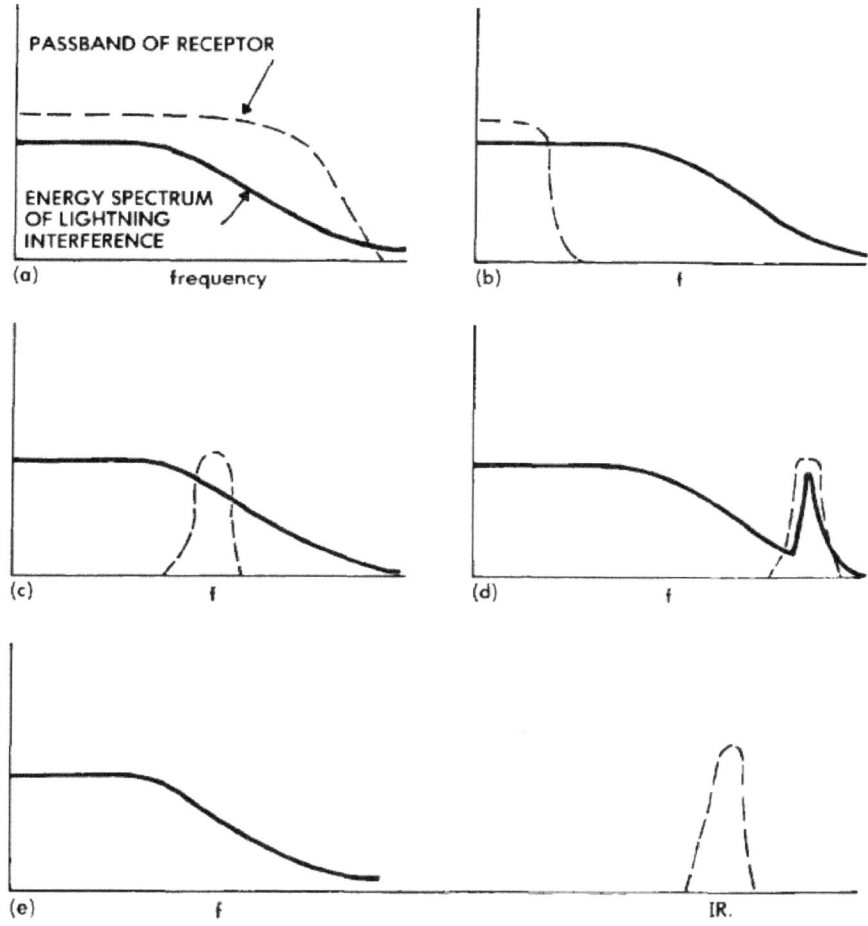

Figure 15.6 Frequency considerations.

to couple lower voltages in the signal path. It must not be forgotten, however, that the use of twisted-pair transmission lines does not eliminate the common-mode voltage to which electronic systems may be subjected. Common-mode voltages applied to the unbalanced transmission path, as in Figure 15.7(b), can lead to line-line voltages which may at times be as high as the common-mode voltage.

Differential transmission and reception devices, shown in Figure 15.7(c), can offer a many-fold improvement in the ability to reject the common-mode voltages produced by lightning.

In general it is preferable that wiring interconnecting two different pieces of electronic equipment not interface directly with the junctions of semiconductors, as shown in Figures 15.7(a) and 15.7(c). Even modest amounts of resistance connected between the junctions and the interfacing wires, shown in

Figure 15.7(d), can greatly improve the ability of semiconductors to resist the transient voltages and currents. Chapter 16 on component damage mechanisms gives examples of the degree of improvement that may be obtained through the use of series resistors. Transmission through balanced transmission lines and transformers, coupled with input protection for semiconductors, probably provides the greatest amount of protection against the transients induced on control wiring.

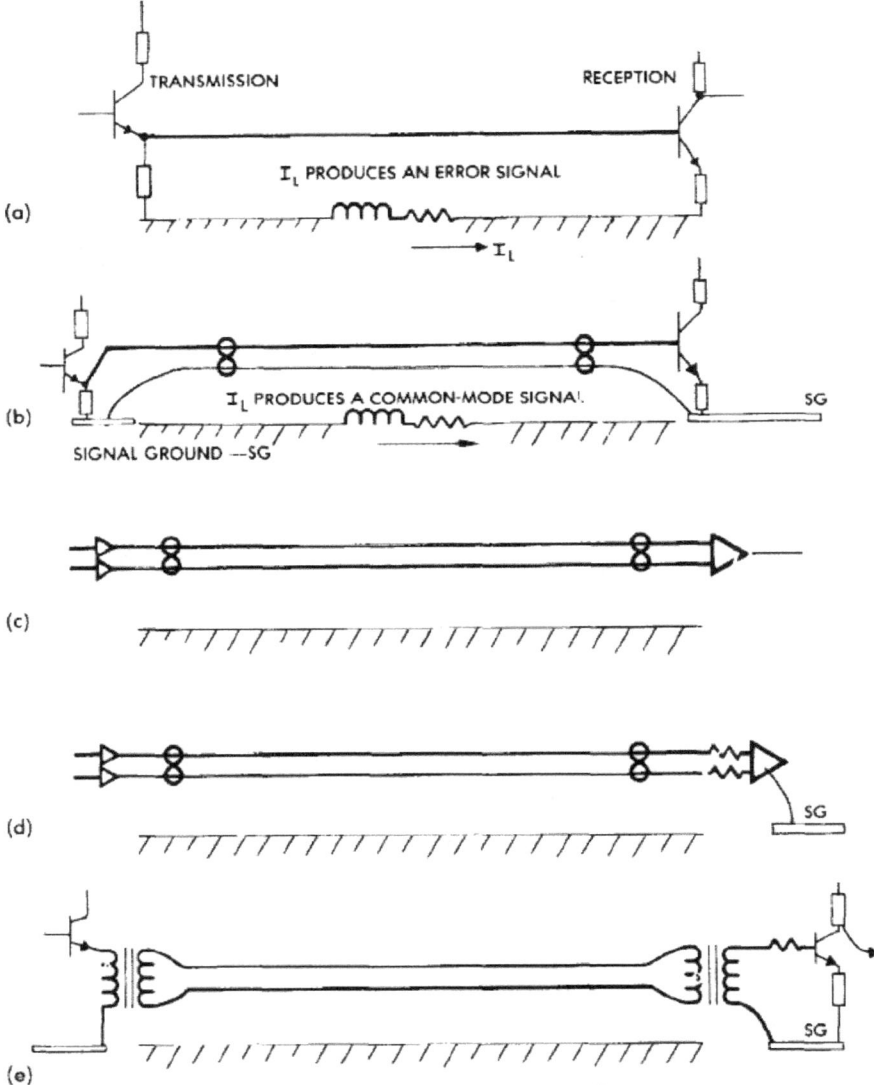

Figure 15.7 Considerations regarding circuit design.

15.6 Circuit Protection Through Use of Protective Devices

Circuit protective devices can sometimes be used to limit the amount of electrical energy that a wire can couple into a piece of electronic equipment. While one can seldom eliminate interference through the use of circuit protective devices, circuit protective devices judiciously used can virtually eliminate physical damage to electronic devices. "Judicious use" usually means that protective devices must be incorporated into a piece of equipment at the time it is built, not added after trouble has been experienced.

There are two basic types of overvoltage or transient protection devices (TPD): those which, on sensing an overvoltage, switch to a low-impedance state and thus cause the impressed voltage across them to collapse to a low value; and those which, on sensing an overvoltage, tend (by virtue of their nonlinear voltage-current relation) to maintain the voltage at that level but do not tend to collapse the voltage. Examples of the first type are spark gaps or arcing dielectric devices. Examples of the second type are Zener diodes and varistors. There are also devices which, on sensing an overvoltage, interrupt the power flow to the load. If this interruption is accomplished by electromechanical means, they should not be considered transient protection devices because they are inherently slow to respond.

Switching devices inherently offer greater surge-power handling capability than do the Zener or varistor types of devices. The instantaneous power dissipated in a transient protective device is the product of the surge current flowing through the device and the voltage across the device. For a constant surge current, a switching device like a spark gap, across which the voltage is low while in the conducting state, will have less power released in it than a device like a Zener diode, across which the surge voltage remains high. For a given surge-power handling capability, a spark gap will thus be smaller physically than a Zener diode or varistor device.

Another fundamental difference between switching devices (spark gaps) and nonswitching devices (Zener diodes or varistors) relates to their recovery characteristics after the surge has passed. If a line is protected by a spark gap and if that line is connected to a source of energy (a power bus, for example), that energy source must be disconnected from the line before the spark gap can switch back from its low-impedance conducting state to its high-impedance nonconducting state. Generally this requires opening a circuit breaker on the line. A Zener diode or varistor effectively ceases to conduct as soon as the voltage returns to its normal value. Operation of remote circuit breakers is not required.

All types of overvoltage protection devices inherently operate by reflecting a portion of the surge energy to its source and by diverting the rest into another path, all with the intention of dissipating the surge energy in the resistance of the ground and interconnecting leads. The alternative to reflecting the energy is to absorb the surge energy in an unprotected load. Reflection and diversion of the surge energy are not without their hazards. Some are the following:

441

- The reflected energy can possibly appear on other unprotected circuits.
- Multiple reflections may cause the transient to last longer than it would otherwise.
- The spectral density of the energy may be changed, either high or low frequencies being enhanced. Interference problems on other circuits may well be increased even though the risk of damage to the protected circuit is reduced.

Most commonly the appropriate type of transient protective device to be used depends on the amount of surge energy to be dealt with. Generally, this energy decreases the further away one gets from the stroke. The surge energy to be expected can also be related crudely to the normal operating power of the circuit involved. One would normally expect lower surge levels on low-voltage signal circuits than on medium-power control circuits, and even lower levels than those on main-power distribution buses. Thus, one might logically use Zener diodes on individual circuit boards, varistors on terminal boards, and spark gaps on leads running to prime entry and exit points.

There are basically four generic types of transient protective devices applicable to lightning hardening of aeronautical subsystems. These types are gas-filled spark gaps, Zener diodes, varistors, and arcing dielectrics. The latter are currently in development, but the others are readily available commercially. Each type has both advantages and disadvantages.

15.6.1 Spark Gaps

Spark gaps are generally composed of two metal electrodes separated by a dielectric and held at a fixed distance from each other. The gap may be sealed in a container. Electrodes may be spherical, but in sophisticated devices they are not. Sparkover voltage is determined by dielectric composition, density, and electrode geometry. Sparkover voltage is also dependent on voltage waveshape. If the voltage wave is increasing rapidly, the sparkover voltage will be higher than it is on a slowly rising wave. Commercially available spark gaps frequently contain minute amounts of tritium or other radioactive elements to reduce the dependence of sparkover voltage on voltage waveshape.

The advantages of spark gaps are as follows:
- They are simple and reliable.
- They have very low-voltage drop during the conducting state. When the gap is carrying maximum current, the voltage across the gap is typically 10 to 20 V. If more current tries to flow, the arc channel increases in diameter and holds the same arc-drop. A low arc-drop indicates relatively low power absorption during the conducting phase.
- They have large power-handling capability. Gas-filled gaps have the highest peak current-handling capabilities of any transient protection device, and almost any gap can handle the maximum surge currents induced by lightning.

442

- They have high impedance and low capacitance. The low-shunt capacity and leakage current characteristics of gas-filled spark gaps minimize insertion problems for operating frequencies below 1 GHz.
- They provide bilateral operation, having the same characteristics on either polarity.

The disadvantages of spark gaps include the following:
- They have relatively high sparkover voltage.
- Simple gaps do not extinguish follow current. This is a most important point to consider if they are to be used on a power circuit. The arc must be extinguished by removing the voltage (circuit breaker of fuse) or by inserting resistance rapidly into the circuit by an additional element, such as a silicon carbide or zinc oxide varistor. Through suitable designing, gaps *can* be made self-extinguishing for applied voltages up to about 100 V. Such self-extinguishing properties require the use of the magnetic blow-out principle.
- They may have a large dependence of sparkover voltage on the waveshape of the voltage. Specifications relating to the impulse ratio or volt-time effect should be carefully considered.
- Since spark gaps reflect more energy than they absorb, external resistive components may be required to minimize ringing. Discrete linear or nonlinear resistors are sometimes used to assist in extinguishing a gap. This component may serve the dual purpose of limiting current and dissipating power.

15.6.2 Zener Diodes

This category includes all single-junction semiconductor devices such as rectifiers, in addition to Zener diodes. While other semiconductor devices, such as PNPN devices and bipolar transistors, may have application as surge arrestors, they will not be covered here because of the limited pertinent data available.

Zener diodes are basically polarized devices which exhibit an avalanche breakdown when the applied voltage in the reverse bias direction exceeds the device's specified breakdown, or Zener voltage of the device. Operated in an opposed series configuration, diodes can be used as effective suppression devices. Since Zener diodes are designed to operate in the breakdown mode, they usually can perform more effectively as terminal protection devices than can signal diodes. While the energy-handling capabilities of Zener diodes are modest when compared with those of spark gaps, they are very well adapted for protection of individual components or circuit boards.

The advantages of Zener diodes include the following:
- They are of small size.
- They are easily mounted.
- They have low "firing" voltage.
- They have low dynamic impedance when conducting.

443

- They are self-extinguishing. When applied voltage drops below the Zener level, they cease conduction.
- They exhibit low volt-time turnup, or impulse ratio.

The disadvantages of Zener diodes include the following:
- They may be expensive.
- They are not bilateral. To protect against both polarities, two diodes in series back-to-back configuration are necessary.
- Diodes have relatively high-junction capacitance; therefore, they may cause significant signal loss at operating frequencies above 1 MHz. Special diode assemblies may extend the useful frequency to approximately 50 MHz.
- They do not switch state between a conducting and a nonconducting mode. The voltage across the diode does not switch to a low value when conducting but remains at the Zener voltage. This characteristic accounts for their ability to cease conduction when the voltage falls below the Zener level, but it has a disadvantage thermally. During conduction, the power absorbed by the diode is the product of the current through the diode and the voltage across the diode. The power absorbed for constant current, thus, is directly proportional to the diode voltage.

  Partially offsetting this disadvantage, however, is the phenomenon that surge energy absorbed in the diode is energy that cannot be reflected back into the system to cause trouble elsewhere.
- They provide lower energy capabilities than do spark gaps. Since the Zener action takes place across a narrow P-N junction, the mass of the protecting junction is small and hence cannot store much energy. As a result, diode networks cannot be used where extremely high transient current or energy is predicted. For most hardening applications, this is not a serious limitation, since the induced surge-current levels are in the 1 to 100 A range at those locations where Zener diodes are most likely to be used.
- They are not available for voltage below about 5 V.
- They are not normally available for voltages above a few hundred volts.

15.6.3 Forward-Conducting Diodes

In a forward-conducting state, a diode conducts little current below about 0.3 V for germanium and 0.6 V for silicon. They can, as a result, be placed directly across a low-voltage line and afford substantial protection.

The advantages of forward-conducting diodes include the following:
- They are of small size.
- They are not costly.
- They provide protection at very low-voltage levels.
- They have excellent surge-current ratings.

444

The disadvantages of forward-conducting diodes include the following:

- They are not bilateral. For protection of both polarities, two diodes in parallel must be used. Some vendors supply dipolar diodes for protection purposes.
- Conduction may occur on normal signals with attendant signal-clipping and frequency-multiplication effects. Diodes must be used in series to raise voltage levels.
- They have relatively high capacitance.

15.6.4 Nonlinear Resistors

This category includes nonlinear resistors (varistors) that may be characterized by the expression

$$I = KV^N$$

where N and K are device constants dependent on the varistor material.

Varistors may be constructed of silicon carbide, selenium, or metal oxide. This section will concentrate on the metal oxide varistor (Reference 15.1).

The advantages of metal oxide-based nonlinear resistors include the following:

- They are bilateral devices.
- They are of small size.
- They are easily mounted. One common configuration is very similar in appearance to a disk ceramic capacitor.
- They are self-extinguishing. When applied voltage drops below the voltage for which the device is rated, they conduct very little current.
- They have an inherently fast response/low-impulse ratio.
- They provide high power-handling capability. The current- and energy-handling capability is second only to certain types of spark gaps. This device gives a higher ratio of energy absorbed to energy reflected than conventional gaps give. Morever, the energy is absorbed throughout the bulk of the material and is not concentrated in a narrow P-N junction.

The disadvantages of nonlinear resistors include the following:

- They have low impedance and high capacitance. The zinc oxide varistor is characterized by high-shunt capacity, limiting its use to frequencies below 1 MHz. While it has a much higher idling current than have either gaps or diodes, the standby power dissipation is in the milliwatt region and is usually not a significant limitation.
- They are not suitable for operating voltages below 20 to 30 V. Operating voltage is proportional to material thickness. Good surge protection of low-voltage circuits would require an impractically thin piece of material.

15.6.5 Surge Protecting Connectors

At least two lines of activity have been pursued with the aim of incorporating surge protective devices directly in electrical connections. One approach (Reference 15.2) incorporated spark gaps between the connector pins and the connector shell. A high dielectric material, rutile, was used to stimulate internal flashover and to achieve a protective level of about 1100 to 1300 V with an impulse ratio nearly unity, even with voltage rates of change of 10 kV/μs. This design withstood extremely large currents (100 to 250 kA). The work on this device was performed at Sandia Laboratories and supported by the U.S. Atomic Energy Commission.

Another line of approach uses zinc oxide varistor material to hold the pins in a connector. This work is being done by General Electric Company for the U.S. Army Harry Diamond Laboratories. Neither of these protective connector designs is yet available commercially.

15.7 Improvement Through Transient Coordination

Transient coordination is a concept which, when reduced to its simplest terms, implies that targets relative to transients will be assigned both to those that design electronic equipment ("black boxes") and those that design wiring to interconnect those black boxes. The task that is to be assigned the designers of black boxes is to produce equipment that will be able to withstand transients on all of the input and output wiring. The targets that form a part of the task will be specifications describing the type of transients that the equipment must withstand and to which it will be subjected as part of an acceptance or proof test. The task that is to be assigned to the groups designing interconnecting wiring is that no external threat, such as lightning or switching of inductive devices, shall produce on the wiring transients larger than those which the black box was designed to withstand. The target numbers to which the wiring designer must work will be maximum amplitudes of current, voltage, and surge energy.

The assignment of such tasks implies that there must be a referee who assigns the appropriate target numbers and oversees the work to ensure that both parties fulfill the tasks assigned. This referee may be called the system integrator. The transient coordinator philosophy is illustrated in Figure 15.8. The aims of transient coordination would be the following:

1. Insure that the actual transient level produced by lightning (or any other source of transient) will be less than that associated with the transient control level number assigned to the cable designer. The cable designer's job would be to analyze the electromagnetic threat that lightning would present and to use whatever techniques of circuit routing or shielding would be necessary to ensure that the actual transients produced by lightning did not exceed the values specified for that particular type of circuit.

2. The transient design level controlling the type of circuit or circuit protection techniques used, and assigned to the avionics designer,

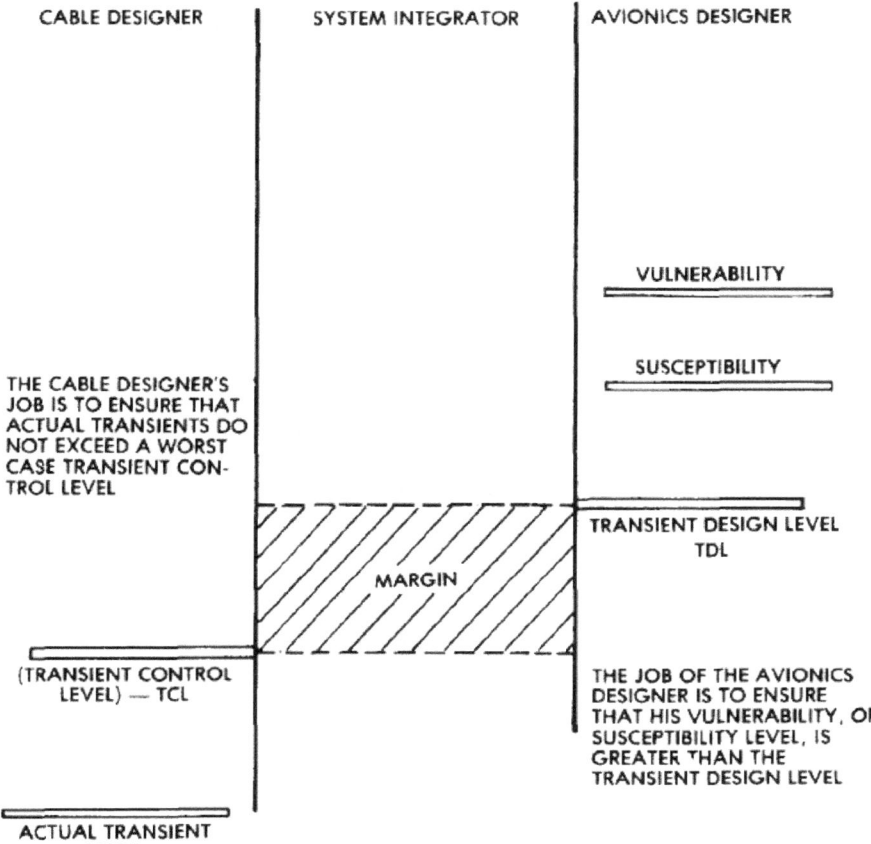

CABLE DESIGNER   SYSTEM INTEGRATOR   AVIONICS DESIGNER

VULNERABILITY

SUSCEPTIBILITY

THE CABLE DESIGNER'S
JOB IS TO ENSURE THAT
ACTUAL TRANSIENTS DO
NOT EXCEED A WORST
CASE TRANSIENT CON-
TROL LEVEL

TRANSIENT DESIGN LEVEL
TDL

MARGIN

(TRANSIENT CONTROL
LEVEL) — TCL

THE JOB OF THE AVIONICS
DESIGNER IS TO ENSURE
THAT HIS VULNERABILITY, OR
SUSCEPTIBILITY LEVEL, IS
GREATER THAN THE
TRANSIENT DESIGN LEVEL

ACTUAL TRANSIENT
LEVEL

Figure 15.8 The transient coordination philosophy.

would be higher than the transient control level by a margin reflecting how important it was that lightning did not in fact interfere with the piece of equipment under design. A margin is necessary because any single lightning flash might produce an actual transient level higher than the assigned transient control level, which would have been derived for a predicted average in spite of the cable designer's good intentions. Prediction of actual transient levels is an imperfect art.

3.   The job of the avionics designer would be to ensure that the vulnerability and susceptibility levels of the equipment that he is to supply would be higher than the assigned transient design level. The vulnerability level is that level of transient which, if applied to the input or output circuit under question, would cause the equipment to be permanently damaged. The susceptibility level is defined as

447

that level of transient that would result in interference with or malfunction of the equipment. The vulnerability level by definition, then, would have to be at least as high as the susceptibility level.

There are several ways in which the levels might be set. In the first, the system integrator would set the desired transient level, then set the required margin, which in turn would set the transient control level. Whatever the rationale by which the system integrator sets the transient design level, that level would become a part of the purchase specifications and would, presumably, not be subject to variation by the vendor of the avionics. As an alternative, the avionics designer might determine by suitable testing the vulnerability and susceptibility levels of his equipment and provide a guarantee as to the level of transients that his equipment could withstand. That level would then be the transient design level. After the system integrator had set the desired safety margin, the appropriate transient control level for the cable designer would have been established. One approach to the setting of margins appears in the *Space Shuttle Lightning Criteria Document* (Reference 15.3).

The numbers that would be assigned to the transient design level probably should be expressed in terms of the maximum voltage appropriate to a high-impedence circuit (open circuit voltage) or the maximum current appropriate to a low-impedance circuit (short circuit current). In order for the transient coordination philosophy to have most impact, there should be a limited number of levels. One set of levels that has been proposed (Reference 15.4) is shown in Table 15.1. With each level there is associated an open circuit voltage and a short circuit current; the two are related by a standard transient-source impedance, shown in Figure 15.9 (Reference 15.5)

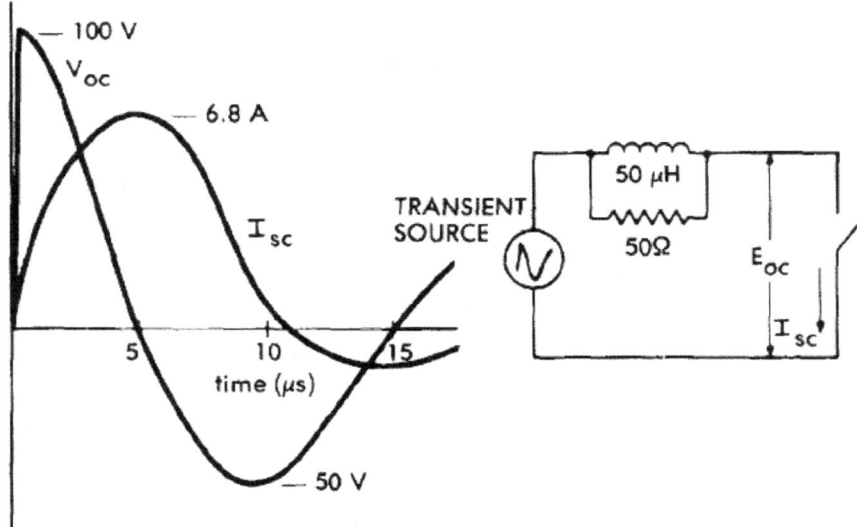

Figure 15.9 Short circuit current (I_{SC}) resulting from a transient source with V_{OC} open circuit voltage and $50\Omega/50\mu Hz$ source impedance.

Table 15.1 PROPOSED TRANSIENT CONTROL LEVELS

Proposed Transient Control Level Number	Open Circuit Voltage Level (volts)	Short Circuit Current Level (amperes)
1	10	0.68
2	25	1.7
3	50	3.4
4	100	6.8
5	250	17
6	500	34
7	1000	68
8	2500	170
9	5000	340

An alternative set of levels for which some voltages are numerically equal to the voltages in existing specifications is presented in Table 15.2.

Table 15.2 ALTERNATIVE TRANSIENT CONTROL LEVELS

Proposed Transient Control Level Number	Open Circuit Voltage Level (volts)	Short Circuit Current Level (amperes)
1	15	1
2	30	2
3	60	4
4	150	10
5	300	20
6	600	40
7	1500	100
8	3000	200
9	6000	400

REFERENCES

15.1 *Transient Voltage Suppression Manual*, Semiconductor Products Department, General Electric Company, Electronics Park, Syracuse, New York (1976).

15.2 J. A. Cooper and L. J. Allen, *The Lightning Arrester Connector: A New Concept in System Electrical Protection*, IEEE International Electromagnetic Compatibility Symposium Record, IEEE 72CH0638-EMC, Institute of Electronic and Electrical Engineers, New York, New York (1972).

15.3 *Space Shuttle Program Lightning Protection Criteria Document*, JSC-07636, Revision A, National Aeronautics and Space Administration, Lyndon B. Johnson Space Center, Houston, Texas (November 4, 1975), pp. D-2 to D-5.

15.4 F. A. Fisher and F. D. Martzloff, "Transient Control Levels: A Proposal for Insulation Coordination in Low-Voltage Systems," *IEEE Transactions on Power Apparatus and Systems*, PAS-95, 1, Institute of Electronic and Electrical Engineers, New York, New York (January/February 1976): 120-29.

15.5 Fisher and Martzloff, "Transient Control Levels," p. 128.

CHAPTER 16
COMPONENT DAMAGE ANALYSIS

16.1 Introduction

The energy coupled into a system by lightning induces large current pulses into the cabling and wiring of the system and resulting voltage pulses across loads. Determining whether or not these voltages and currents cause upset or damage to active or passive components requires a knowledge of the failure thresholds of devices. These failure thresholds were investigated extensively during studies of nuclear electromagnetic pulse (NEMP) effects on electronic equipment.

Assessment of the vulnerability of a system to a transient environment leads to the observation that failure of the system to function properly may result from a degradation in performance of a single active or passive component device when it is subjected to a transient voltage pulse. Systems vulnerability evaluation therefore requires that one know the failure thresholds for devices subjected to transients. Simulation of failure resulting from transients at the component level can be conveniently performed in the laboratory by means of high-voltage pulse generators. Information from such simulation can then be correlated with the original threat (EMP or lightning) by analytical techniques. Semiconductor electronic components are generally more vulnerable under pulse conditions than are nonsemiconductor components; therefore, most of the discussion of this section will be devoted to semiconductors.

16.2 Theoretical Models

The number of semiconductor devices that have been experimentally tested (mostly diodes and transistors) is only a small sample of the total number of devices available. However, the scope of the experimental results has been extended by development of theoretical models which relate significant changes in the properties of a semiconductor PN junction to high temperatures generated in the junction region during application of a high-voltage pulse. Theoretical models based on thermal analysis of the junction region yield a mathematical relation between junction temperature and power dissipation in the junction region. This relation can be used to define a constant characterizing the operation in a given time domain of each device.

16.3 Empirical Models

In addition to the theoretical correlations just described, the experimental data have been used in the development of empirical relations which are obtained from two models of semiconductor junction devices--the junction capacitance model and the thermal resistance model. These models provide a framework from which the power failure threshold of an untested device can be

451

estimated from the quantities listed in a data sheet description prepared by manufacturers for a diode or a transistor.

16.4   Limitations

Certain limits to the accuracy and application of the theoretical and empirical models exist and must be noted in estimating component device vulnerability. The models have been verified only for diodes and bipolar transistors. Other devices, such as FETS and unijunction transistors, have been tested in insufficient numbers for conclusions to be drawn as to their conformance with the model. Also, the results and models which are presented apply strictly only to an isolated device——that is, not to a device in a circuit. In the case of a multiple-terminal device, they apply only to the two terminals connected for test, with any other terminals open.

The assumptions made about junction heating and transfer of heat in the derivation of the models limit their applicability to the region of pulse durations of approximately 0.1 to 20 μs. For longer times appreciable heat transfer may take place away from the junction area during the pulse input. For short pulses the power levels are so high (1 to 10 kW) that very large currents flow; consequently, the joule heating in the bulk material is appreciable. The transition behavior between these three regions of pulse duration (regions that will be more precisely described later in this section) is not well defined and may vary from one device type to another. Examination of available data indicates the transition region generally occurs between 100 ns and 1 μs.

Still another limitation is fundamental to the work summarized in this section: it applies only to junction burnout. Other modes of device failure, such as metallization burnout and internal arcing, are not treated. Based on the results obtained in studies of junction burnout, it would seem that other effects, such as metallization burnout, occur at higher power input levels than those input levels sufficient to damage the junction.

16.5   Failure Mechanisms——Semiconductors

The two principal breakdown modes for semiconductor PN junctions are the following:
1.   Surface damage around the junction as a result of arcing
2.   Internal damage to the junction region as a result of elevated temperatures.
Surface damage refers to the establishment of a high-leakage path around the junction which effectively eliminates any junction action. The junction itself is not necessarily destroyed, since, if it were possible to etch the conducting material away from the surface, the device might be able to return to its normal operating state. This is not practical, of course, in an operational semiconductor. It is equally likely that the formation of any surface leakage path would be the result of excessive heat formation in the bulk of the material, and this would typically be an irreversible phenomenon. It is very difficult to predict

452

theoretically the conditions which will lead to surface damage because they depend upon many variables, such as the geometrical design and the details of the crystal structure of the surface. The theoretical prediction of surface arcing under pulse conditions is not practical (Reference 16.1). It should be cautioned that surface damage, in the general case, may occur in devices at power levels which are orders of magnitude below those sustainable by devices in which *bulk damage* occurs (Reference 16.2).

Bulk damage, which results in a permanent change in the characteristic electrical parameters of the junction, indicates some physical change in the structure of the semiconductor crystal in the region of the junction. The most significant change is melting of the junction as a result of high temperatures. Other types of change may involve the impurity concentrations, the formation of alloys of the crystal materials, or a large increase in the number of lattice imperfections, either crystal dislocations or point defects.

The simplest structure to analyze is a diode. Consider Figure 16.1. In this diode a current is assumed to flow as a result of some outside stimulus and, in doing so, to produce a voltage across the diode. This voltage may be the forward bias voltage (0.5 to 1.5 V), or it may be the reverse breakdown voltage if the outside stimulus has biased the diode in the reverse direction. Assume that I is a square wave and that V is not a function of time, as it might be if V depended upon the junction temperature. The instantaneous power dissipated in the diode is then also a square pulse of magnitude

$$P = IV \qquad (16.1)$$

and the total energy produced is

$$W = \int_0^t Pdt = IVt \qquad (16.2)$$

Assume that the power level is sufficient that the device fails at the end of the pulse. Both experimental and theoretical analyses indicate that the power (or energy) required to cause failure depends on pulse width: the narrower the pulse, the greater the power required to cause failure. Over a broad range of times, typically between 0.1 μs and 100 μs, the power required to cause failure is inversely proportional to the square root of time. For very short pulse durations the power required to cause failure is inversely proportional to the first power of time, and for very long pulse durations the power required to cause failure is a constant threat. These relations may be expressed by the following equations:

$$Pt \quad = C \qquad t < T_0 \qquad (16.3)$$

$$Pt^{1/2} \quad = K \qquad T_0 < t < 100 \, \mu s \qquad (16.4)$$

$$P \quad = \text{Constant} \quad t > 100 \, \mu s \qquad (16.5)$$

where T_0 generally lies between 10 ns and 1 μs.

Figure 16.2 (Reference 16.3) shows an example for a 10 W diode. The most

453

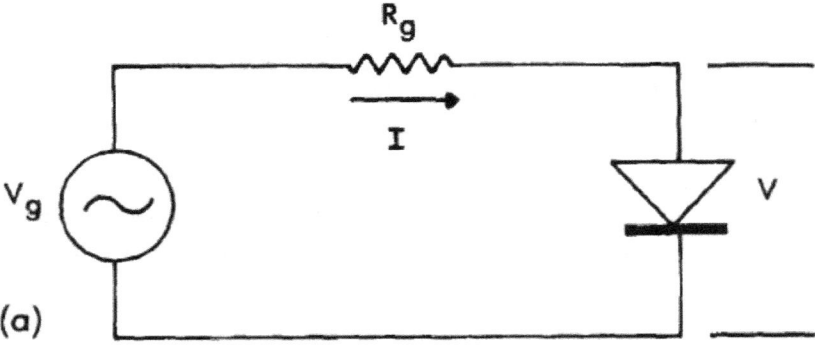

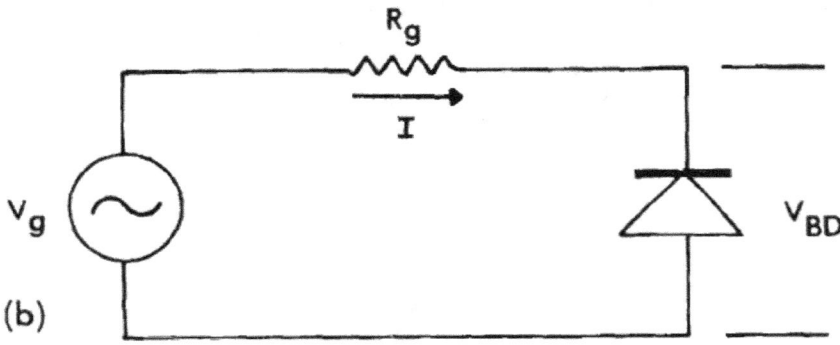

Figure 16.1 Voltage and current through a diode junction.
(a) Forward bias case
(b) Reverse bias case, $V_g > V_{BD}$

important region is the center region, where

$$P = Kt^{-\frac{1}{2}} \qquad (16.6)$$

    Junctions are less susceptible to burnout when operated with forward bias: first, because the power produced by a given current is lower when flowing through the low forward bias voltage than when developed across the higher reverse bias breakdown voltage, V_{BD}, and, second, because the current is more uniformly distributed across the junction in the forward direction. Accordingly, it requires more power in the forward direction to cause failure. An example is shown in Figure 16.3 (Reference 16.4). The 2N2222 transistor is a 0.5 W NPN silicon high-speed switch.

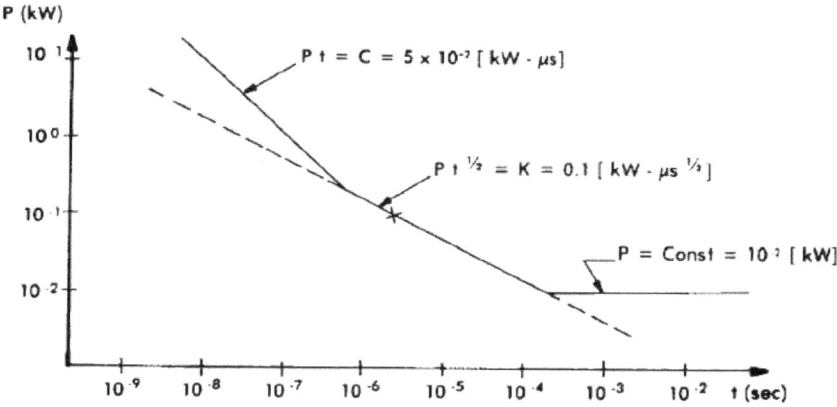

P (kW)

$P t = C = 5 \times 10^{-2} [\, kW \cdot \mu s \,]$

$P t^{1/2} = K = 0.1 [\, kW \cdot \mu s^{1/2} \,]$

$P = Const = 10^{-2} [\, kW \,]$

Figure 16.2 Expected time dependence of pulse power failure: threshold for an example 10 W diode.

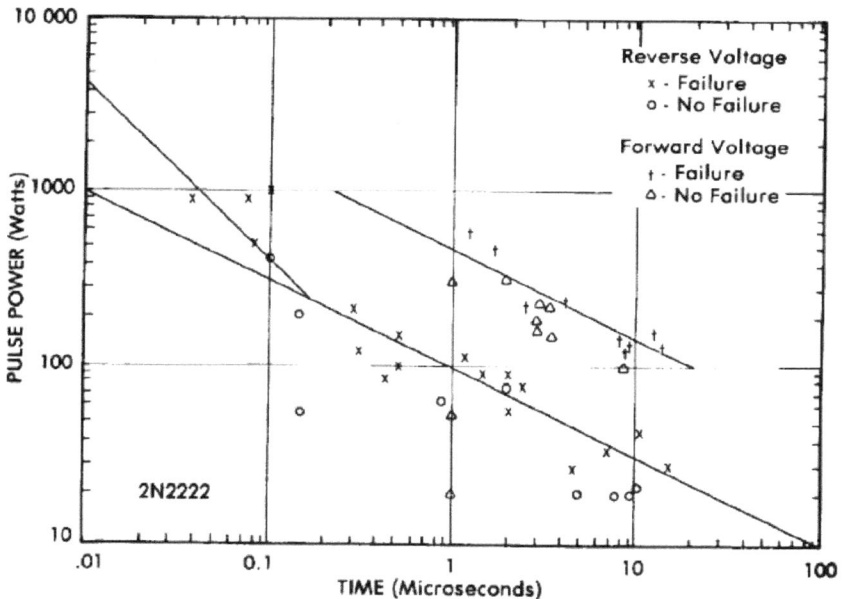

Figure 16.3 Experimental data points for failure of the base-emitter: junction of a 2N-2222 transistor for forward and reverse polarity voltage pulses.

16.6 Damage Constants

From curves, such as those of Figure 16.3, the value of a damage constant, K (or C), can be determined and tabulated for a range of devices. The appropriate threshold damage curve can then be reproduced as desired from a single damage characterization number, the damage constant K. It is convenient

455

to express K in kW • $\mu s^{1/2}$, since the value of K in these units then becomes numerically equal to the power necessary for failure, dissipated by a square pulse of 1 μs duration. If this point is located on a log-log graph (Figure 16.2 or Figure 16.3), then a curve of slope, -1/2, drawn through this point reconstructs the curve fit to the data for a particular device, and the power for failure at other pulse durations can be read directly from such a graph. Ideally, the K factor should be known for both the forward bias and the reverse bias conditions. Generally, only the K factor for the reverse bias condition is known. This limitation gives conservative answers, since K for the reverse bias condition is almost always lower than is K for the forward bias condition.

The magnitude of the damage constant depends upon the type of junction under consideration: broadly speaking, it is larger for large junctions and smaller for small junctions. Figures 16.4 and 16.5 (Reference 16.5) show the range of the damage constant for typical diodes and transistors.

16.6.1   Experimental Determination of K Factor

Experimentally the K factor is determined by injecting power pulses into the semiconductor junction, starting at low levels and increasing the levels until either failure or significant degradation of the junction occurs. Devices would normally be pulsed in both the forward and reverse directions.

The K factors and breakdown voltages for a number of semiconductors are given in Table 16.1 and 16.2 (Reference 16.6). In the case of transistors, the K factor listed generally refers to the base-emitter junction, since this is generally the junction most susceptible to burnout. In all the cases the K factor refers to the reverse bias direction.

16.6.2   K Factor as Determined from Junction Area

If the K factor is not measured, it may be estimated by one or more of three methods. The most accurate of the indirect methods involves a knowledge of the area of the junction. If the area is known, the K factor may be estimated from the following relations:

$$\text{Diodes} - K = 0.56 \text{ A} \qquad (16.7)$$

$$\text{Transistors} - K = 0.47 \text{ A} \qquad (16.8)$$

$$K \text{ in kW} \bullet \mu s^{1/2}$$

$$A \text{ in cm}^2$$

For transistors, the junction area to be used is that of the base-emitter region. This is generally the weaker junction (lower breakdown voltage), and it is that for which the experimental average value for K(A) was obtained.

This method is of course limited by the availability of information on junction area, but where such information is available, the method yields damage

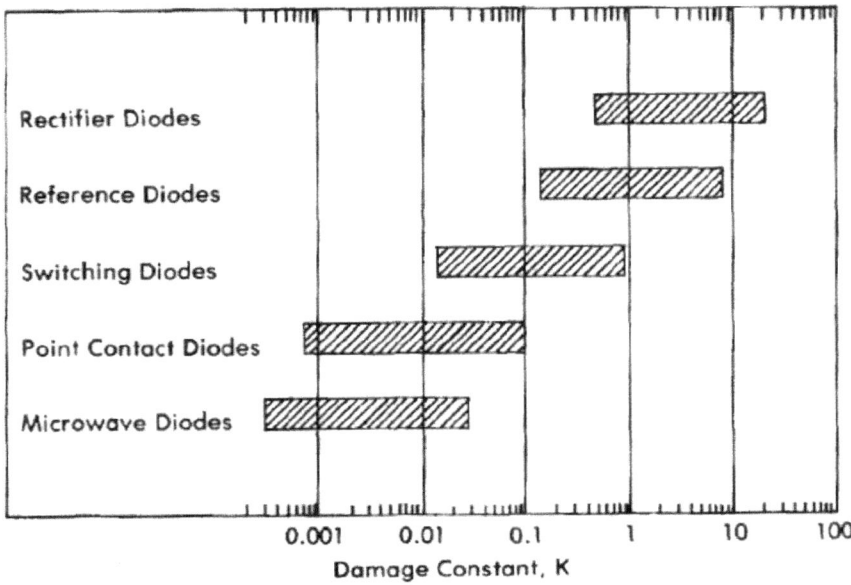

Figure 16.4 Range of pulse power damage constants for representative diodes.

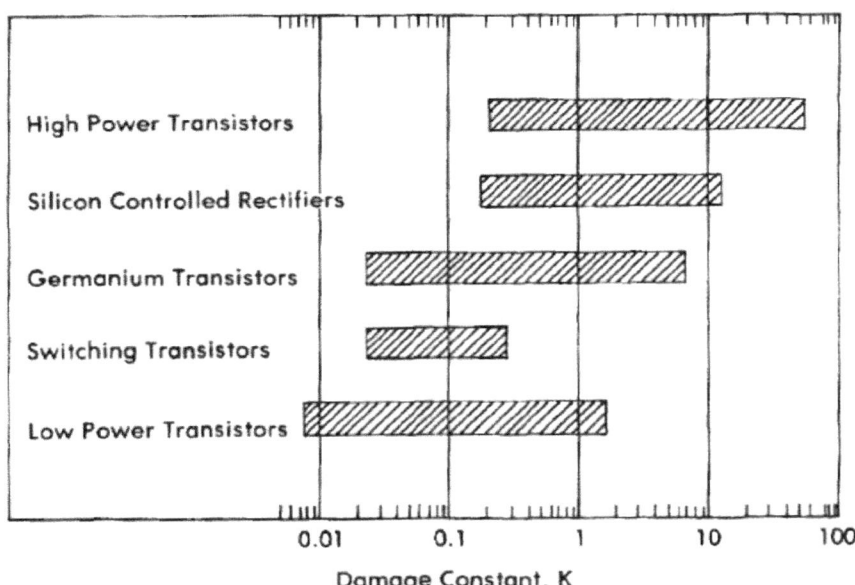

Figure 16.5 Range of pulse power damage constants for representative transistors.

constants accurate to within a factor of two. For planar devices, the junction area can often be measured directly on the silicon chip.

16.6.3 K Factor as Determined from Junction Capacitance

The next most reliable method of determining the damage constant is from a knowledge of the capacitance (C_j) and breakdown voltage (V_{BD}) of the junction. For three different categories of devices, the relations are

Category 1 – insufficient data $\qquad$ (16.9)

Category 2 – $K = 4.97 \times 10^{-3} \, C_j \, V_{BD}^{\,0.57}$ $\qquad$ (16.10)

Category 3 – $K = 1.66 \times 10^{-4} \, C_j \, V_{BD}^{\,0.992}$ $\qquad$ (16.11)

The different categories are

Category 1 – Germanium diodes and germanium transistors

Category 2 – Silicon diodes, all silicon transistor structures except planar and mesa

Category 3 – Silicon planar and mesa transistors

If the junction is a transistor base-emitter junction, the capacitance used should be taken at a reverse bias of approximately 1 V. If it is a collector-base junction or a diode junction, the value should be taken at the reverse bias of approximately 5 to 10 V.

16.6.4 K Factor as Determined from Thermal Resistance

A third, but the least reliable, way of estimating the damage constant is from a knowledge of the thermal resistance of the junction, either the thermal resistance from junction to case (θ_{jc}) or from junction to ambient (θ_{ja}). The empirical expressions for each category for each of the thermal resistances are

Category 1 – insufficient data for an accurate curve fit

Category 2 – $K = 707 \, \theta_{jc}^{\,-1.93} \, (\theta_{jc} > 10.0)$ $\qquad$ (16.12)

$\qquad K = 4.11 \times 10^4 \, \theta_{ja}^{\,-1.7}$ $\qquad$ (16.13)

Category 3 – $K = 707 \, \theta_{jc}^{\,-1.93} \, (\theta_{jc} > 10.0)$ $\qquad$ (16.14)

$\qquad K = 2.74 \times 10^5 \, \theta_{ja}^{\,-2.55}$ $\qquad$ (16.15)

Normally θ_{jc} and θ_{ja} are not given in the transistor data sheets but, rather, must be calculated from the maximum operating junction temperature, ($T_j[\text{max}]$), the total power dissipation (P_d), case temperature (T_c), and ambient temperature (T_{amb}).

$$\theta_{jc} = (T_j[\text{max}] - T_c) / P_d \qquad (16.16)$$

$$\theta_{ja} = (T_{j[max]} - T_{amb}) / P_d \qquad (16.17)$$

Generally, at least one of these thermal resistances may be determined from the manufacturer's data sheet.

The accuracy of the damage constant as determined from either the junction capacitance or the junction thermal resistance is somewhat limited. Table 16.3 (Reference 16.8) gives some estimate of the accuracy within which the damage constant can be calculated.

16.6.5 Oscillatory Waveforms

Equation 16.6 is based on the assumption that the applied voltage, current, and power waves are rectangular in shape. While actual transients are very seldom rectangular, it is possible to derive equivalent rectangular pulses for more common types of transients. One such type of transient typically encountered is the damped oscillatory wave. Based on multiple pulse studies by Wunsch and others (References 16.9 and 16.10), it can be assumed that device damage will occur, if at all, during the first half cycle of a damped sine wave. Therefore, the lower amplitude cycles may be neglected. Two cases may be considered, one in which one of the half cycles of the transient does not exceed the reverse breakdown voltage of the junction, Figure 16.6 (Reference 16.11), and one in which the reverse breakdown voltage is exceeded, Figure 16.7 (Reference 16.12). In either case, one of the half cycles will bias the junction in a forward direction. Treating first the case in which the reverse breakdown voltage is not exceeded (Figure 16.6), the rectangular wave of the same peak amplitude, V_0, and producing the same probability of damage as the sine wave, has a duration τ_p

where

$$\tau_p = \frac{\tau_s}{5} \qquad (16.18)$$

of the sine wave. If the transient does exceed the reverse breakdown voltage, the duration of the equivalent transient depends upon the fraction of the time that the oscillatory transient does exceed the reverse breakdown voltage. The duration of the equivalent rectangular wave is given by the expression

$$\tau_p = \frac{1 - \left(\dfrac{V_{BD}}{V_o}\right)^2}{\pi \cos^{-1}\left(\dfrac{V_{BD}}{V_o}\right)} \tau_s \qquad (16.19)$$

This expression is shown plotted in Figures 16.8 (Reference 16.13) and 16.9 (Reference 16.14). For oscillatory transients whose initial amplitude considerably exceeds the reverse breakdown voltage of the junction, Equation 16.19

459

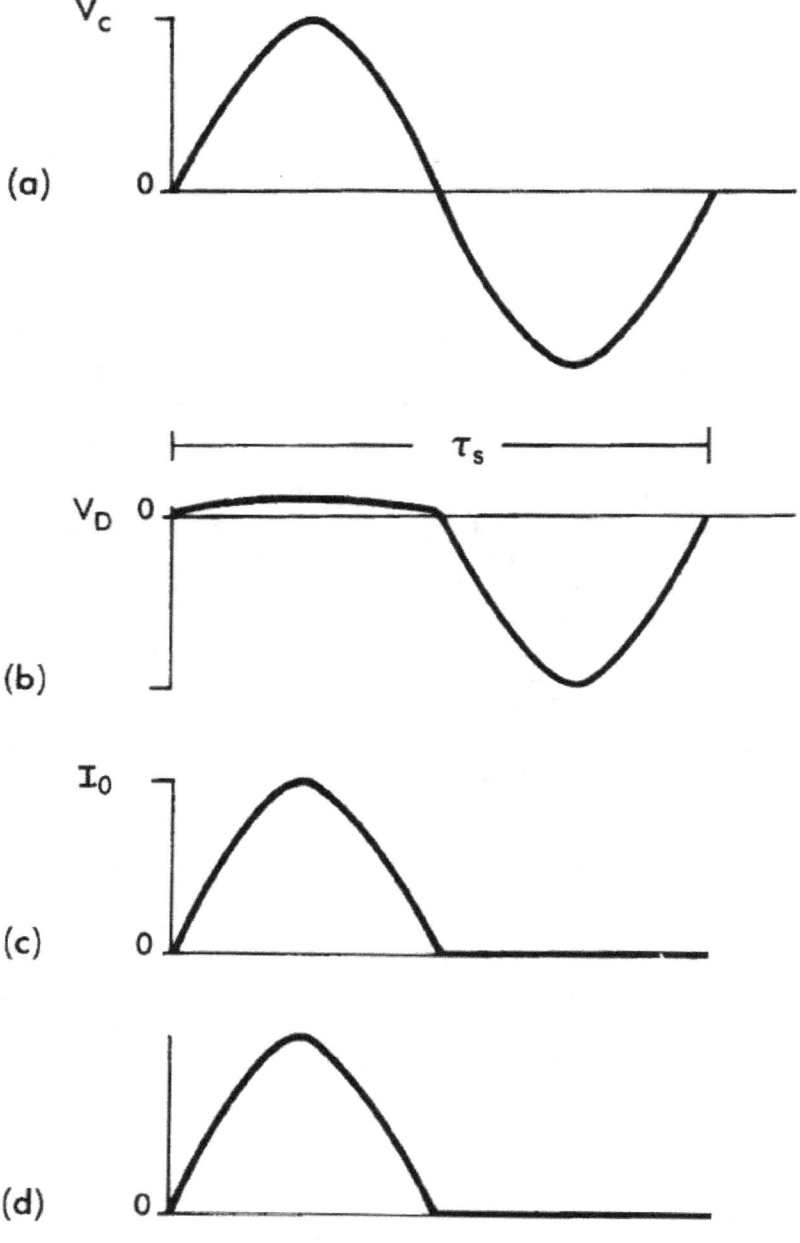

Figure 16.6 Device waveforms for $V_g < V_{BD}$.
  (a) Generator voltage
  (b) Diode voltage
  (c) Junction current
  (d) Junction power

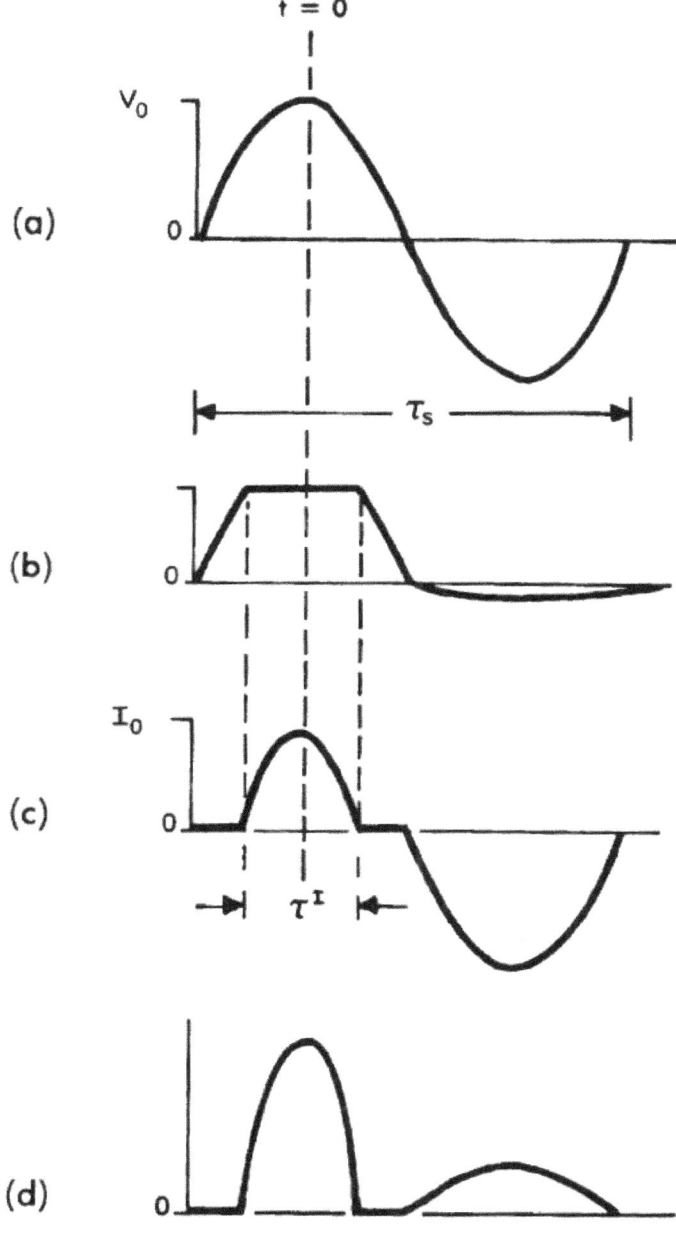

t = 0

(a)

(b)

(c)

(d)

Figure 16.7 Device waveforms for $V_g > V_{BD}$.
    (a) Generator voltage
    (b) Diode voltage
    (c) Junction current
    (d) Junction power

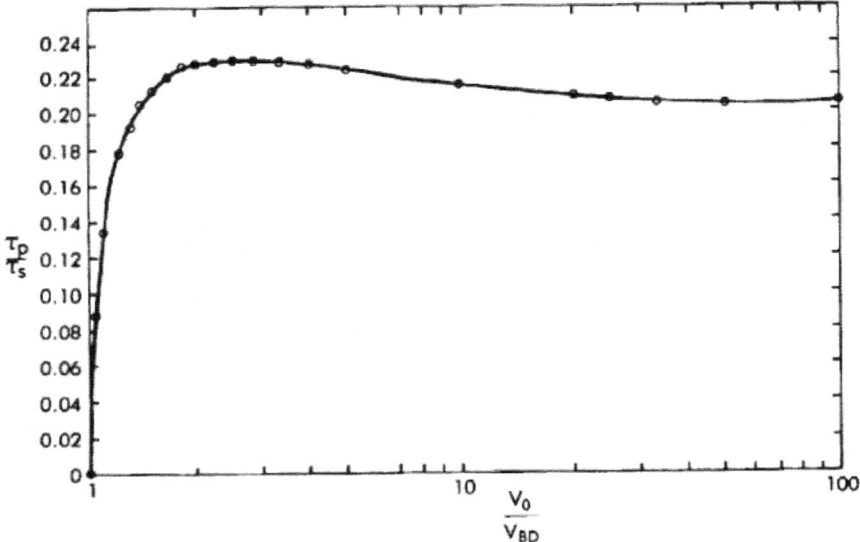

Figure 16.8 Plot of τ_p/τ_s versus V_o/V_{BD}.

approaches a limiting value of 0.2, and thus becomes identical with the forward bias case, Equation 16.18.

A limited amount of data relating voltage and current durations to the breakdown of integrated circuits is available. Figures 16.10 (Reference 16.15), 16.11 (Reference 16.16), and 16.12 (Reference 16.17) show the results of measurements on SN55107 line receivers, SN55109 line drivers, and CD4050 AE hex buffers.

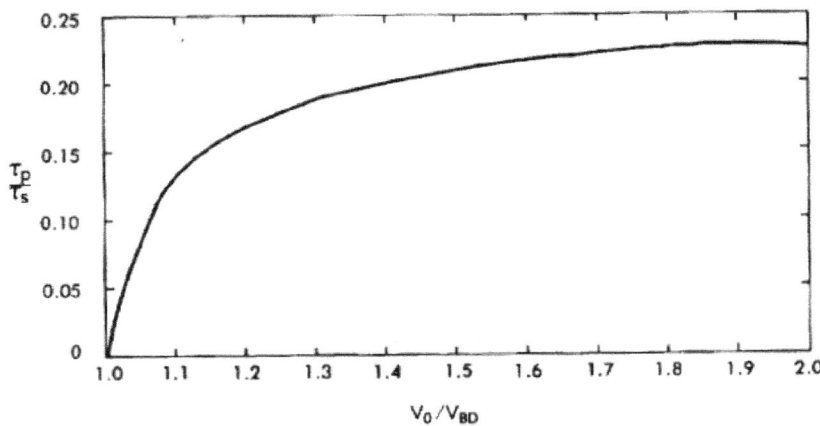

Figure 16.9 Plot of τ_p/τ_s versus V_o/V_{BD} for values of V_o/V_{BD} less than 2.

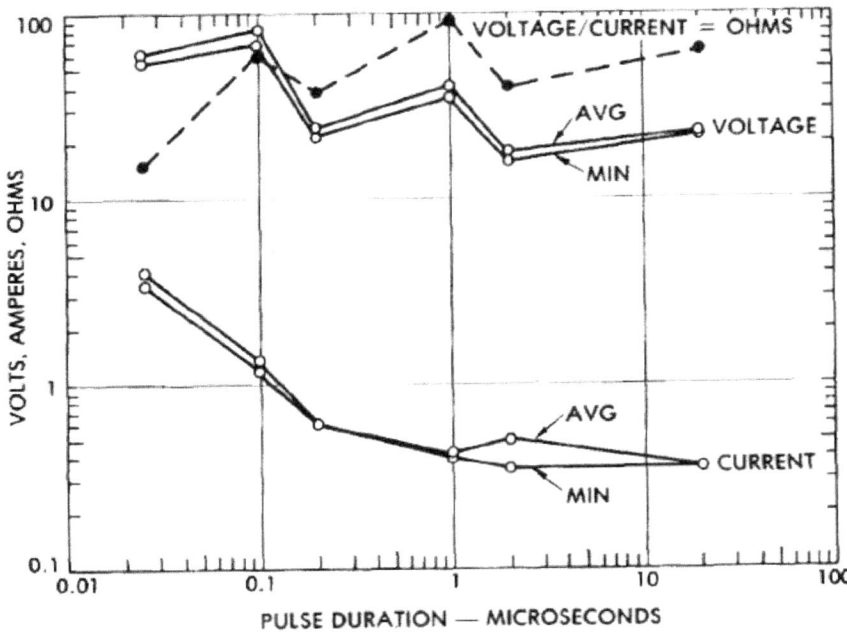

Figure 16.10 Damage thresholds of SN 55107 line receivers.

16.7 Failure Mechanisms––Capacitors

Capacitors fail by a mechanism different from that of semiconductors. The mechanism of capacitor failure depends upon the type of dielectric. Capacitors with solid dielectrics––paper, Mylar, or ceramics––will, when subjected to nonrepetitive transients, either fail by puncture of the dielectric or not fail at all. Typically, a capacitor can withstand short-duration transient voltages several times greater than the dc rating of the insulation. The pulse-breakdown rating of the dielectric, however, is not a constant ratio to the dc voltage rating, nor is it normally part of any manufacturer's specification. Accordingly, it is safest to consider that such a capacitor is in danger of failure if the pulse voltage exceeds the dc rating of the capacitor.

Electrolytic capacitors, on the other hand, do not experience abrupt failure when exposed to short-duration transients. If the voltage across the capacitor exceeds the voltage used to form the dielectric film, the dielectric film begins to conduct. After the pulse has disappeared, the dielectric returns to nearly its normal state. During the transient period the dielectric film can carry substantial transient current without permanent or catastrophic degradation. Transients, however, may lead to increased leakage currents. An example of data that is available relates to a series of tests made on tantalum electrolytic capacitors of value 0.47 μF, 0.047 μF, and 0.0047 μF with a dc voltage rating of 350 V (Reference 16.18). The data indicate that failure (defined as a substantial increase in the leakage current at voltages of less than 350 V) can generally be

463

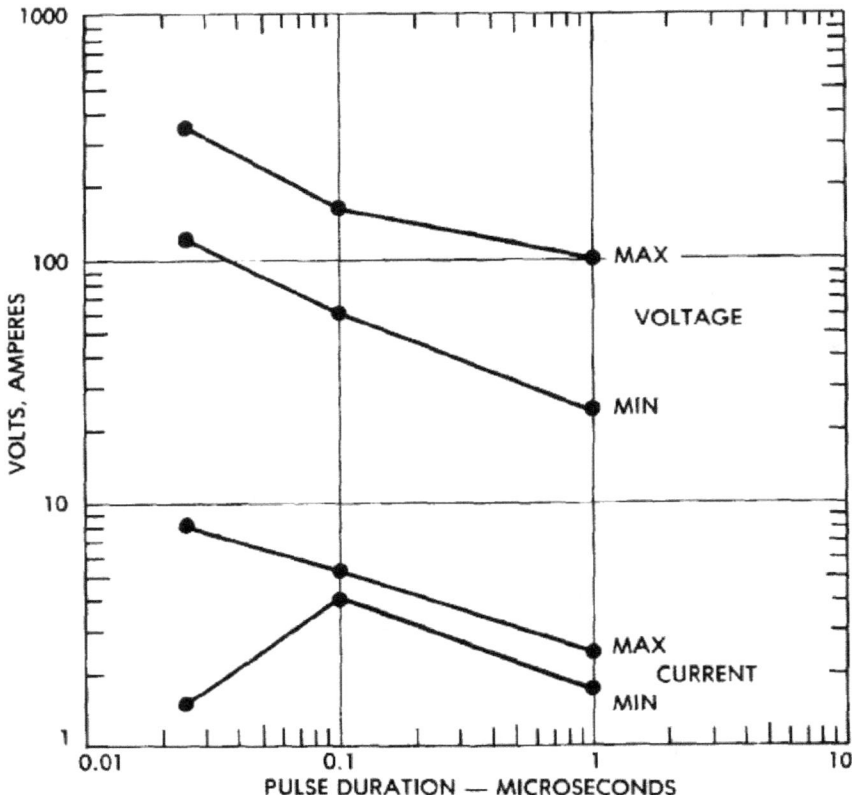

Figure 16.11   Damage thresholds of SN 55109 line drivers.

associated with the time during which internal conduction occurs. For these components, conduction was initiated at 3 to 4 times the voltage rating. Leakage current increased continuously with time of conduction, from initial values of a few nanoamperes through milliamperes. The value of the capacitance determines how quickly the voltage across the capacitor reaches the breakdown voltage range, 90 to 140 V, which then relates to the time of conduction and the extent of damage. Figure 16.13 (Reference 16.19) shows the data for the nine 0.0047 μF capacitors tested. For a particular pulse duration of 5 μs, an increase in leakage current is expected; for pulse voltages of 100 to 150 V and for pulses of 150 to 200 V, an increase in leakage current to milliamperes is possible. It is cautioned that this behavior may not be readily extended to capacitors of different materials or construction. The capacitor data remains insufficient to draw general conclusions as to system implications at this time.

16.8   Failure Mechanisms—Other Components

A limited amount of pulse test data is available for various nonsemicon-

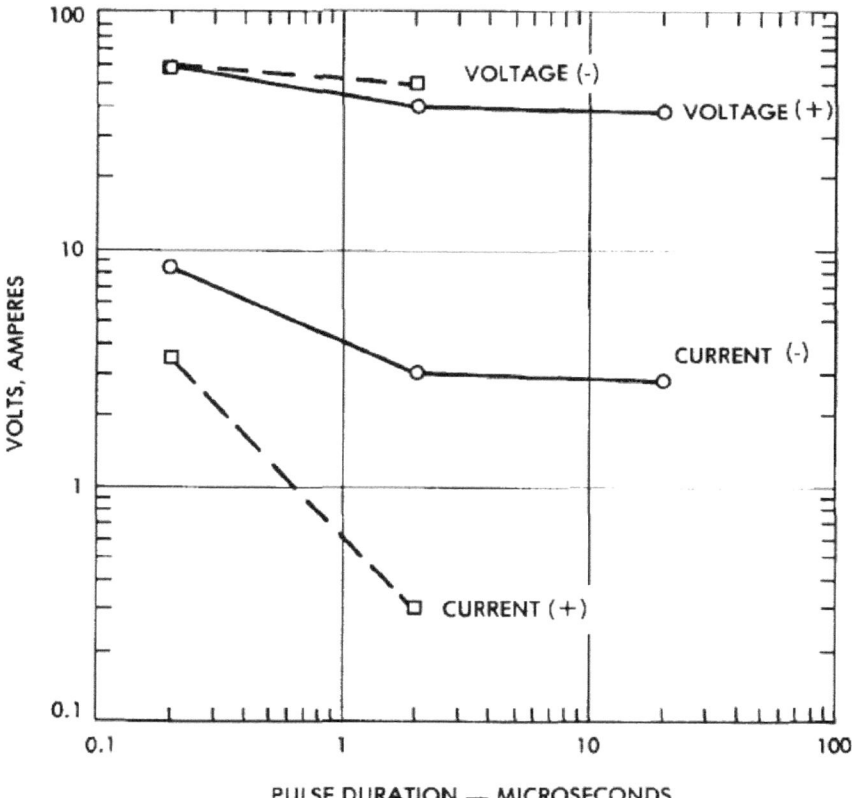

Figure 16.12 Damage thresholds of CD 4050 AE hex buffers.

ductor electronic circuit components. These data were mostly obtained by testing with a square-wave pulse input of 1 to 10 μs duration and up to 1 kV peak. As would be expected, not all such components are invulnerable to pulses of this shape. The test results for several kinds of components are presented in Table 16.4 (Reference 16.20). The test conditions consisted of an 8 μs, 1 kV pulse applied 10 times to each device. In the case of multiterminal components, several pairs of terminals were tested in this manner.

16.9   Examples of Use of Damage Constants

Some examples of how the preceding material may be used to determine whether or not a given transient will cause damage to semiconductors follow.

The first circuit chosen for analysis (Figure 16.14) (Reference 16.21) is a simple remote-controlled relay. Across the terminals of the relay coil there is a diode which would be exposed to the same transients as those to which the coil is exposed. The analysis approach that will be taken is first to calculate the current level that would cause the diode to fail and then to see whether or not

465

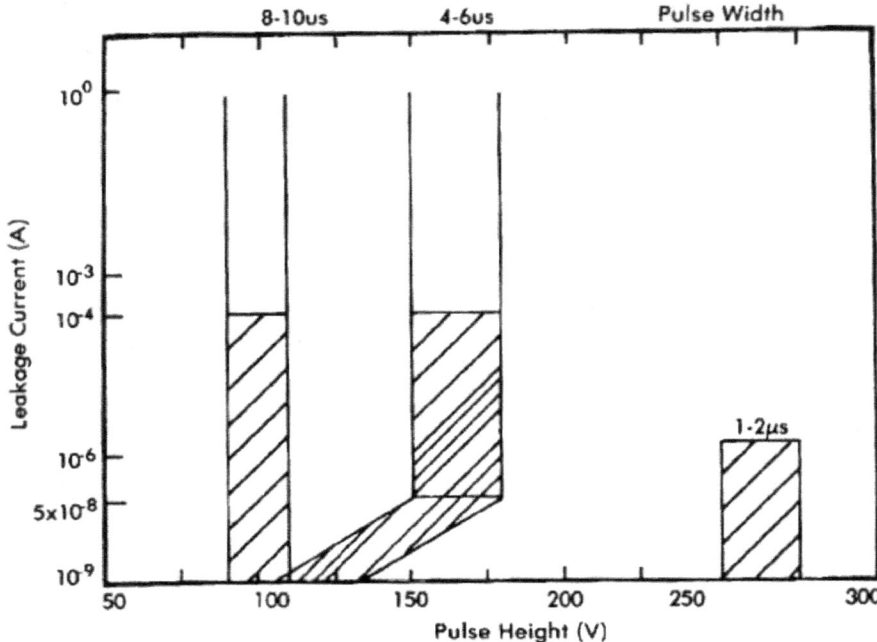

Figure 16.13 Representation of pulse test data for Sprague 0.0047 μF tantalum electrolytic capacitors.

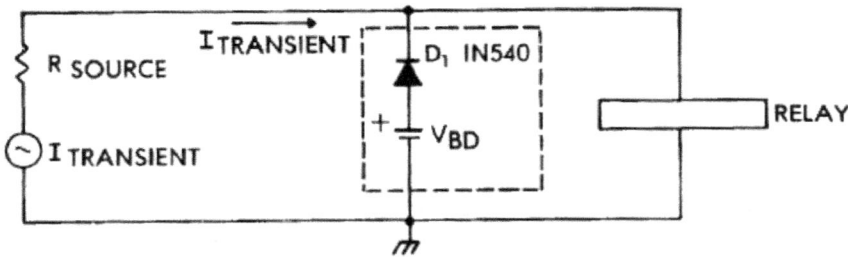

Figure 16.14 Simple remote-controlled relay.

the transient voltage source could supply that current. It will be assumed that the oscillatory pulse is a transient of 1 MHz frequency or 1 μs period. At this frequency the inductive reactance of the relay coil would be sufficiently large that the relay could be neglected. The current required to cause failure at time, t, would be

$$I_F = \frac{P_f}{V_{BD}} = \frac{Kt^{-\frac{1}{2}}}{V_{BD}} \qquad (16.20)$$

466

For a 1N540 diode, the reverse breakdown voltage, V_{BD}, is 400 V and the damage constant, K, is 0.93 (See Table 16.1). If a 200 ns pulse is used to approximate at 1 MHz damped sine wave, the failure current for the diode would be

$$I_F = \frac{0.93\,(2 \times 10^{-7})^{-\frac{1}{2}}}{400} \qquad (16.21)$$

$$I_F = 5.2\ A$$

Assume now that the impedance of the source from which the voltage transient generated is 10 Ω. The voltage required to produce a current of 5.2 A through the diode would be

$$V_{Transient} = V_{BD} + I_{Transient}\,R_{Source} \qquad (16.22)$$

$$V_{Transient} = 452\ V$$

Therefore, a 452 V pulse, 200 ns wide or a 1 MHz damped sine wave having a peak amplitude of 452 V would cause the diode to fail.

The second circuit chosen for analysis is the simple phase-splitter amplifier shown in Figure 16.15 (Reference 16.22). The first step in determining the input current required for damage is to simplify the circuit. Again assume that the voltage source producing the transient is a damped sine wave of 1 MHz frequency. At such a frequency the reactances of capacitors C1 and C2 will be so small that they may be neglected. Likewise the 12 V power supply line can be considered to be at ac ground potential. The resultant circuit after simplification is shown in Figure 16.16 (Reference 16.23). The circuit can be further simplified by determining the equivalent resistances for the base and for collector circuits. The base-emitter junction and the base-collector junction can also be replaced by their diode equivalents to represent operation in the breakdown regions. This simplified circuit is shown in Figure 16.17 (Reference 16.24). Also shown in Figure 16.17 are the breakdown voltages and damage constants for the 2N706B. Note that for this transistor a damage constant for the collector-base junction is available, though not listed in Table 16.2.

The circuit is now simplified to the point where it lends itself easily to hand analysis. The next step is to determine which junction will fail and what the failure mode is. The passive components are generally able to withstand higher energies for short-duration pulses than can transistors. Therefore, the transistor is the element to consider for damage. Failure is also assumed to occur in the reverse biased direction.

Using the Wunsch damage model ($P = Kt^{-\frac{1}{2}}$), a calculation is made to see whether the emitter-base junction or the collector-base junction would fail first.

$$P_{EB} = K_{EB}t^{-\frac{1}{2}} = 17\ W \qquad (16.23)$$

$$P_{CB} = K_{CB}t^{-\frac{1}{2}} = 130\ W \qquad (16.24)$$

467

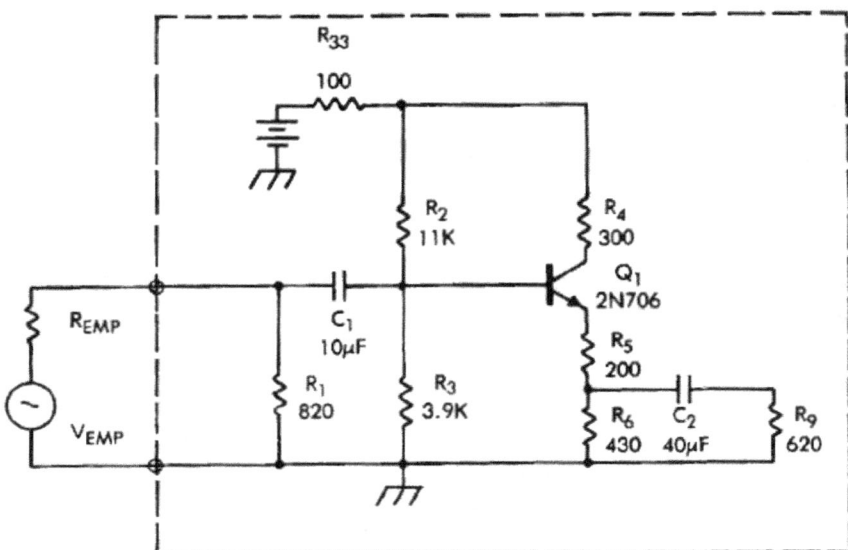

Figure 16.15 Phase-splitter circuit.

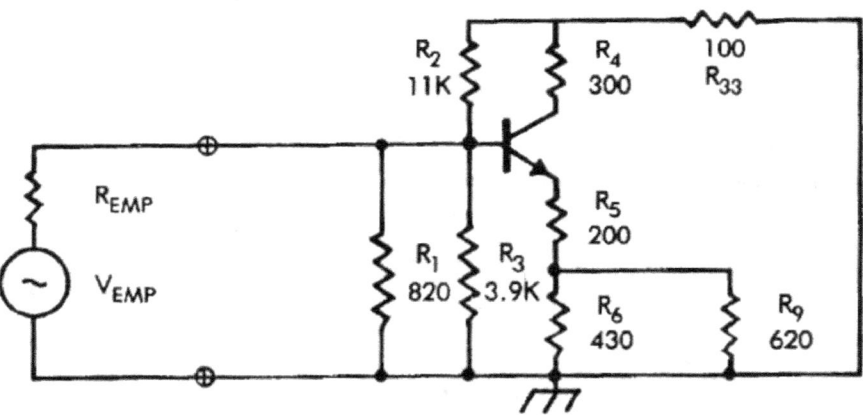

Figure 16.16 Simplified phase-splitter circuit.

This calculation shows that the emitter-base junction is the more susceptible. The current required to fail the emitter-base junction would be

$$I_{jF} = \frac{P_{EB}}{V_{BD}} = 3.4 \text{ A} \qquad (16.25)$$

The voltage from the base to ground is

$$V_{BASE} = BV_{EBO} + I_{jF} R_{EQ2} = 1.5 \text{ kV} \qquad (16.26)$$

468

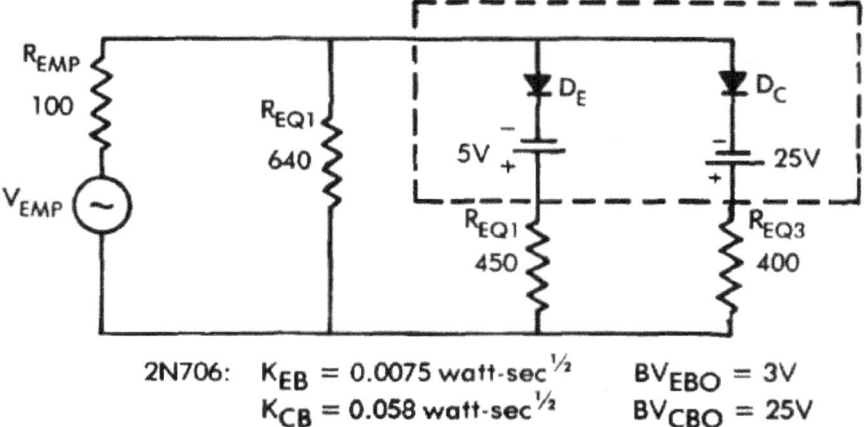

$$2N706: \quad K_{EB} = 0.0075 \text{ watt-sec}^{1/2} \quad BV_{EBO} = 3V$$
$$K_{CB} = 0.058 \text{ watt-sec}^{1/2} \quad BV_{CBO} = 25V$$

Figure 16.17 Further simplification of phase-splitter circuit.

The current through the collector-base junction is

$$I_{CB} = \frac{V_{BASE} - BV_{CBO}}{R_{EQ3}} = 3.7 \text{ A} \tag{16.27}$$

The power dissipated in the collector-base junction is

$$P_{CB} = BV_{CBO} I_{CB} = 93 \text{ W} \tag{16.28}$$

which is below its failure-threshold power.

The total current into the circuit is then

$$I_{Transient} = I_{jF} + I_{CB} + \frac{V_{BASE}}{R_{EQ1}} = 9.4 \text{ A} \tag{16.29}$$

and the $I_{Transient}$ voltage required to cause failure is

$$V_{Transient} = V_{BASE} + I_{Transient} R_{Source} = 2.5 \text{ kV} \tag{16.30}$$

Therefore (assuming a 100 Ω source impedance) a 2.5 kV pulse, 200 ns wide, will cause the transistor to fail.

Table 16.1 DIODE DAMAGE DATA

Device Number	K	V_{BD}	Reference
IN23B, C	.0009		SP
IN23RF	.00094		DX
IN23WE	.00029		DX
IN25	.026		DX
IN34A	.014	60.	DX
IN39A	.006	230.	SP
IN39B	.006	200.	SP
IN43B	.005	70.	CM
IN64	.041	25.	DX
IN67A	.003	80.	SP
IN69, A	.005	70.	CM
IN81	.003	10.	SP
IN82A	.0007	5.	DX
IN91	.0055	100.	CM
IN128	.005	40.	CM
IN191	.005	90.	SP
IN198	.024	80.	SP
IN248A	40.	50.	SP
IN249	40.	100.	SP
IN249B	40.	100.	SP
IN250	40.	200.	SP
IN250B	80.	200.	SP, DE
IN251	.03	40.	SP
IN253	86.	95.	DX
IN254	3.5	190.	SP
IN260	.0027	30.	CM

Table 16.1 DIODE DAMAGE DATA (CONT.)

Device Number	K	V_{BD}	Reference
IN270	.022	100.	CM
IN276	.0055	100.	CM
IN277	.027	125.	DX
IN295, A	.005	40.	SP
IN320	1.2	500.	SP
IN332	3.5	400.	SP
IN333	1.5	400.	SP
IN335	1.5	300.	SP
IN337	1.5	200.	SP
IN338	18.3	100.	DE
IN341	3.5	400.	SP
IN342	1.5	400.	SP
IN346	1.5	200.	SP
IN429	.6	6.2	DX
IN457	.12	70.	DX
IN458	.5	150.	SP
IN459	.59	200.	DX
IN459A	.96	200.	DX
IN461	.24	35.	SP
IN462	.05	80.	SP
IN466	.78	3.5	SP
IN467	.78	4.1	SP
IN468	.78	4.9	SP
IN470	.78	7.1	SP
IN474A	.219	5.8	CM
IN482A	.96	36.	DX

Table 16.1   DIODE DAMAGE DATA (CONT.)

Device Number	K	V_{BD}	Reference
IN483, A	.3	70.	SP
IN483, B	.3	80.	SP
IN484A	.45	130.	DX
IN484B	.3	130.	SP
IN485	.3	180.	CM
IN486, B	.29	225.	SP
IN487, Z	.3	300.	SP
IN488	.3	380.	SP
IN536	1.	50.	DE
IN537	.51	100.	DX
IN538, M	1.	200.	SP
IN539	1.	300.	SP
IN540	.93	400.	DX
IN547	12.1	600.	DX
IN560	.625	800.	CM
IN561	.625	1000.	CM
IN562	1.8	800.	SP
IN619	.36	10.	SP
IN622	.347	150.	CM
IN625	.164	30.	CM
IN625A	.045	20.	CM
IN643	.44	200.	SP
IN643A	.1	200.	DX
IN645	2.8	225.	SP
IN646	2.29	300.	DX

Table 16.1   DIODE DAMAGE DATA (CONT.)

Device Number	K	V_{BD}	Reference
1N647	2.8	400.	SP
1N648	2.8	500.	SP
1N649	2.9	600.	DX
1N658	.92	120.	DX
1N660	.44	100.	SP
1N661	.41	200.	DX
1N662	.29	100.	SP
1N663	.44	100.	SP
1N676	.27	100.	SP
1N689	1.1	600.	SP
1N691	.418	80.	SP
1N692	.5	100.	SP
1N702, A	1.	2.6	SP, DX
1N703A	1.	3.5	SP
1N704, A	1.	4.1	SP
1N705, A	.91	4.8	SP
1N706	.288	5.8	CM
1N709, A	.78	6.2	SP
1N710	.78	6.8	SP
1N711A	2.1	7.5	DX
1N712	.78	8.2	SP
1N714A	.78	10.	SP
1N715A	.78	11.	SP
1N718A	.1	15.	SP
1N719A	.1	16.	SP
1N721, A	.35	20.	SP

Table 16.1   DIODE DAMAGE DATA (CONT.)

Device Number	K	V_{BD}	Reference
1N725A	.349	30.	CM
1N729	.06	43.	SP
1N746, A	1.1	3.3	SP
1N747, A	1.1	3.6	SP
1N748A	1.1	3.9	SP
1N749	1.1	4.3	SP
1N750A	1.1	4.7	SP
1N751, A	1.1	5.1	SP
1N752, A	1.1	5.6	SP
1N753, A	1.2	6.2	SP, DX
1N754, A	.63	6.8	SP
1N755, A	.63	7.5	SP
1N756, A	.63	8.2	SP
1N757, A	.63	9.1	SP
1N758, A	.63	10.	SP
1N759, A	.63	12.	SP
1N761	1.8	4.9	SP
1N762	1.8	5.8	SP
1N763	1.8	7.1	SP
1N763-2	3.	7.0	DX
1N766A	1.8	12.8	SP
1N767	1.8	15.8	SP
1N769A	1.8	23.5	SP
1N769-3	2.	26.	DX
1N816, W	1.5	26.	DX
1N817	.46	200.	SP

Table 16.1   DIODE DAMAGE DATA (CONT.)

Device Number	K	V_{BD}	Reference
1N821	.577	6.2	CM
1N823	1.8	6.2	DX
1N845	.365	200.	CM
1N890	.357	60.	CM
1N914	.85	100.	DX
1N916	.44	100.	SP
1N933	.014	100.	DX
1N933J	.1	100.	DX
1N936	.14	9.	DX
1N936A, B	7.	9.	SP
1N937	.824	9.	CM
1N938, B	7.	9.	SP
1N939	.824	9.	CM
1N939B	7.0	9.	DE
1N960B	1.0	9.	SP
1N961B	1.0	10.	SP
1N963B	1.	12.	SP
1N964B	1.	13.	SP
1N965B	1.	15.	SP
1N967B	.73	18.	DX
1N968B	1.	20.	SP
1N969B	1.	22.	SP
1N970B	1.	24.	SP
1N972B	1.	30.	SP
1N973B	1.	33.	SP
1N974B	1.	36.	SP

Table 16.1 DIODE DAMAGE DATA (CONT.)

Device Number	K	V_{BD}	Reference
IN975B	1.	39.	SP
IN976B	1.	43.	SP
IN977B	1.	47.	SP
IN979B	1.	56.	SP
IN981B	1.4	68.	DX
IN983A	1.	82.	SP
IN987A, B	1.	120.	SP
IN1095	.9	500.	DX
IN1096	.9	600.	SP
IN1118	11.392	400.	CM
IN1124A	7.985	250.	CM
IN1126A	14.	500.	SP
IN1184	31.5	100.	CM
IN1199, A	15.	50.	SP
IN1200	62.32	100.	CM
IN1201	62.32	150.	CM
IN1202	21.	200.	SP
IN1204A	46.106	400.	CM
IN1206	62.32	600.	CM
IN1217	5.8		SP
IN1222B	2.563	400.	CM
IN1317A	.19	19.	SP
IN1319A	.19	28.	SP
IN1342A	38.4	100.	DE
IN1348A	1.827	200.	CM
IN1367	34.	47.	SP

Table 16.1 DIODE DAMAGE DATA (CONT.)

Device Number	K	V_{BD}	Reference
IN1583	11.391	200.	CM
IN1585	3.5	400.	DE
IN1614	.38	200.	SP
IN1615	.666	480.	CM
IN1693	3.2	200.	SP
IN1695	3.2	400.	SP
IN1731	3.2	1500.	CM
IN1733A	11.3	3000.	DE
IN1770A	14.2	9.1	DE
IN1773A	1.9	12.	SP
IN1780A	1.9	24.	SP
IN1783	21.3	33.	DE
IN1818RA	4.3	16.	SP
IN1823C, A	4.3	27.	SP
IN1828C	4.3	43.	SP
IN1834	33.8	75.	CM
IN1835A	4.3	82.	SP
IN1836C	4.3	91.	SP
IN1904	28.	100.	SP
IN1909	6.8	200.	SP
IN2037	.05	12.8	SP
IN2154	20.	50.	SP
IN2158	21.5	400.	CM
IN2164	2.3	9.4	SP
IN2483	2.1	400.	SP
IN2610	2.6	100	SP

Table 16.1   DIODE DAMAGE DATA (CONT.)

Device Number	K	V_{BD}	Reference
IN2611	2.6	200.	SP
IN2613	2.6	400.	SP
IN2615	2.6	600.	SP
IN2789	40.	400.	SP
IN2795	40.	150.	SP
IN2796	40.	200.	SP
IN2808	249.	10.	CM
IN2818	249.	20.	CM
IN2823B	249.	30.	CM
IN2824	156.	33.	SP
IN2826B	249.	39.	CM
IN2844B	15.	160.	SP
IN2846B	15.	200.	SP
IN2862	2.8	400.	SP
IN2864	2.8	600.	SP
IN2929A	.073	1.	DX
IN2930	.196	.74	CM
IN2970B	15.0	6.8	SP
IN2976B	15.	12.	SP
IN2979B	15.	15.	SP
IN2984, B	15.	20.	SP
IN2985, B, RB	15.	22.	SP
IN2986B	15.	24.	SP
IN2987B	15.	25.	SP
IN2988B	15.	27.	SP
IN2989B	15.	30.	SP

478

Table 16.1 DIODE DAMAGE DATA (CONT.)

Device Number	K	V_{BD}	Reference
IN2991B	15.	36.	SP
IN2995, B	15.	47.	SP
IN2997B	15.	51.	SP
IN3001B	15.	68.	SP
IN3008B	15.	120.	SP
IN3015B	33.84	200.	CM
IN3016B	19.5	6.8	DE
IN3017B	1.9	7.5	SP
IN3019B	1.9	9.1	SP
IN3022B	1.9	12.	SP
IN3024B	1.9	13.	SP
IN3025B	1.9	16.	SP
IN3026B	1.9	18.	SP
IN3027B	1.9	20.	SP
IN3028, B	1.9	22.	SP
IN3029B	1.9	24.	SP
IN3030B	1.9	27.	SP
IN3031B	1.9	30.	SP
IN3033B	1.9	36.	SP
IN3035B	1.9	43.	SP
IN3037B	1.9	51.	SP
IN3040B	1.9	68.	SP
IN3041, B	1.9	75.	SP
IN3051B	1.9	200.	SP
IN3064	.02	75.	SP
IN3070	.365	200.	CM

Table 16.1   DIODE DAMAGE DATA (CONT.)

Device Number	K	V_{BD}	Reference
IN3157	.625	8.4	CM
IN3189	10.	200.	SP
IN3190	4.1	600.	CM
IN3560	.038	.475	CM
IN3561	.038	.475	CM
IN3582A	.35	11.7	SP
IN3600	.18	50.	SP
IN3821	1.947	3.3	CM
IN3828A	1.95	6.2	CM
IN3893	6.41	400.	CM
IN3976	132.	200.	CM
IN4241	33.84	6.	CM
IN4245	2.4	200.	SP
IN4249	2.4	1000.	SP
IN4312	.116	150.	CM
IN4370A	.625	2.4	CM
IN4816	6.8	50.	DE
IN4817	6.8	100.	DE
IN4820	10.	400.	DE
IN4823	.208	'00V	CM
IN4989	14.33	200.	CM
AM2	1.4	50.	SP
D4330	.001		SP
FD300	.18	125.	SP
SG22	.23		SP
SLD10EC		10,000.	CM

Table 16.1 DIODE DAMAGE DATA (CONCLUDED)

Device Number	K	V_{BD}	Reference
SV1035	1.71	26.	CM
SV2092	2.6		SP
SV2183	2.6		SP
TM7	20.	70.	SP
TM21	18.	200.	SP
TM27	20.	200.	SP
TM84	11.	800.	SP
TM124	11.	1200.	SP
UT 242	2.6	200.	SP

SP — SAP-1 Computer Listing from *SAP-1 Computer Code Manual* (unpublished), U.S. Air Force Weapons Laboratory, 1972.

DX — Experimental data from *DASA* (Defense Atomic Support Agency) *Handbook* (See Reference 16.3)

DE — Estimated data from *DASA Handbook* (See Reference 16.3)

CM — Computed data (Reference 16.7)

Table 16.2 TRANSISTOR DAMAGE DATA

Device Number	K	BV EBO	BV CBO	BV CEO	Reference Source
2N43, A	.28	5.	45.	30.	SP
2N117	.15	1.	45.	45.	SP
2N118	.15	1.	45.	45.	SP
2N128	.017	10.	10.	4.5	SP
2N158	.499	30.	60.	60.	CM
2N176	.46		40.	30.	CM
2N189	.17		25.	25.	SP
2N190	.58		25.	25.	DX
2N243	.05	1.	60.	60.	SP
2N244	.05	1.	60.	60.	SP
2N263	.38	1.	45.	30.	SP
2N264	.36		45.	30.	SP
2N274	.0076	.5	35.	40.	CM
2N279A	.047		45.	30.	CM
2N297A	.499	40.	60.	40.	CM
2N329, A	.21	20.	50.	30.	SP
2N332	.45	1.	45.	30.	SP
2N333	.32	1.	45.	30.	SP
2N335, A	.55	1. (4.-2N335A)	45.	45.	SP
2N336	.55	1.	45.	30.	SP
2N337	.12	1.	45.	30.	SP
2N338	.12	1.	45.	30.	SP
2N339	2.	1.	55.	55.	SP
2N341	1.	1.	125.	85.	SP

Table 16.2 TRANSISTOR DAMAGE DATA (CONT.)

Device Number	K	BV EBO	BV CBO	BV CEO	Reference Source
2N343	.047	1.	60.	60.	SP, DX
2N343A	.05	1.	60.	60.	SP
2N357	.05	20.	20.	15.	SP
2N359	.04	6.	25.	18.	SP
2N375	1.02	40.	80.	60.	DX
2N388	.084	15.	25.	20.	CM
2N389	2.14	10.		60.	DX
2N395	.09	20.	30.	15.	SP
2N404	.05	12.	25.	24.	CM
2N424A	10.	10.	80.	80.	SP
2N463	6.6	50.	60.	60.	DX
2N480	.132	2.	45.	45.	CM
2N490	1.	60.*	58.*	--	SP
2N491	1.	60.*	58.*	--	SP
2N495, A	.7	20.	25.	25.	SP
2N497	.8	8.	60.	60.	SP
2N498	.8	8.	100.	100.	DX
2N525	.3	15.	45.	30.	SP
2N526	.39	15.	45.	30.	DX
2N527	.3	15.	45.	30.	SP
2N537	.012	1.	30.		CM
2N538	.5285	28.	80.	60.	CM
2N539, A	6.	28.	80.	55.	SP
2N540	.5285	28.	80.	55.	CM
2N542	.18	2.	30.	30.	SP

Table 16.2 TRANSISTOR DAMAGE DATA (CONT.)

Device Number	K	BV EBO	BV CBO	BV CEO	Reference Source
2N551	1.6	6.	60.	60.	SP
2N576A	.023	15.	40.	20.	DX
2N587	.14	40.	40.	30.	SP
2N595	.012		20.	15.	CM
2N618	.88	40.	80.	60.	DX
2N652A	.118	30.	45.	30.	CM
2N656	.2	8.	60.	60.	DX
2N657	.66	8.	100.	100.	DX
2N657A	1.07	8.	ϱ00.	100.	DX
2N682	.33	--	--	50.(A-C)	CM
2N685	1.4	--	--	200.(A-C)	DX
2N687	11.7	--	--	300.(A-C)	DX
2N690	3.1	--	--	800.(A-C)	CM
2N696	1.0	5.	60.	40.	CM
2N697	.2	5.	60.	40.	SP
2N699	.25	5.	120.	80.	DX
2N703	.08	5.	25.	25.	SP
2N706, B	.0075	3. (5.-2N706B	25.	20.	DX
2N708	.03	5.	40.	15.	DX
2N717	.13	5.	60.	40.	SP
2N718	.13	5.	60.	40.	SP
2N718A	.35	7.	75.	32.	SP
2N726	.021	5.	25.	20.	CM
2N730	.165	5.	60.	40.	CM
2N736	.1	5.	80.	60.	DX

Table 16.2 TRANSISTOR DAMAGE DATA (CONT.)

Device Number	K	BV EBO	BV CBO	BV CEO	Reference Source
2N756A	.32	6.	60.	60.	SP
2N757	.032	6.	45.	45.	CM
2N760A	.034	8.	60.	60.	DX
2N834	.03	5.	40.	30.	SP
2N859	.18	25.	40.	40.	DX
2N869A	.009	5.	25.	18.	CM
2N910	.218	7.	100.	60.	CM
2N912	.07	7.	100.	60.	SP
2N914	.04	5.	40.	15.	SP
2N916	.043	5.	45.	25.	CM
2N917	.004	3.	30.	15.	SP
2N918	.004	3.	30.	15.	CM
2N927	.1	70.	70.	60.	DX
2N930	.046	5.	45.	45.	DX
2N930A	.02	6.	60.	45.	DX
2N1016, B	1.6	25.	100.	100.	DX
2N1039	1.4	20.	60.	60.	DX
2N1045-1	.55	20.	100.	60.	SP
2N1048	3.9	6.	120.	120.	SP
2N1049	3.9	6.	80.	80.	SP
2N1050	6.082	6.	120.	120.	CM
2N1069	9.3	9.	60.	45.	SP
2N1099	1.	40.	80.	60.	DX
2N1115	.38		20.	15.	DX
2N1116A	.98	6.	60.	60.	DX

485

Table 16.2 TRANSISTOR DAMAGE DATA (CONT.)

Device Number	K	BV EBO	BV CBO	BV CEO	Reference Source
2N1118	.19	10.	25.	25.	DX
2N1132	.23	5.	50.	35.	DX
2N1136B	18.4		100.	65.	SP
2N1150	.18	1.	45.		SP
2N1154	21.	1.	50.	28.	SP
2N1156	18.	1.	120.	68.	SP
2N1184	.471	20.	45.	20.	CM
2N1212	13.129	10.	60.	60.	CM
2N1303	.087	25.	30.		CM
2N1308	.084	25.	25.	15.	CM
2N1309	.087	25.	30.		CM
2N1445	.5	8.	120.	120.	SP
2N1458	.5285	15.	80.	65.	CM
2N1469	.65	40.	40.	35.	DX
2N1480	5.5	12.	100.	55.	SP
2N1481	2.2	12.	60.	40.	SP
2N1483	3.633	12.	60.	40.	CM
2N1485	4.1	12.	60.	40.	SP
2N1486	5.	12.	100.	55.	SP
2N1489	12.3	10.	60.	40.	SP
2N1490	12.3	10.	100.	55.	SP
2N1564	.56	5.	80.	60.	SP
2N1565	.11	5.	80.	60.	SP
2N1566	.11	5.	80.	60.	SP
2N1596	.94	--	—	100.	DX

486

Table 16.2 TRANSISTOR DAMAGE DATA (CONT.)

Device Number	K	BV EBO	BV CBO	BV CEO	Reference Source
2N1602	.40	--	--	200.	DX
2N1613	.27	7.	75.	50.	SP
2N1615	.553	8.	100.	100.	CM
2N1642	.13	30.	30.	6.	DX
2N1700	4.134	6.	60.	40.	CM
2N1701	4.5	6.	60.	40.	SP
2N1711	.36	7.	75.	50.	SP
2N1722	54.5	10.	175.	80.	DE
2N1751	1.05	2.5	80.	80.	CM
2N1753	.039	.5	30.	18.	CM
2N1772A	.651	--	--	100.	CM
2N1776A	1.584	--	--	300.	CM
2N177A	C-G C-A 2.0 .46	--	--	400.	DX
2N1871A	1.1	--	--	60.	CM
2N1890	.27	7.	100.	60.	SP
2N1893	.4	7.	120.	80.	DX
2N1916W	2.22	--	--	400.	CM
2N2015	26.462	10.	100.	50.	CM
2N2035	3.633	10.	80.	60.	CM
2N2060	.21	7.	100.	60.	SP
2N2102	.77	7.	120.	65.	SP
2N2156	.471	25.	45.	30.	CM
2N2218A	.264	6.	75.	40.	CM
2N2219	.3	5.	60.	30.	SP
2N2219A	.264	6.	75.	40.	CM

Table 16.2 TRANSISTOR DAMAGE DATA (CONT.)

Device Number	K	BV EBO	BV CBO	BV CEO	Reference Source
2N2222	.1	5.	60.	30.	DX
2N2222A	.1	6.	75.	40.	SP
2N2223, A	.21	7.	100.	60.	SP
2N2270	.5	7.	60.	45.	SP
2N2346	3.2	--	--	100.	DX
2N2369A	.03	4.5	40.	15.	SP
2N2417	.549	30.*	35.*	--	CM
2N2432	.189	15.	30.	30.	CM
2N2481	.099	5.	40.	15.	CM
2N2509	.126	7.	125.	80.	CM
2N2516	.209	8.	80.	60.	CM
2N2563	.55	20.	100.	100.	SP
2N2646	.72	30.*	35.*	--	SP
2N2708	.018	3.	35.	20.	CM
2N2857	.018	2.5	30.	15.	CM
2N2894, A	.03	4. (4.5-2N2894A)	12.	12.	SP
2N2904A	.221	5.	60.	60.	CM
2N2905	.221	5.	60.	40.	CM
2N2906	.044	5.	60.	40.	DX
2N2906A	.221	5.	60.	60.	CM
2N2907, A	.1	5.	60.	40.(60.-A)	DX
2N2920	.04	6.	60.	60.	DX
2N2996	.01	.3	15.	10.	SP
2N3014	.02	5.	40.	20.	CM
2N3050	.01	5.	25.	20.	CM

Table 16.2 TRANSISTOR DAMAGE DATA (CONT.)

Device Number	K	BV EBO	BV CBO	BV CEO	Reference Source
2N3053	.721	5.	60.	40.	CM
2N3054	3.633	7.	90.	60.	CM
2N3055	20.084	7.	100.	70.	CM
2N3118	.53	4.	85.	60.	SP
2N3217	.126	15.	15.	10.	CM
2N3235	20.	7.	65.	55.	SP
2N3240	1.5	8.	160.	160.	SP
2N3251	.143	5.	50.	40.	CM
2N3308	.12	3.	30.	25.	SP
2N3384	.094	--	30.†	--	CM
2N3436	.488	--	50.†	--	CM
2N3440	1.75	7.	300.	250.	SP
2N3585	5.278	6.	440.	300.	CM
2N3708	.507	6.	30.	30.	CM
2N3777	2.	8.	100.	100.	SP
2N3785	.012	.5	50.	12.	SP
2N3819	.22	25.	25.†	--	CM
2N3823	.228	30.	30.†	--	CM
2N3902	43.35	5.	400.	400.	CM
2N3907	.165	6.	60.	45.	CM
2N4037	.045	7.	60.	40.	CM
LN75497	1.9				DX
LN75638	2.3				DX
MIS17331	.1				DX
T1482	.21	5.	20.	20.	SP

Table 16.2 TRANSISTOR DAMAGE DATA (CONCLUDED)

Device Number	K	BV EBO	BV CBO	BV CEO	Reference Source
T1487	4.5	6.	80.	60.	SP
TIXM101	.01	.3	15.	7.	SP
SW3042	.1	--	--		DX

SP — SAP-1 Computer Listing from *SAP-1 Computer Code Manual* (unpublished), U.S. Air Force Weapons Laboratory, 1972.

DX — Experimental data from *DASA* (Defense Atomic Support Agency) *Handbook* (See Reference 16.3)

DE — Estimated data from *DASA Handbook* (See Reference 16.3)

CM — Computed data (Reference 16.7)

Table 16.3 ACCURACY OF K FACTOR AS DETERMINED BY INDIRECT METHODS (CATEGORY 2 SEMICONDUCTORS)

	Conditions	Accuracy
K_{amb}	$50 < \theta_{ja} < 200$	Factor of 2
	$200 < \theta_{ja} < 500$	Factor of 10
	$\theta_{ja} > 500$	Factor of 30
K_{case}		Factor of 3
K_{cj}	$V_{bd} < 10$	Factor of 30
	$10 < V_{bd} < 200$	Factor of 10
	$200 < V_{bd} < 2000$	Factor of 2

K_{amb} = K as determined from θ_{ja}

K_{case} = K as determined from θ_{jc}

K_{cj} = K as determined from junction capacitance

Table 16.4 NONSEMICONDUCTOR DAMAGE TEST RESULTS

Device Type	Manufacturer	Manufacturer's Part Number	Properties	Test Results
Capacitor	Cornell-Dublier	C100K	10 pF	No change in capacity or leakage resistance
Capacitor	Cornell-Dublier	CK62 Series	4700 pF, 500 Vdc	
Capacitor	Sprague	96P Series	1 µF, 200 Vdc	
Capacitor	WES CAP	KF223KM	0.022 µF, 600 Vdc	
Coil	Collins	240-2524-00	220 µH	No change in inductance or resistance
Coil	Collins	542-0916-002	2 µH 10³ Hz	
Filter	Bundy	21-0526-00	Notch 400 and 1200 Hz	No change in frequency response
Filter	Varo	954-0429-400	Bandpass 400 Hz	
Potentiometer	Computer Insts.	M18-178105	400 Ω	No changes
Potentiometer	Ohmite	51927-1	250 Ω	
Potentiometer	Ohmite	51927-3	25 Ω	
Relay	Babcock	RP11573-G2	Armature	Resistance increase: 625 Ω > 629 Ω
Relay*	C. P. Clare	A5245-1	Armature	Resistance increase: 418 Ω > 425 Ω
Relay	Hathaway	63862	Magnetic Reed	Resistance increase: 2.5%
Relay	Potter Brumfield	FLB4002	Magnetic Latching	Resistance decrease: <1%
Relay	Struthers Dunn	FC6-365	Armature	No changes
Transformer	Dektronics	D78Z222	Audio Frequency	No change in resistance or voltages, no arcing during pulse
Transformer	Dektronics	D78Z225	Power, Isolation	
Transformer	Freed	667-0386-00	Audio Frequency	
Transformer	Varo	950-1622-200	Power, Isolation	
Transformer	Varo	999-0197-200	Power, Isolation, Stepdown	
Vacuum Tube		6BX7	Med µ Twin	No change in characteristics
Vacuum Tube		5876	UHF High µ	Transconductance decrease, 35%
Vacuum Tube		6BC4	Med µ Twin	

*Arcing present continually for 700 V pulses.

491

REFERENCES

16.1   R. L. Davies and F. E. Gentry, "Control of Electric Field at the Surface on PN Junctions," *IEEE Transactions on Electron Devices*, ED-11, Institute of Electronic and Electrical Engineers, New York, New York (July 1964): pp. 313-23.

16.2   H. S. Velorie and M. P. Prince, "High-Voltage Conductivity-Modulated Silicon Rectifier," *The Bell System Technical Journal*, July 1957, pp. 975-1004.

16.3   *DNA EMP* (Electromagnetic Pulse) *Handbook*, Vol. 2, Section 13: Analysis and Testing (Confidential), DNA 2114H-2, Defense Nuclear Agency, Washington, D.C. (November 1971).

16.4   *DNA EMP Handbook*, p. 13-34.

16.5   *DNA EMP Handbook*, p. 13-53.

16.6   *EMP Susceptibility Threshold Handbook*, (Appendix D), prepared by the Boeing Company for the U.S. Air Force Weapons Laboratory, Kirtland Air Force Base, Albuquerque, New Mexico (July 21, 1972).

16.7   *EC-135 Pretest EMP Analysis*, I, BDM/A-701-705, prepared by Braddock, Dunn and McDonald, Inc., for the Air Force Special Weapons Center, Albuquerque, New Mexico (March 1971).

16.8   *DNA EMP Handbook*, p. 13-47.

16.9   J. B. Singletary and D. C. Wunsch, *BDM Final Report, Semiconductor Damage Study, Phase II*, Report BDM/A-66-70-TR, prepared by Braddock, Dunn and McDonald, Inc., for the U.S. Army Mobility Equipment Research and Development Center, Fort Belvoir, Virginia (June 1970).

16.10   D. C. Wunsch and L. Marzitelli, *BDM Final Report, Semiconductor and Nonsemiconductor Damage Study*, I, Report BDM-375-69-F-0168, prepared by Braddock, Dunn and McDonald, Inc., for the U.S. Army Mobility Equipment Research and Development Center, Fort Belvoir, Virginia (April 1969).

16.11   *EMP Susceptibility Threshold Handbook*, p. 212.

16.12   *EMP Susceptibility Threshold Handbook*, p. 217.

16.13   *EMP Susceptibility Threshold Handbook*, p. 223.

16.14   *EMP Susceptibility Threshold Handbook*, p. 224.

16.15   E. Keuren, R. Hendrickson, and R. Magyarics, "Circuit Failure Due to Transient-Induced Stresses, *First Symposium on Electromagnetic Compatibility, Montreux, Switzerland, May 20-22, 1975*, IEEE EMC Conference Record 75 CH10124, Institute of Electronic and Electrical Engineers, New York, New York (1975), pp. 500-505.

16.16   Keuren, Hendrickson, and Magyarics, "Circuit Failure," p. 504.

16.17   Keuren, Hendrickson, and Magyarics, "Circuit Failure."

16.18   *DNA EMP Handbook*, pp. 13-19 to 13-101.

16.19   *DNA EMP Handbook*, p. 13-101.

16.20   *DNA EMP Handbook*, p. 13-100.

16.21 *EMP Susceptibility Threshold Handbook*, p. 144.

16.22 *EMP Susceptibility Threshold Handbook*, p. 146.

16.23 *EMP Susceptibility Threshold Handbook*, p. 148.

16.24 *EMP Susceptibility Threshold Handbook*, p. 148.

16.24 *EMP Susceptibility Threshold Handbook*, p. 148.

16.25 *Electromagnetic Susceptibility of Semiconductor Components*, AFWL-TR-74-280, Air Force Weapons Laboratory, Kirtland Air Force Base, Albuquerque, New Mexico (1974).

16.26 B. D. Faraudo and L. C. Martin, *Review of Factors for Application in Component Damage Analysis*, Protection Engineering and Management Note PEM-52, Lawrence Livermore Laboratory, University of California, Livermore, California (September 1976).

16.27 D. M. Tasca, *Theoretical and Experimental Studies of Semiconductor Device Degradation Due to High Power Electrical Transients*, 735D4289, Corporate Research and Development, General Electric Company, Schenectady, New York (December 1973).

16.28 *Determination of Upset and Damage Circuit Thresholds*, PREMPT Program TN-25, the Boeing Company, Seattle, Washington (September 15, 1975).

CHAPTER 17
TEST TECHNIQUES FOR EVALUATION OF INDIRECT EFFECTS

17.1 Introduction

There are as yet few official standard tests or even generally recognized test techniques for evaluation of how well electrical and electronic equipment withstands the indirect effects of lightning. The purpose of this section is to discuss some of the existing test techniques, to discuss some of the philosophy behind some of the existing or evolving standards, and to suggest possible avenues of improvement in test techniques.

The lightning current to which an aircraft may be exposed cannot be predicted beforehand except in statistical terms. Aircraft are thus designed to some type of model lightning flash——one which duplicates the essential characteristics of lightning, even though the chances of any particular lightning current actually having all those characteristics is vanishingly small. A design model that has strongly influenced test techniques is the severe composite lightning model developed for the *Space Shuttle* Program (Reference 17.1). This model, with some revisions, seems to be emerging as a standard for aircraft design, testing, and certification. The model, as incorporated in present test practice (Reference 17.2), is shown as Figure 1.36.

As regards indirect effects, only the first and possibly the second return stroke are likely to be of consequence. The wave fronts of the initial return stroke and the restrike are not specified in Figure 17.1 (Reference 17.2), since they are not significant for the evaluation of direct effects. In the *Space Shuttle Program Lightning Protection Criteria Document* and in the earlier MIL specification *Bonding, Electrical and Lightning Protection for Aerospace Systems* (Reference 17.3), the current is taken to rise linearly to crest in 2 μs. Dividing the peak amplitude of 200 000 A by a front time of 2 μs yields an average rate of change of 100 kA/μs, a figure frequently quoted with respect to indirect effects.

Ideally, for proof tests to determine indirect effects, a simulated lightning current having this waveshape (or at least the waveshape of the initial stroke would be circulated through the aircraft and the electronic equipment observed to insure that no harmful effects would be caused by the current. In practice this is not done, since available surge generators do not have the stored energy required to circulate this amount of current through the inductance that would be associated with a complete aircraft. Compromises with perfection have to be made, and these take the following three forms.

- Tests may be made on a complete aircraft, during which tests currents of high amplitude are circulated through the aircraft and the effects noted. In such tests the waveshape of the current may differ radically from the waveshapes of actual lightning currents. Tests of this nature might be called *proof tests* and are discussed in Section 17.3.1.

495

- Tests may be made on a complete aircraft, during which tests currents of low amplitude but correct waveshape are circulated through the aircraft. Such tests are sometimes called *transient analysis tests* and are discussed in Section 17.3.2.

- Tests may be made upon individual pieces of electronic equipment or upon interconnected electronic subsystems. During such tests it is assumed, explicitly or implicitly, that the waveshapes of the current or voltage surges injected into the pieces of equipment or subsystems are representative of those produced by natural lightning. Some of the aspects of such tests are discussed in Section 17.5.

17.2 Waveform Definitions

Before discussing test techniques in detail or discussing the advantages and disadvantages of different types of tests, it would be well to discuss general types of waveforms and the terms used to describe these waveforms.

Waveforms of either voltage or current tend to fall into one of the categories shown in Figure 17.1. The waveforms shown in Figure 17.1 are described in two ways: in terms of the natural type of waveform produced (1) by physically realizable test circuits and (2) by their triangular approximation. When any test waveform is described in terms of its triangular approximations, it must be understood that there is no intent that an actual test be made with triangular waveforms, since such waveforms can never be produced with physically realizable hardware. Any test that is to be done should be performed with natural waveforms, of which a specified triangular wave is only a convenient approximation.

Perhaps the most common waveform encountered during indirect effects testing is the damped oscillatory waveform of Figure 17.1(a). The major characteristics of importance for such a waveform are the peak amplitude, A, and the period, τ, or alternatively, the frequency, f. Since an oscillatory waveform may be either a basically cosinusoidal waveform (one starting at the peak amplitude A) or a sinusoidal waveform (one starting at zero), the initial time to crest, t_1, should generally be specified in addition to the period, or natural, frequency. In addition, specifications may be required as to the decrement of the wave, or the time required for the oscillations to damp out to a negligible amplitude.

The second most commonly encountered waveform is the exponential wave of Figure 17.1(b). This waveform describes the characteristics of a capacitor discharging through a resistor. The most common specifications relate to the peak amplitude, A, the time to crest, t_1, and the duration of the tail, or decay, t_2. The front time is described most commonly in terms of its triangular approximation, t'_1, the time required for a tangent line through the major portion of the leading edge to intercept the projection of the peak amplitude. Surge generators frequently have enough inherent inductance that the leading edge is fairly linear over a reasonable length of time, in which case the best fit tangent line may be drawn by inspection. To cover those situations in which the

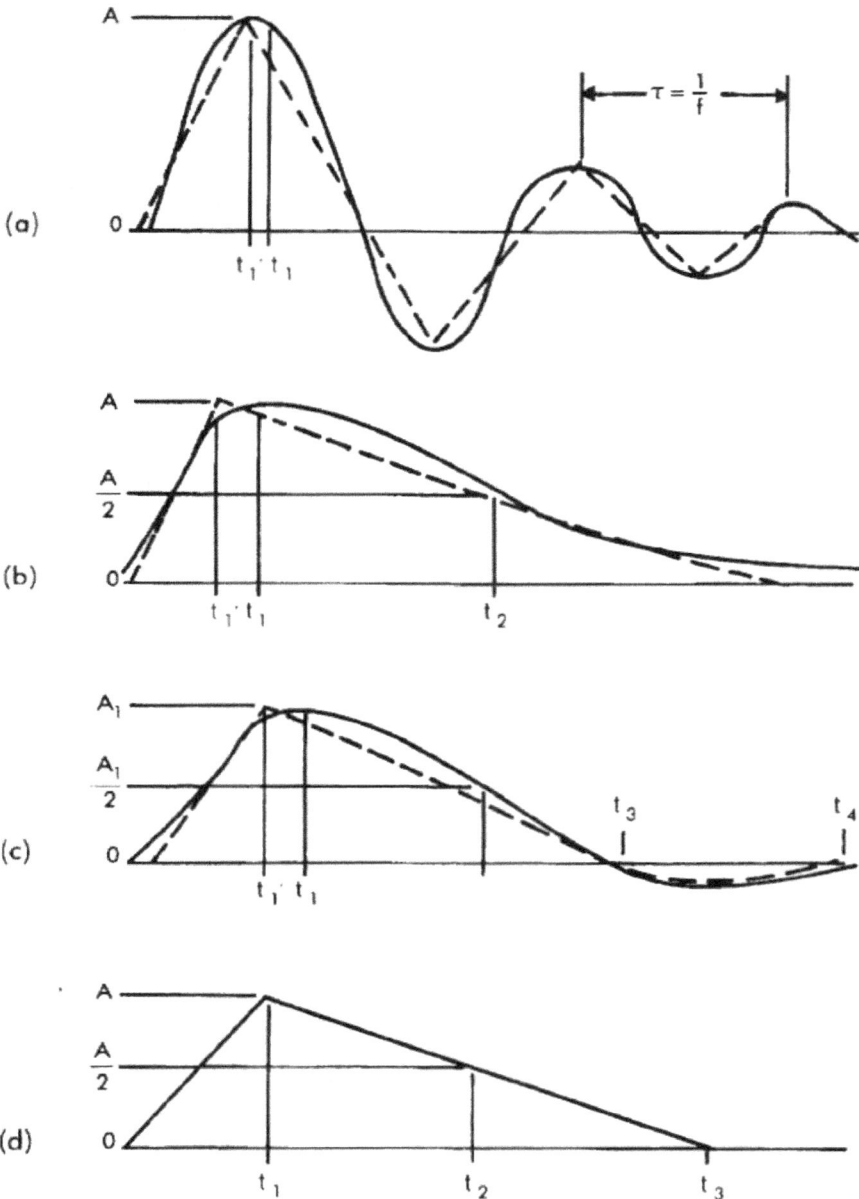

Figure 17.1 Description of waveforms.

(a) Damped oscillatory wave
(b) Exponential wave
(c) Exponential wave with backswing
(d) Generalized triangular wave

497

tangent line is not obvious by inspection, specifications frequently require that the straight line defining the front be that passing through the 10% and 90% points. Sometimes specifications require that the straight line be drawn through the 30% and 90% points. The duration of the tail, t_2, is most frequently defined as the time required for the wave to decay to one-half its initial amplitude. Some alternate specifications may define the tail in terms of the e-folding time, the time required for the wave to decay to $1/e$, or approximately 37% of its initial amplitude.

With some types of test circuits the characteristic wave produced is that of an exponential wave with a relatively small backswing, as shown in Figure 17.1(c). Technically, such a wave is a damped oscillatory wave with a high degree of damping. Specifications made with regard to the peak amplitude of such a wave, A_1, and to the time to crest, t_1 or t'_1, will be the same as those for the exponential wave. Specifications relating to the decay time may refer to either the time required for the wave to decay to one-half (or alternatively 37%) of its initial amplitude or the time required for the wave to decay to zero, t_3. Specifications which relate to an exponential wave with backswing frequently require that the maximum amplitude of the backswing, A_2, be less than some specified fraction of the initial amplitude. The duration of the backswing is frequently unspecified. In summary, then, the characteristics required to specify the generalized triangular wave of Figure 17.1(d) are the peak amplitude, A, the time to crest, t_1, and a measure of the duration of the wave. This latter may be either the time required to decay to half value, t_2, or the time required to decay to zero, t_3. If t_1 is small with respect to either t_2 or t_3, t_2 may be assumed to be equal to one-half of t_3.

17.3   Tests on Complete Vehicles

Tests on complete vehicles may be intended either as full-scale proof tests of the aircraft and all its internal electrical and electronic systems, or they may be transient analysis tests to determine only how the wiring responds to the electromagnetic fields produced by the simulated lightning current.

17.3.1   Proof Tests Techniques and Waveforms

Proof tests on a complete vehicle are intended primarily to identify system incompatibility problems––systems which may operate properly by themselves but which may fail when the complete vehicle is subjected to lightning currents. For example, redundant systems may provide no redundancy when they are simultaneously affected by the same induced voltages. Typical test setups are shown in Figure 17.2 (Reference 17.4).

In a test of this nature a high-power surge generator is placed adjacent to the aircraft; current is passed into one extremity of the aircraft and returned to the surge generator through an external wire or wires––preferably wires sufficiently well insulated that the test current is confined to the test circuit and does not enter the ground network of the facility where the tests are made.

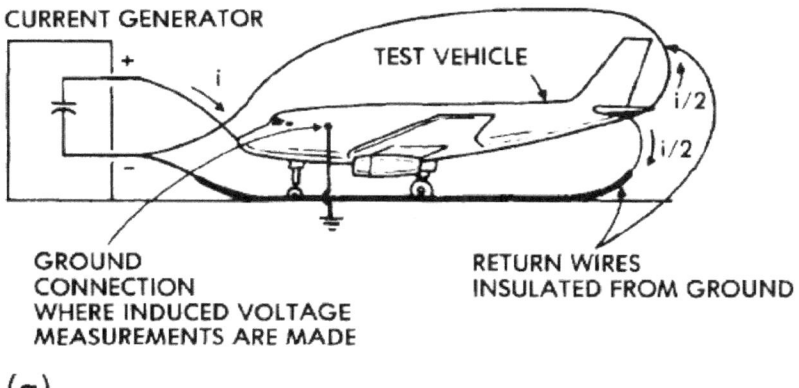

CURRENT GENERATOR

TEST VEHICLE

i

+

−

i/2

i/2

GROUND
CONNECTION
WHERE INDUCED VOLTAGE
MEASUREMENTS ARE MADE

RETURN WIRES
INSULATED FROM GROUND

(a)

CURRENT
GENERATOR

TEST VEHICLE

i

+

−

i/2

i/2

GROUND
CONNECTION

WHEELS ON
ELECTRICAL INSULATION

RETURN WIRES
INSULATED FROM GROUND

(b)

Figure 17.2 Typical setups for complete vehicle tests.
　　　　　　(a) Vehicle under test grounded
　　　　　　(b) Low side of current generator grounded

There should be a ground on the system. If the low side of the surge generator can be insulated from ground, it is generally preferable that the vehicle under test be grounded as shown in Figure 17.2(a). If this is not possible, the test vehicle must be insulated from ground and the ground made at the low side of the surge generator, as shown in Figure 17.2(b).

Since present surge generators are not sufficiently powerful to circulate the full-design lightning current through the aircraft, it is accepted practice to perform the proof test with two or more damped oscillatory currents having as waveforms one which provides the fast rate of rise characteristic of a natural lightning stroke wave front and the other a long-duration period characteristic of natural lightning-stroke duration. The two different components are necessary, since the mechanism of coupling electromagnetic fields into the interior of a vehicle is different at different frequencies. The high-frequency waveform is intended primarily for evaluation of aperture effects and has a minimum amplitude of 10 kA and a fundamental frequency of 100 kHz. The low-frequency waveform is primarily for evaluation of diffusion effects and has a minimum amplitude of 20 kA and a fundamental frequency of 2 kHz. As a result of the inductance of the test circuit, the waveforms are typically of a sinusoidal nature, and no independent definition of the front time, t_1, need be made.

During the tests the responses of the electrical equipment should be measured. Since neither of the test current waves is of the full amplitude associated with lightning, the test results must be extrapolated to the full lightning-threat level, 200 kA. For the high-frequency test current wave, the average rate of rise of the first quarter cycle of the test waveform should be determined by measurement or calculation and the test results extrapolated to 100 kA/μs. Measurements made with the low-frequency current wave must be extrapolated from the measured peak amplitude of the current wave to the 200 kA level.

The use of these oscillatory currents is one of the necessary compromises with perfection previously mentioned. These waves are used for proof tests simply because test facilities capable of subjecting a complete aircraft to a full-threat lightning current do not exist. For purposes of simulation the oscillatory waves have both merits and drawbacks. These are summarized below.

The fast oscillatory wave has the following merits:

- Over a narrow band of frequencies it simulates one important aspect of indirect effects—the aperture coupling of electromagnetic fields.
- High amplitude is available from relatively low-energy-surge generators, since little energy is lost in waveshaping resistance.
- It can be nearly a full-threat-level test because of the higher frequency coupling effects.

The fast oscillatory wave has the following drawbacks:

- It overemphasizes (and deliberately so) one of the component frequencies of lightning currents.
- It underemphasizes low-frequency coupling effects, particularly the effects of vehicle resistance and the diffusion penetration of magnetic fields.
- It is fundamentally different in waveshape from the waveshapes of natural lightning currents.
- It is potentially damaging to avionic equipment. In this respect, however, it must be remembered that if avionic equipment can be damaged by a simulated lightning proof test, it can also be damaged by an actual lightning flash.
- Bulky and expensive test equipment is required to make the test.

Some of the drawbacks of the fast oscillatory test wave are offset by its companion, the slow oscillatory wave. The slow oscillatory wave has the following merits:

- It simulates over a narrow band of frequencies one important aspect of indirect effects——the resistive coupling of electromagnetic fields.
- High amplitude is available from relatively low-energy-surge generators, since little energy is lost in waveshaping resistance.
- It can approach a full-threat-level test of lower frequency resistive coupling effects, at least for lightning strokes of low to average amplitude.

The slow oscillatory wave has drawbacks parallel to those of the high-frequency test wave:

- It overemphasizes (and deliberately so) one of the component frequencies of lightning currents.
- It may not correctly simulate diffusion penetration of magnetic fields. One can take refuge, however, in the observation that imperfect simulation is better than no simulation.
- It does not constitute a full-threat-level test for the higher amplitude lightning flashes. Data obtained at low amplitudes require extrapolation to give effects that would be present at full lightning-current amplitude.
- It is fundamentally different in waveshape from the waveshapes of natural lightning currents.
- It is potentially damaging to avionic equipment. The observation made previously about the high-frequency test waveform applies equally well to the same drawback in the low-frequency test.
- It too requires bulky test equipment.

17.3.2 Transient Analysis Tests

A major virtue of the fast oscillatory test wave is that it can simulate fairly well one of the important aspects of natural lightning——the electromagnetic effects associated with the initial rate of change of lightning current. Because of the emphasis that is placed on rate of change, it is often believed that only rate of change of current is important. Likewise, it is often believed that a nondamaging test for indirect effects analysis can be made by applying a current that initially rises at the full-threat-level rate, 100 kA/μs, but not allowing that current to flow for very long or not allowing it to increase to a high amplitude. Such concepts are fallacious.

Consider Figure 17.3 (Reference 17.5) in which two different current waves are applied to an equivalent circuit of the type shown in Figure 17.3(a). One current rises at the rate of 100 kA/μs to a value of 10 kA, with t_1 equal to 0.1 μs, and the second rises at the rate of 100 kA/μs to a value of 100 kA, with t_1 equal to 1 μs. The responses of the circuit, shown in Figure 17.4 (Reference 17.6), are fundamentally different.

The most nearly correct method of making a nondestructive test of indirect effects is to keep the basic waveshape the same, reducing only the amplitudes. Such scaling is shown in Figure 17.5 (Reference 17.7). Frequently, concern is expressed about the fact that reducing amplitude but keeping front

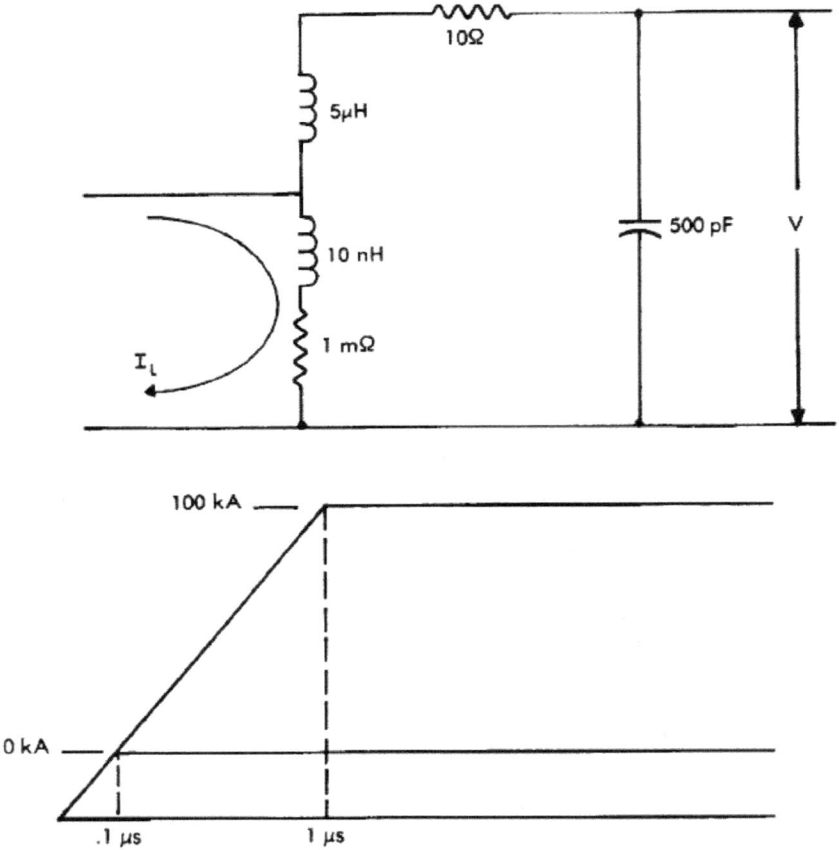

Figure 17.3 Why duration of rate of rise is as important as rate of rise.
   (a) Circuit analyzed
   (b) Current waveshapes

time the same reduces the rate of change of current. The initial rate of change is indeed reduced, but it is reduced in exactly the same ratio as that of the peak amplitude.

   Such considerations of scaling have led to the development and use of the transient analysis technique in which currents of waveshape similar to those produced by natural lightning, but of lower amplitude, are circulated through the aircraft. During such a test many measurements would be made of the transients induced on the wiring of the aircraft. These measurements, when scaled up to full-threat level, provide a measure of the transients to which lightning would subject the individual pieces of electronic equipment. An assessment of whether or not such transients might present a hazard to the equipment might involve additional tests on the individual pieces of equipment. Some of the considerations in such tests are given in Section 17.5.

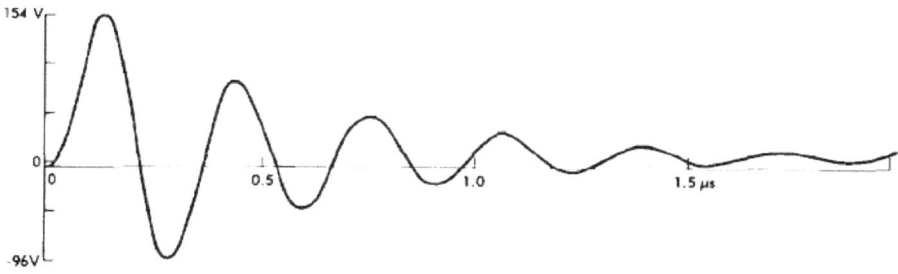

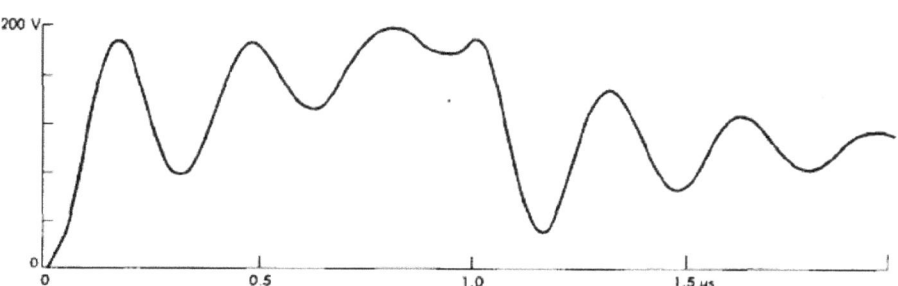

Figure 17.4 Responses of circuit.
(a) Response to a 10 kA current
(b) Response to a 100 kA current

The techniques of making transient analysis tests have been described in other literature (References 17.8, 17.9, and 17.10) and will not be discussed here. A summary of their advantages and disadvantages follows.

The transient analysis type of test has, among others, the following advantages:

- The tests may be made with current surges of waveshape nearly representative of those of natural lightning.
- The waveshapes have the correct ratio of high- and low-frequency components.
- The test correctly simulates frequency-dependent resistance and diffusion-coupled electromagnetic effects.
- The test is nondamaging to avionic equipment if it is made at low current levels.
- Relatively small and inexpensive equipment is required to make the tests.

The transient analysis type of test has, among others, the following drawbacks:

- Because a considerable amount of energy is lost in resistive circuit elements, a given size test generator produces less current than it would if it were used to produce damped oscillatory current waves.
- Data obtained at low amplitudes must be extrapolated to full-threat

503

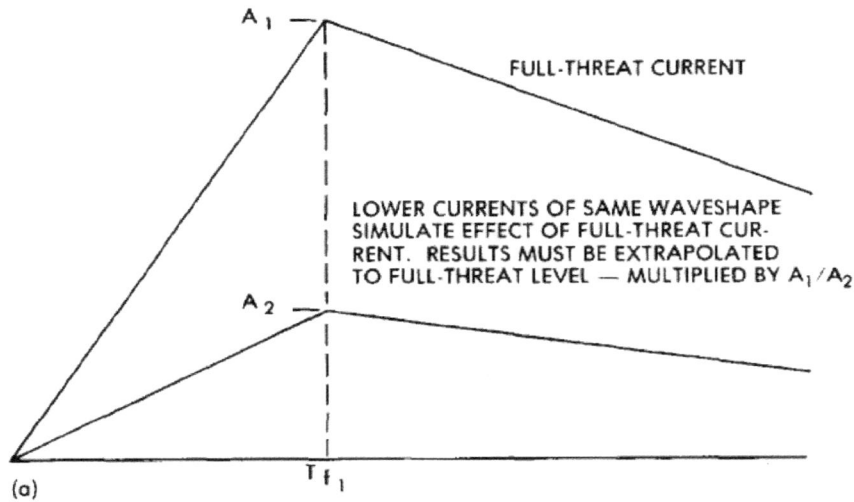

FULL-THREAT CURRENT

LOWER CURRENTS OF SAME WAVESHAPE
SIMULATE EFFECT OF FULL-THREAT CUR-
RENT. RESULTS MUST BE EXTRAPOLATED
TO FULL-THREAT LEVEL — MULTIPLIED BY A_1/A_2

(a)

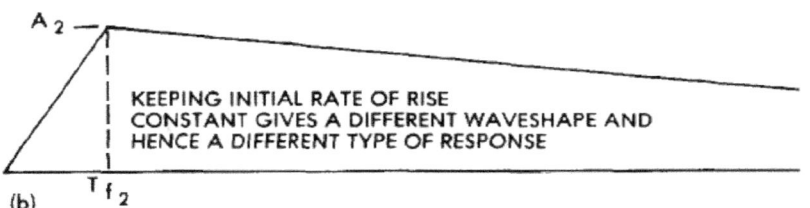

KEEPING INITIAL RATE OF RISE
CONSTANT GIVES A DIFFERENT WAVESHAPE AND
HENCE A DIFFERENT TYPE OF RESPONSE

(b)

Figure 17.5 Considerations regarding amplitude scaling.
(a) Correct scaling
(b) Incorrect scaling

level. Such extrapolation requires an assumption of linearity——that is, 10 A produces 10 times the response of 1 A. While such linearity can usually be relied upon, there are some nonlinear effects involved in the indirect effects of lightning.

• It is not a proof test and hence has no pass or fail criteria.

17.4 Tests on External Electrical Hardware

The test techniques described so far have dealt with determining how the wiring and equipment within the aircraft respond to the passage of lightning current through the aircraft. There are, however, some types of equipment containing electrical hardware, or containing wiring that connects to electrical hardware, that are frequently mounted on the exterior of the aircraft. Some of these are likely to be struck by the lightning flash. A pitot boom containing a deicing heater and the electrical wiring to supply power to that heater is an excellent example. An externally mounted navigation light is another example.

Tests are frequently required to determine what voltages and currents might be induced on the electrical wiring by a lightning stroke to such hardware. These tests are needed because the currents and voltages may be conducted to other electrical systems in the vehicle. The basic principle by which one of such tests is made is shown in Figure 17.6 (Reference 17.11). The test object should be mounted on a shielded test chamber so that access to the electrical connectors of the object can be obtained in an area relatively free from the electromagnetic fields produced by the surge generator. The test object should be mounted to the shielded enclosure in a manner similar to the way in which it is mounted on the aircraft, since normal bonding impedances may contribute to the voltages induced on circuits. If the shielded enclosure is large enough, the measurement oscilloscope may be contained within it. If not, a suitably shielded instrument cable may be used to transfer the induced voltage signal from the shielded enclosure to the oscilloscope. Measurements of the voltage induced on the electrical circuits within the test objects should be made on both a line-ground and a line-line mode. Since the direct contact of the lightning flash with the test object frequently produces quite high voltages, dividers or attenuators external to the oscilloscope may be needed. If used, these should be checked to ensure that the signals displayed on the oscilloscope are not the result of either spurious pickup on the attenuator or overload effects in the amplifiers used in the oscilloscope.

The appropriate waveforms of the surge currents to be injected into the test object depend on where the object will be located on the aircraft and whether or not it will be likely to intercept the initial lightning current stroke or whether it is likely only to intercept a stroke blown back along the surface of the vehicle by the swept-stroke mechanism. The literature on test requirements discusses in considerable detail the waveforms appropriate to the different zones on the aircraft, and for any such test the reader is referred to that literature (Reference 17.12).

17.5 Tests on Internal Equipment

The second major category of tests involves tests to determine how the avionic equipment within the aircraft responds to the transients induced on the wiring by the flow of lightning current. Equipment for and specifications relating to these tests are in an embryonic state of development at the present time. In the following sections we will attempt to discuss some of the available test techniques, the goals they are intended to accomplish, and some of the factors to which attention must be given during the conduct of such tests.

17.5.1 Waveforms and Amplitudes

Tests and measurements have shown that the transient voltages and currents to which avionic equipment is subjected are most commonly of a damped oscillatory nature. Voltages are almost always of this nature and currents usually are as well, although the currents flowing on the shields of shielded conductors have more of a tendency to be of a unidirectional nature

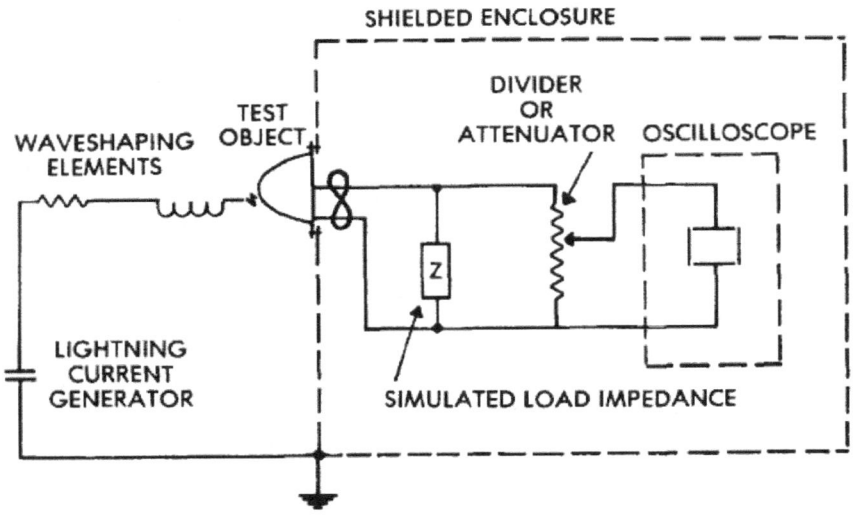

Figure 17.6 Typical electrical hardware indirect effects, test and measurement circuit.

like that of the external lightning current than do the induced voltages.

Some examples of transients found in aircraft during transient analysis tests are shown in Figure 17.7 (Reference 17.13). Whether the measurement is of an open circuit voltage or of a short circuit current is indicated in this figure.

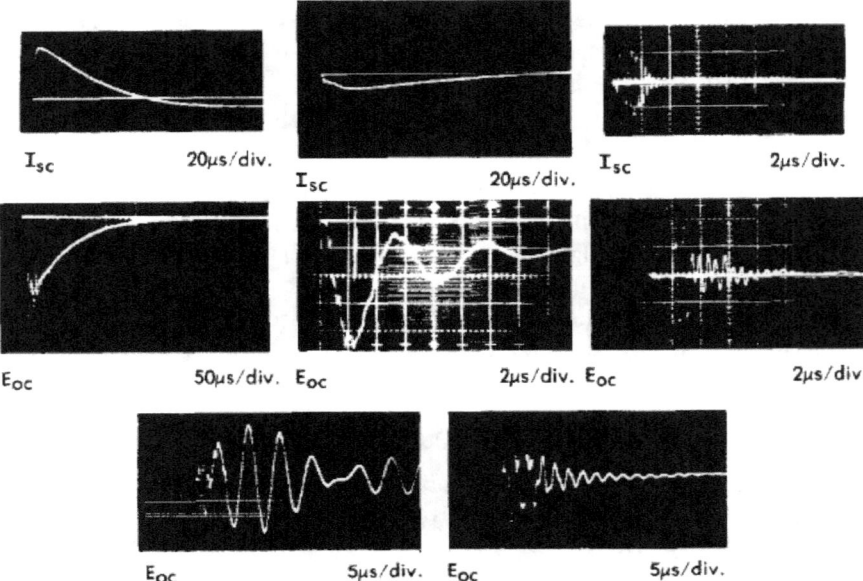

Figure 17.7 Some examples of lightning-induced transients found in aircraft.

506

Information relating to amplitude and to the conditions under which the transient was measured has deliberately been omitted, since the intent of the figure is merely to illustrate the broad diversity of waveforms that might be encountered.

The question of what waveforms to specify for transients used during tests related to indirect effects is another subject for which there are as yet only evolving standards. Some of those standards, or evolving standards, that have been used outside as well as inside the aircraft community are summarized in Tables 17.1 (Reference 17.14) and 17.2 (Reference 17.15). Specifications relating waveforms that are basically unidirectional will be discussed first.

TABLE 17.1 SPECIFICATIONS RELATING TO UNIDIRECTIONAL TRANSIENTS

	Peak Amplitude A_1	Crest Time T_1	Decay Time T_2	Amplitude of Backswing A_2
Space Shuttle — Component 1				
I_{SC}	10 A	2 μs	100 μs	$A_2 < 0.25\ A_1$
E_{OC}	50 V	2 μs	100 μs	$A_2 < 0.25\ A_1$
Space Shuttle — Component 2				
I_{SC}	5 A	300 μs	600 μs	$A_2 < 0.25\ A_1$
E_{OC}	0.5 V	300 μs	600 μs	$A_2 < 0.25\ A_1$
Standard ANSI Test Waves				
I_{SC}	(depends on	4 μs	10 μs	NA
E_{OC}	application)	1.2 μs	50 μs to half value	NA
Fisher and Martzloff				
I_{SC}	(depends on	5 μs	not specified	not specified
E_{OC}	application)	0.25 μs	5 μs	$A_2 < 0.5\ A_1$

The waveforms of longest standing are the American National Standards Institute (ANSI) waves relating to lightning effects on high-voltage apparatus (Reference 17.16). These call for current waves of 4 μs to crest and 10 μs to half value, and voltage waves of 1.2 μs to crest and 50 μs to half amplitude. Although

507

other waveforms are specified for other purposes, these are the most common waveforms.

TABLE 17.2 SPECIFICATIONS RELATING TO OSCILLATORY TRANSIENTS

	Peak Amplitude	Period	Frequency	Damping Definition
Space Shuttle — Component 1				
I_{SS}	10 A	8 μs	125 kHz	$A_2 > 0.25\,A_1$ on fifth cycle
E_{OC}	50 V	8 μs	125 kHz	$A_2 > 0.25\,A_1$ on fifth cycle
Space Shuttle — Component 2				
I_{SS}	5 A	1200 μs	833 Hz	$A_2 > 0.25\,A_1$ on fifth cycle
E_{OC}	0.5 V	1200 μs	833 Hz	$A_2 > 0.25\,A_1$ on fifth cycle
ANSI SWC Test				
I_{SS}	NA	NA	NA	NA
E_{OC}	2.5-3.0 kV	0.67-1 μs	1.0-1.5 MHz	Envelope decay to 50% in no less than 6 μs
Martzloff and Howell				
I_{SS}	NA	NA	NA	NA
E_{OC}	0.8 kV	10 μs	100 kHz	$A_{n+1}/A_n > 0.6$

Some of the older American literature refers to a voltage wave having a time to crest of 1.5 μs and a decay time to half amplitude of 40 μs. For all practical purposes, this is the same as the 1.2 x 50 μs waveshape, the 1.2 x 50 μs wave representing a fusion of slightly different American and European standards. The 1.2 x 50 μs figure is often viewed as representing the waveshape of the lightning current. This view, however, is incorrect; it represents the voltage produced at high-voltage equipment, such as transformers, by the effects of a lightning stroke at some remote point. The 4 x 10 μs current wave

508

represents the lightning-related current passing through a lightning arrester after it has been caused to spark over internally by a voltage surge of different waveshape. Both the voltage and current waveshapes represent transients that may be produced by basically simple laboratory test equipment. They do not necessarily represent the shape of the current or voltage transients produced by natural lightning.

At one stage in the development of the *Space Shuttle Criteria Document*, there was an allowance made for two basically unidirectional transient test waves. While those test waves had some deficiencies and have been largely superseded, discussion of them is still appropriate, since that discussion will illustrate some of the problems inherent in waveform specifications.

The first of these specifications dealt with a transient rising to crest in 2 μs and decaying to zero in 100 μs. The intent of this test wave was to duplicate in some manner the effects produced by magnetic flux leaking through apertures. The waveform had its basis in the ANSI test waves. These waveforms permitted a backswing, a feature characteristic of transformer-coupled surges. The backswing amplitude was required to be less than 25% of the initial amplitude, but the duration of the backswing was left uncontrolled. The test waveform was not intended to be interpreted as requiring a backswing. An overdamped wave was perfectly satisfactory. If an overdamped waveform was used, there would be no clearly defined time to zero. In such cases the decay time was intended to be taken as one-half the indicated value (50 μs instead of 100 μs) and measured to the time at which the wave had decayed to 50% of its initial amplitude. The waveform was thus similar to the standard ANSI test waveform derived for tests on high-voltage apparatus. The front time of 2 μs reflected the 2 μs front time of the basic lightning design current. The transients themselves were specified as having amplitudes of 50 V open circuit and 10 A short circuit.

The second specification related to a long-duration transient representing the effects produced by magnetic flux diffusing through the walls of cavities. Such flux would have rise and decay times much longer than those of the lightning current. The specification called for a short circuit current of 5 A and an open circuit voltage of 0.5 V, both taking 300 μs to reach crest and another 300 μs for decay to zero. The specification of equal times to crest and from crest back to zero is incompatible with the response of real physical elements. In practice, any waveform with a rise time of 300 μs would have a decay time longer than 600 μs.

One common deficiency of the above specifications was that both the short circuit current and open circuit voltage had the same waveshape. In practice the open circuit voltage would be of a duration shorter than that of the short circuit current.

In the original paper in which the concept of transient control levels was first presented (Reference 17.17), Fisher and Martzloff proposed a test wave that was primarily unidirectional. The open circuit voltage was characterized by a fast rise to crest and then a decay to zero in 5 μs, or greater. To allow for transformer coupling of the transient, a backswing was allowed after the transient had decayed to zero. The character of the backswing was not specified——only that its amplitude should be less than 50% of the initial amplitude.

The rationale behind this waveform concept included the following several aspects.

- It should be in some measure proportional to the derivative of the magnetic field produced by a lightning current.
- The duration of the transient should be long enough that possible failures of semiconductors would not be strongly affected by the waveshape of the transient, since, with transients of duration shorter than about a microsecond, the failure levels of semiconductors are strongly affected by waveshape.
- The duration of the transient should be roughly comparable to the duration of clock cycles in digital equipment.
- The transient should include a rapidly changing phase to excite inductively coupled circuit elements.
- The transient should be one that could be produced by and coupled to equipment by relatively simple test equipment.

Directly associated with the open circuit voltage transient was a short circuit current transient. The short circuit current was that current which would flow from a source the internal impedance of which could be represented as 50 μH in parallel by 50 Ω. The test waves taken together as a set were thus more consistent than were those relating to the *Space Shuttle*. Amplitudes of neither current nor voltage were specified, since the test levels would be part of transient control level specification.

The most common type of transient encountered would be oscillatory. Some existing specifications relating to oscillatory transients are summarized in Table 17.2. The first two shown deal with the *Space Shuttle* and were essentially oscillatory versions of the unidirectional transients shown in Table 17.1 As presented at that time there was no specification on the initial front time; however, the maximum front time would be governed by the oscillatory frequency. The decrement of the transient was defined by requiring that the amplitude of the fifth half-cycle be more than 25% of the amplitude of the initial amplitude.

The low-frequency component 2 was ultimately dropped from further consideration, since there was little evidence from transient analysis tests of such low-frequency oscillations being excited by diffusion-coupled magnetic flux.

A specification of somewhat longer standing is the ANSI (SWC) Surge Withstand Capability Test (Reference 17.18). This test was derived from considerations of the transients found on control equipment in high-voltage substations. The specification makes no reference to short circuit currents. The open circuit voltages are specified as being between 2.5 and 3 kV, with the oscillatory frequency being from 1.0 to 1.5 MHz. Decrement is specified by requiring that the envelope decay to 50% in no less than 6 μs.

The test was intended for simulation of the transients found in high-voltage electrical substations and to which equipment in such substations might be exposed. The voltage range specified is reasonable for such apparatus

510

but may be high for most electronic equipment, particularly for equipment located in shielded locations. The waveshape, while not unreasonable, is sometimes difficult to inject into circuits by transformer coupling. This test does not treat in any way the short circuit current that would flow on a shorted conductor connected to a low-impedance load or flow upon the shield of a cable. The specification that the envelope decay to 50% in no less than 6 μs implies a lower loss test circuit than may be found in practice. The recommended test generator circuit is somewhat uncontrolled, using an untriggered spark gap firing virtually at random at the crest of an ac charging voltage, though the waveform itself does not imply any particular generator test circuit. The specification does call for a 150 Ω source impedance. The question of source impedance deserves serious consideration and 150 Ω may not be the optimum value.

This test (and waveform) is one of the few that have been formally approved by an industry standardizing group.

Martzloff and Howell (Reference 17.19) proposed a test wave with a frequency of 100 kHz and a voltage range of 0 to 8 kV. The wave was intended for the duplication of transients found in residential circuits on 120 V ac lines. The damping was specified such that the ratio of successive half-cycles should be greater than 0.6. This also may be somewhat high, particularly if the transient is to be coupled into equipment through transformers. No specification was made about the magnitude or waveshape of the short circuit current. Martzloff and Howell show a test circuit capable of injecting the transient onto 120 V ac lines where, since the output impedance of the circuit is basically 150 Ω resistive, the shape of the short circuit current would be about the same as that of the open circuit voltage.

This waveform and this test circuit have been widely accepted in some fields. One example is their use in relation to ground fault interrupters (Reference 17.20).

In answer to response from readers and users of the original Fisher and Martzloff paper (Reference 17.21) on the transient control level philosophy, Crouch, Fisher, and Martzloff have proposed a test wave somewhat different from the original one (Reference 17.22). The revised voltage wave, shown in Figure 17.8, emphasizes the oscillatory nature of the wave, rather than deemphasizing it, as did the original test wave. The front time was raised to 0.5 μs, the course of the wave after crest being defined in terms of its oscillatory frequency of 100 kHz and the decrement specified by requiring that the ratio of successive half-cycles be greater than 0.6. The voltage wave thus becomes nearly identical with that proposed by Martzloff and Howell.

The proposed waveshape should be viewed in relationship to its companion short circuit current. The current, similar to that from natural transient sources, is essentially the integral of the voltage.

17.5.2 Equipment for Direct Injection of Transients

Surge generators capable of producing the types of transients described

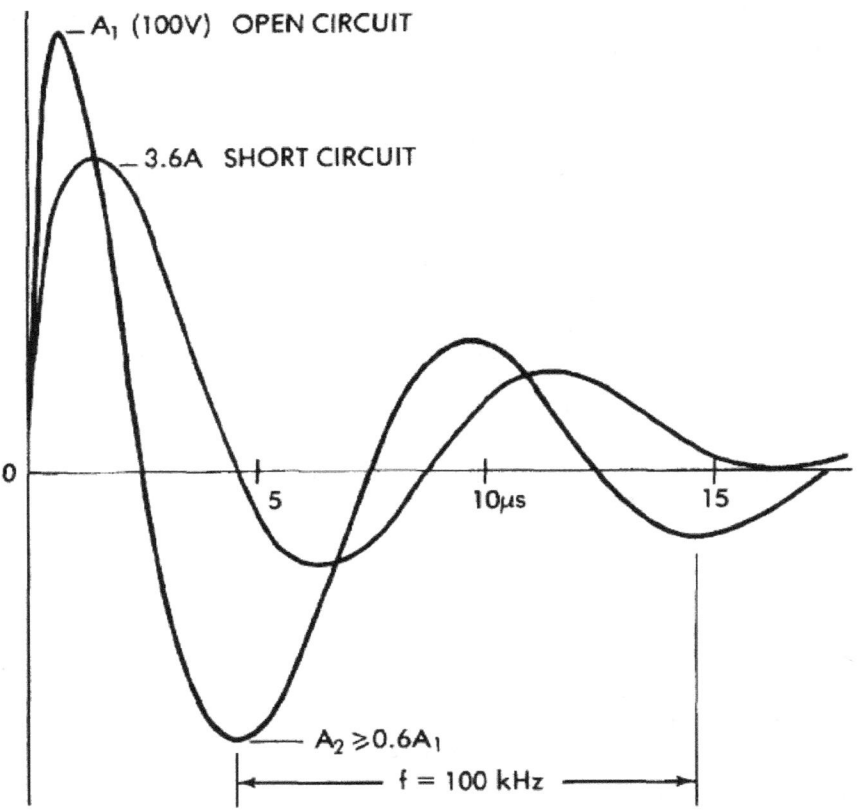

Figure 17.8 Revised proposals for TCL test waves.
(a) Open circuit voltage
(b) Short circuit current

here generally employ capacitors that are discharged into a wave-shaping circuit. That circuit would be resistive if unidirectional transients are to be produced, and inductive if oscillatory transients are to be produced. Some representative examples of circuits are shown in Figure 17.9 (Reference 17.23). The switching gap may be either triggered or untriggered depending on the degree of sophistication desired. Components should be chosen and laid out in such a way as to minimize undesired residual inductance. Circuit voltages are typically high enough to be hazardous; therefore, appropriate safety precautions should be taken in both their construction and their use. Figure 17.10 (Reference 17.24) shows an example of the types of transients that are produced by the revised TCL test circuit shown in Figure 17.9(b).

One way in which the generators may be used is to connect the output directly to the terminals of the device under test. When generators are used in such a manner, one may determine the level of transient that causes a device to fail or, alternatively, determine that a given transient does not cause the device to fail. The test may be applied in a common mode, as it is in Figure 17.11(a) (Reference 17.25), or in a differential mode, as it is in Figure 17.11(b). Depending on the nature of the circuits under test, it may be appropriate to couple the transient into the terminals of the device through coupling capacitors.

Direct injection of transients into the terminals of a device has the drawback that the device becomes connected in a manner different from its normal connections. As a result, it may not be possible to check the device to see whether the transient interferes with the normal operation of the device. One way the operation of the device may be checked is to discharge the test generator into shields placed over interconnecting cables, if there are such shields, or into the cases housing electronic equipment. In either case, the generator would be used to generate a short circuit current, and that current,

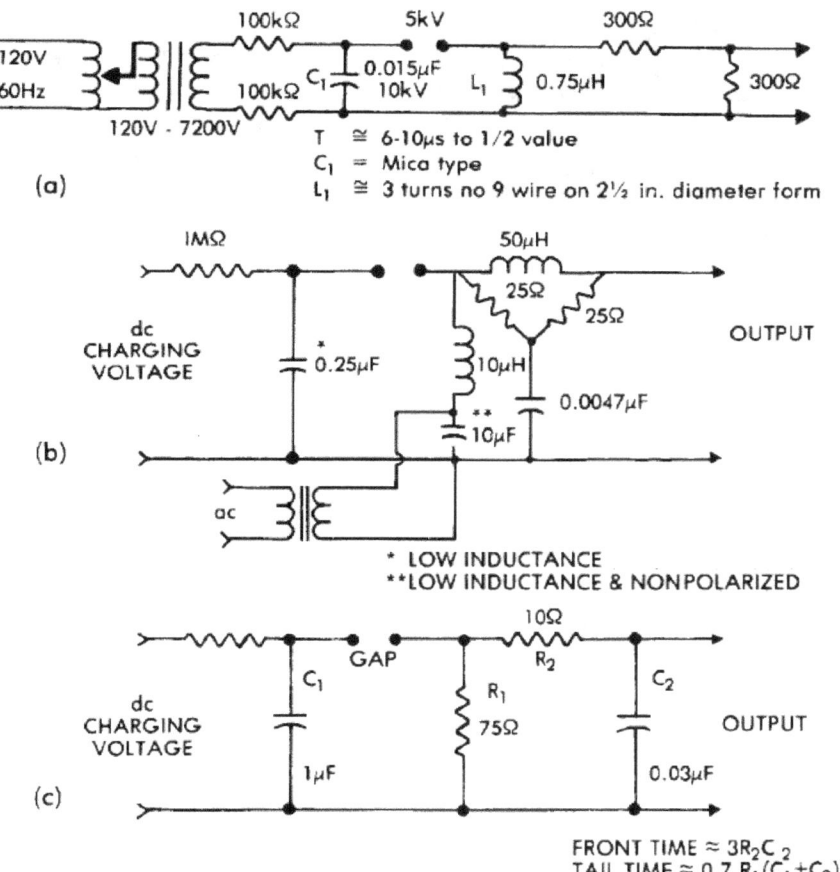

(a)

$T \cong$ 6-10μs to 1/2 value
C_1 = Mica type
$L_1 \cong$ 3 turns no 9 wire on 2½ in. diameter form

(b)

* LOW INDUCTANCE
**LOW INDUCTANCE & NONPOLARIZED

(c)

FRONT TIME $\approx 3R_2C_2$
TAIL TIME $\approx 0.7 R_1(C_1+C_2)$

Figure 17.9　Representative transient generator circuits.
　　　　(a) SWC test circuit – 1.5 MHz oscillatory
　　　　(b) Revised TCL test circuit – 100 kHz oscillatory
　　　　(c) Unidirectional – 1.2 x 50 μs

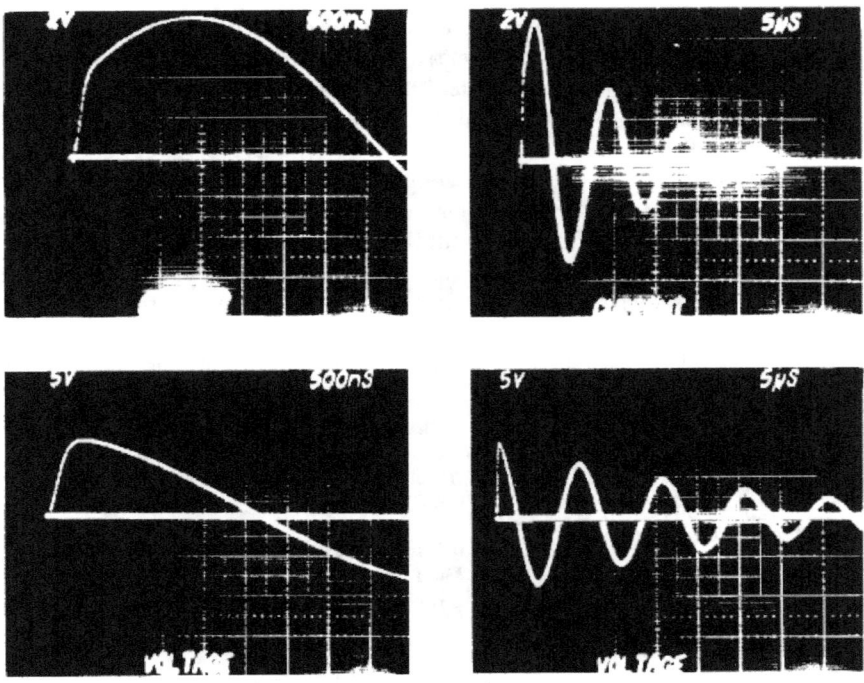

Figure 17.10 Output characteristics of TCL generator shown in Figure 17.9(b).
Current — 40 A/major div.
Voltage — 1000 V/major div.

flowing through whatever transfer impedances might be present, would induce
transients upon the internal circuitry. Presumably one would have previously
determined in some manner what would be an appropriate amount of current to
inject into the shields.

Direct injection of current is an appropriate test technique if the system
under test will be exposed to a severe electromagnetic field threat, since any
successful system will almost certainly be fitted with shields on the interconnecting wires.

17.5.3 Equipment for Transformer Injection of Transients

Another powerful test technique by which transients may be injected into
the electrical wiring interconnecting different pieces of electronic equipment is
through the use of a pulse-injection transformer and a suitable pulse generator.
The elements of such a system are shown in Figure 17.12. Current that is passed
from the pulse generator through a primary winding on a magnetic core sets up a

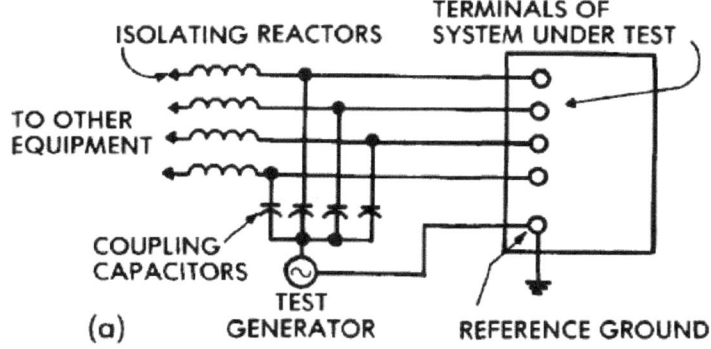

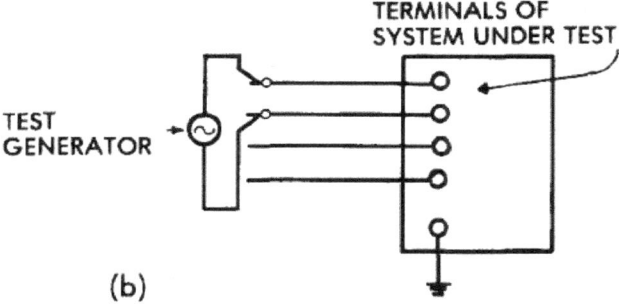

Figure 17.11   Direct injection of transients.
           (a)   Common mode
           (b)   Differential mode

magnetic field in that core. If the core is placed around a cable interconnecting two pieces of electronic equipment, the magnetic field contained in that core induces current or voltage in the interconnecting wiring in a manner very similar to that produced by magnetic flux from an external lightning source. This type of equipment has been used for NEMP effects testing on the B-1 (Reference 17.26) and has also been specified for use in the *Space Shuttle Lightning Criteria Document* (Reference 17.27), though only in principle, not in detail. An example of the type of test waves that may be produced by a transformer pulse-injection circuit is shown in Figures 17.13 and 17.14 (Reference 17.28). The inclusion of more complex waveshaping elements in the pulse generator (shown dotted in Figure 17.12) would allow other waveshapes to be produced.

     Some considerations regarding the capabilities of such transient injection follow. First, it should be remembered that the circuit under test is primarily exposed to the changing electromagnetic field confined in the core of the pulse injection transformer. If the circuit under test is a high-impedance circuit, the natural effect of that changing magnetic field is to induce an open circuit voltage, the magnitude and duration of which will depend upon the amplitude

515

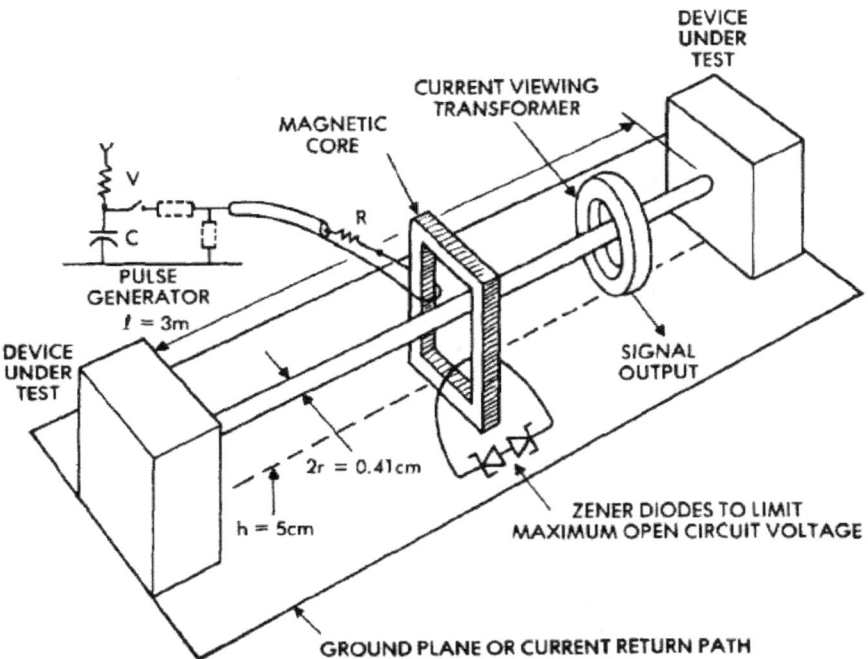

DEVICE
UNDER
TEST

CURRENT VIEWING
TRANSFORMER

MAGNETIC
CORE

V

C

PULSE
GENERATOR
l = 3m

R

DEVICE
UNDER
TEST

2r = 0.41cm

h = 5cm

SIGNAL
OUTPUT

ZENER DIODES TO LIMIT
MAXIMUM OPEN CIRCUIT VOLTAGE

GROUND PLANE OR CURRENT RETURN PATH

Figure 17.12 Transformer pulse injection.

and rate of change of the magnetic field in the core but which will not be significantly affected by the length of the circuit under test. If the circuit is of low impedance, however, the magnetic field will induce a circulating current in the circuit. The shape of this current will be of a duration longer than that of the open circuit voltage, and will be nearly the same as the duration of the current produced by the pulse generator in the primary of the transformer. The magnitude of this current will depend upon the impedance of the circuit under test, and this impedance will in turn depend upon the height of the circuit above a ground plane and upon the diameter of the circuit or bundle of wires in which current is injected, since these latter parameters affect the self-inductance of the circuit. A complete specification of the test circuit must deal both with the maximum open circuit voltage that may be produced and with the maximum short circuit current that may be produced. The inductance of the circuit under test is almost directly proportional to the length of the circuit, but proportional only to the logarithm of the diameter of the circuit or its height above the ground plane. In the absence of any other specifications, it is suggested that the pulse generator and core be such that the specified short circuit current may be induced on a conductor 3 m long (about 10 ft) and having a diameter of 0.41 cm and spaced 5 cm above a ground plane. Such a conductor may be provided by the shield of RG-58 coaxial cable. Furthermore, it is proposed that the open circuit voltage be measured when the pulse generator is set at the amplitude producing the specified short circuit current. Specifications have yet to be

516

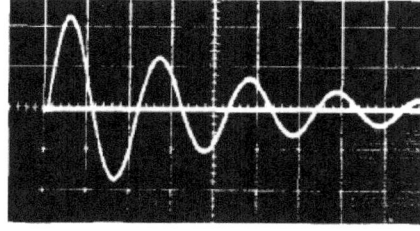

VERTICAL — 10 A/div.
HORIZONTAL — 2 µs/div.
(f = 123 kHz)

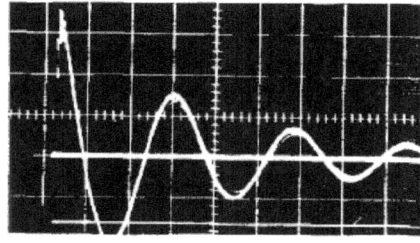

VERTICAL — 60 V/div.
HORIZONTAL — 2 µs/div.
(f = 91 kHz)

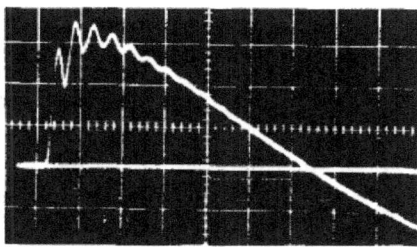

VERTICAL — 60V/div.
HORIZONTAL — 0.2µs/div.

C = 0.1µF
V = 520V
R = 0Ω

Figure 17.13 Oscillatory current injected.

    (a) I_{SC}

    (b) E_{OC}

    (c) E_{OC} (different time base)

developed as to what this open circuit voltage should be in terms of the corresponding short circuit current, although suggestions are given by Fisher and Martzloff and by Howell and Martzloff (Reference 17.29). If the open circuit voltage is too high, there are two alternatives: (1) to modify the pulse generator so as to reduce the rise time of the current pulse in the primary of the transformer, or (2) to use a Zener diode or other clipping device on an auxiliary winding on the transformer core. After the required short circuit current was

517

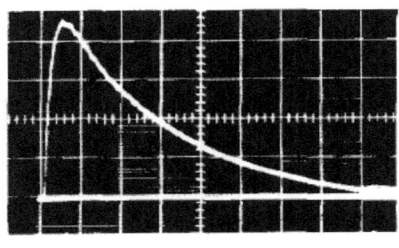

VERTICAL — 1 A/div.
HORIZONTAL — 1 μs/div.

(a)

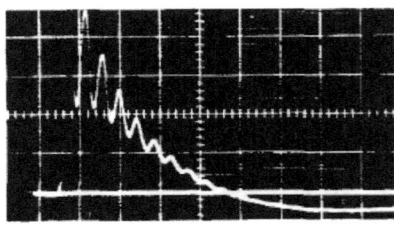

VERTICAL — 20 V/div.
HORIZONTAL — 0.2 μs/div.

(b)

$C = 0.1 \ \mu F$
$V = 520 \ V$
$R = 27 \ \Omega$

Figure 17.14   Unidirectional current injected
(a) I_{SC}
(b) E_{OC}

shown to be developed on the standard conductor (Figure 17.12), that standard conductor would be replaced by the actual cable into which transients were to be injected. The actual current induced in the cable could be monitored by the current-viewing transformer. Recording it as well should probably be one of the requirements of the test plan. The actual current, of course, may be lower than the standard short circuit current if the actual cable is terminated in impedances higher than zero. Conversely, if the actual cable under test is provided with an overall shield and that shield is connected to the ground structure at both ends and if it has a diameter larger than 0.41 cm, the actual short circuit current may be larger than the standard or specified current.

Whether or not it is appropriate to measure the voltage actually induced on the individual circuits would depend upon the characteristic of those circuits. Connection of a voltage probe would certainly be more likely to disturb the circuit and cause spurious response than would the insertion of a current-viewing transformer.

During the demonstration tests in which the waves shown in Figures 17.13 and 17.14 were measured, the conductor was a 15-foot section of RG-58 coaxial cable lying directly on a metal-covered floor. One end of the shield of this cable was connected to the floor, and the other end could be left either open circuited or shorted to the floor. Measurements were made only of the open circuit voltage or the short circuit current on the shield of the cable. Measurements were not made of either the current or the voltage induced on the signal conductor within the shield. The short circuit current is seen to rise to crest in about 1 μs and to oscillate at a frequency of 123 kHz. The peak current obtained with this configuration was about 23 A.

If the conductor was left open at one end, the magnetic field that was then produced in the core of the transformer induced an open circuit voltage of about 210 V. The voltage rose to crest in about 0.2 μs and oscillated at a frequency of 91 kHz. The frequency was different because the inductance seen by the capacitor in the pulse generator was different, depending on whether or not there was a short circuit on the secondary of the transformer. The higher frequency oscillations on the front of the open circuit voltage were caused by the stray inductance and capacitance of the conductor under test.

Significant points to observe here are that the open circuit voltage will invariably rise to crest faster than does the short circuit current and that the open circuit voltage will typically oscillate at a lower frequency than does the short circuit current. When the conductor under test is shorted, the inductance seen by the pulse generator is lower than it is when the secondary is open. This difference in oscillatory frequencies should be either recognized and allowed for in test specifications or corrected by adjusting the capacitance of the pulse generator used to excite the transformer.

When a unidirectional current was injected into the transformer (Figure 17.14), the short circuit current rose to a crest of about 5.5 A in about 0.5 μs. The rise time was greater and the amplitude less because of the damping resistor that was required to be in series with the primary winding. The open circuit voltage again rose to crest more rapidly than did the short circuit current. If an equal number of turns are used on the primary and secondary windings, there will be less current in the secondary winding than in the primary. With the transformer core used, the secondary current was about 65% of the primary current. The factors that affect the performance of pulse-injection transformers are discussed in more detail in paragraph 17.5.4.

As an alternative to transformer coupling, it is possible to discharge the pulse generator into a wire run parallel to the cable interconnecting the two pieces of equipment. This technique is less efficient in coupling low-frequency energy into the cable under test than is the transformer technique, but it may be more effective in coupling high-frequency energy.

17.5.4 Equipment for Generation of Electric and Magnetic Fields

The test equipment described so far has been aimed at the injection of current or voltage pulses into the terminals of electronic equipment. There is sometimes a need for subjecting an entire piece of electronic equipment to an engulfing electric or magnetic field, principally to check for magnetic field leakage of a cabinet. Sometimes these effects are checked with a wire-wrap technique in which a coil of wire is wrapped around the item to be tested and the coil then excited from a pulse generator. The type of magnetic field so produced is perhaps not the best that may be derived for indirect effects testing. In the study of nuclear electromagnetic pulse (NEMP) effects, use is often made of strip line simulators in which a pulse generator is matched to a large, open transmission line. In such simulators attention is given to ensuring that there is a suitable transition from the small geometry of the pulse generator to the large geometry of the test chamber, with the intent of producing plane wave fields in the working chamber. The pulse generators and working chambers are also carefully matched to each other to allow the production of electromagnetic fields having very fast fronts.

For testing the electromagnetic effects resulting from lightning, the physical design of such chambers can frequently be simplified. The rise times involved are not as fast as those in NEMP studies, and the requirement of plane wave propagation in the test chamber is not as important. A test chamber suitable for many types of equipment might be like that shown in Figure 17.15. It would basically consist of a Helmholtz coil of rectangular cross section about 1 m^3 in volume. Such a coil, when excited by the pulse generator sketched, would produce in its working volume a magnetic field of approximately 1000 A/m or an electric field intensity of about 20 kV/m. Magnetic fields of this nature might frequently be encountered in typical equipment bays of an aircraft unless these bays were deliberately built to provide electromagnetic shielding. The impedance (E field intensity/H field intensity) would be lower than that of free space or lower than the impedance of an NEMP strip line simulator. This would be appropriate, since the electromagnetic field impedance inside an aircraft equipment bay would also be lower than the impedance of free space. The duration of the pulses produced by such a simulator with the generator and load resistor constants shown would be about 100 μs. The front times would be determined by the inductance of the chamber and the values of any waveshaping elements, L and C, employed in the pulse generator. Longer duration magnetic fields, typical of those that might be found in an aircraft equipment bay, might be produced by short circuiting the 20 Ω termination resistor shown in the figure and, possibly, increasing the value of the capacitance and inductance of the pulse generator.

17.6 Design Notes for Transformer Pulse-Injection Equipment

This section will discuss some of the detailed considerations in the design of equipment suitable for pulse-injection equipment. The pulse generator and transformers described should not be viewed as representing the only designs that might be suitable or even the best that are suitable. They are intended only to provide a starting point for those who wish to initiate a test program at minimum cost.

17.6.1 Pulse Generator

A simple single-polarity current injection-pulse generator could be made as

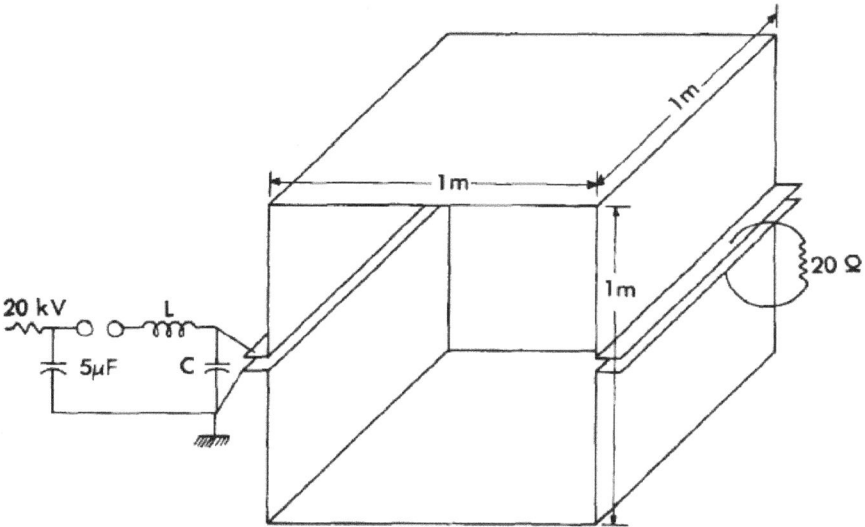

Figure 17.15 Possible test chamber for evaluating electric and magnetic field effects.

shown in Figure 17.16 (Reference 17.30). This pulse generator is basically the one used to produce the waveforms shown in Figures 17.13 and 17.14. The energy storage element is capacitor C, which is charged through D_2 to a voltage determined by the ratings of transformer T_2 and the setting of the variable autotransformer T_1. Charging of the capacitor would take place on the positive half-cycle (b) of the supply voltage. During this time the discharge switch, RY1, a relay with mercury-wetted contacts, would be unenergized and resting on contact b. During the negative half-cycle (a) RY1 will be energized through diode D_1 and closed to position a, thus discharging the capacitor during the half-cycle when there is no voltage being applied through diode D_2. Because of mechanical inertia, the switch armature does not necessarily operate in phase with the ac drive signal, and, consequently, during a portion of the positive charging half-cycle, the contact may remain closed to point a. In order to prevent excessive power drain from the transformer or excessive power dissipation in R_3, the series resistor R_2 is used to limit the charging current drawn from the transformer.

The capacitor would be discharged into a single-turn primary on a transformer core. If a unidirectional pulse were desired, switch S_3 would be open and resistor R_2 inserted in the discharge circuit to provide damping. If S_3 were closed, the capacitor would oscillate through the inductance of the transformer. A parts list for the pulse generator is shown in Table 17.3.

Higher voltages could be obtained by increasing the ratings of T_2, D_2, and C_1. If the charging voltage were appreciably higher, the mercury-wetted contact relay RY1 used for switching the charged capacitor into the transformer would have to be replaced by a different device, either a higher voltage relay or some type of discharge tube.

TABLE 17.3 PARTS LIST — SINGLE-POLARITY PULSE GENERATOR

B_1	—	Yellow (power on)
B_2	—	Red (pulse on)
C_1	—	0.1 μF, 600 V
D_1	—	IN4003 (1 A, 200 V)
D_2	—	IN4007 (1 A, 1000 V)
R_1	—	6800 Ω, 1 W
R_2	—	15 KΩ, 1 W
R_3	—	27 Ω, 2 W
S_1	—	SPST (power and relay on)
S_2	—	SPST (pulse on)
S_3	—	SPST (unidirectional or oscillatory)
T_1	—	Variable autotransformer (120 V, 2 A)
T_2	—	120/350 V, 40 mA
		(Any of a variety of plate transformers is satisfactory. Half or all of a center-tapped secondary may be used).
T_3	—	Ferroxcube 1F53C5 U-core (2 required)
F	—	120 V, 1 A
RY1	—	Mercury-wetted contact relay Clare HG 1002 or Potter and Brumfield JML 1200-81

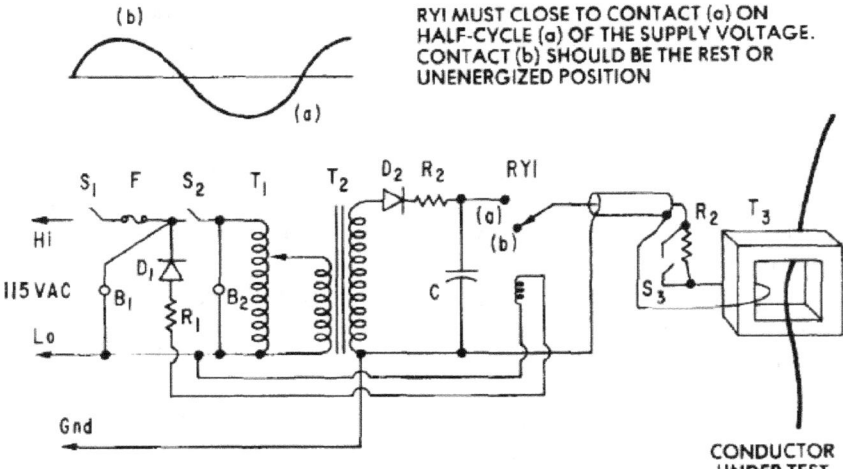

RYI MUST CLOSE TO CONTACT (a) ON HALF-CYCLE (a) OF THE SUPPLY VOLTAGE. CONTACT (b) SHOULD BE THE REST OR UNENERGIZED POSITION

Figure 17.16 Single-polarity current injection-pulse generator.

522

17.6.2 Current-Injection Transformers

Some of the parameters affecting circuit response are shown in Figure 17.17 (Reference 17.31). These include the capacitance of the storage capacitor; the inductance of the lead connecting the storage capacitor to the transformer; the primary, secondary, and leakage inductances of the transformer; and the inductance and resistance of the conductor under test.

The characteristics of the pulse transformer are among the most important properties. These parameters relate to the primary and secondary inductances, the flux in the transformer core, and the proportion of flux produced by current

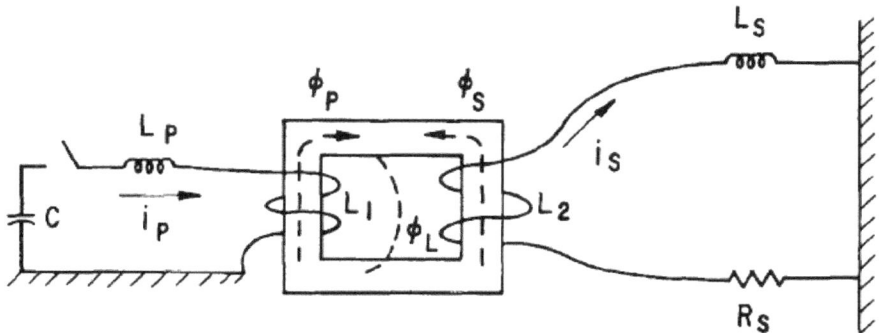

C = STORAGE CAPACITOR

L_P = INDUCTANCE OF PULSE GENERATOR AND CONNECTING LEAD (0.77 μH/ft FOR 50 Ω COAXIAL CABLE)

L_I = PRIMARY INDUCTANCE OF TRANSFORMER

L_2 = SECONDARY INDUCTANCE OF TRANSFORMER

L_S = INDUCTANCE OF LEAD UNDER TEST

R_S = RESISTANCE OF LEAD UNDER TEST

ϕ_P = FLUX DUE TO CURRENT IN PRIMARY

ϕ_S = FLUX DUE TO CURRENT IN SECONDARY

ϕ_L = LEAKAGE FLUX, FLUX ORIGINATING IN PRIMARY, BUT NOT LINKING SECONDARY

Figure 17.17 Parameters determining circuit response.

in the primary that links the secondary winding. The difference in these two fluxes is the leakage flux, and it is the leakage flux that prevents the secondary current from being as large as the primary current.

For a given transformer, the primary and secondary inductances may either be measured on an inductance bridge or be calculated by discharging a known capacitor through the winding and observing the frequency of oscillation. Mutual inductance, which is related to the degree of coupling between windings, can be determined by connecting the primary and secondary windings in series, first in series so that the two magnetic fields set up are in the same direction (series aiding) and then so the magnetic fields are in opposite directions (series bucking).

Knowing the primary, secondary, and mutual inductances, one can produce an equivalent circuit of an injection transformer. One such circuit is the Pi circuit of Figure 17.18(b) (Reference 17.32). If the number of turns on the primary winding is equal to the number on the secondary winding (the usual case), the circuit reduces to that of Figure 17.18(c).

The amount of open circuit voltage or short circuit current that can be transformed by a given transformer depends upon the magnetization curve of the core. A typical curve for a core without any air gap in the magnetic path is shown in Figure 17.19(a) (Reference 17.33). The operation of the core depends on the type of pulse to be produced. The most difficult operation involves a train of pulses, all of the same polarity. If the core is initially demagnetized, the first pulse will follow path a on the magnetization curve, whereas successive pulses will follow curve b. Since the amount of open circuit voltage that may be produced depends upon the change in magnetic field, operation along curve b cannot support as high a voltage as would be expected if one considers only path a, nor one lasting for as long a period of time. If alternate positive and negative pulses are applied, then operation along paths c and d will occur, and a much higher output voltage may be produced. If the first pulse carries the magnetic field along path a and then lets it relax to point 1, the second pulse will carry the magnetization along path c and let it relax to point 2. The third pulse would then carry it along path d and let it relax back to point 1 again.

If there is an air gap in the magnetic path, the B-H curve is flattened out and the remnant magnetic flux in the core becomes very small. In such a case, the core can support a series of pulses all of the same polarity and amplitude, since operation along path e is virtually the same as that along path f.

The B-H curve of a typical ferrite core suitable for a current injection transformer is shown in Figure 17.20 (Reference 17.34). The core dimensions are shown in the figure. The core was a ferrite U-core with a removable end yoke. If the end yoke butts closely against the pole pieces of the U-core, the core shows a clearly defined B-H loop. The initial magnetic path, corresponding to curve a of Figure 17.19(a), is identified as the *zero-mil gap*. With gaps of 1.8 or 3.3 mils (.0018 or .0033 inches) on the two-pole faces, the curve flattens out, or becomes less steep. In each case the core begins to saturate when the magnetic field reaches 1.8×10^{-5} W, corresponding to about 3500 gauss (lines per square centimeter).

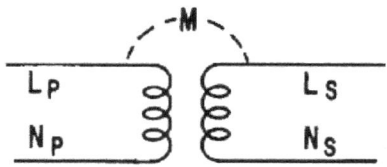

$L_A = L_P \ \& \ L_S \ \text{IN SERIES AIDING}$

$L_B = L_P \ \& \ L_S \ \text{IN SERIES BUCKING}$

$M = \frac{1}{4}(L_A - L_B)$

$K = \dfrac{M}{\sqrt{L_P L_S}}$

(a)

$n = N_S / N_P$

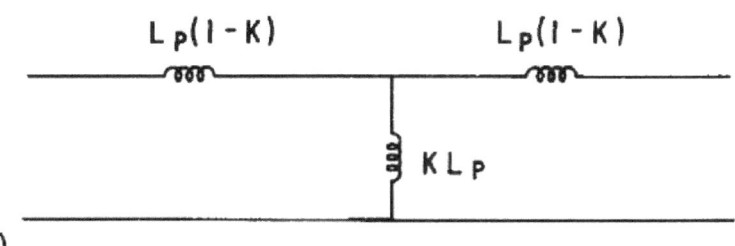

(b)

(c)

Figure 17.18 Equivalent circuits.
(a) Basic transformer
(b) Equivalent circuit referred to secondary
(c) Equivalent circuit if $N_P = N_S$ and $L_P = L_S$

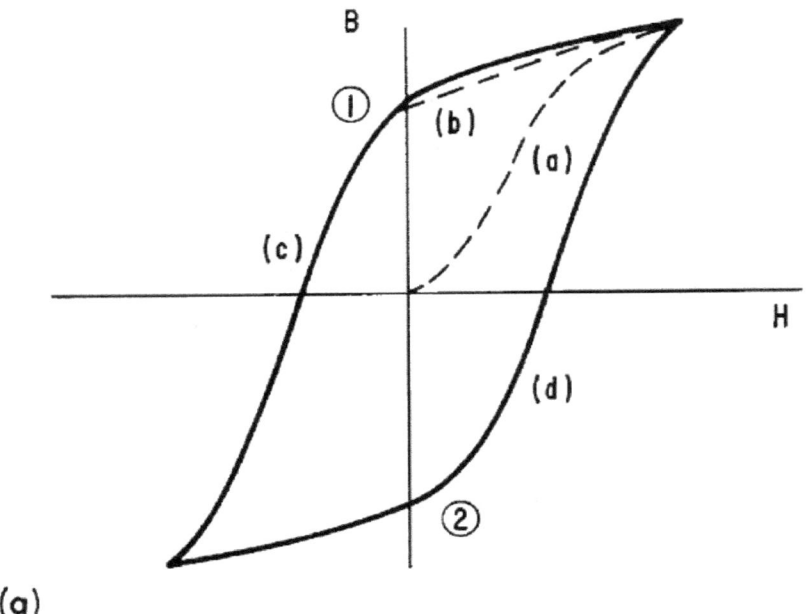

(a)

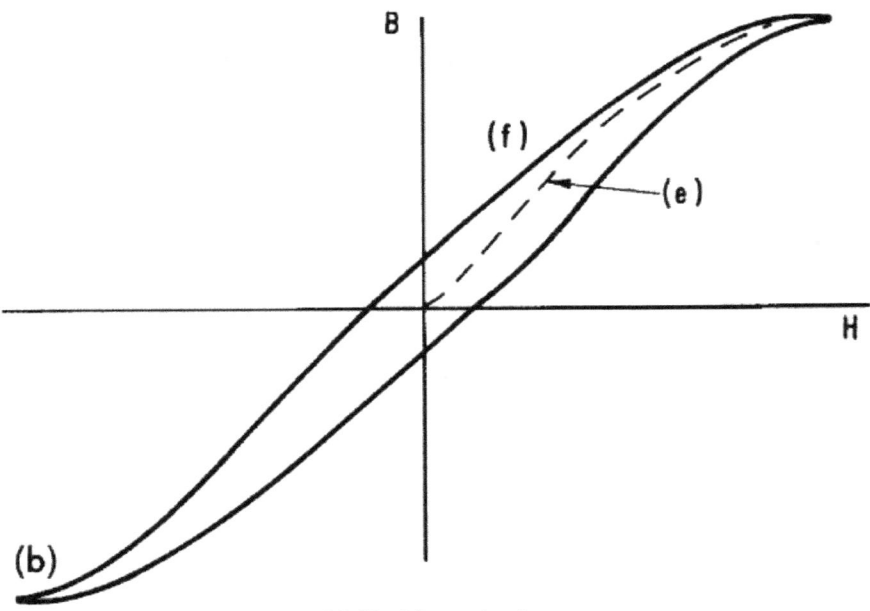

(b)

Figure 17.19 Magnetization curves.
(a) No gap
(b) With gap

526

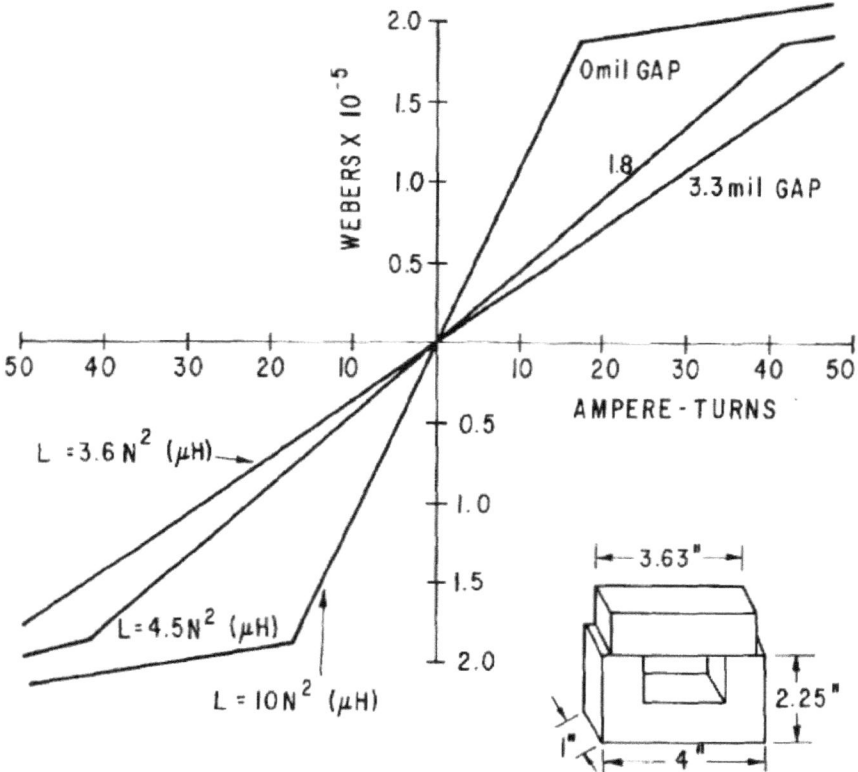

Figure 17.20 Saturation characteristics of ferrite current injection transformer.

The primary or secondary inductance is defined as the slope of the B-H curve:

$$L_P = \frac{\Delta\theta_P}{\Delta I_P} \qquad (17.1)$$

$$L_S = \frac{\Delta\theta_S}{\Delta I_S} \qquad (17.2)$$

The inductance is proportional to the square of the number of turns in the winding. For the three curves the inductance would be as indicated.

In the absence of a controlled gap in the magnetic path, the inductance of a winding depends largely on the amount of pressure exerted on the pole piece, since with more pressure any small residual gap is reduced. When the B-H curve shown in Figure 17.20 was obtained by exciting the core with 60 Hz ac, the magnetic forces involved held the poles closely together. The result is that the inductance that would be calculated from the zero-mil gap curve in Figure

17.20 would be higher than one that would be encountered if the core were excited by a short-duration pulse, one lasting only a few microseconds. With short pulses the magnetic field may be high and may exert a significant pressure on the pole pieces, but the inertia of the pole piece prevents the poles from moving together and reducing the residual air gap.

The transformers are not constrained to be operated with only one turn on

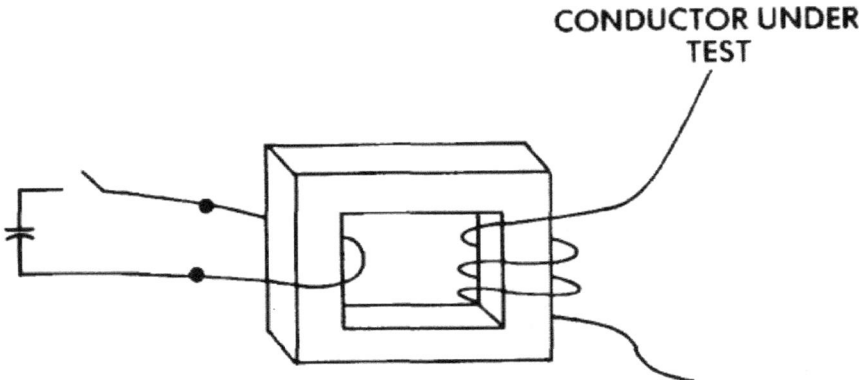

Figure 17.21   Effect of more turns on secondary.
- Less short circuit current
- More open circuit voltage
- A decrease in the frequency of the natural oscillatory mode of the cable under test

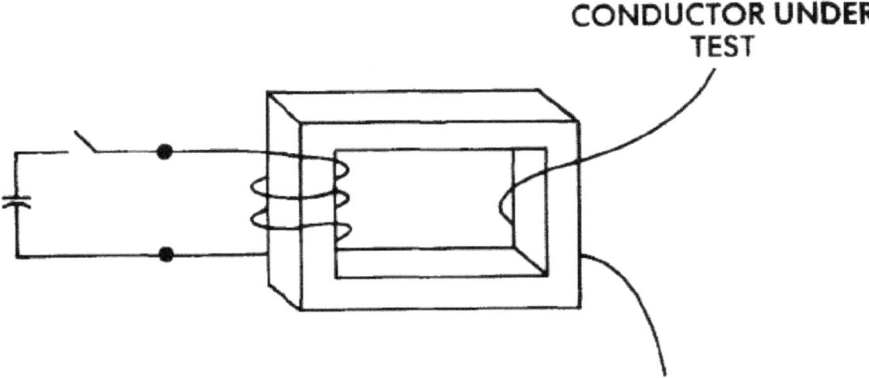

Figure 17.22   Effect of more turns on primary.
- More short circuit current
- Less open circuit voltage
- A decrease in the frequency of the natural oscillatory mode of the pulse generator
- A longer rise time of the current pulse

528

each winding or, for that matter, with the same number of turns on the two windings. Figures 17.21 and 17.22 (Reference 17.35) give qualitatively some of the ways in which turns on one winding or the other affect the results if the number of turns on the two windings are not equal. For instance, if the wire under test is looped through the core several times, a given excitation on the transformer will produce less short circuit current, more open circuit voltage, and a decrease in the frequency of the natural oscillatory mode of the cable under test. If there are more turns on the primary, there will be more short circuit current, less open circuit voltage, and a decrease in the frequency of the natural oscillatory mode of the pulse generator. There will also be a longer rise time of the current and voltage pulses.

REFERENCES

17.1 *Space Shuttle Program Lightning Protection Criteria Document*, JSC-07636, Revision A, National Aeronautics and Space Administration, Lyndon B. Johnson Space Center, Houston, Texas (November 4, 1975), p. A-5, and *Lightning Test Waveforms and Techniques for Aerospace Vehicles and Hardware*, Society of Automotive Engineers, Warrendale, Pennsylvania (5 February 1976), p. 9.

17.2 *Space Shuttle Program Lightning Protection Criteria Document*, p. 7-6.

17.3 *MIL-B-5087-B: Bonding, Electrical and Lightning Protection for Aerospace Systems* (August 31, 1970).

17.4 *Lightning Test Waveforms and Techniques for Aerospace Vehicles and Hardware*, Society of Automotive Engineers, Warrendale, Pennsylvania (5 February 1976), p. 21.

17.5 F. A. Fisher, *Some Observations Regarding Amplitude Scaling for Lightning Currents*, Aircraft Lightning Protection Note 74-1, High Voltage Laboratory, Environmental Electromagnetics Unit, Corporate Research and Development, General Electric Company, Pittsfield, Massachusetts (14 June 1974).

17.6 Fisher, *Some Observations Regarding Amplitude Scaling*, (n.p.).

17.7 Fisher, *Some Observations Regarding Amplitude Scaling*, (n.p.).

17.8 L. C. Walko, *A Test Technique for Measuring Lightning-Induced Voltages on Aircraft Electrical Currents*, NASA CR-2348, National Aeronautics and Space Administration, Lewis Research Center, Cleveland, Ohio (February 1974).

17.9 J. A. Plumer, F. A. Fisher, and L. C. Walko, *Lightning Effects on the NASA F-8 Digital Fly-By-Wire Airplane*, NASA-2524, National Aeronautics and Space Administration, Flight Research Center, Edwards, California (March 1975).

17.10 J. A. Plumer, *Guidelines for Lightning Protection of General Aviation Aircraft*, FAA-RD-73-98, Federal Aviation Administration, Department of Transportation, Washington, D.C. (October 1973).

17.11 *Lightning Test Waveforms and Techniques*, p. 20.

17.12 *Space Shuttle Program Lightning Protection Criteria Document and Lightning Test Waveforms and Techniques*.

17.13 F. A. Fisher, *Equipment for Indirect Effects Testing*, Aircraft Lightning Protection Note 75-3, High Voltage Laboratory, Environmental Electromagnetics Unit, Corporate Research and Development, General Electric Company, Pittsfield, Massachusetts (4 August 1975), p. 6.

17.14 *Space Shuttle Program Lightning Protection Criteria Document; Techniques for Dielectric Tests*, ANSI C 68.1-1968/IEEE No. 4 (April 1969), the Institute of Electrical and Electronic Engineers, New York, New York (1969); F. A. Fisher and F. D. Martzloff, "Transient Control Levels: A Proposal for Insulation Coordination in Low-Voltage Systems," *IEEE Transactions on Power Apparatus and Systems*, PAS-95, 1, (January/February 1976): 120-29.

530

17.15 *Space Shuttle Program Lightning Protection Criteria Document; Surge Withstand Capability (SWC) Tests*, ANSI C 37.90a-1974/IEEE STD 472-1974; F. D. Martzloff and E. K. Howell, *Hi-Voltage Impulse Testers*, 75 CRD 039, Corporate Research and Development, General Electric Company, Schenectady, New York (March 1975).

17.16 ANSI C 68.1-1968/IEEE No. 4 (April 1969).

17.17 Fisher and Martzloff, "Transient Control Levels: A Proposal for Insulation Coordination in Low Voltage Systems."

17.18 ANSI C 37.90a-1974.

17.19 Martzloff and Howell, *Hi-Voltage Impulse Testers*.

17.20 *Proposed Requirements for Surge Tests on Ground Fault Circuit Interrupters*, Underwriters Laboratories, Inc., Melville, New York (July 1975).

17.21 Fisher and Martzloff, "Transient Control Levels: A Proposal for Insulation Coordination."

17.22 K. E. Crouch, F. A. Fisher, and F. D. Martzloff, *Transient Control Levels: A Better Way to Voltage Ratings in Power Converter Applications*, TIS 76CRD154, Electronic Power Conditioning and Control Laboratory, Corporate Research and Development, General Electric Company, Schenectady, New York (July 1976), p. 2.

17.23 Part (a): ANSI C 37.90a-1974; Part (b): Crouch, Fisher, and Martzloff, *Transient Control Levels: A Better Way to Voltage Ratings*, p. 3.

17.24 Crouch, Fisher, and Martzloff, *Transient Control Levels: A Better Way to Voltage Ratings*, p. 3.

17.25 Adapted from ANSI C 37.90a-1974.

17.26 *EMP Susceptibility Threshold Handbook*, Draft Report submitted by the Aerospace Group of the Boeing Company for the U.S. Air Force Weapons Laboratory, Kirtland Air Force Base, Albuquerque, New Mexico (July 21, 1972), pp. 18-25.

17.27 *Space Shuttle Program Lightning Protection Criteria Document*, Appendix D.

17.28 Fisher, *Equipment for Indirect Effects Testing*, pp. 20, 22.

17.29 "Transient Control Levels: A Proposal for Insulation Coordination" and *Hi-Voltage Impulse Testers*.

17.30 Fisher, *Equipment for Indirect Effects Testing*, p. 37.

17.31 Fisher, *Equipment for Indirect Effects Testing*, p. 24.

17.32 Fisher, *Equipment for Indirect Effects Testing*, p. 26.

17.33 Fisher, *Equipment for Indirect Effects Testing*, p. 28.

17.34 Fisher, *Equipment for Indirect Effects Testing*, p. 29.

17.35 Fisher, *Equipment for Indirect Effects Testing*, p. 34.

APPENDIX 1
SR-52 CALCULATIONS – CONDUCTOR

**SR-52
User Instructions**

TITLE ___Conductor Temperature Rise and Elongation_____ PAGE _1_ OF _2_

	◄A►			
∫I²dt	Area	Density	Sp. Ht.	ΔT
Resistivity	T. coeff.	L. coeff.		ΔL

	◄B►			

STEP	PROCEDURE	ENTER	PRESS			DISPLAY
1	Enter Program A		CLR	•Read		0
2	Enter ∫I²dt (A²-s)		A			
2	Enter Area (cm²)		B			
3	Enter Density (g/cm³)		C			
4	Enter Sp. Ht. (g-cal/g-°C)		D			
5	Enter Resistivity (Ω-cm)		•A			
6	Enter Temp. coeff. (1/°C)		•B			
7	Enter Exp. coeff. (1/°C)		•C			
8	Calculate ΔT (°C)		E			ΔT (°C)
9	Calculate ΔL (cm/cm)		•E			ΔL (cm/cm)

    Part No. 1220479

SR-52 CALCULATIONS – CONDUCTOR (CONCLUDED)

TITLE Conductor Temperature Rise and Elongation   PAGE 2 OF 2
PROGRAMMER J. A. Plumer   DATE 25 October 1976

SR-52 Coding Form

LOC	CODE	KEY	COMMENTS	LOC	CODE	KEY	COMMENTS	LOC	CODE	KEY	COMMENTS	LABELS
000 112	46	*LBL			53	(			55	+		A $\int I^2\,dt$
	11	A			93	.			53	(		B Area
	42	STO		040 152	02	2			01	1		C Density
	00	0			03	3			75	−		D Sp. Ht.
	00	0			08	8		080 192	43	RCL		E ΔT
005 117	81	HLT			09	9			00	0		A' Resistivity
	46	*LBL			65	X			05	5		B' T. coeff.
	12	B		045 157	43	RCL			65	X		C' L. coeff.
	42	STO			00	0			43	RCL		D'
	00	0			00	0		085 197	00	0		E' ΔL
010 122	01	1			65	X			06	6		**REGISTERS**
	81	HLT			43	RCL			54	)		00 $\int I^2\,dt$
	46	*LBL		050 162	00	0			95	=		01 Area
	13	C			04	4			42	STO		02 Density
	42	STO			54	)		090 202	00	0		03 Sp. Ht.
015 127	00	0			55	+			07	7		04 Resistivity
	02	2			53	(			81	HLT		05 T. coeff.
	81	HLT		055 167	43	RCL			46	*LBL		06 K
	46	*LBL			00	0			18	*C		07 ΔT
	14	D			03	3		095 207	42	STO		08 L. coeff.
020 132	42	STO			65	X			00	0		09
	00	0			43	RCL			08	8		10
	03	3		060 172	00	0			81	HLT		11
	81	HLT			02	2			46	*LBL		12
	46	*LBL			65	X		100 212	10	*E		13
025 137	16	*A			53	(			53	(		14
	42	STO			43	RCL			43	RCL		15
	00	0		065 177	00	0			00	0		16
	04	4			01	1			07	7		17
	81	HLT			40	*x²		105 217	65	X		18
030 142	46	*LBL			54	)			43	RCL		19
	17	*B			95	=			00	0		**FLAGS**
	42	STO		070 182	42	STO			00	0		0
	00	0			00	0			08	8		1
	05	5			06	6		110 222	54	)		2
035 147	81	HLT			43	RCL			95	=		3
	46	*LBL			00	0			81	HLT		4
	15	E		075 187	06	6						

TEXAS INSTRUMENTS INCORPORATED

534

APPENDIX 2
SR-52   CALCULATIONS – TUBE

SR-52
User Instructions

TITLE ___ Grounding Tube Temperature Rise and Elongation _____ PAGE __1__ OF __3__

	◄A◄			
ʃI² dt	O.D	Wall	Density	ΔT
Sp. Ht.	Resistivity	T. coeff.	L. coeff.	ΔL

	◄B◄			

STEP	PROCEDURE	ENTER	PRESS			DISPLAY
1	Enter Program A		CLR	*Read		0
2	Enter Program B		*Read			0
3	Enter ʃI² dt (A²-s)		A			
4	Enter Outside Diameter (in)		B			
5	Enter Wall Thickness (in)		C			
6	Enter Density (g/cm³)		D			
7	Enter Sp. Ht. (g-cal/g-°C)		*A			
8	Enter Resistivity (Ω-cm)		*B			
9	Enter T. coeff. (1/°C)		*C			
10	Calculate ΔT (°C)		E			ΔT (°C)
11	Enter L. coeff. (1/°C)		*D			
12	Calculate ΔL		*E			ΔL (in/ft)

© 1975 Texas Instruments Incorporated   Part No. 1220479

535

TITLE __Grounding Tube Temperature Rise and Elongation__   PAGE __2__ OF __3__

PROGRAMMER __J. A. Plumer__   DATE __16 March 1976__

SR-52 Coding Form

LOC	CODE	KEY	LOC	CODE	KEY	LOC	CODE	KEY
000/112	46	*LBL		42	STO		02	2
	11	A		00	0		93	•
	42	STO	040/152	06	6		05	5
	00	0		81	HLT		04	4
	00	0		46	*LBL	080/192	54	)
005/117	81	HLT		15	E		40	*x²
	46	*LBL			(		54	)
	12	B	045/157		*π		95	=
	42	STO			+		42	STO
	00	0			4	085/197	00	0
010/122	01	1			)		07	7
	81	HLT			X		43	RCL
	46	*LBL	050/162	53	(		00	0
	13	C		53	(		05	5
	42	STO		43	RCL	090/202	55	+
015/127	00	0		00	0		43	RCL
	02	2		01	1		00	0
	81	HLT	055/167	65	X		07	7
	46	*LBL		02	2		95	=
	14	D		93	•	095/207	42	STO
020/132	42	STO		05	5		00	0
	00	0		04	4		08	8
	03	3	060/172	54	)		53	(
	81	HLT		40	*x²		93	•
	46	*LBL		75	-	100/212	02	2
025/137	16	*A		53	(		03	3
	42	STO		53	(		08	8
	00	0	065/177	43	RCL		09	9
	04	4		00	0		65	X
	81	HLT		01	1	105/217	43	RCL
030/142	46	*LBL		75	-		00	0
	17	*B		02	2		00	0
	42	STO	070/182	65	X		65	X
	00	0		43	RCL		43	RCL
	05	5		00	0	110/222	00	0
035/147	81	HLT		02	2		08	8
	46	*LBL		54	)			
	18	*C	075/187	65	K			

LABELS

A	$\int I^2\,dt$
B	O.D.
C	Wall Thick.
D	Density
E	Calc. ΔT
A'	Sp. Ht.
B'	Resistivity
C'	T. coeff.
D'	L. coeff.
E'	Calc. ΔL

REGISTERS

00	$\int I^2\,dt$
01	O.D.
02	Wall
03	Density
04	Sp. Ht.
05	Resistivity
06	T. coeff.
07	Wall Area
08	R/cm
09	K
10	ΔT
11	L. coeff.
12	
13	
14	
15	
16	
17	
18	
19	

FLAGS

0
1
2
3
4

TEXAS INSTRUMENTS INCORPORATED

536

TITLE Grounding Tube Temperature Rise and Elongation PAGE 3 OF 3
PROGRAMMER J. A. Plumer DATE 16 March 1976

SR-52 Coding Form

LOC	CODE	KEY	COMMENTS	LOC	CODE	KEY	COMMENTS	LOC	CODE	KEY	COMMENTS	LABELS
000/112	54	)			81	HLT						A
	55	+			46	*LBL						B
	53	(		040/152	19	*D						C
	43	RCL			42	STO						D
	00	0			01	1		080/192				E
005/117	04	4			01	1						A
	65	X			81	HLT						B
	43	RCL		045/157	46	*LBL						C
	00	0			10	*E						D
	03	3			53	(		085/197				E
010/122	65	X			43	RCL						REGISTERS
	43	RCL			01	1						00
	00	0		050/162	00	0						01
	07	7			65	X						02
	54	)			01	1		090/202				03
015/127	95	=			02	2						04
	42	STO			65	X						05
	00	0		055/167	43	RCL						06
	09	9			01	1						07
	43	RCL			01	1		095/207				08
020/132	00	0			54	)						09
	09	9			95	•						10
	55	÷		060/172	81	HLT						11
	53	(										12
	01	1						100/212				13
025/137	75	-										14
	43	RCL										15
	00	0		065/177								16
	06	6										17
	65	X						105/217				18
030/142	43	RCL										19
	00	0										FLAGS
	09	9		070/182								0
	54	)										1
	95	•						110/222				2
035/147	42	STO										3
	01	1										4
	00	0		075/187								

TEXAS INSTRUMENTS
INCORPORATED

537

APPENDIX 3
PROGRAM *MAGFLD*

Table A3.1   LISTING OF PROGRAM *MAGFLD*

- 1 -

MAGFLD   10/26/76

```
100 REM FA FISHER GENERAL ELECTRIC COMPANY ENVIRONMENTAL
110 REM ELECTROMAGNETICS UNIT BLDG 9-209 PITTSFIELD,MA 01201
120 REM PHONE (413)-494-4380 OR DIAL COMM 8-236-4380
130 REM THIS PROGRAM CALCULATES THE MAGNETIC FIELD INTERNAL OR EXTERNAL
140 REM TO A GROUP OF CURRENT CARRYING CONDUCTORS.  ALL CONDUCTORS
150 REM ARE ASSUMED LONG ENOUGH THAT END EFFECTS ARE NEGLIGIBLE.
160 REM IT ALSO CALCULATES THE DISTRIBUTION OF CURRENT AMONG THE
170 REM CONDUCTORS. THE CONDUCTORS ARE ALL ASSUMED TO BE EQUIDISTANT FROM
180 REM THE RETURN PATH.  THE DISTANCE TO TO THE RETURN PATH IS ASSUMED TO BE 300
190 REM METER.  THE PROGRAM DOES NOT CALCULATE ANY PROXIMITY EFFECTS.
200 REM
210 REM DATA IS STORED IN A FILE WHICH IS GIVEN THE NAME DATFIL.
220 REM THE FIRST RECORD OF THE FILE SHOULD GIVE THE TOTAL NUMBER OF CONDUCTORS (N1)
230 REM SUCCESSIVE RECORDS OF THIS FILE SHOULD GIVE THE LOCATION OAND
232 REM SIZE OF THE CONDUCTORS.  THEY SHOULD BE IN THE FORM "CONDUCTOR
234 REM NUMBER" (1 THROUGH N1), "X COORDINATE" (IN METERS), "Y COORDINATE"
236 REM (IN METERS) AND  "CONDUCTOR DIAMETER" (IN METERS).
238 REM
239 REM
240 REM A SAMPLE DATA FILE IS SHOWN BELOW.
241 REM THIS FILE DESCRIBES FOUR CONDUCTORS, EACH 0.01 METERS DIAMETER,
242 REM ARRANGED AT THE CORNERS OF A SQUARE 1 METER ON A SIDE AND
243 REM CENTERED AT X=Y=0.
244 REM
245 REM
246 REM       100: 4
247 REM       101: 1,.5,.5,.01
248 REM       102: 2,.5,-.5,.01
252 REM       103: 3,-.5,-.5,.01
254 REM       104: 4,.5,-.5,.01
256 REM
258 REM
260 REM THE LENGTH OF THE CONDUCTORS IS NOT ENTERED.  THEY ARE ASSUMED TO
262 REM BE OF INFINITE LENGTH.
270 REM
280 REM THE PROGRAM IS PRESENTLY DIMENSIONED FOR A MAXIMUM OF
290 REM 50 CONDUCTORS
300 REM
310 REM THE OUTPUT IS NORMALIZED TO A TOTAL CURRENT OF ONE AMPERE
320 REM FLOWING THROUGH THE GROUP OF CONDUCTORS.
345 FILES DATFIL
350 DIM M(50,50),V(50,1),I(50,1),D(50,4),N(50,50)
360 DIM J(50),K(50)
365 READ#1,N1
370 MAT READ #1,D(N1,4)
380 MAT V=ZER(N1,1)
390 FOR I=1 TO N1
400 LET V(I,1)=1
410 NEXT I
420 MAT M=ZER(N1,N1)
```

539

---------- ------------------------------------- -------- --

MAGFLD   10/26/76

```
430 MAT I=ZER(N1,1)
440 MAT N=ZER(N1,N1)
450 PRINT"CONDUCTOR","X COORDINATE","Y COORDINATE","DIAMETER-METERS"
455 :####         #.###^^^^
460 MAT PRINT D
470 PRINT
480 PRINT
490 LET P2=6.2831d
500 LET E2=2.7128
510 LET R1=300
520 FOR I=1 TO N1
530 FOR J=1 TO N1
540 IF I=J THEN 580
550 LET R2=SQR((D(I,2)-D(J,2))^2 +(D(I,3)-D(J,3))^2)
560 LET M(I,J)=LOG(R1/R2)
570 GOTO 590
580 LET M(I,J)=LOG(R1*2/D(I,4))
590 NEXT J
600 NEXT I
610 MAT N=INV(M)
620 MAT I=N*V
630 LET I1=0
640 FOR I=ITON1
650 LET I1=I(I,1)+I1
660 NEXT I
670 FOR I=1 TO N1
680 LET J(I)=I(I,1)/I1
690 NEXT I
700 PRINT "CONDUCTOR","FRACTIONAL CURRENT"
710 PRINT
720 FOR I=1 TO N1
730 PRINT USING 455,I,J(I)
740 NEXT I
750 PRINT
760 PRINT
770 PRINT"THE FIRST CHARACTER OF ANY KEYBOARD INPUT DATA IS A CONTROL"
780 PRINT"CHARACTER. WHEN AN INPUT QUESTION ASKS FOR 'CONTROL', ENTER"
790 PRINT"1  (ONE) IF YOU WISH TO CONTINUE WITH THAT TYPE OF CALCUL-"
800 PRINT"ATION AND ENTER 0  (ZERO) IF YOU WISH TO GO ON TO THE"
810 PRINT"NEXT TYPE OF CALCULATION"
820 PRINT
830 REM THIS STARTS CALCULATION OF FIELD AT A DEFINED POINT
835 PRINT"DO YOU WANT A POINT BY POINT CALCULATION OF THE FIELD?"
840 PRINT"YES=1   NO=0":
870 INPUT A1
880 IF A1=0 THEN 1022
890 PRINT
900 REM THIS STARTS CALCULATION OF THE FIELD AT A DEFINED POINT
910 PRINT"   X         Y         H-X       H-Y       H-TOT   ANGLE-DEG"
920 PRINT
```

Table A3.1 LISTING OF PROGRAM *MAGFLD* (CONT.)

```
---------------------------------------------------------------------------

                                  - 3 -

MAGFLD      10/26/76

930 PRINT
940 REM X1 AND Y1 ARE COORDINATES OF POINT P1
950 PRINT"CONTROL,X1,Y1"!
960 INPUT Q,X1,Y1
965 PRINT
970 IF  Q=0 THEN 1020
980 GOSUB 1048
990 PRINT USING 1021 ,X1,Y1,B2,B3,B4,B5
1000 PRINT
1010 GOTO 950
1020 REM THIS ENDS THE CALCULATION OF THE FIELD AT A DEFINED POINT
1021 !#.###^^^^   #.###^^^^   #.###^^^^   #.###^^^^   ####.##
1022 REM THIS STARTS CALCULATION OF THE FIELD AT A SERIES OF POINTS
1024 PRINT"DO YOU WANT A CALCULATION OF THE FIELD AT A SERIES OF POINTS"
1025 PRINT"COVERING AN AREA?  YES=1,NO=0"!
1026 INPUT Q
1027 IF Q=0 THEN 1280
1028 PRINT"WHAT ARE START,STOP AND INCREMENT VALUES FOR X"!
1029 INPUT S1,S2,S3
1030 PRINT"WHAT ARE START,STOP AND INCREMENT VALUES FOR Y"!
1031 INPUT S4,S5,S6
1032 PRINT
1033 PRINT
1036 PRINT"    X          Y          H-X        H-Y        H-TOT    ANGLE-DEG"
1037 FOR X1=S1 TO S2 STEP S3
1038 FOR Y1=S4 TO S5 STEP S6
1039 GOSUB 1049
1040 PRINT USING 1021 ,X1,Y1,B2,B3,B4,B5
1041 NEXT Y1
1042 NEXT X1
1043 REM THIS ENDS THE CALCULATION OF FIELD AT A SERIES OF POINTS
1044 PRINT
1045 PRINT
1046 GOTO 1280
1048 REM THIS STARTS THE SUBROUTINE FOR CALCULATION OF FIELD AT A POINT
1049 LET B2=0
1050 LET B3=0
1060 FOR I=1 TON1
1070 REM X3=X COMPONENT BETWEEN POINT I AND POINT 1
1080 REM DITTO FOR Y3
1090 REM R3= TOTAL DISTANCE BETWEEN POINT I AND POINT 1
1100 LET X3=X1-D(I,2)
1110 LET Y3=Y1-D(I,3)
1120 LET R3=SQR(X3*X3+Y3*Y3)
1130 LET B1=J(I)/(P2*R3)
1135 REM C2=INCREMENTAL COMPONENT IN X DIRECTION
1137 REM C3=INCREMENTAL COMPONENT IN Y DIRECTION
1140 LET C2=B1*Y3/R3
1150 LET B2=B2+C2
1160 LET C3=-B1*X3/R3
```

541

- 4 -

MAGFLD   10/26/76

```
1170 LET B3=B3+C3
1180 NEXT I
1190 REM B2= X COMPONENT OF TOTAL FIELD
1200 REM B3=Y COMPONENT OF TOTAL FIELD
1210 REM B4= TOTAL FIELD
1220 REM B5=ANGLE OF TOTAL FIELD
1230 LET B4=SQR(B2*B2+B3*B3)
1231 REM THE FOLLOWING STATEMENTS GUARD AGAINST DIVISION BY ZERO IN LINE 1240
1232 IF B2<>0 THEN 1240
1233 IF B3<0 THEN 1236
1234 LET B5=90
1235 GOTO 1260
1236 LET B5=-90
1237 GOTO 1260
1240 LET B5=ATN(B3/B2)
1245 LET B5=B5*57.29583
1246 REM IF B2>0 ANGLE IS IN FIRST OR FOURTH QUADRANTS AND LINE 1240 IS OK
1247 REM IF B2<0 AND B3<0 THE ANGLE IS IN THE THIRD QUADRANT
1248 REM IF B2<0 AND B3>0 THE ANGLE IS IN THE SECOND QUADRANT
1252 IF B2>0 THEN 1260
1254 IF B3<0 THEN 1258
1256 LET B5=180+B5
1257 GOTO 1260
1258 LET B5=-180+B5
1260 RETURN
1270 REM THIS ENDS THE SUBROUTINE FOR CALCULATION OF THE FIELD AT A POINT
1280 PRINT"DO YOU WANT THE FLUX INTEGRATED OVER A PATH BETWEEN"
1290 PRINT"TWO POINTS. YES=1  NO=0"!
1320 INPUT A2
1325 IF A2=0 THEN 1400
1330 PRINT
1349 REM X4 AND Y4 ARE THE COORDINATES OF P1
1350 REM X5 AND Y5 ARE THE COORDINATES OF P2
1360 PRINT"CONTROL,X-P1,Y-P1,X-P2,Y-P2"!
1370 INPUT Q,X4,Y4,X5,Y5
1380 IF Q=1 THEN 1470
1390 PRINT
1400 PRINT"DO YOU WANT TO GO BACK AND DO MORE POINT BY POINT CALCULATIONS"
1410 PRINT"(ENTER 2), DO MORE CALCULATIONS OF FLUX OVER AN AREA (ENTER 1)"
1420 PRINT"OR STOP (ENTER 0)  "
1430 INPUT A2
1440 IF A2=2 THEN 890
1450 IF A2=1 THEN 1330
1460 IF A2=0 THEN 1720
1470 REM THIS STARTS CALCULATION OF FLUX OVER AN AREA
1480 LET X5=X5-X4
1490 LET Y6=Y5-Y4
1500 LET R6= SQR(X6*X6+Y6*Y6)
1510 REM 01 IS THE ANGLE BETWEEN THE LINE JOINING P1 AND P2
1520 REM AND THE X AXIS. SIN(01)=Y6/R6
```

542

Table A3.1 LISTING OF PROGRAM *MAGFLD* (CONCLUDED)

MAGFLD   10/26/76

```
1530 REM AND COS(01)=X6/R6
1540 REM THIS STARTS THE INTEGRATION OVER THE AREA BETWEEN P1 AND P2
1542 LET 03=50
1545 LET 04=R6/03
1550 FOR J=0 TO 03
1560 LET 02=J/03
1570 LET X1=X4+02*X6
1580 LET Y1=Y4+02*Y6
1590 GOSUB 1049
1600 REM D1 AND D2 ARE THE COMPONENTS OF H-X AND H-Y NORMAL TO THE
1610 REM LINE JOINING P1 AND P2
1620 LET D1=-B2*Y6/R6
1630 LET D2=B3*X6/R6
1640 LET K(J)=D1+D2
1660 NEXT J
1670 LET B7=0
1675 FOR J=0 TO 03-1
1680 LET B7=(K(J)+K(J+1))*04/2+B7
1685 NEXT J
1687 LET B7=2E-7*P2*B7
1690 PRINT"TOTAL FLUX=";B7;"WEBERS"
1700 PRINT
1710 GOTO 1360
1720 END
```

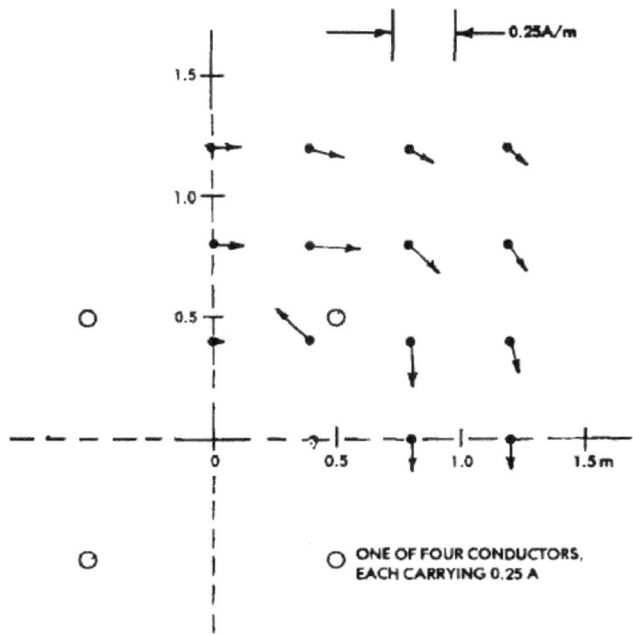

Figure A3.1 Pattern of field around the conductors described in lines 246 to 254 of Table A3.1.

543

Table A3.2 SAMPLE OUTPUT FROM PROGRAM *MAGFLD*

```
MAGFLD      09:27EDT    10/26/76

CONDUCTOR      X COORDINATE    Y COORDINATE    DIAMETER-METERS

    1             -0.5            0.5              0.01
    2              0.5            0.5              0.01
    3             -0.5           -0.5              0.01
    4              0.5           -0.5              0.01


CONDUCTOR       FRACTIONAL CURRENT

    1             .250E+00
    2             .250E+00
    3             .250E+00
    4             .250E+00
```

THE FIRST CHARACTER OF ANY KEYBOARD INPUT DATA IS A CONTROL
CHARACTER. WHEN AN INPUT QUESTION ASKS FOR 'CONTROL',ENTER
1 (ONE) IF YOU WISH TO CONTINUE WITH THAT TYPE OF CALCUL-
ATION AND ENTER 0 (ZERO) IF YOU WISH TO GO ON TO THE
NEXT TYPE OF CALCULATION

DO YOU WANT A POINT BY POINT CALCULATION OF THE FIELD?
YES=1   NO=0? 1

```
    X           Y          H-X        H-Y        H-TOT     ANGLE-DEG


CONTROL,X1,Y1? 1,,3,.3

  .300E+00    .300E+00   -.395E-01   .395E-01    .559E-01    135.00

CONTROL,X1,Y1? 1,1,1

  .100E+01    .100E+01    .849E-01  -.849E-01    .120E+00    -45.00

CONTROL,X1,Y1? 0,0,0
```

DO YOU WANT A CALCULATION OF THE FIELD AT A SERIES OF POINTS
COVERING AN AREA? YES=1,NO=0? 1
WHAT ARE START,STOP AND INCREMENT VALUES FOR X? 0,1.2,.4
WHAT ARE START,STOP AND INCREMENT VALUES FOR Y? 0,1.2,.4

```
    X           Y          H-X         H-Y        H-TOT     ANGLE-DEG
  .000E+00    .000E+00    .140E-08   -.186E-08   .233E-08    -53.13
  .000E+00    .400E+00    .370E-01    .140E-08   .370E-01      .00
  .000E+00    .800E+00    .124E+00    .198E-08   .124E+00      .00
  .000E+00    .120E+01    .118E+00    .640E-09   .118E+00      .00
  .400E+00    .000E+00   -.186E-08   -.370E-01   .370E-01    -90.00
  .400E+00    .400E+00   -.138E+00    .138E+00   .195E+00    135.00
  .400E+00    .800E+00    .184E+00   -.120E-01   .184E+00     -3.73
  .400E+00    .120E+01    .119E+00   -.279E-01   .122E+00    -13.22
  .800E+00    .000E+00   -.233E-08   -.124E+00   .124E+00    -90.00
  .800E+00    .400E+00    .120E-01   -.184E+00   .184E+00    -86.27
  .800E+00    .800E+00    .117E+00   -.117E+00   .166E+00    -45.00
  .800E+00    .120E+01    .983E-01   -.596E-01   .115E+00    -31.24
  .120E+01    .000E+00   -.466E-09   -.113E+00   .113E+00    -90.00
  .120E+01    .400E+00    .279E-01   -.119E+00   .122E+00    -76.78
  .120E+01    .800E+00    .596E-01   -.983E-01   .115E+00    -58.76
  .120E+01    .120E+01    .684E-01   -.684E-01   .967E-01    -45.00
```

DO YOU WANT THE FLUX INTEGRATED OVER A PATH BETWEEN
TWO POINTS. YES=1 NO=0? 1

CONTROL,X-P1,Y-P1,X-P2,Y-P2? 1,0,0,.4,.4
TOTAL FLUX=-4.87611E-9 WEBERS

CONTROL,X-P1,Y-P1,X-P2,Y-P2? 0,0,0,0,0

DO YOU WANT TO GO BACK AND DO MORE POINT BY POINT CALCULATIONS
(ENTER 2), DO MORE CALCULATIONS OF FLUX OVER AN AREA (ENTER 1)
OR STOP (ENTER 0)
? 0

544

INDEX

A

Access doors, effects on magnetic
    field, 327
    sparking at, 159-164
Action integral, 82, 188, 190
Airworthiness regulations, 105-113, 150
Aluminum foil, as diverter strip, 222
    as protective coating, 246
    shielding effectiveness, 241, 243, 244
Anodizing, effects on conductivity,
    163, 166
    effects on couplings, 168
Aperture coupling – Electric fields
    equations, 343-344
    shape factor, 341
Aperture coupling – Magnetic fields
    dipole approximations, 329-332,
    335
    equations, 333-334, 336-337
    equivalence of aperture shapes, 340
    examples, 342-343
    patterns of field intensity, 333
    reflecting surfaces, 337, 338
    shape factor, 332
Arcing (See *sparking*)
Attachment of flash (See *attachment
    zones*), as influenced by diverter
    strips (See *diverter strips*)
    initial attachment, 48, 49, 53
    on nonmetallic surfaces, 207-209
    to fuel tanks (See *fuel tanks*)
    to radomes (See *radomes*)
Attachment zones, 50-53, 105-108,
    132-134
    defined, 51, 106
    determined by scale models, 106
    field experience, 107, 109
    lightning environment in, 110-111
    swept-stroke zone, 50-53, 106, 107,
    108, 109
Avoidance, of storms, 59, 65-71
    of strikes, 69-70

B

Bonding, jumpers, 204
    magnetic force effects, 201-203

    of access doors, 163, 164
    of composites, 240
    of fuel caps, 160
    of hinges, 205
    of pipe couplings, 169, 170
    on fuel dump pipes, 135
    via fasteners, 165
Boys camera, 8, 9, 149
Breakdown, attachment point tests,
    211-213
    of air, 7, 174
    of canopies, 88
    of electrical insulation, 91, 93
    of fiberglass, 87, 216
    of spark gaps (See *protective devices*)

C

Canopies, puncture of, 88
    scorching of, 90
    streamers inside, 88
Capacitance, aircraft, 40
    sphere, 39
Channel of lightning, initial develop-
    ment, 5-9
    pressure of, 83
    return stroke, 9-14
    temperature of, 83
Charge, delivered by stroke com-
    ponents, 16, 19, 25, 27
    in leader, 7, 41
    on aircraft, 40, 41
    produced by engines, 48-49
    to burn through metals, 136, 143,
    152
Circulating currents, 304, 306, 307,
    311,
Coatings, 224, 245-250
Composite materials, boron, 230-233
    dielectric coatings for, 248-249
    electrical models of, 231, 232
    energy absorption, 229
    examples of damage, 234, 235, 237
    extent of damage, 236, 268, 239
    graphite, 232, 237-238
    invisible damage, 233, 235
    joints, 240
    mechanism of failure, 87-89, 230, 233

protective coatings, 245-247
shielding properties, 241-245
temperature rise, 230
Conductivity, at access doors, 163-164
at joints, 165-167, 240
of composites, 229-230, 233-238
of metals, 190, 303
through couplings and interfaces,
171-172
Connectors, grounding of, 394, 400, 404
transfer characteristics, 394, 399
Continuing current, 18
effects of, 78, 136-154, 186-188
Corona, 3, 4, 44, 49
Coulomb ignition thresholds, 140-143,
150, 152
Covers, access (See *access doors*)
effects of internal fields, 325
sparking at, 163
Current distribution, 280, 283, 287, 291
in internal conductors, 160-171
time constant of redistribution,
305-308

on composites (See *composites*)
on metallic surfaces (See *metallic
surfaces, fuel systems*)
magnetic force, 78-79, 81
melting and burnthrough, 78, 80
pitting, 79
resistance heating, 82
Diverters, application of, 213-220
electromagnetic interference from,
223
foil, 222-223
inductance of, 222
internal, 221
metallic coatings, 224
on composite materials, 247
on wing tips, 132, 156
required size, 220
segmented, 225-226
solid bar, 220-221
tests on, 218-219
thermal elongation, 194, 195
types of, 220-228
woven wire, 224
Dwell time (See *swept strokes*)

D

Damage   Direct effects, from shock
waves, 85, 86
to composites, 88, 89, 230-240
to fiberglass, 83, 87, 92, 206-213
to metal surfaces, 78, 136-154
to radomes, 83, 85, 87
to structures, 79-86, 199-201,
206-213
Damage – Indirect effects, 98
constants, 453, 455-462, 465-492
examples of, 92, 99-101
to capacitors, 463-464
to semiconductors, 451-462, 465-492
Dart leader, 18
Design to minimize indirect effects, circuit design, 438-440
in fuel systems, 172-179
location of equipment, 429-430
location of wiring, 430-433
premises for design, 429
protective devices, 441-446
shielding, 176-179, 433-437
transient coordination, 446-448
Diffusion (See *pulse penetration*)
Direct effects, defined, 76, 77
on fuel systems (See *fuel systems* and
fuel tanks)

E

Eddy currents, 299, 303, 309
Electric field, as influenced by aircraft, 37, 39
at aircraft surface, 38-41
at surface of a sphere, 40
coupling through apertures, 341-
342
effects on circuits, 353-357
external intensity, 353
on interior of cylinders, 298-302
on interior of shield, 369
Electron density, 48
Elliptical conductors, current distribution, 274-276
internal magnetic field, 288, 289,
309-320
tangential field intensity, 274-276,
287
Engines, lightning effects on, 75, 96,
97, 98
Explosion, in fuel vent systems, 127-
132
in fuel tanks, 115-116, 162
of conductors, 82, 84, 188-193
of radomes, 83

546

G

H

I

549

1. Report No. NASA RP-1008	2. Government Accession No.	3. Recipient's Catalog No.
4. Title and Subtitle LIGHTNING PROTECTION OF AIRCRAFT		5. Report Date October 1977
		6. Performing Organization Code
7. Author(s) Franklin A. Fisher and J. Anderson Plumer		8. Performing Organization Report No.
		10. Work Unit No.
9. Performing Organization Name and Address General Electric Company 100 Woodlawn Avenue Pittsfield, MA 01201		11. Contract or Grant No.
		13. Type of Report and Period Covered Reference Publication
12. Sponsoring Agency Name and Address National Aeronautics and Space Administration Washington, D.C. 20546		
		14. Sponsoring Agency Code

15. Supplementary Notes

16. Abstract

This handbook summarizes the current knowledge concerning the potential lightning effects on aircraft and the means that are available to designers and operators to protect against these effects. The impetus for writing this book comes from two sources - the increased use of nonmetallic materials in the structure of aircraft and the constant trend toward using electronic equipment to handle flight-critical control and navigation functions. Nonmetallic structures are inherently more likely to be damaged by a lightning strike than are metallic structures. Nonmetallic structures also provide less shielding against the intense electromagnetic fields of lightning than do metallic structures. These fields have demonstrated an ability to damage or cause upset of electronic equipment. The persons who can best use information on aircraft protection from lightning are the aircraft designers and operators, but generally they are not among those who produced this information. Moreover, they are often unaware of its existence, and they seldom have the background to distill from it the important facts that can and should be applied to achieve safer designs. The purpose of this book is to present the most important parts of this body of knowledge in a manner most useful to the designer and the operator.

17. Key Words (Suggested by Author(s)) Lightning; Thunderstorms; Electronic equipment; Aircraft equipment; Aircraft instruments; Aircraft safety; Aircraft structures		18. Distribution Statement Unclassified - unlimited STAR Category 03		
19. Security Classif. (of this report) Unclassified		20. Security Classif. (of this page) Unclassified	21. No. of Pages 560	22. Price* A24

* For sale by the National Technical Information Service, Springfield, Virginia 22161

www.ingramcontent.com/pod-product-compliance
Lightning Source LLC
Chambersburg PA
CBHW081426170526
45166CB00008B/2114